# Achieving sustainable cultivation of ornamental plants

It is widely recognised that agriculture is a significant contributor to global warming and climate change. Agriculture needs to reduce its environmental impact and adapt to current climate change whilst still feeding a growing population, i.e. become more 'climate-smart'. Burleigh Dodds Science Publishing is playing its part in achieving this by bringing together key research on making the production of the world's most important crops and livestock products more sustainable.

Based on extensive research, our publications specifically target the challenge of climate-smart agriculture. In this way we are using 'smart publishing' to help achieve climate-smart agriculture.

Burleigh Dodds Science Publishing is an independent and innovative publisher delivering high quality customer-focused agricultural science content in both print and online formats for the academic and research communities. Our aim is to build a foundation of knowledge on which researchers can build to meet the challenge of climate-smart agriculture.

For more information about Burleigh Dodds Science Publishing simply call us on +44 (0) 1223 839365, email info@bdspublishing.com or alternatively please visit our website at www.bdspublishing.com.

**Related titles:**

Achieving sustainable greenhouse cultivation
Print (ISBN 978-1-78676-280-1); Online (ISBN 978-1-78676-282-5, 978-1-78676-283-2)

Advances in horticultural soilless culture
Print (ISBN 978-1-78676-435-5); Online (ISBN 978-1-78676-437-9, 978-1-78676-438-6)

Chapters are available individually from our online bookshop:
https://shop.bdspublishing.com

climate•SMART•publishing
IN AGRICULTURAL SCIENCE

BURLEIGH DODDS SERIES IN AGRICULTURAL SCIENCE
NUMBER 82

# Achieving sustainable cultivation of ornamental plants

Edited by Emeritus Professor Michael Reid,
University of California-Davis, USA

burleigh dodds
SCIENCE PUBLISHING

Published by Burleigh Dodds Science Publishing Limited
82 High Street, Sawston, Cambridge CB22 3HJ, UK
www.bdspublishing.com

Burleigh Dodds Science Publishing, 1518 Walnut Street, Suite 900, Philadelphia, PA 19102-3406, USA

First published 2020 by Burleigh Dodds Science Publishing Limited


Library of Congress Control Number: 2020934453

British Library Cataloguing in Publication Data
A catalogue record for this book is available from the British Library

ISBN 978-1-78676-328-0 (Print)
ISBN 978-1-78676-331-0 (PDF)
ISBN 978-1-78676-330-3 (ePub)
ISSN 2059-6936 (print)
ISSN 2059-6944 (online)

DOI 10.19103/AS.2020.0066

Typeset by Deanta Global Publishing Services, Dublin, Ireland

# Contents

# Series list

| Title | Series number |
|---|---|
| **Achieving sustainable cultivation of maize - Vol 1**<br>From improved varieties to local applications<br>Edited by: Dr Dave Watson, CGIAR Maize Research Program Manager, CIMMYT, Mexico | 001 |
| **Achieving sustainable cultivation of maize - Vol 2**<br>Cultivation techniques, pest and disease control<br>Edited by: Dr Dave Watson, CGIAR Maize Research Program Manager, CIMMYT, Mexico | 002 |
| **Achieving sustainable cultivation of rice - Vol 1**<br>Breeding for higher yield and quality<br>Edited by: Prof. Takuji Sasaki, Tokyo University of Agriculture, Japan | 003 |
| **Achieving sustainable cultivation of rice - Vol 2**<br>Cultivation, pest and disease management<br>Edited by: Prof. Takuji Sasaki, Tokyo University of Agriculture, Japan | 004 |
| **Achieving sustainable cultivation of wheat - Vol 1**<br>Breeding, quality traits, pests and diseases<br>Edited by: Prof. Peter Langridge, The University of Adelaide, Australia | 005 |
| **Achieving sustainable cultivation of wheat - Vol 2**<br>Cultivation techniques<br>Edited by: Prof. Peter Langridge, The University of Adelaide, Australia | 006 |
| **Achieving sustainable cultivation of tomatoes**<br>Edited by: Dr Autar Mattoo, USDA-ARS, USA & Prof. Avtar Handa, Purdue University, USA | 007 |
| **Achieving sustainable production of milk - Vol 1**<br>Milk composition, genetics and breeding<br>Edited by: Dr Nico van Belzen, International Dairy Federation (IDF), Belgium | 008 |
| **Achieving sustainable production of milk - Vol 2**<br>Safety, quality and sustainability<br>Edited by: Dr Nico van Belzen, International Dairy Federation (IDF), Belgium | 009 |
| **Achieving sustainable production of milk - Vol 3**<br>Dairy herd management and welfare<br>Edited by: Prof. John Webster, University of Bristol, UK | 010 |
| **Ensuring safety and quality in the production of beef - Vol 1**<br>Safety<br>Edited by: Prof. Gary Acuff, Texas A&M University, USA & Prof. James Dickson, Iowa State University, USA | 011 |
| **Ensuring safety and quality in the production of beef - Vol 2**<br>Quality<br>Edited by: Prof. Michael Dikeman, Kansas State University, USA | 012 |
| **Achieving sustainable production of poultry meat - Vol 1**<br>Safety, quality and sustainability<br>Edited by: Prof. Steven C. Ricke, University of Arkansas, USA | 013 |
| **Achieving sustainable production of poultry meat - Vol 2**<br>Breeding and nutrition<br>Edited by: Prof. Todd Applegate, University of Georgia, USA | 014 |

# Reviewers to acknowledge

We wish to acknowledge the following for their help in reviewing particular chapters:

- Chapter 1: Dr Tony Stead, Royal Holloway University of London, UK
- Chapter 2: Professor John Ruter, University of Georgia, USA
- Chapter 5: Professor Thomas Debener, Institute of Plant Genetics - Hannover University, Germany; and Dr Jan De Riek, ILVO, Belgium
- Chapter 6: Dr Yulong Guo, Southwest University, China

We would also like to thank Professor Michelle Jones of Oregon State University, USA for her help in reviewing material for this book.

# Introduction

Ornamental plants are plants grown for decorative purposes such as gardening and landscape design and include cut flowers, bulbs, potted plants, shrubs and trees. Like other crops, ornamentals face challenges such as biotic and abiotic stresses as well as the need to develop more sustainable, 'climate-smart' methods of cultivation. This collection reviews the rich range of research addressing these challenges. Part 1 discusses advances in understanding plant physiology, genetic diversity and breeding techniques. Chapters cover recent research on how plants respond to abiotic stress, ways of exploiting genetic diversity to improve target traits, advances in both conventional and marker-assisted breeding techniques, as well as their use to produce abiotic stress-resistant varieties. Part 2 surveys advances in cultivation techniques in such areas as nutrition, irrigation, protected cultivation and post-harvest storage. The book also includes chapters on developments in integrated disease and pest management.

## Part 1 Physiology and Breeding

As Chapter 1 points out, growers produce hundreds of species from all over the world to market often at a specific time of year. To achieve this, growers must have a sophisticated understanding of the environmental physiology of plant growth and development. This chapter reviews key research on the physiology of sexual and asexual propagation, vegetative and reproductive development, photosynthesis and dormancy.

Chapter 2 discusses ways of exploiting genetic diversity in ornamentals. This chapter reviews collecting, conserving and utilizing ornamental germplasm. Topics include collection strategies, *in situ* and *ex situ* conservation methods. The chapter also summarises the management of ornamental genetic resources in key collections around the world. It concludes with a case study on utilizing ornamental germplasm to exploit genetic diversity in roses.

Chapter 3 reviews developments in breeding programs using conventional techniques such as classical cross breeding and selection, polyploidization and interspecific hybridization. The review summarises advances in key breeding techniques in ornamental crops and discusses their application in such areas as: drought tolerance, low temperature tolerance, elevated pH tolerance, resistance and tolerance to biotic stress, nutrient use efficiency and postharvest life.

Advances in tissue culture techniques for ornamental plant propagation is the focus for Chapter 4. It reviews research on key techniques and factors affecting their effectiveness. Techniques include: clonal propagation and organogenesis, embryo rescue, thin cell layer (TCL) technology, somatic embryogenesis, the use of tissue culture in mutagenesis and mutation breeding, slow-growth conservation techniques, bioreactor/liquid culture, plant improvement through genetic engineering and the use of molecular markers in tissue culture.

Chapter 5 reviews advances in molecular breeding of ornamental plants such as marker-assisted selection. To illustrate developments it includes examples of breeding to improve flower colour and flowering. The chapter also reviews developments in genetic modification using the examples of carnations, roses and chrysanthemums. The chapter also discusses trends in genome sequencing and gene editing. Building on the previous chapter, Chapter 6 examines how innovation in breeding has increased with the advent of genome editing and its application in horticulture. These techniques use nucleases to guide site-directed mutagenesis such as the Clustered Regularly Interspaced Short Palindromic Repeats nuclease-associated systems (CRISPR). To illustrate the potential of the technique, the chapter focuses on the example of reducing ethylene sensitivity in the potted plant *Campanula portenschlagiana*.

Recently there has been more focus in ornamental breeding on the importance of tolerance to abiotic stresses, in response to more extreme weather events related to climate change such as drought and flooding. The final chapter in Part 1, Chapter 7, reviews ways of improving resistance to a wide range of stresses, including: low and high temperatures, drought, waterlogging and saline conditions as well as poor soil quality.

## Part 2 Cultivation techniques

Chapter 8 reviews nutrient management in different production systems for ornamentals. Proper nutrition in containerized and field grown plants is essential for maximized growth and profitability. The chapter discusses the use of macro- and micro-nutrients along with substrate or soil amendments to ensure that plants remain healthy and continue to grow optimally in both mineral soils and soilless substrates. The chapter reviews irrigation and water quality, fertilizer types and usage, the effect of soil and substrates on fertilizers, and the impact of production systems on fertilizer longevity.

Two issues will likely dominate irrigation and water management for ornamental cultivation in the coming years, namely the decreasing availability of fresh water for agricultural use and environmental concerns that will increasingly potentially contaminate the runoff of water from commercial operations. Chapter 9 discusses how both issues are the subject of increasing scrutiny and regulation in the US and Europe, and how many operations are addressing these issues through an array of better management practices. It discusses ways of increasing the precision of irrigation systems and how we can better time the frequency and duration of irrigation events, not only to optimize plant growth but also to limit leaching and runoff.

Chapter 10 addresses key developments in protected cultivation systems for ornamentals. The advent of LED technology has created new and unique opportunities for growing plants, while vertical propagation systems and the ability to precisely control the light spectra have the potential to revolutionize the industry in the coming years. Labour issues have also spurred the engineering of automated and robotic systems to improve production efficiency. The drive to reduce the industry's dependency on synthetic pesticides, non-renewable resources and plastics has also led to the exploration of alternative production techniques and materials. This chapter explores these recent advances as well as future directions for this dynamic industry.

The final chapter in the book addresses the key challenge of diseases affecting ornamentals. It focuses on the example of geophytes which represent a significant segment of the world floriculture industry, particularly in cut flower production. Chapter 11 reviews the main biotic and abiotic diseases affecting geophytes and summarizes current best practice in integrated disease management to reduce fungicide use to address the problem of resistance. To illustrate integrated disease management in practice, it focusses on *Botrytis* infection of ornamental geophytes.

# Part 1

## Physiology and breeding

# Chapter 1

## Environmental physiology of ornamental crops

*John Erwin, University of Maryland, USA*



## 1 Introduction

Ornamental horticulture is one of the most challenging sectors of agriculture. Growers produce hundreds of species from all over the world to market at a specific time of a year or for a holiday. To achieve this, growers must have a sophisticated understanding of the environmental physiology of plant growth and development including propagation and vegetative and reproductive development. These issues will be reviewed in this chapter along with identifying key areas that are gaps in our knowledge that have practical implications. Useful background on this important topic can be found in Dole and Wilkins (2004) and Acquaah (2008).

## 2 Propagation: sexual propagation

The degree of sophistication of seed production, storage, sowing, and germination technologies increased greatly with the advent of the 'plug' or seedling production industry in the 1980s. Seed producers developed better seed harvesting, maturing, and storage procedures that increased the rates of germination success and uniformity. Simultaneously, the seedling production industries installed seed storage facilities, environmentally controlled germination chambers and automated seeders and irrigation booms to ensure

http://dx.doi.org/10.19103/AS.2020.0066.01

uniform and precise environmental and moisture conditions to maximize germination success. Yet seed and seedling producers still have challenges, often around providing optimal conditions for germination. Current knowledge about seed physiology and factors affecting dormancy, germination, and propagation is reviewed in Copeland and McDonald (2001), Basra (2006), Bewley et al. (2013), Geneve (2013), Finch-Savage and Bassel (2016), and Davies et al. (2017).

## 2.1 Seed storage

Improper temperature and/or humidity during seed storage can reduce the percentage of germination, uniformity in germination, and early growth. In general, optimal seed storage conditions for many species indigenous to temperate climates include low humidity and temperatures above 2°C. A general 'rule of thumb' used in the USA is that % humidity + temperature (°F) should be <70. Put in another way, seed longevity doubles when storage temperatures decrease to 5-6°C, and seed longevity doubles when seed moisture content is reduced to 1-2.5% (Roberts, 1991).

Seed viability over time varies with species. Seed viability of *Aquilegia*, *Callistephus*, *Catharanthus*, *Centauria*, *Cleome*, *Consolida*, *Delphinium*, *Linum*, *Lunaria*, *Phlox*, *Salvia*, *Torenia*, *Verbena*, and *Viola* decreases after the first year (Dole and Wilkins, 1999); in addition, the vigor of germinated seedlings decreases as the seed age increases (Corbineau and Come, 1991). Species in the genera *Anemone*, *Asparagus*, *Callistephus*, Calceolaria, *Delphinium*, *Dahlia*, *Gloxinia*, *Phlox*, *Salivia*, and *Viola* are generally viable for only 1-2 years. In contrast seed of species in the *Begonia*, *Dendranthema*, *Centaurea*, *Cyclamen*, and *Tropaeolum* genera can remain viable for 3-15 years (Corbineau and Come, 1991).

## 2.2 Temperature and germination

Delivering the optimal temperature for seed germination is still one of the greatest challenges in commercial production; this is particularly the case if growers are germinating seed in a greenhouse. In general, seedling producers manage air temperatures to maintain a media temperature of 22-24°C to maximize germination (Erwin, personal observation); this temperature range is optimal for many ornamental plant species (Table 1). In contrast to most species, foxglove (*Digitalis purpurea*), *Phlox*, and sweet pea (*Lathyrus splendens*), seed germination is optimal at 16-18°C. Vinca (*Catharanthus roseus*) and tomato (*Lycopersicon esculentum*) seed germination is optimal at 24-27°C.

Fluctuating temperatures can stimulate the germination of some species (Aragino, 1981). This can be the case with species from areas with large day/

**Table 1** Optimal temperature, light, and moisture requirements for seed germination of different ornamental crop species

| Species | Temperature | Light | Cover | Moisture | Storage |
|---|---|---|---|---|---|
| *Abutilon* | 22-24 | - | Yes | - | - |
| *Achillea* | 18-21 | Light | No | - | - |
| *Agastache* | 15-21 | - | Yes | - | - |
| *Ageratum houstonianum* | 24-25 | - | No | M | M |
| *Ammi majus* | 22-24 | Light | - | - | - |
| *Anchusa capensis* | 20-21 | Light | - | - | - |
| *Anemone coronaria* | 18-21 | - | Yes | - | S |
| *Angelonia* | 22-24 | Light | No | - | - |
| *Antirrhinum majus* | 18-21 | Light | No | M | M |
| *Aquilegia caerulea* | 21-24 | - | Yes | - | - |
| *Asclepias* | 21-24 | - | Yes | - | - |
| *Asclepias tuberosa* | 18-24 | - | - | - | - |
| *Asparagus densiflorus* | 29 day/24 night | Dark | Yes | M | S |
| *Astilbe* x *arendsii* | 21-24 | - | No | - | - |
| *Astrophytum myriostigma* | 20-25 | - | - | - | - |
| *Bacopa* | 20-23 | Light | No | - | - |
| *Begonia* 'Dragon Wing' | 24-25 | Light | No | - | - |
| *Begonia* x *hiemalis* | 24-26 | Light | - | - | - |
| *Begonia semperflorens* | 24-27 | Light | No | W | S |
| *Begonia tuberhybrida* | 24-26 | Light | No | W | S |
| *Bellis perennis* | 21-24 | Light | | | |
| *Brachycome iberidofolia* | 21-22 | Light | - | M | - |
| *Browallia speciosa* | 24-25 | - | No | M | S |
| *Calceolaria herbeohybrida* | 18-21 | Light | - | M | - |
| *Calendula officinalis* | 20-21 | Dark | Yes | D | - |
| *Callistephus chinensis* | 18-21 | - | Yes | D | S |
| *Campanula isophylla* | 15-18 | Light | No | - | - |
| *Canna* x *generalis* | 21-24 | Dark | - | M | - |
| *Catharantus roseus* | 25-26 | Dark | Yes | W | S |
| *Celosia argentea* | 24-25 | - | No | W | M |
| *Centaurea americana* | 21-24 | - | Lightly | M | - |
| *Clarkia amoena* | 20-22 | Dark | - | - | - |
| *Cleome hassleriana* | 26 day/21 night | Light | No | M | |
| *Cobea scandens* | 21 | Cover lightly | - | - | - |
| *Cordyline indivisa* | 22-26 | Light | Lightly | M | - |

(Continued)

**Table 1** (Continued)

| Species | Temperature | Light | Cover | Moisture | Storage |
|---|---|---|---|---|---|
| *Coreopsis grandiflora* | 18-24 | Light | No | D | - |
| *Cosmos bipinnatus* | 18-21 | - | Yes | D | - |
| *Crossandra infundibuliformis* | 26 day/21 night | - | Yes | - | - |
| *Cuphea platycentra* | 21 | - | Cover lightly | M | - |
| *Cyclamen persicum* | 18-20 | Dark | - | W | M |
| *Cynoglossum amabile* | 18-21 | Dark | - | - | - |
| *Dahlia x hybrida* | 20-21 | - | Yes | D | M |
| *Delphinium grandiflorum* | 24-26/18-21 | - | Yes | - | S |
| *Dendranthema x grandiflorum* | 15-21 | Light | - | M | - |
| *Dianthus barbaratus* | 21-24 | - | Cover lightly | M | M |
| *Dianthus chinensis* | 18-20 | - | Cover lightly | M | M |
| *Diascia* | 18-21 | - | Yes | - | - |
| *Dichondra* | 22-24 | - | Yes | - | - |
| *Digitalis purpurea* | 21-24 | Light | No | - | - |
| *Dolichos lablab* | 21-22 | Dark | - | - | - |
| *Echinacea purpurea* | 20-22 | - | Yes | - | - |
| *Eschscholzia californica* | 21-22 | - | - | - | - |
| *Euphorbia marginata* | 15-20 | Dark | - | - | - |
| *Eustoma grandiflorum* | 20-25 | Light | Lightly | - | - |
| *Exacum affine* | 22-25 | - | Cover lightly | M | - |
| *Freesia* spp. | 15-20 | Dark | Yes | - | - |
| *Gaillardia x grandiflora* | 21-24 | Light | No | M | - |
| *Gaura lindheimeri* | 21-22 | - | Yes | - | - |
| *Gazania rigens* | 20-21 | Dark | Yes | D | - |
| *Gerbera jamesonii* | 20-22 | Light | Yes | M | - |
| *Gomphrena globosa* | 22-24 | - | Yes | M | - |
| *Helianthus annuus* | 18-25 | - | Yes | M | - |
| *Helichrysum bracteatum* | 21-24 | - | Yes | M | - |
| *Heliotrope* | 24-26 | - | - | M | - |
| *Hibiscus moscheutos* | 21-26 | - | Yes | - | - |
| *Hypoestes phyllostachya* | 21-24 | - | No | M | - |
| *Iberis amara* | 15-18 | Light | Cover lightly | W | - |
| *Impatiens balsamina* | 22-25 | Light | - | W | S |
| *Impatiens walleriana* | 22-24 | Light | No | W | S |
| *Impatiens hawkeri* | 22-25 | Light | No | W | S |
| *Lathyrus splendens* | 13-20 | - | Yes | - | L |

| Species | Temperature | Light | Cover | Moisture | Storage |
|---|---|---|---|---|---|
| *Lavendula angustifolia* | 18–24 | – | Yes | M | – |
| *Lavatera trimestris* | 22 | Dark | Yes | – | – |
| *Leucanthemum lacustre* | 20–25 | – | Cover lightly | M | L |
| *Limonium sinuatum* | 21–25 | Dark | – | M | – |
| *Linum perenne* | 21–25 | – | Cover lightly | – | – |
| *Lisianthus* | 20–22 | Light | No | M | M |
| *Lobelia erinus* | 21–24 | Light | No | M | M |
| *Lobularia maritima* | 20–22 | Light | – | M | M |
| *Lupinus polyphyllus* | 18–21 | – | Yes | M | – |
| *Mathiola incana* | 17–21 | – | Yes | – | L |
| *Melampodium divaricatum* | 18–20 | – | Cover lightly | M | – |
| *Mimosa pudica* | 26 | Dark | Yes | – | – |
| *Mimulus* x *hybridus* | 15–21 | Light | No | M | – |
| *Mirabilis jalapa* | 22 | Dark | Yes | – | – |
| *Nemesia strumosa* | 20–23 | – | Yes | – | – |
| *Nemophila menziesii* | 18 | Dark | Lightly | – | – |
| *Nicotiana alata* | 21–24 | Light | No | M | – |
| *Nierembergia hippomanica* var. *violacea* | 22–24 | – | Cover lightly | M | – |
| *Nigella damascene* | 18–21 | – | Cover lightly | – | – |
| *Opuntia* spp. | 25–35 | – | – | – | – |
| *Osteospermum* | 21–24 | – | Yes | – | – |
| *Papaver nudicaule* | 15–18 | Light | No | M | – |
| *Papaver orientale* | 18–24 | Light | No | M | – |
| *Pelargonium* x *hortorum* | 21–24 | – | Yes | W | M |
| *Pelargonium peltatum* | 21–24 | – | Yes | W | M |
| *Pentas* | 23–25 | – | No | M | – |
| *Pericallis* x *hybrida* | 20–24 | Light | – | M | – |
| *Petunia* x *hybrida* | 24–26 | Light | No | M | M |
| *Phlox drummondi* | 18–20 | Dark | Yes | D | S |
| *Platycodon grandifloras* | 20–22 | Light | No | M | – |
| *Portulaca grandiflora* | 25–26 | – | No | M | M |
| *Primula vulgaris* | 18–20 | Light | Lightly | W | – |
| *Ranunculus asiaticus* | 15–20 | – | Yes | M | – |
| *Rudbeckia fulgida* | 28–31 | – | Cover lightly | M | – |
| *Rudbeckia hirta* | 20–21 | – | Yes | M | – |
| *Salpiglossus sinuate* | 21–22 | Dark | No | – | – |

*(Continued)*

**Table 1** (Continued)

| Species | Temperature | Light | Cover | Moisture | Storage |
|---|---|---|---|---|---|
| *Salvia coccinea* | 24-26 | Dark | Yes | W | S |
| *Salvia farinacea* | 21-25 | - | Yes | W | S |
| *Salvia splendens* | 21-25 | - | Yes | W | S |
| *Scabiosa atropurpurea* | 21-24 | Light | No | - | - |
| *Senecio cineraria* | 22-24 | - | No | M | M |
| *Sinningia speciosa* | 18-21 | Light | No | - | - |
| *Solenostemon scutellarioides* | 22-24 | - | Yes | M | M |
| *Stenocereus griseus* | 24-25 | Yes | No | - | - |
| *Streptocarpus* x *hybridus* | 21 | Light | No | - | - |
| *Tagetes erecta* | 21-22 | - | Yes | M | M |
| *Tagetes patula* | 21-22 | - | Yes | M | M |
| *Thunbergia alata* | 21-24 | - | Yes | M | - |
| *Tithonia* | 20-21 | - | Yes | - | - |
| *Torenia fournieri* | 22-24 | Light | No | M | - |
| *Tropaeolum majus* | 18-21 | Dark | Yes | M | - |
| *Verbena* x *hybrida* | 24-26 | - | Yes | D | M |
| *Verbena bonariensis* | 24 | - | Cover lightly | D | M |
| *Viola tricolor* | 18-24 | - | Yes | W | S |
| *Viola wittrockiana* | 20-21 | - | Cover lightly | M | S |
| *Zinnia angustifolia* | 20-22 | - | Yes | D | L |
| *Zinnia elegans* | 20-22 | - | Yes | D | L |

*Source*: developed from Styer and Koranski (1997), Dole and Wilkins (1999), Ball Horticultural Germination Fact Sheet, Pimiento-Barrios and del Castillo (2002), Erwin, personal observation.

night temperature (DT/NT) fluctuations such as deserts and mountainous regions (Tudela-Isanta et al., 2018). For instance, *Rumex retroflexus* will not germinate when maintained at constant 20°C or 30°C, but 100% germination occurs when temperature is fluctuated between 20°C and 30°C diurnally (Toole et al., 1955). *Cucumis anguria* germination is inhibited by light, but it will germinate in the light if temperatures are fluctuated between 15°C and 35°C diurnally (Felippe, 1980). *Rumex obtusifolius* requires light to germinate, but it will germinate in the dark if temperature is fluctuated between 15°C and 35°C (Toole et al., 1955). Interestingly, recent research on cacti suggest that many species germinate better with constant 20–25°C rather than with fluctuating temperatures; authors suggest this may encourage germination under plants or in crevices where survival may be maximized (Lindow-Lopez et al., 2018).

Interestingly, short-term low- or high-temperature pulses can stimulate germination in some cases (Takaki et al., 1981). *Cucumis* imbibed at 25°C can be stimulated to germinate by a 0°C exposure for 2 h compared to seed germinated at constant temperature (Felippe, 1980). The seed of *Rumex*, *Nicotiana*, and *Nigella* can be stimulated to germinate by a short-term high-temperature exposure (35°C; Aragino, 1981). Similarly, a 33°C exposure for 30 min can stimulate lettuce seed germination compared to seed germinated at constant temperature (Hendricks and Taylorson, 1976).

The seed of a number of perennial and woody species must be 'stratified' to germinate (VanDeWoude and Toole, 1980). Stratification is a cool (4–6°C), moist (seed must be imbibed) treatment for 6–12 weeks that promotes germination (Hartmann and Kester, 1959). Interestingly, the effectiveness of a stratification treatment in promoting germination can be modified by light; a red-light exposure after imbibition, but prior to stratification, can reduce the duration of the stratification treatment needed to promote germination (Bevington, 1981; Toole et al., 1962; Duke et al., 1977). The ability of red light to reduce stratification is reversible if the red-light exposure is followed immediately by a far-red light exposure (Taylorson and Hendricks, 1976; Cone and Kendrick, 1986).

The basis for how stratification promotes germination is unclear. During the cooling process abscisic acid (ABA) levels decrease and gibberellin (GA) levels increase (Taiz and Zeiger, 1998; Oracz and Karpinski, 2016). ABA content in *Fraxinus americana*, *Juglans regia*, and *Corylus avellana* decreases as the chilling duration increases (Wareing and Saunders, 1971). In contrast, the application of GA4+7 overcomes the stratification requirement of a number of species (Felippe, 1980).

## 2.3 Light and germination

Light affects the germination of some species, but not others. Borthwick and Hendricks showed that lettuce (*Lactuca sativa*) seed germinated only after the exposure to red light (660–680 nm) and that the promotion of germination was reversible if a far-red light (720–740 nm) exposure occurred immediately after the red-light exposure (Borthwick et al., 1952a,b). Sunlight has a red:far-red light ratio of 1.1 (Taiz and Zeiger, 1998; Erwin et al., 2006). The red:far-red light ratio of sunlight decreases under plant canopies as leaves preferentially filter red versus far-red light; the red:far-red ratio under a canopy of 42 herbaceous ornamental crop species (single leaf) varied from 0.16 to 0.58 (Erwin et al., 2006). The low-red:far-red light ratio under a canopy may be a mechanism to promote germination only when the seed survival is maximized when direct the sunlight hits the ground (red:far red > 1.0). Light exposure ensures seed germination only when the seed is not covered deep in soil/media in which condition the survival would be reduced. In contrast, some species have a dark requirement for seed germination, presumably to ensure the seed is buried deep enough to maximize the survival (Table 1).

A light requirement for seed germination is evident in many small-seeded species (Erwin, personal observation; Table 1). For instance the seed germination of *Begonia*, *Calceolaria*, and *Sinningia* is promoted by red light; all are small seeded (Table 1). In contrast, a dark requirement for seed germination occurs in a number of large-seeded species; darkness promotes the germination of *Dahlia*, *Tagetes*, *Calendula*, *Cyclamen*, and *Tropaeolum* (Table 1). Interestingly, some more primitive species grown as ornamental crops, such as ferns, have different light germination requirements; spore germination of different fern species is promoted by red light, but can be inhibited by far-red light, blue light (440–480 nm) in or dark promoted germination (Suo et al., 2015). Red-light promotion of germination is impacted by temperature. VanDeWoude and Toole (1980) found that temperatures below 18°C increased the sensitivity of lettuce seed germination to red-light.

The intensity of light required to promote, or inhibit, seed germination is low (Borthwick et al., 1952a); generally, <2 μmol $m^{-2}$ $s^{-1}$ is provided in ornamental species germination chambers (Dole and Wilkins, 1999; Erwin, personal observation). Historically, seedling producers deliver red light to light-requiring species to promote germination by germinating seed in the greenhouse under sunlight (red:far-red ratio of approximately 1.1) or in germination chambers with vertically mounted fluorescent lights (red:far-red ratio of approximately 15). Advances in recent lighting technology have resulted in light-emitting diodes (LEDs) that can emit nearly monochromatic red light with a higher electrical efficiency than fluorescent lamps. Although not yet commonplace, red LEDs will likely become commonplace in germination chambers moving forward.

## 2.4 Moisture and germination

Either limited, or excess, moisture can reduce seed germination (Styer and Koranski, 1997). Precisely managing the moisture environment in the microenvironment surrounding a seed can be challenging. Sufficient moisture must be present to imbibe seed, but not so much that gas exchange is limited. Vinca (*C. roseus*), impatiens (*Impatiens wallerana*), and gerbera (*Gerbera jamesonii*) germination can be inhibited by excess moisture and/or limited media air/pore space. Media pore space can be reduced by fine particle size, over-compression of media prior to sowing, or overwatering (Erwin, personal observation). In some cases the seedling radicle has difficulty initially penetrating saturated/compressed media. Species in which radicle penetration can reduce seedling establishment include vinca, gerbera, and impatiens (Erwin, personal observation). Optimal moisture levels for the seed germination of different species are outlined in Table 1.

## 2.5 Seed dormancy

Germination of some species is contingent on an environmental cue being fulfilled or a period of time passing (Taiz and Zeiger, 1998). For instance, seed of temperate perennial species may require a stratification treatment before germination is possible, presumably to ensure germination occurs after winter (Corbineau and Come, 1991). In contrast, species from hot, dry climates can require a prolonged moisture exposure before germination can occur, presumably to inform the seed that a dry season is done. As mentioned earlier, some seed require a red-light exposure before germination can occur. *Helianthus* germination is cold insensitive, and ethylene or dry storage promotes germination (Corbineau et al., 1990). Many species simply require the passage of time before germination is possible, which results in a decrease in the endogenous germination inhibitor ABA (Taiz and Zeiger, 1998). In some cases, dormancy can be overcome early by submerging seed in a solution that has GA (Corbineau and Come, 1991).

## 2.6 Special factors impacting seed germination

In some cases, we are just now understanding that many species can have very specific requirements that are often remnants from their indigenous habitat. Two examples of special requirements are listed as follows:

- Orchid seed lack an endosperm. In some cases, seedling survival will not occur unless germination occurs in which naturally occurring fungi/bacteria are present that develop a synergistic relationship with the seed to provide needed carbohydrates. In a commercial setting, growers circumvent this process by germinating seed on a carbohydrate-rich agar-based medium.
- Some species from regions with periodic fires must be exposed to a solution from burned foliage to promote germination (Keeley and Pausas, 2018).

## 2.7 Seedling storage

In some cases, ornamental crop finishers need to hold seedlings (plugs) for a period of time (with minimal leaf unfolding) to allow for planting into finished containers over time. Alternatively, storing plugs allows a plug producer to hold them and ship at a desired time to a finisher and open up valuable greenhouse space for additional production. A storage environment should be cool to reduce leaf unfolding, elongation, and respiration. Light in the storage environment generally enables the plugs to be stored longer (Heins et al., 1994). In addition, if light is not present, storage temperature is often decreased (Heins et al., 1994). Storage requirements for different seed-propagated ornamental species are shown in Table 2.

### 2.8 Gaps in knowledge

There are several interesting areas that represent 'gaps' in our knowledge of seed germination that offer research opportunities, including:

- *Epigenetics*: There is increasing evidence that the environmental conditions of the 'mother plant' from which the seed is harvested impart stress tolerance on seedling progeny (Cendan et al., 2013). The question arises as to whether mother plants can be stressed to create more resilient seedlings.
- *Temperature fluctuations*: Little work has been conducted on the impact of diurnally fluctuating temperatures, as well as on the impact of short-term high- or low-temperature exposures, on the seed germination of ornamental crop species. This may be an interested area to study on species with high seed viability but poor germination percentage.
- *Shipping - temperature and fertility*: Seedlings are routinely shipped long distances. Increasingly, during the shipping process there are significant losses; often the basis for losses is not known. In particular, how temperature, light, media fertility, and ethylene during shipping interact to affect seedling/plant survival is of interest.
- *Media inoculation with fungi/bacteria to improve stress tolerance*: There is increasing evidence that inoculation with biological organisms can improve the stress tolerance of roots, above-media growth, and/or disease tolerance. This area could benefit finishers, shippers, retailers, and consumers.

## 3 Propagation: asexual propagation

Asexual propagation allows for the rapid introduction and commercialization of a plant with a new trait with no consideration as to whether that new trait is heritable. Asexual propagation techniques commonly used in ornamental crops production include division, propagation from cuttings, and tissue culture. Division is primarily used to propagate herbaceous perennial species that do not propagate readily from cuttings or tissue culture. However, propagation via division is becoming increasingly less common because of high production, harvesting, and shipping costs. Cutting propagation is used predominantly with species that root easily from vegetative stem cuttings. When propagating from cuttings, 'stock' or 'mother plants' must be grown either on site or at a cutting supplier's facility (and shipped). Tissue culture of ornamental species is

**Table 2** Storage conditions and duration for plugs of several seed-propagated ornamental plant species with and without lighting in storage

| Species | Light | Temperature (°C) | Duration (weeks) |
|---|---|---|---|
| *Ageratum houstonianum* | Yes | 5-13 | 6 |
| | No | 7 | 6 |
| *Begonia semperflorens - cultorum* | Yes | 2-7 | 6 |
| | No | 5-7 | 6 |
| *Begonia tuberhybrida* | Yes | 5-7 | 6 |
| | No | 5-7 | 4 |
| *Catharanthus roseus* | Yes | 7-13 | 6 |
| | No | 7-13 | 3 |
| *Celosia argentea* | Yes | 10 | 1 |
| | No | 10 | 1 |
| *Cyclamen persicum* | Yes | 2-5 | 6 |
| | No | 2-5 | 6 |
| *Dahlia x hybrida* | Yes | 5-10 | 3 |
| | No | 5-7 | 1 |
| *Impatiens hawkeri* | Yes | 13 | 3 |
| | No | 13 | 1 |
| *Impatiens walleriana* | Yes | 7 | 6 |
| | No | 7 | 6 |
| *Lobelia erinus* | Yes | 5 | 6 |
| | No | 5 | 4 |
| *Lobularia maritima* | Yes | 2 | 6 |
| | No | 2 | 2 |
| *Pelargonium x hortorum* | Yes | 2 | 4 |
| | No | 2 | 4 |
| *Petunia x hybrida* | Yes | 2 | 6 |
| | No | 2 | 6 |
| *Portulaca grandiflora* | Yes | 10-13 | 6 |
| | No | 5-7 | 6 |
| *Salvia splendens* | Yes | 7-13 | 6 |
| | No | 7 | 4 |
| *Tagetes patula* | Yes | 5 | 6 |
| | No | 5 | 3 |
| *Verbena x hybrida* | Yes | 7-13 | 4 |
| | No | 7-13 | 1 |
| *Viola x wittrockiana* | Yes | 2-7 | 16 |
| | No | 0-2 | 16 |

*Source*: adapted from Heins et al. (1994).

becoming more common with orchids, species that produce few cuttings, and/or high-value crops (Aparecida de Oliveira et al., 1995; Starling and Dodds, 1983). Cutting and tissue culture propagation will be reviewed in this section. Current research on asexual propagation is reviewed in De Kerck (2002), Oinam et al. (2011), and Davies et al. (2017).

## *3.1 Cutting propagation*

Cutting propagation is commonplace with potted plants such as chrysanthemum, poinsettia, kalanchoe, and African violets, as well as many foliage plants (Dole and Wilkins, 1999). Stock plants are managed using photoperiod or by the application of ethylene (via ethephon, Florel, etc.) to cause early flower bud abortion to produce cuttings that are 'vegetative' or 'flowerless' that will root more readily.

Asexual propagation of flowering ornamental annuals and perennials, other than geraniums, is a more recent development. Historically, stock plants were produced by a finisher who harvested cuttings to root and finish. In some cases, spring annual finishers grew geranium stock plants, harvested cuttings from those stock plants, and rooted and finished flowering geraniums for sale. Over time, poinsettia, chrysanthemum, and geranium cuttings were produced by domestic suppliers who grew stock plants, harvested cuttings, and shipped to finishers. More recently, stock plant (and cuttings) production has moved to warm locations in other countries where stock can be grown outdoors such as Central America, Africa, and Guatemala where input costs are lower.

Ornamental species differ in the optimal conditions for cutting shipping and storage. Generally, optimal cutting storage temperatures vary from 4°C to 16°C (Table 3). Some species, such as *Ipomea*, vinca, or *Celosia*, have optimal cutting storage temperatures above 13°C; temperatures below 13°C can reduce and delay rooting. In contrast, chrysanthemum (a temperate perennial species) cuttings are shipped at temperatures from 2°C to 6°C without deleterious effects. Generally, as storage temperature increases, cutting respiration increases, thus reducing the carbohydrate status, which can decrease subsequent rooting (Enfield, 2011).

The length of time a cutting can be stored and successfully rooted differs among species. In some cases, unventilated cutting storage duration is limited by ethylene sensitivity; ethylene concentrations as low as 100 ppb for as short as 2 h can damage plants (Nell, 1993). In all cases, boxed cuttings should be unpacked upon arrival to release ethylene and excess moisture and placed in a cooler room at the appropriate temperature. Cuttings should be prioritized for unpacking, sticking, and placement in the propagation environment, with the most sensitive species being stuck first (Table 3). What determines the rapidity a cutting should be stuck is likely a combination of ethylene sensitivity and respiration rate.

### 3.2 Rooting temperature

Optimal media temperature for root cutting is generally between 22°C and 24°C for many herbaceous species. However, some tropical species may root more rapidly or completely with warmer media temperatures (25–35°C). Woody species can often be more tolerant of varying media temperatures and will eventually root, particularly if they are dormant wood cuttings.

### 3.3 Rooting photoperiod

The photoperiod during which stock plants are grown, and the photoperiod in the rooting environment, can affect how, or if, cuttings take root. In general, stock plants and cuttings of photoperiodic species are grown in an environment to promote vegetative growth; root cutting is more rapid and complete when plants are maintained in the vegetative state. For instance, *Dahlia*, *Tarragon* (*Artemesia dracunculus*), and tuberous begonia (*Begonia* x *tuberhybrida*) root poorly under short-day (SD) conditions because SDs promote the storage root formation in these crops (Erwin, personal observation, Table 3). Therefore, stock plants are grown and cuttings are rooted under long-day (LD) conditions (often with night interruption (NI) lighting; see the later discussion). In contrast, *Petunia, Lobelia, and Verbena* naturally flower under LDs; therefore, stock plants are grown and cuttings are rooted under SDs (Table 3). Similarly, as SDs promote the flowering of chrysanthemum (*Dendranthema grandiflorum*) and poinsettia (*Euphorbia pulcherrima*), stock plants of these crops are grown and cuttings of these crops are rooted under LD conditions (Table 3). We note that it is common for stock plants and cuttings to be sprayed with compounds that release ethylene (such as ethephon) to cause early flower bud abortion and to enable them grow under LD conditions to maximize photosynthesis and mass (see the later discussion).

### 3.4 Rooting environment irradiance

Although cuttings are severed from roots, they are still capable of conducting photosynthesis if hydrated. Providing adequate irradiance to maximize the carbohydrate status of cuttings results in earlier and more complete rooting (Lopez and Runkle, 2008). Irradiance is managed in a cutting propagation facility to minimize leaf transpiration while maximizing photosynthesis. Optimal irradiance and daily light integral levels vary among species. Lopez and Runkle (2008) showed that the degree of rooting increased and crop time decreased as

**Table 3** Optimal cutting storage temperatures and 'sticking' priority of different common vegetative annuals propagated from cuttings

| Plant species | Optimal storage temperature (°C) | Sticking priority | Ethylene sensitivity |
|---|---|---|---|
| *Achimenes* | - | - | Very |
| *Aeschynanthus* | - | - | Very |
| *Ageratum* | - | 3 | Moderate |
| *Alternanthera* | - | 3 | |
| *Angelonia* | 10-12 | 3 | - |
| *Anthirrinum* | - | - | Moderate |
| *Arctotis* | - | 2 | - |
| *Argyranthemum* | 7-10 | 3 | Slight |
| *Artemesia* | - | 2 | - |
| *Asparagus* | - | - | Slight |
| *Bacopa* | 7-10 | 2 | - |
| *Basil* | 13-15.5 | - | - |
| *Begonia boliviensis* | - | 2 | Very |
| *Begonia semperflorens - cultorum* | - | - | Very |
| *Bidens* | 7-10 | 3 | - |
| *Bouganvillea* | - | - | Very |
| *Brachyscome* | 7-10 | 3 | - |
| *Bracteantha* | 7-10 | 3 | Slight |
| *Browallia* | - | - | Very |
| *Calibrachoa* | 7-10 | 2 | Slight |
| *Celosia* | - | 4 | - |
| *Chrysocephalum* | - | 3 | - |
| *Coleus* | 10-13 | 2 | - |
| *Clerodendrum* | - | - | Very |
| *Cuphea* | 7-10 | 3 | - |
| *Dahlia* | 10-13 | 1 | Moderate |
| *Dendranthema* | - | - | Moderate |
| *Diascia* | - | 2 | Slight |
| *Erysimum* | 10-13 | 2 | - |
| *Evolvulus* | - | 2 | - |
| *Exacum* | - | - | Slight |
| *Felicia* | - | 3 | - |
| *Fuchsia* | - | 2 | Very |
| *Gaura* | - | 4 | - |
| *Geranium* (ivy) | - | 1 | Moderate |

| Plant species | Optimal storage temperature (°C) | Sticking priority | Ethylene sensitivity |
|---|---|---|---|
| *Geranium* (zonal) | - | 1 | Moderate |
| *Gypsophila* | - | - | Slight |
| *Hedera* | - | 4 | Slight |
| *Helichrysum* | - | 3 | - |
| *Heliotrope* | 10-13 | 1 | - |
| *Hibiscus* | - | 1 | Very |
| *Impatiens* (double) | 10-13 | 2 | |
| *Impatiens* (interspecific) | 10-13 | 3 | |
| *Impatiens* (walleriana) | 10-13 | 2 | Moderate |
| *Impatiens* (hawkeri) | - | 3 | Very |
| *Ipomea* | 13-15.5 | 1 | - |
| *Iresine* | 10-13 | 3 | - |
| *Kalanchoe* | - | - | Very |
| *Lantana* | - | 1 | Very |
| *Lamium* | - | 3 | - |
| *Lobelia* | 7-10 | 1 | Slight |
| *Lobularia* | - | 2 | - |
| *Lophospermum* | - | 4 | - |
| *Lysimachia* | - | 4 | - |
| *Nemesia* | 7-10 | 2 | Moderate |
| *Nierembergia* | - | 3 | Slight |
| *Osteospermum* | 7-10 | 2 | Moderate |
| *Petunia* | 7-10 | 2 | Slight/very |
| *Perilla* | 10-13 | 2 | - |
| *Penstemon* | - | - | Moderate |
| *Petchoa* | - | 2 | - |
| *Phlox* | - | 3 | Moderate |
| *Plectranthus* | 7-10 | 2 | - |
| *Portulaca* | - | - | Very |
| *Purslane* | 13-15.5 | 1 | Very |
| *Poinsettia* | 10-13 | 1 | Moderate |
| *Rosa* | 2-4 | - | Slight |
| *Salvia* | 7-10 | 2 | Moderate |
| *Sanvitalia* | - | 4 | - |
| *Strobilanthes* | 10-13 | - | - |

(*Continued*)

**Table 3** (Continued)

| Plant species | Optimal storage temperature (°C) | Sticking priority | Ethylene sensitivity |
|---|---|---|---|
| *Streptocarpus* | - | 4 | - |
| *Scaevola* | 7-10 | 3 | Slight |
| *Schlumbergera* | - | - | Very |
| *Sutera* | - | - | Moderate |
| *Thunbergia* | 10-13 | 1 | - |
| *Torenia* | - | 3 | Moderate |
| *Verbena* | 7-10 | 2 | Slight |
| *Vinca Vine* | - | 4 | - |
| *Viola* | - | 2 | - |

Sticking cutting is defined as highest (1) to lowest priority (4).
*Source*: adapted from Ball Horticultural Fact Sheet - Step up to the plate - Grow a better liner (2015), Mayers et al. (1997), Dole and Wilkins (1999), Leatherwood and Dole (2007), and Williams et al. (2006).

daily light integral increased from 1.2 to 7.5 mol $m^{-2}$ $d^{-1}$ and 10.7 mol $m^{-2}$ $d^{-1}$ for petunia and New Guinea impatiens, respectively. Currey et al. (2012) showed that the total, shoot, and root dry mass of *Argyranthemum*, *Diascia*, and *Nemesia* increased significantly when daily light integral was increased from 1.2 to 12.3 mol $m^{-2}$ $d^{-1}$. In general, 100–200 μmol $m^{-2}$ $s^{-1}$ of light is provided initially after sticking the cutting in media and during callus and early root formation (root length < 1 cm). Irradiance is increased to 300–400 μmol $m^{-2}$ $s^{-1}$ when roots are >1/2 cm in length and to >500 μmol $m^{-2}$ $s^{-1}$ on rooted cuttings (roots hit the edge of the container). In some cases, such as with tissue culture cuttings/rooted cuttings, initial irradiance levels are as low as 50 μmol $m^{-2}$ $s^{-1}$ (Erwin, personal observation).

## 3.5 Rooting and moisture

In most cases, additional moisture, in the form of mist or fog, is applied to foliage and media or in the rooting environment to minimize transpiration and promote rooting. Therefore, ornamental cutting propagators periodically mist or fog root cuttings during the day; plants do not receive mist during the night. Cuttings are commonly propagated in an open greenhouse with stationary misting systems or with a mobile mist boom that can often mist an entire greenhouse bay. The most common herbaceous ornamental plant species such as chrysanthemum, poinsettia, geranium (*Pelargonium zonale*), and petunia are propagated in this way.

In contrast, some species root best when they do not receive overhead water; these species are often propagated in mist tents where high humidity can be maintained without regular overhead watering or with fog. For instance, *Hibiscus*, *Campanula*, *Rhododendron*, and *Jasmine* root better under humidity tents than under mist and are often propagated in this way (Erwin, personal observation). A number of species with silver foliage indigenous areas with low rainfall prefer wet-dry cycles to promote rapid and complete rooting; providing continuous overhead mist slows rooting and can result in the rotting of foliage (Erwin, personal observation).

Rooting of some species is inhibited by any media moisture at all. Most notably, rooting of some succulents is promoted by little or no media moisture, no misting, and no humidity tent. This phenomenon is perhaps most obvious with severed *Echeveria* leaf cuttings that initiate and grow roots from their base in the open air (Cabahug et al., 2018). Similarly, *Bryophyllum* plantlets form on the edge of mature leaves and drop after they unfolded a single leaf set and after forming roots in the air. Even in the open air, *Echeveria* leaf cuttings have an optimal temperature for rooting (Erwin et al., 2015). Succulents that are stem propagated are often stuck in barely moistened media. For instance, kalanchoe (*Kalanchoe blossfeldiana*), *Echeveria*, and *Aloe* root cutting is also done in barely moistened media (Cabahug et al., 2018; Erwin, personal observation).

## 3.6 Plant growth regulators and rooting

Synthetic auxins are applied to the base of stem cuttings in a dry talc or liquid form to promote early, uniform, and increased rooting (Hartmann and Kester, 1959). For instance, indolebutyric acid (IBA) is applied (1000–3000 ppm) to cuttings of numerous herbaceous ornamental species in a talc form to promote rooting. IBA rates are increased as the difficulty in rooting a stem cutting of a species increases. IBA rates of 3000 to 16 000 ppm are applied to the base of semi-hardwood or hardwood cuttings of woody ornamentals to promote rooting. In some cases, napthalacetic acid (NAA) is applied alone, or in combination with IBA to promote rooting of more-difficult-to-root woody plant species such as *Camellia*.

More recently, water-soluble forms of IBA (typically, potassium salts; K-IBA) are sprayed on foliage after sticking cuttings in media. The foliarly applied IBA is then transported basipetally where it promotes rooting at the base of the cutting. Generally, concentrations of 100–200 ppm are sprayed on the foliage of herbaceous cuttings. This is now a common practice in summer poinsettia propagation. A higher concentration, 200–700 ppm, is sprayed on the foliage of more-difficult-to-root or woody species. Alternatively, a second spray application can be made 7 to 14 days later on difficult-to-root species. This technique of applying rooting hormone is gaining popularity as it allows for reduced cutting handling (that can spread pathogens), reduces labor costs, and results in earlier, and more complete, rooting compared to historical basal talc or liquid application techniques. It is also likely that more of the IBA is able to enter the plant than with dry talc, for instance.

## 3.7 Tissue culture (TC)

Tissue culture (TC) generally involves the promotion of undifferentiated callus cells on an agar-based medium in a laboratory setting. That agar-based medium used in this process often includes key plant growth regulators (GA, cytokinins, and auxin) as well as some carbohydrates, often in the form of sucrose (Fujiwara et al., 1992). Growth of callus, and differentiation of callus into shoots, is promoted by altering the hormonal concentration in the agar media. Shoot tips that develop from callus are removed and rooted either in another tissue culture vessel or in a cutting propagation environment (Amoo et al., 2009).

Tissue culture vessels are typically opened only in a sterile laminar flow hood, and cultures are grown under low irradiance levels (often <50 μmol $m^{-2}$ $s^{-1}$). Early growth can be rapid and is supported by carbohydrates (usually sucrose; 30 mg $l^{-1}$) in the agar medium, rather than through autonomous callus/cutting photosynthesis. However, as plantlets (leaves and stems) mature and media carbohydrate levels decrease, plants in tissue culture vessels are increasingly carbohydrate limited (Fujiwara et al., 1992). In fact, the question of whether sucrose should be added to media at all is being discussed as its addition appears to delay the onset of autonomous photosynthesis and may decrease plantlet fresh weight (Martins et al., 2015). Therefore, optimal temperature, irradiance, and carbon dioxide levels for photosynthesis need to be optimized. Current systems appear to be carbon dioxide and irradiance limited (Fujiwara et al., 1987; Erwin and Guenthner, 2018).

Perhaps one of the most challenging parts of TC propagation is the successful establishment of cuttings in greenhouse media after removal from TC vessels. Tender shoots generally have little cuticle, can desiccate easily, and may not be conducting photosynthesis. As a result, these weak shoots can easily die unless a grower has a high humidity-low irradiance environment to acclimate these shoots to new conditions. As a result, these cuttings are generally rooted in humidity tents under low irradiance levels initially (Misic et al., 2006; Erwin, personal observation).

## 3.8 Gaps in knowledge

Gaps in knowledge of asexual propagation revolve around identifying optimal conditions for non-traditional propagation techniques. Specific areas of opportunity include:

- Identification of 'direct stick' propagation protocols to reduce crop time, input costs, and labor. By 'direct stick', I mean propagation by cuttings rooted in finished containers, often not under a typical propagation environment with traditional mist.

- Identification of optimal tissue culture vessel environment conditions to maximize photosynthesis (irradiance, $CO_2$ level, and temperature) and ways to safely modify that environment (with and without sucrose in media) to optimize photosynthesis to reduce tissue culture tub time.
- Identification of protocols to successfully acclimate tissue culture produced cuttings/plantlets to environments outside of the tissue culture vessel.

## 4 Vegetative development

Plants have a diversity of shapes and phenotypic plasticity that enables them to adapt to some extent to their surroundings (Teichmann and Muhr, 2015). That plasticity primarily centers on the rate of development, stem elongation, and branching. Axillary meristems initiate in leaf axils where they initiate a few leaves and form a 'bud' that either goes dormant or continues to grow (Rameau et al., 2014). Endogenous and environmental factors impact whether a bud goes dormant or grows. The shoot meristem, itself, suppresses branching in many cases; this suppression is often referred to as 'apical dominance' (Leyser, 2005). Auxins synthesized in the shoot tip and young leaves suppress continued axillary bud growth and branching (Ljung et al., 2001). Horticulturalists modify plant architecture by pinching, or pruning, to encourage branching and 'shape' a plant. That 'pinching' removes the apical meristem, or auxin source, thus allowing axillary branch development (Thimann and Skoog, 1933). Application of auxin to the end of a decapitated shoot tip restores apical dominance and suppresses branching. We now better understand that after shoot decapitation, auxin concentration is reduced; there is a subsequent flux of auxin out of axillary buds, releasing a bud from dormancy. Those new apical meristems eventually reestablish apical dominance, and branching is again suppressed. An overview of vegetative development can be found in Leyser and Day (2002) and Dole and Wilkins (2004).

Aside from auxin, a secondary messenger is involved in axillary bud break and branching as the application of auxin to axillary buds does not, in itself, inhibit bud break; both cytokinins and strigolactones (SLs) have been implicated. Cytokinins are produced in roots and stems and transported upward in xylem (Sachs and Thimann, 1964). Cytokinins are antagonistic to auxin-suppression of axillary bud growth, and auxin can suppress cytokinin synthesis (Nordstrom et al., 2004; Tanaka et al., 2006) and promote cytokinin oxidases that inactivate cytokinins (Shimizu-Sato et al., 2009; Werner et al., 2001). In addition, shoot tip removal increases cytokinin biosynthesis (Bangerth, 1994), particularly in axillary buds (Turnbull et al., 1997). Yet cytokinins are not, in themselves, the only branching promoter as transgenic roots overproducing cytokinins grafted to a wild-type stem did not branch more (Faiss et al., 1997). SLs were subsequently identified as control agents in long-distance branching control (Gomez-Roldan et al., 2008; Umehara et al., 2010). Auxin increases SL synthesis (Sorefan et al., 2003), and SLs are synthesized primarily in roots and in the shoot (Mashiguchi et al., 2009).

The transcription factor BRANCHED1 plays a central role in integrating both internal and external cues to impact branching (Wang et al., 2019). In wild-type *Arabidopsis*, less than 40% of axillary buds break and grow out; in contrast, the *brc1* knock-out mutants have rosette branching and 100% of axillary buds elongate (Aguilar-Martinez et al., 2007). BRC1 expression is downregulated transcriptionally by SLs (Aguilar-Martinez et al., 2007). However, SL synthesis mutant branching still responds to changes in the environment, suggesting SLs alone do not control branching (Beveridge et al., 2009; Ferguson and Breveridge, 2009).

GAs inhibit shoot branching (Murfet and Reid, 1993), and brassinosteroids (BRs) stimulate it. Overexpressing GA catabolism genes to reduce GA levels increased branching (Agharkar et al., 2007; Lo et al., 2008). Pea double mutants (GA and SL) displayed an additive response. In contrast to GA mutants, BR-deficient pea and rice mutants have less branching (Murfet and Reid, 1993; Tong et al., 2009). GA mutants can also have modified branching (Murfet and Reid, 1993). Interestingly, SL mutants suggest that these responses may play a role in another pathway than the before-mentioned SL-dependent pathway (Ferguson and Breveridge, 2009), suggesting that GAs act independently of SL on branching (de Saint Germain et al., 2013).

Light plays a major role in determining whether axillary buds grow or go dormant (Leduc et al., 2014). Additionally, genes that regulate photoperiodic responses can have a strong influence on shoot branching (Lejune-Henaut et al., 2008; Klein et al., 2014). Interestingly, sucrose has also been implicated as a factor in branching (Mason et al., 2014). The before-mentioned branching suppressor, BRANCHED1, is downregulated after sucrose treatment.

## 4.1 Plant stem elongation

In general, it is desirable to produce compact plants that have shorter internodes in ornamental crop production. Plant stem elongation is often controlled in ornamental crop production chemically or environmentally. Chemically, plant growth retardants are applied that reduce GA biosynthesis.

Plant stem elongation responses to DT and NT were first described in depth by Went (1944, 1957). However, ornamental crop growers did not actively use temperature to manage plant height until the advent of automated and computerized environmental control systems for greenhouses were developed and distributed in the 1980s.

Another way to manage stem elongation is to manage DT and NT diurnally to reduce elongation (Berghage and Heins, 1991; Erwin, 1991a,b; Erwin and

Heins, 1990; Erwin et al., 1989a; Heins and Erwin, 1990). Stem elongation increases as DT increases and NT decreases (Karlsson, 1988; Tageras, 1979; Went, 1944, 1953, 1957). Subsequent plant stem elongation was highly correlated with the difference (DIF) between DT and NT (DIF = DT − NT; Erwin, 1991a,b; Erwin et al., 1989a,b; Heins and Erwin, 1990, 1991; Moe et al., 1991). Anatomically, cellular elongation (stem parenchyma and epidermal cells), but not division, was impacted by DIF (Erwin et al., 1991a,b).

Subsequent work identified diurnal sensitivity of stem elongation to temperature fluctuations; stem elongation is most sensitive to a temperature change at the end of the night and the beginning of the day (Erwin, 1991a,b; Erwin et al., 1989a). As a result, commercial growers often decrease 3-5°C temperature at the end of the night and at the first 3 h of the day to reduce elongation while not substantially slowing crop development. Dropping temperatures at the end of the night has been referred to as 'DIP' or 'DROP'. In contrast to DROP/DIP, there appears to be a sensitivity to temperature increase during the photoperiod with respect to stimulating stem elongation. Mortensen and Moe (1992) reported that a high temperature at the end of the photoperiod increased the elongation of some species, but not others. In contrast, Erwin (1991a,b) reported that if temperature was increased at the beginning of the photoperiod, *Impatiens* elongation can be increased.

The impact of DT/NT on plant morphology was referred to as 'thermomorphogenesis' (Erwin et al., 1989b). DIF effects on stem elongation may be elicited through impacts on GA synthesis and/or response (Erwin, 1991a,b; Erwin et al., 1989b; Moe et al., 1991; Pinthus and Mieri, 1979; Zieslin and Tsujita, 1988). Alternatively, other factors such as temperature impacts on cell turgor, cell wall elasticity, and phytochrome photoequilibria or cycling interact with DIF and likely play a role in DIF impacts on stem elongation (Erwin, 1991a,b; Erwin et al., 1991a,b; Lecharny et al., 1985; Moe et al., 1991). DIF is used extensively worldwide to control elongation in European greenhouse production (Roberts, 1991), North American greenhouse production (Erwin, personal observation), tissue culture facilities (Kozai et al., 1992), and growth chamber research (M. Bates, personal communication).

The responsiveness of different species and varieties within a species can differ. For instance, tomato and cucumber are very responsive to DIF, but pea and bean much less so (Erwin et al., 1994a). Poinsettia cultivars differ in their response to DIF (Moe et al., 1991).

The physiological basis for DIF responses appears to be associated with differences in GA synthesis and/or responses. Application of GA3 to *Triticum* grown with a −DIF overcame the suppression of elongation due to −DIF (Pinthus and Mieri, 1979). Similarly, applying GA4+7 to *Lilium* bulbs before planting overcame the inhibition of elongation due to −DIF (Zieslin and Tsujita, 1988). Moe et al. (1991) and Erwin and Pierson (1992) showed that spray applications of GA3 overcame the inhibition of *Campanula* and *Lycopersicon* internode elongation, respectively, when plants were grown with a −DIF. *Lycopersicon* gib-1 mutants (reduced ability to synthesize GA) responded more on an absolute and relative basis to a GA3 application when plants were grown with a −DIF versus +DIF (Erwin, personal observation).

Photoperiod, irradiance, and light quality interact with diurnal temperature to affect stem elongation (Berghage and Heins, 1991; Erwin, 1991a,b; Erwin et al., 1991a,b; Moe et al., 1991). In general, stem elongation responses to DIF decrease as photoperiod length increases (Erwin et al., 1991a). On an absolute basis, stem elongation decreases as irradiance increases. For instance, the internode length of potato (*Solanum tuberosum*) decreased as irradiance increased (Kozai et al., 1992). However, stem elongation responses to DIF on a percent basis increase as irradiance increases (Erwin et al., 1991a,b; Heins and Erwin, 1990; Kaczperski, 1989; Ludolph, 1992; Myster and Moe, personal communication).

DIF interacts with red and far-red light, implicating an interaction with phytochrome. Continuous far-red light exposure during the night overcame the reduction in stem elongation by reducing DIF on *Campanula* (Moe et al., 1991) and *Fuchsia* (Erwin, 1991a,b). In contrast, continuous red light exposure during the night enhanced the inhibition of stem elongation by a −DIF in *Campanula* (Moe et al., 1991). In contrast, supplementing daylight with 10 μmol $m^{-2}$ $s^{-1}$ of red light enhanced the response of *Fuchsia* stem elongation to a −DIF (Erwin, personal observation); supplementing daylight with blue light did not impact stem elongation responses to DIF (Erwin, personal observation).

Thermomorphogenic stem elongation responses to DIF, DROP, and DIP are now used broadly in commercial greenhouse production and, in some cases, eliminated the need for chemical growth retardants (Roberts, 1991). The effectiveness of DIF in controlling elongation varies in different locations geographically and seasonally. For instance, experiments and growers in Europe have reported less response of many plant species to DIF than that of those same species in the USA (Hendricks, 1991; Langton et al., 1992; Vogelezang et al., 1992a). Variation in plant responses to DIP/DROP between European and American growers has also been reported (Vogelezang et al., 1992b). Erwin (personal observation) also reported that DIF responses are greater in early than in late spring. Taken together, the before-mentioned interaction between DIF and irradiance and daylength likely explains these observed differences.

A non-chemical way to reduce internode elongation is through touching plants which elicits a thigmotropic response (Biddington, 1986; Jaffe, 1973). Touching or shaking a plant reduces internode elongation by stimulating ethylene synthesis (Mitchell et al., 1975; Latimer 1991, 1993; Latimer and Thomas, 1991; Miller, 1991). Drawbacks of utilizing touching or shaking to limit elongation include labor requirements and/or mechanical damage to plants (Latimer, 1991, 1993; Latimer and Thomas, 1991).

Interestingly, the effectiveness of moving, touching, or shaking plants appears to have a diurnal response where plants are most sensitive to touching/shaking when plants are elongating most (early morning; Van Gaal and Erwin, 2005). Therefore, previous studies that reported little or no response to touching may have seen this only because plants were touched or shaken at a time other than early in the morning.

### 4.2 Leaf expansion

How temperature affects leaf expansion varies with species. Some species appear to have the greatest leaf expansion when temperatures are held constant. For instance, *Cucumis* leaf expansion was greatest when plants are grown at constant 20D (Milthorpe, 1959). In contrast, leaf expansion in *Fuchsia* (Erwin et al., 1991b), *Solanum* (Kozai et al., 1992), and *Brassica* expansion increased as DIF increased. Lastly, the leaf length of *Lilium longiflorum* was affected only by NT (Erwin et al., 1989b).

### 4.3 Plant development rate

Temperature effects on the plant development rate have traditionally been quantified by recording the days to flower. Time to flower generally decreases as temperature increases to some optimal temperature and then increases as temperature further increases. For instance, *Kalanchoe*'s (*Kalanchoe blossfeliana*) days to flower decreased 19 days when temperature increased from 18°C to 24°C but then increased 2 days when temperature was further increased from 24°C to 26°C (Carvalho et al., 2006). Thanksgiving cactus's (*Schlumbergera truncata*) days to flower decreased from 100 to 52 as temperature increased from 12°C to 20°C and was unchanged when temperature was further increased from 20°C to 24°C (Erwin et al., 1990).

More recently, the plant development rate has been quantified by calculating the node-unfolding rate per day (Karlsson et al., 1988, 1989; Berghage, 1998; Roberts and Sommerfield, 1987; Vaid and Runkle, 2013). The node-unfolding rate is primarily dependent on the average daily temperature (ADT) a plant is grown within a limited temperature range. The impact of temperature on the leaf-unfolding rate is a curvilinear response where the leaf-unfolding rate increases as temperature increases to some optimal temperature. Generally, that temperature range is between the base temperature (leaf-unfolding rate = 0) and the optimal temperature (maximum leaf-unfolding rate).

For many species, the 'base temperature' for leaf unfolding ranges from 0°C to 10°C and the optimum temperature generally ranges from 25°C to 30°C. Sweet pepper (*Capsicum annuum*; Yaping and Heins, 1996), poinsettia (*E. pulcherrima*; Berghage, 1998), dahlia (*Dahlia pinnata*; Brondum and Heins, 1993), and Boston fern (*Nephrolepis exaltata*, Erwin et al., 1993) have optimal node/leaf-unfolding rates at temperatures of 26°C to 30°C. The optimum leaf-unfolding rate occurs at temperatures from 30°C to 35°C for Easter lily (*Lilium longiflorium*; Karlsson et al., 1988) and vinca (*C. roseus*; Pietsch et al., 1995). In contrast, the leaf-unfolding rate of some succulents (*Aeonium*, spp., *Adromischus* spp. for example) occurs at an optimal temperature of 22°C (Erwin et al., 2017). Leaf-unfolding rate of some other succulents, such as *Haworthia*, *Aloe*, and *Agave* sp., increased from 10°C to 28°C (it had not reached an optimum at 28°C; Erwin et al., 2017).

Irradiance and photoperiod may appear to impact leaf-unfolding rate. Leaf-unfolding rate reportedly increases as irradiance increases (Pietsch et al., 1995). Kozai et al. (1992) reported leaf-unfolding rate also increased as daylength increased. In both cases, higher irradiance and longer photoperiod increase leaf-unfolding rate by increasing leaf temperatures; high irradiance levels can increase leaf temperatures 2–5°C.

### 4.4 Gaps in knowledge

Opportunities in vegetative development revolve around using DIF/DIP and leaf-unfolding responses to maximize non-chemical height control and increased heat tolerance. Specifically:

- Producers commonly use temperature manipulation (DIF, DIP) to limit stem elongation. Breeding programs can likely collect germplasm with high or low DIF responses based on indigenous habitats.
- Maximum temperatures for leaf unfolding have been identified only for a small number of ornamental species. We note that production temperatures often exceed these maximum temperatures, thereby slowing leaf unfolding. Breeding programs could emphasize the use of maximum temperature for leaf unfolding as selection criteria for the identification of greater heat tolerance.

## 5 Increasing plant mass

### 5.1 Photosynthesis

In general, ornamental crop photosynthetic rate increases as irradiance increases from 0 to some maximum irradiance (varies from 300 to 1650 μmol $m^{-2}$ $s^{-1}$), above which the photosynthetic rate is unchanged. The photosynthetic rate increases as carbon dioxide ($CO_2$) concentration increases from 0 to some maximum value (typically around 1200 ppm) and then can decrease when $CO_2$ levels are increased above this level.

As with $CO_2$ levels, the photosynthetic rate increases as temperature increases to some maximum temperature. When temperatures are increased above this level, photosynthetic rates decrease. The optimal temperature for photosynthesis can vary significantly among ornamental crops. For instance, some leaf greens, crops, and pansy (*Viola wittrockiana*) have optimal temperatures for photosynthesis ranging from 16°C to 20°C. *Gerbera* and tomato have optimal temperatures for photosynthesis between 25°C and 27°C, whereas poinsettia and Aloe can have the optimal temperature at or around 30°C.

Increasingly, we are appreciating that short-term high-temperature exposures can depress photosynthesis for a longer period of time. For instance, exposure of *Impatiens hawkeri* and *V.* x *wittrockiana* to high temperatures (35°C) for as short as 2 h reduced photosynthetic rates for up to 3 days compared to 'un-stressed' plants (Warner and Erwin, 2002). Photosynthetic recovery following a high-temperature stress event varied between *I. hawkeri* and V. x *wittrockiana*; *V. wittrockiana* recovered more quickly.

## 5.2 Carbon partitioning

In general, the total plant dry weight increases when photosynthesis is maximized and respiration is minimized. Karlsson and Heins (1992) showed the total plant dry weight increased as DT increased (20–30°C) and NT decreased from 30°C to 20°C in *Dendranthema*. Total shoot dry weight of *Fuchsia* was highest when plants were grown with a 25/15°C or 25/20°C DT/NT (Erwin et al., 1990, 1991a,b). In contrast, Kozai et al. (1992) showed the dry weight of *Solanum* plantlets was unaffected by DIF in tissue culture; this would be consistent with minimal photosynthesis due to the presence of sucrose in the media.

Partitioning to *Dendranthema* roots differed than that of shoots in that partitioning decreased as DT increased (Karlsson and Heins, 1992). In contrast, the dry weight of *Impatiens* and *Fuchsia* leaves and flowers had an optimal temperature that ranged from 20°C to 24°C (Erwin and Heins, 1995; Erwin et al., 1991a,b). Similarly, the number and weight of Thanksgiving cactus flower were highest at constant 20°C (Erwin et al., 1990). In contrast, chrysanthemum flower partitioning was correlated with DT, NT, and DIF (Karlsson and Heins, 1992).

## 5.3 Gaps in knowledge

Specific obvious gaps in knowledge include:

- Identification of optimal light, irradiance, and $CO_2$ levels (and interactions between them) for maximum photosynthetic rates for ornamental species.
- How optimal conditions for photosynthesis change with plant age.
- Impact of single and repeated abiotic stresses on plant photosynthesis.

# 6 Reproductive development

## 6.1 Juvenility vs. maturity

The process whereby a shoot meristem produces flowers opposed to leaves is referred to as 'floral evocation'. The signal that a meristem receives that promotes floral evocation is often referred to as the 'floral induction' signal. The formation of flower buds after induction is referred to as 'flower initiation'. The process after flower initiation until anthesis is referred to as 'flower development'. A meristem is 'competent' to flower when flower evocation occurs in response to an autonomous or external environmental stimulus. A meristem is referred to as 'determined' when flower development continues even after an inductive signal is removed. In some species, flower induction, initiation, and initial flower development occurs but flower development is arrested until some environmental condition is fulfilled such as cold or time. Such a suspension in flower development until such condition is fulfilled is referred to as 'flower dormancy'.

The complexity of control mechanisms associated with flowering is enormous. If a single factor is promotive, it is possible for flowering to be inhibited if other conditions are not met. Yet plants have a built-in redundancy in processes that can result in flowering to ensure a number of cues that can enable flowering to compensate for environmental fluctuations. There are six pathways that converge to regulate the 'floral integrator genes' that determine flowering time. Of particular interest to ornamental plant growers are the autonomous, photoperiod, and vernalization pathways elaborated in the following section. Overviews of the topic include Dole and Wilkins (2004), McDonald and Kwong (2005), Anderson (2007), and Erwin (2009).

## 6.2 Autonomous flowering pathway

Sexually propagated organisms pass through a series of developmental phases during growth and maturation. In plants, three developmental phases occur in the apical meristem: the juvenile, mature vegetative (competent), and mature reproductive (determined) phases. The distinction between the 'juvenile' and 'mature' phases is that a juvenile meristem is incapable of flowering (incompetent) and a mature meristem is. The distinction between the 'mature vegetative' and 'mature reproductive' phase is simply whether a meristem is induced to flower (mature reproductive) or not (mature vegetative). Often, there are no phenotypic changes that show whether a plant has transitioned from a juvenile to mature phase.

Whether a plant is mature and competent to flower can be evaluated using grafting experiments (McDaniel et al., 1992). If a non-flowering scion is grafted onto a mature rootstock induced to flower and the scion flowers, then the scion must have been mature. *Betula verrucosa* J.F. Ehrh. tissue harvested from the base of the tree (juvenile) grafted onto a mature rootstock remains juvenile or vegetative; that is the scion was not competent to flower, whereas mature tissue collected from the top of the tree was competent to flower after 2 years (Longman, 1976). Alternatively, placement of a plant in inductive conditions and failure to flower is an indication that that plant is incompetent to flower. The transition from the juvenile to mature phase is a continuous process, and it is not uncommon for a population of seedlings to transition phases at different times.

As a plant ages, there is an increased tendency to transition from a juvenile to mature phase and to flower. For example, the percentage of *Lunaria biennis* L. (Wellensiek, 1958), Brussels sprouts (Stokes and Verkerk, 1951), and beets (Wellensiek and Hakkaart, 1955) plants induced to flower by cold temperatures (vernalization) increases as plants age. Similarly, as successive populations of cuttings are harvested from *E. pulcherrima* Willd. Ex Klotzsch stock, or mother plants, the percentage of those cuttings that spontaneously produce floral organs (cyathia) under non-inductive LD conditions increases with each successive cutting generation (Siraj-Ali et al., 1990; Carver and Tayama, 1992).

Once the mature phase is achieved, it is relatively stable. For instance, the juvenile or adult phase is maintained through vegetative propagation and grafting (Longman, 1976). *H. helix* cuttings taken from juvenile tissues will remain juvenile after propagation and vice versa (Poethig, 1990). However, sexual or apomictic reproduction will result in reversion back to the juvenile phase (Hackett, 1980).

Species vary considerably in juvenile phase length. 'Longer-lived' species often have a longer juvenile phase. For instance, nut trees can have approximately a 25-year juvenile phase. In contrast, more 'shorter-lived' herbaceous perennial species can have a juvenile phase that can range from 4 to 9 months. Herbaceous annual species can have a juvenile phase length that can be 1 to 4 weeks. Plant age and a way to quantify when and whether a plant transitions from a juvenile to mature state are indicated in years with longer-lived species and leaf number with shorter-lived herbaceous perennial and annual species. Example developmental time (measured in years or the number of leaves) required for the transition from the juvenile to mature phase for some woody versus herbaceous species is shown in Tables 4 and 5.

The association between leaf number and phase change may be related to meristem size more than leaf number. *Chrysanthemum* x *morifolium* Ramat. flower initiation does not occur until after a minimum meristem apex size has been attained regardless of leaf number (Cockshull, 1985). The importance of meristem size as opposed to leaf number was also noted by Hackett and Srinivasani (1983) with *H. helix*. Pomologists also note the transition of fruit trees from the juvenile to mature phase may be associated with meristem size

**Table 4** Juvenile phase length across woody plant species

| Species | Juvenile period length |
|---|---|
| Rose (*Rosa* x *hybrida*) | 20–30 days |
| Grape (*Vitis* spp.) | 1 year |
| Orchid spp. | 4–7 years |
| Apple (*Malus* spp.) | 4–8 years |
| Orange (*Citrus* spp.) | 5–8 years |
| English ivy (*Hedera helix*) | 5–10 years |
| Redwood (*Sequioa sempervirens*) | 5–15 years |
| Sycamore maple (*Acer pseudoplatanus*) | 15–20 years |
| English oak (*Quercus robur*) | 25–30 years |
| European beech (*Fagus sylvatica*) | 30–40 years |

*Source*: adapted from Clark (1983) and Goh and Arditti (1985).

**Table 5** Juvenile period lengths as identified by leaf number

| Species | Leaf number plants transition from juvenile to mature (nodes) |
|---|---|
| *Aquilegia* 'McKana's Giant' | 12 |
| *Aquilegia 'Fairyland'* | 15 |
| *Calceolaria herbeohybrida* | 5 |
| *Callistephus chinensis* | 4 |
| *Coreopsis grandiflora* 'Sunray' | 8 |
| *Gaillardia* x *grandiflora* 'Goblin' | 16 |
| *Heuchera sanguinea* 'Bressingham' | 19 |
| *Lavandula angustifolia* 'Munstead' | 18 |
| *Rudbeckia fulgida* 'Goldstrum' | 10 |

*Source*: adapted from Yuan (1995), Whitman (1995) and Sheldron and Weiler (1982a,b).

(Visser and DeVries, 1970). Singer and McDaniel (1986) showed tobacco plants produced 37 leaves from terminal buds prior to flowering and concluded, through grafting experiments, that node number prior to flowering was a function of floral stimulus strength and the competence of a meristem to respond to that signal.

Environmental conditions that retard growth can delay the transition from juvenile to adult phase. For instance, low irradiance can delay the transition from juvenile to mature phase or even cause reversion from the mature to juvenile phase. In contrast, increasing irradiance is used as a technique to reduce juvenile phase length and hasten flowering in seed-propagated ornamental crops. Irradiance can affect the earliness of flowering of many species (Armitage and Tsujita, 1979; Armitage et al., 1981; Dole and Wilkins,

1999; Erwin et al., 1997; Zhang et al., 1996). For instance, *Achillea millefolium* L. 'Summer Pastels' grown under a 16-h photoperiod flowered after 57, 45, and 37 days when grown under 100, 200, or 300 μmol $m^{-2}$ $s^{-1}$, respectively (Zhang et al., 1996). Armitage and Tsujita (1979) grew four *Pelargonium* x *hortorum* Bailey cultivars under natural daylight conditions (February-March 1977, Guelph, Ontario, Canada) plus 27 or 54 μmol $m^{-2}$ $s^{-1}$; the addition of 27 μmol $m^{-2}$ $s^{-1}$ hastened flowering for some cultivars but not for others. *Asclepias curassavica* L. node number below the inflorescence decreased from 21 to 16 nodes when plants were grown under continuous lighting versus NI lighting (Nordwig, 1999). Juvenile phase length can be reduced in a number of other herbaceous ornamental crops by increasing irradiance (Mattson and Erwin, 2003; Warner and Erwin, 2001b; Table 6).

Daily light integral (DLI) may be more important than irradiance on impacting juvenile phase length. Adams et al. (1999) showed that decreased DLI lengthened the time to flower in *P. hybrida*. Pearson et al. (1993) noted that increased DLI shortened the time to flower for *D. grandiflora* Ramat; the reduction in time to flower was related to DLI rather than to maximum irradiance. Erwin and Warner (2002) reported 11 species flowered earlier developmentally when DLI was increased. Hedley (1974) showed *Anthirrhinum majus* 'Orchid Rocket' flowering was hastened developmentally (leaf number below the flower decreased from 80 to 34 leaves) under LDs when irradiance increased from 115 to 500 μmol $m^{-2}$ $s^{-1}$. Erickson et al. (1980) found a linear relationship between DLI and days to flower for *P.* x *hortorum* until a threshold DLI between 6.89 and 9.01 μmol $m^{-2}$ $s^{-1}$ was reached. In contrast, Fausey et al. (2001) noted 100% flowering of *Digitalis purpurea* L. 'Foxy' was reached only with DLI greater than 11 μmol $m^{-2}$ $s^{-1}$. Supplemental 90 μmol $m^{-2}$ $s^{-1}$ lighting on *Gerbera jamesonii* H. Bolus hastened flowering by 23 days in the winter, but only 11 days during spring (leaf number below the first flower not reported; Gagnon and Dansereau, 1990). Therefore, the benefit of supplemental irradiance on stimulating earlier flowering can be dependent on ambient light conditions. In our experiment, average DLIs (Table 2) under all lighting treatments were greater than the threshold levels found by Erickson et al. (1980) and Fausey et al. (2001).

The term 'facultative irradiance' (FI) was coined to describe the developmental hastening of flowering by increased irradiance or DLI. Species that exhibit an FI response display decreased leaf number below the first flower when irradiance or DLI is increased (Erwin and Warner, 2002). In contrast, plants in which flowering is not hastened developmentally by increased irradiance or DLI are referred to as 'irradiance indifferent' (Erwin and Warner, 2002). Species exhibiting an FI response include *Convolvulus*, *Dianthus*, *Gazania*, *Lavatera*, *Limnanthes*, *Linaria*, *Nemophila*, *Nicotiana*, oregano, *Silene*, snapdragon, petunia, pansy, and seed geranium (Table 6). Species with an II response

**Table 6** Optimal conditions for rooting common herbaceous ornamental plants

| Species | Mist or tent | Rooting hormone | Pinch stick | Daylength | Direct |
|---|---|---|---|---|---|
| *Achillea* | Mist | Y | N | SD | - |
| *Ageratum houstonianum* | Mist/Tent | Y/N | N | SD | Y |
| *Ajuga repens* | Mist | N | N | SD | - |
| *Angelonia* | Mist | N | Y | LD | - |
| *Argyranthemum* | Mist | N | N | LD | Y |
| *Bacopa* | Mist | N | Y | LD | Y |
| *Calibrachoa* | Mist | N | Y | SD | - |
| *Coleus* | Mist | N | Y | LD | Y |
| *Coreopsis* | Mist | Y/N | Y | SD | - |
| *Cuphea* | Mist | N | Y/N | LD | - |
| *Dahlia* | Mist/Tent | Y | Y | LD | Y |
| *Diascia* | Mist | N | Y/N | LD | - |
| *Evolvulus* | Mist/Tent | Y/N | Y | LD | - |
| *Gaillardia* | Mist | N | Y/N | SD | - |
| *Gaura* | Mist | N | Y | LD | - |
| *Geranium* (ivy) | Mist | N | Y/N | LD | - |
| *Geranium* (zonal) | Mist | N | N | LD | Y |
| *Iberis* | Mist | Y | Y | LD | - |
| *Impatiens* (double) | Mist | N | Y | LD | Y |
| *Impatiens* (New Guinea) | Mist | N | N | LD | Y |
| *Lantana* | Mist | Y | Y | LD | - |
| *Lavender* | Mist/tent | Y | Y | LD | - |
| *Lithodora* | Mist | Y | Y | LD | - |
| *Lobelia* | Mist | Y | Y | SD | Y |
| *Osyeospermum* | Mist | Y | N | LD | Y |
| *Petunia* | Mist | Y | N | SD | - |
| *Phlox* (creeping) | Mist | Y | Y | LD | - |
| *Rosemary* | Mist/tent | Y | Y | LD | Y |
| *Salvia* (greggi) | Mist/tent | Y | Y | SD | Y |
| *Scaevola* | Mist | Y | Y | LD | - |
| *Verbena* | Mist | Y | Y | SD | - |
| *Vinca* (vine) | Mist/tent | Y | Y | LD | - |

include *Ageratum*, *Celosia*, *Cleome*, *Cosmos*, *Gomphrena*, *Statice*, *Lobelia*, *Mimulus*, *Nigella*, *Salvia*, *Tithonia*, and *Zinnia* (Table 6).

In our experiment, ADT at the plant level was 1–2°C greater under the highest lighting treatments (data not presented). Faust and Heins (1998) reported that supplemental irradiance increased shoot-tip temperatures of *C. roseus* L. by 1.2, 1.5, and 1.7°C, with HPS light additions of 50, 75, and 100 μmol $m^{-2}$ $s^{-1}$, respectively. Pietsch et al. (1995) concluded that increased development rates of *C. roseus* under supplemental light were a result of increased shoot-tip temperatures.

Temperature can play some role in the earliness of flowering as well, but to a lesser extent. The node where the first flower cluster appeared on *L. esculentum* increased when plants were grown at 27 versus 10–16°C (Calvert, 1957).

The physiological basis for phase change in plants is not well understood. Application of GAs to *Cupressus arizonica* Greene caused male cone formation (indicator of transition to adult phase) when plants were 2 months old (Pharis and King, 1985). Similarly, other treatments that increase endogenous GAs in conifers can reduce the juvenile phase length. In contrast, the application of GAs to *H. helix* can cause reversion from the mature to juvenile phase (Hackett and Srinvasani, 1985). In contrast, inhibition of the transition from juvenile to adult phases following GA application was noted on *Citrus* (Cooper and Peynado, 1958), deciduous fruit trees, (Luckwill, 1970) and *Ipomoea caerula* L. (Njoku, 1958). The application of Amo 1618, a GA-synthesis inhibitor, did not promote early transition from the juvenile to mature phase in *H. helix* (Frydman and Wareing, 1974); the application of another GA-synthesis inhibitor, ancymidol, prevented *H. helix* reversion to the juvenile phase under low irradiance (Rogler and Hackett, 1975). A variety of GA-synthesis inhibitors are commonly applied to seed-propagated ornamental crops without any impact on flowering time, suggesting these compounds generally do not affect phase transition in commercial ornamental crops (Erwin, personal observation).

The impact of plant age on precociousness to flower is likely related to the accumulation of SQUAMOSA promoter binding-like (SPL) transcription factors with time. SPL transcription factors promote flowering. SPL protein synthesis is repressed by the microRNA miR-156; levels of this microRNA decrease as a plant ages (Table 7).

## 6.3 Photoperiodic flowering pathway

'Photoperiodism' refers to the ability of an organism to detect daylength (or more correctly night length) and elicit a response. In animals, hibernation, egg laying, and migration are controlled by photoperiod. In plants, day/night length can impact flowering, tuber and rhizome formation, and the onset of dormancy. Flowering responses to day/night length were first observed at the U.S. Department of Agriculture laboratories in Beltsville, Maryland, by Wightman Garner and Henry Allard in the 1920s. Garner and Allard (1920) noted a mutant of a tobacco variety (Maryland Mammoth) did not flower outside during the

**Table 7** Photoperiodic and irradiance classifications based on leaf number below the first open flower

| Species | Photoperiod | Irradiance |
|---|---|---|
| *Abutilon* x *hybridum* | DNP | - |
| *Achimenes hybrida* | DNP | - |
| *Agastache* hybrids | FLDP | - |
| *Ageratum houstonianum* L. 'Blue Danube' | FLDP | II |
| *Alcea rosea* | FLDP | - |
| *Amaranthus caudatus* | FSDP | - |
| *Amaranthus hybridus* L. 'Pygmy Torch' | DNP | II |
| *Ammi majus* L. | OLDP | II |
| *Anemone coronaria* | DNP/OLDP | - |
| *Anethum graveolens* L. 'Mammoth' | OLDP | II |
| *Anigozanthos flavidus* | FLDP | - |
| *Anigozanthos manglesii* | FSDP | - |
| *Anigozanthos pulcherrimus* Hook. | DNP | - |
| *Anigozanthos rufus* Labill. | DNP | - |
| *Anisodontea* x *hypomandarum* K. Presl. | FLDP | - |
| *Antirrhinum majus L.* | FLDP | FI |
| *Asclepias curassavica* L. | DNP | FI |
| *Aquilegia caerulea* | DNP/FLDP | - |
| *Asclepias tuberosa* L. | OLDP | - |
| *Asperula arvensis* L. 'Blue Mist' | OLDP | II |
| *Asilbe* x *arendsii* | DNP/FLDP | - |
| *Begonia* x *hiemalis* Fotsch | O/FSDP | - |
| *Begonia tuberhybrida* | FLDP | - |
| *Begonia semperflorens* | DNP | FI |
| *Bougainvillea* spp. | FSDP | FI |
| *Browallia speciosa* | - | - |
| *Calceolaria herbeohybrida* | FLDP | - |
| *Calendula officinalis* 'Calypso Orange' | O/FLDP | II |
| *Calibrachoa* hybrids | FLDP | - |
| *Callistephus chinensis* L. | FLDP | - |
| *Campanula isophylla* | O/FLDP | - |
| *Canna* x *generalis* | DNP | - |
| *Catharanthus roseus* | DNP | - |
| *Catananche caerulea* L. Per. 'Blue' | OLDP | FI |
| *Carpanthea pomeridiana* L. 'Golden Carpet' | DNP | II |
| *Celosia argentea* | O/FSDP | - |

| Species | Photoperiod | Irradiance |
|---|---|---|
| *Celosia plumose* L. 'Flamingo Feather Purple' | OSDP | II |
| *Centaurea cyanus* L. 'Blue Boy' | OLDP | II |
| *Centranthus macrosiphon* Boiss. | DNP | FI |
| *Clarkia amoena* | DNP/OLDP | - |
| *Cleome hasslerana* Chodat 'Pink Queen' | FLDP | - |
| *Cleome hasslerana* Chodat 'Rose Queen' | DNP-FSDP | FI |
| *Clerodendrum thomsoniae* | DNP | - |
| *Clerodendrum* x *speciosum* | FSDP | - |
| *Cobea scandens* | DNP | - |
| *Coleus* spp. | FSDP/FLDP | - |
| *Cobaea scandens* Cav. | DNP | II |
| *Collinsia heterophylla* Buist | FLDP | II |
| *Convolvulus tricolor* L. 'Blue Enchantment' | DNP | FI |
| *Coreopsis grandiflora* | O/FLDP | - |
| *Cosmos bipinnatus* Cav. Ann. 'Diablo' | FSDP | II |
| *Cosmos bipinnatus* Cav. Ann. 'Sensation White' | FSDP | FI |
| *Cosmos sulphureus* Cav. | OSDP | - |
| *Crossandra infundibuliformis* L. | DNP | - |
| *Cuphea platycentra* | DNP | - |
| *Cyclamen persicum* Mill. | DNP | FI |
| *Dahlia* x *hybrida* | FSDP | - |
| *Delphinium grandiflorum* | DNP | - |
| *Dendranthema* x *grandiflorum* | F/OSDP | - |
| *Dianthus barbatus* L. | DNP | - |
| *Dianthus chinensis* L. 'Ideal Cherry Picotee' | FLDP | II |
| *Dimorphotheca sinuata* DC. 'Mixed Colors' | DNP | II |
| *Dolichos lablab* L. | OSDP | II |
| Echinacea purpurea | F/OLDP | - |
| *Eschscholzia californica* Cham. 'Sundew' | FLDP | II |
| *Euphorbia marginata* | OSDP | - |
| *Euphorbia pulcherrima* Willd. Ex Klotzsch. | OSDP | - |
| *Eustoma grandiflorum* | DNP/FLDP | - |
| *Exacum affine* Balf. F. | DNP | - |
| *Fuchsia* x hybrida | OLDP | - |
| *Fuchsia* 'Gartenmeister' | DNP | - |
| *Gallardia* x *grandiflora* | OLDP | - |
| *Gaura lindheimeri* | FLDP | - |

(*Continued*)

**Table 7** (Continued)

| Species | Photoperiod | Irradiance |
|---|---|---|
| *Gazania rigens* L. 'Daybreak Red Stripe' | OLDP | FI |
| *Gerbera jamesonii* | FSDP | - |
| *Gomphrena globosa* L. 'Bicolor Rose' | FSDP | II |
| *Gypsophila* spp. | ?LDP | - |
| *Hatoria gaertneri* Reg. | F/OSDP | - |
| *Helianthus annus* L. 'Vanilla Ice' | DNP/FSDP | II |
| *Helipterum roseum* Hook. | OLDP | II |
| *Hibiscus cisplatinus* | DNP | - |
| *Hibiscus laevis* | OLDP | - |
| *Hibiscus moscheutos* | OLDP | FI |
| *Hibiscus radiatus* | OSDP | |
| *Hibiscus rosa-sinensis* L. | DNP | - |
| *Hibiscus trionum* | FLDP | - |
| *Hydrangea hybrids* | FSDP | - |
| *Iberis amara* | DNP | - |
| *Impatiens balsamina* | DNP | - |
| *Impatiens hawkeri* Bull. | DNP | - |
| *Impatiens wallerana* Hook. f. | DNP | - |
| *Ipomoea* x *multifida* Shinn. 'Scarlet' | FSDP | II |
| *Ipomopsis rubra* Wherry 'Hummingbird Mix' | OLDP | II |
| *Kalanchoe blossfeldiana* Poelln. | OSDP | - |
| *Lathyrus odoratus* L. 'Royal White' | OLDP | FI |
| *Lathyrus splendens* | OLDP | - |
| *Lavatera trimestris* L. 'Silver Cup' | OLDP | FI |
| *Lavendula angustifolia* | OLDP | - |
| *Leucanthemum lacustre* | F/OLDP | - |
| *Legousia speculum-veneris* Chaix | OLDP | II |
| *Leonotis menthaefolia* R. Br. | DNP | - |
| *Leptosiphon hybrida* | OLDP | II |
| *Lilium* soo. | FLDP | - |
| *Limnanthes douglasii* R. Br. | OLDP | FI |
| *Limonium sinuata* Mill. 'Fortress Deep Rose' | FLDP | II |
| *Limonium sinuata* Mill. 'Heavenly Blue' | FLDP | II |
| *Linaria maroccana* Hook. f. | FLDP | FI |
| *Linum perenne* L. | DNP/OLDP | FI |
| *Lobelia erinus* L. 'Crystal Palace' | OLDP | II |
| *Lobularia maritima* | DNP | - |

| Species | Photoperiod | Irradiance |
|---|---|---|
| *Matthiola longipetala* Venten. 'Starlight Scentsation' DNP | II | |
| *Mimulus* x *hybridus* L. 'Magic' | OLDP | II |
| *Mina lobata* Cerv. | OSDP | II |
| *Mirabilis jalapa* L. | OLDP | II |
| *Nemophila maculate* Benth. 'Pennie Black' | DNP | FI |
| *Nemophila menziesii* Hook. & Arn. | DNP | II |
| *Nicotiana alata* Link & Otto 'Domino White' | DNP/FLDP | FI |
| *Nierembergia hippomanica var. violacea* | OLDP | - |
| *Nigella damascene* L. 'Miss Jekyll' | OLDP | II |
| *Oenothera pallida* Lindl. 'Wedding Bells' | OLDP | II |
| Osteospermum hybrids | FLDP | - |
| *Oxypetalum caerulea* D. Don 'Blue Star' | DNP | FI |
| *Papaver nudicaule* | DNP | - |
| *Pelargonium* x *domesticum* L.H. Bail. | FLDP | - |
| *Pelargonium* x hortorum L.H. Bail. | DNP | FI |
| *Pelargonium peltatum* L. | DNP | - |
| *Perilla frutescens* | F/OSDP | - |
| *Petunia* x *hybrida* | FLDP | - |
| *Petunia* x *hybrida* 'Purple Wave' | OLDP | - |
| *Phacelia campanularia* A. Gray. | DNP | II |
| *Phacelia tanacetifolia* Benth. | FLDP | II |
| *Pharbitis nil* | FSDP | - |
| Phlox paniculata | OLDP | - |
| *Polemonium viscosum* Nutt. | OLDP | II |
| *Primula malacoides* Franch | OSDP | - |
| *Primula obconica* Hance | DNP | - |
| *Primula* x *polyantha* | DNP | FI |
| *Rhododendron* spp. | OSDP | - |
| *Rosa* x *hybrida* spp. | DNP | FI |
| *Rudbeckia hirta* 'Indian Summer' | OLDP | - |
| *Saintpaulia ionantha* Wendl. | DNP | FI |
| *Salpiglossus sinuata* | OLDP | - |
| *Salvia farinacea* 'Strata' | FLDP | FI |
| *Salvia splendens* F. Sellow 'Vista Red' | FSDP/FLDP | II |
| *Sanvitalia procumbens* Lam. | FSDP | II |
| *Schlumbergera truncata* Haw. | OSDP | II |

*(Continued)*

**Table 7** (Continued)

| Species | Photoperiod | Irradiance |
|---|---|---|
| *Silene armeria* L. 'Elektra' | OLDP | FI |
| *Sinningia speciosa* Lodd. | DNP | - |
| *Solenostemon scutellarioides* | FSDP | - |
| *Solidago* L. spp. | SDP | - |
| *Stephanotis floribunda* | DNP | - |
| *Streptocarpus* x *hybridus* Voss. | DNP | FI |
| *Streptocarpus nobilis* Clarke | FSDP | - |
| *Tagetes erecta* L. | F/OSDP | - |
| *Tagates patula* L. | DNP | - |
| *Tagetes tenuifolia* Cav. | FSDP | - |
| *Thunbergia alata* Bojer. | DNP | II |
| *Tithonia rotundifolia* Mill. 'Fiesta Del Sol' | FLDP | II |
| *Tithonia rotundifolia* Mill. 'Sundance' | FSDP | FI |
| *Verbascum phoeniceum* L. | DNP | II |
| *Verbena* x *hybrida* | OLDP | - |
| *Viguiera multiflora* S.F. Blake | FLDP[z] | II |
| *Viola tricolor* L. | F/O LDP | II |
| *Viola* x *wittrockiana* Gams. | FLDP | FI |
| *Zea mays* H. | DNP | - |
| *Zinnia angustifolia* Kurth. | DNP | - |
| *Zinnia elegans* Jacq. 'Exquisite Pink' | FSDP | II |
| *Zinnia elegans* Jacq. 'Peter Pan Scarlet' | FSDP | II |

Data presented here were derived from the following references: Armitage (1996), Dole and Wilkins (1999), Erwin and Warner (2002), Motum and Goodwin (1987a,b), Nordwig (1999), Mattson and Erwin (2003), Seeley (1985), and Zanin and Erwin (2003). Photoperiod classifications: FSDP (facultative short-day plant); FLDP (facultative long-day plant); OSDP (obligate short-day plant) OLDP (obligate long-day plant); DNP (day-neutral plant). Irradiance classifications: 'FI' (facultative irradiance response: supplemental irradiance hastened induction developmentally); 'II' (irradiance indifferent response; increasing irradiance did not hasten flowering developmentally) (Erwin and Warner, 2002; Mattson and Erwin, 2003). A question mark identifies an uncertain photoperiodic classification.

summer but flowered in the greenhouse during the winter. They subsequently covered outdoor-grown plants during the summer with black cloth to shorten the daylength and noted that plants flowered; they concluded daylength controlled flowering in tobacco.

Manipulation of photoperiodism in the commercial floriculture industry started after Gus Poesch noted street lights outside of a chrysanthemum cut flower facility in Michigan (USA) inhibited flowering in that crop (Poesch, 1931). That observation ultimately resulted in daylength manipulation being used to promote or inhibit flowering in chrysanthemum and potted and cut flower production worldwide. The application of photoperiod to spring annual production is relatively recent and started in the late 1990s and early 2000s after research at the University of Maryland and the University of Minnesota showed effects on spring annuals, and after research at Michigan State University on herbaceous perennial crops.

Photoperiodic flowering responses are divided into five 'response groups': short-day plants (SDPs), long-day plants (LDPs), day-neutral plants (DNPs), intermediate-day plants, and ambiphotoperiodic day plants (Thomas and Vince-Prue, 1997):

- SDP - plants in which night length must be greater than some critical length, generally, species from low latitudes (i.e. coffee, cotton, and rice) or species that flower in the late summer (Cockshull, 1984).
- LDP - plants in which night length must be shorter than some critical length, generally, temperate species (Deitzer, 1984).
- DNP - flower induction is not affected by night length, usually species with a wide latitudinal distribution, such as tomato or corn (Halevy, 1984).
- Intermediate-day plants - plants that require a night length between 12 and 14 h.
- Ambiphotoperiodic day plants - plants in which flower induction occurs under long or short nights but not at intermediate nights.

Within photoperiodic response groups, plants exhibit facultative (quantitative) or obligate (qualitative) responses. Plants with a facultative or quantitative response flower under any photoperiod but flower faster developmentally under the described photoperiod. Plants with an obligate or qualitative response will not flower unless they receive the described photoperiod.

Many herbaceous ornamental species are photoperiodic (Table 6). Shillo (1976) showed *Limonium sinuata* Mill. flowering was delayed 6 weeks and leaf number below the first flower increased when plants were grown under ambient daylength (10–12 h photoperiod) versus ambient daylength plus a 4-h NI with incandescent lights; *L. sinuate* is a facultative LDP. *Eschscholzia californica* Cham. flowered only when grown under 9-h daylength plus a 4-h NI (3–4 μmol $m^{-2}$ $s^{-1}$ from incandescent lamps); *E. californica* was classified as an obligate LDP by Lyons and Booze-Daniels (1986). Keatinge et al. (1998) noted *Dolichos lablab* L. flowered after 69 and 172 days when grown under an 11.5-h versus 14.5-h daylength (respectively, leaf number below the first flower not reported); *D. lablab* was classified as a facultative SDP. Armitage and Garner (1999) reported *Catananche caerulea* L. Per. was an obligate LDP, flowering after 164 days when grown under LD but not SD. Photoperiodic flowering responses of over 50 herbaceous flowering and herb crops were subsequently determined by Erwin and Warner (2002) and Mattson and Erwin (2003).

Species within genera can vary widely in photoperiodic requirements. For instance, Warner and Erwin (2003b) showed all photoperiodic response groups were evident in the genus *Hibiscus*. Cumming (1969) showed *Chenopodium rubrum* L. strains collected from different latitudes expressed a wide range of photoperiodic flowering responses. In addition, flowering responses to photoperiod can vary with varieties within a species. Ross and Murfet (1985) reported three photoperiodic response groups among *Lathryus odoratus* L. varieties: a facultative (spring) or obligate (summer) LD response. Variation in photoperiodic responses among *Salvia splendens* and *Petunia* x *hybrida* varieties has also been observed (Erwin, personal observation). Introduction of new germplasm (often wild in origin) can increase the photoperiod dependence of new varieties. An example of this was observed in the increase in LD dependence for the flowering of *P.* x *hybrida* 'Purple Wave' and 'Fantasy' series (Erwin et al., 1997).

The time required for a crop to be 'determined' to flower is essential for commercial producers. *Begonia* x *hiemalis* flower induction required 2–3 weeks (Karlsson and Heins, 1992). Non-reversible *P.* x *hybrida* 'Purple Wave' flowering occurred when plants were exposed to 20 LDs (Mattson, 2002).

Photoperiod can interact with temperature to affect the rapidity of flower induction and initiation, or if it occurs at all, under inductive conditions (Salisbury and Ross, 1969). When temperatures exceed 22–23°C, at night flowering of many SDP is delayed, often called 'heat delay'. Heat delay was historically observed on poinsettia and chrysanthemum. More recently, heat delay was observed on the SDP *Gomphrena globosa* L. when NT exceeded 25°C (Warner et al., 1997). In general, as NT increases above 22°C, the length of inductive long night for SDPs increases. Therefore, heat delay can be overcome by sufficiently increasing night length. The SDPs most sensitive to high temperatures in the night are most sensitive 8 h after the onset of darkness.

In contrast, some species are photoperiodic at warmer temperatures but not at lower temperatures. *Begonia* x *hiemalis* is a facultative SDP at temperatures over 24°C, but a DNP when grown at temperatures below 24°C (Sandved, 1969). *Cineraria* is a DNP when grown below 17°C, but a SDP when grown above 17°C. In contrast, some species are sensitive to high DTs (*Fuchsia* x *hybrida*, *A. majus*), and some species are sensitive to high DTs or NTs (*I. hawkeri*, *C. morifolium*, *S. truncata*, many *Pelargonium* spp.) (Erwin and Warner, 1999). Higher irradiance levels overcome an LD requirement in *H. moscheutos* when grown under SD conditions (Warner and Erwin, 2003b).

As previously mentioned, plants perceive daylength by measuring night length. SDPs flower when night length exceeds some critical duration regardless of daylength. In contrast, LDP flower when night length is less than some critical duration regardless of the daylength. The importance of night length in photoperiodism was demonstrated by Hamner and Bonner (1938) and Hamner (1940) where flowering of *Xanthium strumarium* and *Glycine max* (L.) Merrill (both SDP) occurred only when night length exceeded 8.5 and 10 h, respectively. Flowering of the SDPs *X. strumarium* or *P. nil* was inhibited by NI lighting of as little as a few minutes even when total night length was sufficient to promote flowering. In contrast, NI lighting can stimulate flowering in LDP even when night length is too long for flowering. In general, the length of NI required to inhibit flowering in SDP is less than that required to promote flowering in LDP. Harder and Bode (1943) showed the inhibition of flowering of the SDP *K. blossfeldiana* by NI was most effective when it occurred 5 to 7 h after the onset of darkness. Salisbury and Bonner (1956) showed *X. strumarium* flowering was most sensitive to NI 8 h after the onset of darkness. Therefore, SDP and LDP appear to be most sensitive to NI approximately 8 h after the onset of darkness (Vince-Prue, 1975; Salisbury, 1963; Papenfuss and Salisbury, 1967). These observations are the basis for the commercial practice of providing NI from 10 pm to 2 am with low-intensity lighting (2 μmol $m^{-2}$ $s^{-1}$) to promote or inhibit flowering in photoperiodic species in ornamental crop production. The effectiveness of a red-light NI decreases as temperature increases from 18.5°C to 25°C (Takimoto and Hamner, 1964).

Separate from flower evocation and initiation, species differ in their photoperiod requirement for flower development. For instance, *C.* x *morifolium* flower development has a photoperiod requirement shorter than that for flower induction and initiation. Therefore, if chrysanthemum (SDP) plants are induced during spring and placed under LD conditions following induction, flower development would be inhibited, resulting in 'crown bud' formation (Dole and Wilkins, 1999). Similarly, if LDP is induced very early in spring with NI lighting and placed back under natural SD conditions too early after induction, flower bud development can be arrested.

Light action spectra for the inhibition of flowering by NI in the SDPs *Glycine* and *Xanthium* were similar to that for the stimulation of flowering in the LDPs *Hordeum vulgare* and *H. niger*, indicating a similar photoreceptor involved (Borthwick et al., 1948, 1952b; Borthwick and Hendricks, 1960). Hendricks et al. (1956) demonstrated that red NI lighting effects on flowering were reversible by a subsequent exposure to far-red light that restored the flowering response (Downs, 1956). This reversibility was a result of two different forms of a single pigment, a red and far-red light absorbing form; the pigment was named phytochrome (Butler et al., 1959). Ultimately, action spectra for the inhibition of flowering and for the reversal of that inhibition identified peaks at 660 nm and 720–740 nm, respectively (Saji et al., 1983). Separate from phytochrome, there is also evidence that a blue light receptor is involved in flowering although these data are not clear (Millar et al., 1995; Hicks et al., 1996; Zagotta et al., 1996). Interestingly, some species differ in the required 'mixture' of light quality needed for optimal stimulation or inhibition of flowering. For instance, many

SDPs are most sensitive to red light (660 nm) at night, whereas the stimulation of flowering in some LDPs by NI requires both red and far-red light (Runkle and Heins, 2001). It is for this reason that the induction of flowering in the LDP *Fuchsia* x *hybrida* requires a light source that contains both red and far-red light. Therefore, care must be taken when utilizing new LEDs to ensure they emit the light quality desired for the plant photoperiodic group.

When a single leaf of the SDP *X. strumarium* was exposed to SDs but the remainder of the plant was grown under LDs, the plant flowered (Hamner and Bonner, 1938). Zeevaart (1969) demonstrated a single excised *Perilla crispa* (L.) Britt. (SDP) leaf exposed to SDs was capable of repeatedly inducing plants grown under LDs to flower after being grafted to those plants. The degree to which an induced leaf can induce a non-induced donor plant to flower can vary with species and the degree a leaf is induced. *X. strumarium* leaves remain induced for a few days and are competent in inducing plants to which they are grafted (Carr, 1959). Imamura (1953) showed a SDP *P. nil* leaf was induced for a few days following a single inductive long night but could remain in an induced state for many days when the leaves received numerous SDs compared to one. Similarly, *P. crispa* leaves could be inductive for a few days or many months depending on the number of inductive SDs received prior to removal from the donor plant (Zeevaart, 1957).

Grafting between species from different photoperiodic classes showed the floral induction stimulus was similar across species. Grafting an induced leaf from the LDP *Hyoscyamus niger* L. to the SDP *Nicotiana tabacum* L. grown under LD caused the SD-grown tobacco to flower (Melchers and Lang, 1941). Although more difficult, inducing flowering in an LDP by grafting an induced SDP leaf can also induce flowering (Lang and Melchers, 1943; Zeevaart, 1957). Flowering of parasitic plants such as dodder can be stimulated when parasitizing induced plants from any photoperiodic class (Frattianne, 1965).

Stimulation of flowering occurred when a leaf of the DNP plant *G. max* 'Agate' was grafted onto the SDP *G. max* var. 'Biloxi' grown under LDs suggest a similar factor inducing flowering in both species (Heinze et al., 1942). Such grafting experiments were repeated by others (Lona, 1946; Zeevaart, 1962; Wellensiek, 1966). Handro (1977) demonstrated that only *Streptocarpus nobilis* L. leaf tissue was capable of subsequent flowering in vitro under inductive conditions, suggesting leaves were critical in perceiving daylength. In contrast, removal of all leaves of the LDP *H. niger* resulted in flowering, suggesting leaves produced an inhibitor and photoperiod (LD) acted to depress the synthesis of that inhibitor (Lang and Melchers, 1943). Guttridge (1959) also showed flowering following leaf removal on the SDP strawberry. Leaves from a photoperiodic tobacco (non-inductive conditions) are capable of inhibiting the flowering of a DNP tobacco cultivar (Lang et al., 1977). Together, these experiments suggested a mobile flowering promoter was produced in induced leaves and was subsequently translocated to the shoot tip where flowering induction was promoted. The term used to describe this promoter was 'florigen'. Early work focused on an association with flower induction and GAs. A 100 ppm application of GA3 to a facultative SDP *Cosmos* was able to substitute for SD when plants were grown under an 18-h photoperiod (Wittwer and Bukovac, 1959). *Lolium temulentum* plants (SDP) are committed to flower after the exposure to a single LD. Excised terminal meristems exposed to a single LD will flower in tissue culture in the presence of GAs. SD-exposed meristems in tissue culture will not flower in the presence of GAs. Therefore, LDs are required for the *determination* of *L. temulentum* L., and GAs are required for the *expression* of the determined state (McDaniels and Harnett, 1996).

Daylength is perceived in the leaves, and a stimulus is translocated to the shoot meristem. Imamura and Takimoto (1955) estimated the timing of the flowering stimulus by developing a two-branched *P. nil* system and identifying the node at which flowering was induced. Subsequent experiments showed the translocation of the compound from the leaf to a receptive bud is comparable or somewhat slower than that for the translocation of sugars (2–4 mm $h^{-1}$) (Evans and Wardlaw, 1966; King et al., 1968). We now know that signaling molecules (RNAs and proteins) are transported from leaves via the phloem to a meristem where those molecules are integrated into complex networks that regulate growth and flower development (Deeken et al., 2008; Giakountis and Coupland, 2008). Inductive LD signals (via phytochrome and circadian clock networks) promote the stable expression of CONSTANS (CO), particularly at the end of LD (Turck et al., 2008; Fornara et al., 2009). CO, in turn, activates the transcription of FLOWERING LOCUS T (FT) and the homolog TWIN SISTER OF FT (TSF) in phloem companion cells (Corbesier et al., 2007; Mathieu et al., 2007; Turck et al., 2008). FT and TSF are small (approximately 175 amino acids), globular, phosphatidylethanoamine-binding proteins and are graft transmissible. FT/TSF encodes the phloem-mobile 'florigen' protein. Flowering is, therefore, transcriptionally regulated by FT/TSF synthesis.

FT and TSF are transported via the phloem to the shoot apical meristem where they induce flowering (Abe et al., 2015; Wigge et al., 2005; Turck et al., 2008). Export of FT and TSF in phloem is mediated by phloem-specific proteins FT-INTERACTING PROTEIN 1 (FTIP1) and ALTERED PHLOEM DEVELOPMENT (APL) that transport FT to phloem sieve elements by an endoplasmic reticulum membrane protein FTIP1 (Liu et al., 2012; Abe et al., 2015). FT expression will occur only if trehalose 6-phosphate (T6P) signaling occurs; T6P is associated with carbohydrate availability signaling (Wahl et al., 2013). Taken together, exclusive of meristem-specific genes, flowering regulation is directly impacted by the expression of CO, FT, TSF, FTIP1, APL, and MRF1.

Recent work showed that phloem companion cells play an integral role where photoperiod upregulates chromatin modifications H3K4me3 and a previously uncharacterized MORN-motif repeat protein (MORN-MOTIF REPEAT PROTEIN REGULATING FLOWERING 1 - MRF1). The *mrf1* mutation delayed flowering, and MRF1 overexpression accelerated flowering, indicating that flowering is regulated not only by the promoter FT but also by the process of moving FT to the meristem (You et al., 2019). Therefore, posttranscriptional regulation of FT and flowering can occur as well.

## 6.4 Vernalization flowering pathway

'Vernalization' is defined as the low-temperature induction of flowering in imbibed seed or a growing plant or bulb (Chouard, 1960; Taiz and Zeiger, 1998). As with photoperiodism, plants can be vernalized only if they are mature. Interestingly, winter annuals can perceive vernalization as imbibed seed (Napp-Zinn, 1969). For instance, imbibed seed of *R. sativus* 'Chinese Jumbo Radish Scarlet' can be vernalized in as little as 6–8 days (Engelen-Eigles and Erwin, 1997; Erwin et al., 2002). Beet and *Oenothera* (Thomas and Vince-Prue, 1997) can also be vernalized as imbibed seed. Yet most plants must be mature before a meristem is capable of being vernalized. *H. niger* must grow for 10 days, and *L. biennis* must grow for 7 weeks before plants can respond to vernalizing temperatures (Thomas and Vince-Prue, 1997). Carrot cv 'Chantenay Red Cored' requires 8–12 leaves before plants are capable of being vernalized (Atherton et al., 1990). Herbaceous ornamental perennial species have a minimum leaf requirement before plants can be induced to flowering via vernalization (Lopes and Weiler, 1977; Sheldron and Weiler, 1982a,b; Iverson and Weiler, 1994; Armitage et al., 1996; Whitman et al., 1996). Varieties of a given species can vary in the leaf number that must unfold before vernalizing temperatures can be perceived and the length of time required for complete vernalization (Yui and Yoshikawa, 1991; Wurr et al., 1994).

Vernalization requirements are characterized into three categories: no requirement, an obligate requirement (required for flowering), or a facultative requirement (promotes earlier flowering). Plants that have a short vernalization requirement (<30 days) such as *Brassica campestris pekinensis* L. (Suge, 1984) and *A. thaliana* L. (Bagnall, 1993; Martinez-Zapater and Somerville, 1990) often have a facultative vernalization requirement, and photoperiod (LD) is often the primary flower induction stimulus. In contrast, plants that require a long vernalization period for induction (6–12 weeks) often have an obligate vernalization requirement.

The site of perception of vernalization is the shoot tip, and the stimulus is not graft transmissible. When different portions of a plant were cooled, only the shoot tip was capable of being vernalized (Curtis and Chang, 1930; Metzger, 1988, 1996). Gregory and de Ropp (1938) showed excised winter rye embryos, or even a fragment thereof, could be vernalized in tissue culture. Wellensiek (1961, 1962) reported young expanding leaves of *L. biennis* were also capable of being vernalized. In all cases, active cell division is required for vernalization (Metzger et al., 1992). Once tissue has been vernalized, all growth from that tissue is vernalized (Schwabe, 1954).

A wide range of cool temperatures (−6°C and 17°C) are capable of promoting vernalization. Chouard (1960) reported plants can be vernalized with temperatures ranging from −6°C to 14°C. Both Lang (1952) and Chouard (1960) reported the optimal temperature for vernalization was between 6°C and 10°C. In general, it is now accepted that vernalization occurs between 0°C and 17°C, with an optimum temperature between 4°C and 6°C.

The time required for complete vernalization varies with species. *Lunaria biennis* L. requires 9 weeks (Wellensiek, 1958), and *Apium graveolens* L. (celery) and *Secale cereale* L. (winter rye) require 6 weeks (Ramin and Atherton, 1991; Friend, 1965) for complete vernalization, respectively. *Alstroemeria* L. hybrids require 6 weeks at 5°C (Healy and Wilkins, 1981, 1982).

The vernalization process, once saturated, can be very stable (Table 3) (Yeh et al., 1997). For instance, once *Hyoscyamus* is vernalized (requires both vernalization and LDs for flowering), the vernalized state can persist for several months under SD conditions: flowering occurs when plants are returned to LD conditions (Thomas and Vince-Prue, 1997). *L. longiflorum* bulbs exposed to vernalizing temperatures retained that information even when several warm temperature periods occurred prior to the completion of a vernalization treatment (Erwin and Engelen-Eigles, 1997). No loss of the vernalization stimulus was evident on henbane after 190 days. In fact, there was no measurable loss in the degree of vernalization on henbane until 300 days had passed (Thomas and Vince-Prue, 1997). Similarly, some cereal seeds can be moistened, vernalized, and then re-dried and maintained for months without loss of the vernalization status (Purvis and Gregory, 1952).

The period between the completion of a vernalization and flower initiation can be divided into two phases; phase I is a period immediately after vernalization when flower induction can be reversed by exposure to warm temperatures, low irradiance, and/or SD. The reversal of vernalization by environmental conditions is referred to as 'devernalization' (Purvis and Gregory, 1952; Lang, 1965). Phase II is that period after Phase I when flower induction is stable and not reversible. The most common cause of devernalization is high temperature (25–30°C). However, devernalization by SD and/or low irradiance during or after vernalization was demonstrated in some species (Thomas and Vince-Prue, 1997). Devernalization treatments are most effective immediately after vernalization, if a plant is not completely vernalized (Table 1), when light intensity during and/or after vernalization is low, and/or when plants are exposed to SD conditions (Thomas and Vince-Prue, 1997). Exposure of *Beta* to SD caused devernalization at any stage of development, even when stem elongation had been initiated. *Oenothera biennis* L. and *C. allionii* L. can be devernalized by transfer to SDs (Wellensiek, 1965). In fact, nearly all vernalizing-requiring plants can hypothetically be devernalized. For instance, Petkus winter rye (Purvis and Gregory, 1952), *C. allionii* (Barendse, 1964), carrot (Hiller and Kelly, 1979), cauliflower (Fujime and Hirose, 1980), kohlrabi (Wiebe et al., 1992), celery (Booiju and Meurs, 1993), *L. longiflorum* Thunb. (Miller and Kiplinger, 1966), and cineraria (Yeh et al., 1997) can all be devernalized. The first 5 days after the vernalization appear to be the most critical period; that is non-optimal temperatures or irradiance after this period have a little impact on induction (Erwin, personal observation).

Photoperiod can interact with vernalization to affect flower induction (Yui and Yoshikawa, 1991; Thomas and Vince-Prue, 1997; Yeh et al., 1997). In some cases, as with winter cereals, SDs can substitute for vernalization entirely. *Coreopsis grandiflora* Hogg ex Sweet. Per. is a SDP–LDP plant; however, the SD requirement can be substituted with a vernalization treatment (Ketellapper and Barbaro, 1966). In contrast, LDs can be additive with vernalization in hastening the flowering of some species including *L. longiflorum* (Wilkins et al., 1968). Many vernalization-requiring plants require LDs after cooling. The number of LDs required and/or the critical daylength required for flowering after vernalization decrease as the length of the vernalization treatment increases (Lang, 1965).

Light during vernalization is required for some species. *Apium graveolens* L. var. dulce. does not perceive vernalization unless plants are lit during vernalization (Ramin and Atherton, 1994). *Dianthus carophyllus* L. and *Ajuga reptans* L. flowering was hastened as irradiance during vernalization increased (Eltzroth and Link, 1970). Low red:far-red light ratio was antagonistic to vernalization of *R. sativus* (Engelen-Eigles, 1996) and *L. longiflorum* (Erwin and Engelen-Eigles, unpublished data), and high red:far-red ratio appears to hasten vernalization. For instance, sprouted *L. longiflorum* bulbs were induced to flower with vernalization for 4 weeks under light with a high red:far-red ratio but required 8 weeks for flower induction when sprouted bulbs were lit with light with a low red:far-red ratio (Erwin and Engelen-Eigles, unpublished data).

Early work on the physiology of vernalization focused on the identification of a specific, cold-induced GA responsible for flowering (Lang, 1965). GAs interact with vernalization to affect the flowering of a number of species (Metzger and Dusbabek, 1991). The application of GAs to some biennial species, including wild-type *A. thaliana* grown under unfavorable conditions for flowering, promoted flowering and bolting (Lang, 1957; Zeevaart, 1983). Conversely, the application of 2-chloroethyltrimethylammonium chloride (CCC

**Table 8** Progressive stabilization of vernalization with increasing duration of exposure to cold in winter rye 'Petkus'

| Cold treatment (weeks) | 2 | 3 | 4 | 5 | 6 | 8 |
|---|---|---|---|---|---|---|
| Plants remaining vernalized after 2 days | | | | | | |
| At 25°C (devernalizing) (%) | 0 | 42 | 44 | 75 | 84 | 97 |

*Source*: adapted from Thomas and Vince-Prue (1984).

- Cycocel), a GA-synthesis inhibitor, delayed flowering in several wild-type *A. thaliana* lines and suggested GA was essential for flower induction (Napp-Zinn, 1985). Similarly, low temperatures and exogenous application promoted flowering in Chinese cabbage (*Brassica campestris* L. ssp. *Chinensis* var. *itilis* Tsen et Lee) in an additive fashion, and the application of uniconazle (GA biosynthesis inhibitor) delayed flowering on low-temperature exposed plants (Shi-wei et al., 2018). Exogenous application of specific GAs can overcome a vernalization requirement for some species, but not for others (Ketellapper and Barbaro, 1966).

Vernalizing temperatures result in the demethylation of plant DNA (Burn et al., 1993). The application of 5-azacytidine, a demethylator, promoted the flowering of non-vernalized cold requiring *Thlaspi* and *A. thaliana* (Metzger, 1996). Transformed *A. thaliana* with reduced DNA methylation because of the presence of a methyltransferase (METI) antisense gene flowered earlier than untransformed plants (Finnegan et al., 1998). However, a 70% reduction of methylation in antisense plants where the promotion of flowering was comparable to that of a vernalization plant resulted in only a 15% decrease in DNA methylation. These data suggest that other methyltransferases may be affected by vernalization, since METI affected only one.

We now know that vernalization acts through FLOWERING LOCUS C (FLC), which acts as an inhibitor of flowering. Vernalization progressively silences FLC. FLC repression typically requires histone modifications at the chromatin of the FLC locus. Such modifications involve two other proteins: VERNALIZATION INSENSITIVE 3 (VIN3) and VERNALIZATION 2 (VIN2) (Table 8).

## *6.5 Stress induction of flowering*

Short-term plant stress can shorten the juvenile phase in some species. Perhaps the best example of this phenomenon is the induction of early flowering in celosia after a short-term water stress (Erwin, personal observation). Similarly, it appears as though short-term heat stress can cause earlier flowering on newer cultivars of garden mums (Erwin, personal communication). Additionally, some plug producers note that *Impatiens walleriana* flowering is hastened by nutritional stress (R. Wagner, B. Gibson, personal communication).

## 6.6 Chemical promotion of flowering

Exposure to ethylene induces flowering in some plant species, but not in others. Most notably, ethylene exposure induces flowering in Bromeliads. Clark and Kerns (1942) showed that the application of auxin to pineapple could stimulate early flowering. Subsequent research by Burg and Burg (1966) demonstrated that the induction of flowering by auxin was associated with auxin-induced stimulation of ethylene. Commercially, pineapple is now stimulated to flower by applying ethephon, which releases ethylene on plant leaves. The application of ethephon also stimulates early flowering in mango (Chacko et al., 1976). Similar ethylene-induced flower induction was observed on *Euphorbia splendens* (Erwin, personal observation).

## 6.7 Temperature inhibition of flowering

Following complete induction of flowering, temperature appears to play a role in flower development. Excessively cool or warm temperatures can delay flower development or cause flower bud abortion. *P. zonale* (DNP) flower number per inflorescence decreased as ADT increased. *P. peltatum* L. 'Nicole' flower number per inflorescence decreased from 9 to 4 as ADT increased from 12°C to 29°C (Erwin, 1999). Similarly, *P.* x *domesticum* L.H. Bailey flower number per inflorescence decreased as ADT increased and flowering was inhibited altogether above constant 17°C (Erwin and Engelen, 1992). *P.* x *hortorum* 'Veronica' flower number per inflorescence decreased from 52 to 15 flowers as ADT increased from 12°C to 29°C (Erwin and Heins, 1992). *Fuchsia hybrida* L. 'Dollar Princess' flower number per node increased from 2.3 to 6.0 as ADT increased from 12°C to 15°C and then decreased to 2.3 flowers per node as ADT was further increased to 25°C (Erwin and Kovanda, 1990). Flower development temperature sensitivity can vary diurnally. *Schlumbergera truncata* Lem. and *E. pulcherrima* flower development is most sensitive to NT. For instance, flower number was greatest on *S. truncata* 'Madisto' when NT was 20°C (Erwin et al., 1990). In contrast, *A. majus* flower development was most sensitive to DT. Flower development in *I. hawkeri* Bull., *D. grandiflora*, and *F. hybrida* was most sensitive to DT and NT.

Short- or long-term temperature stress can reduce flowering. Exposure of *I. hawkeri* and *V.* x *wittrockiana* to high temperatures (35°C) for as short as 2 h reduced photosynthetic rates for up to 3 days compared to 'un-stressed' plants (Warner and Erwin, 2002). Photosynthetic recovery following a high-temperature stress event varied between *I. hawkeri* and *V.* x *wittrockiana*, with *V. wittrockiana* recovering more quickly. Prolonged high-temperature exposure reduces the flowering of numerous flower crops. Exposure of *A. majus*, *Calendula officinalis*, *I. wallerana*, *Mimilus* x *hybridus* Hort. Ex Siebert & Voss, and

*Torenia fournieri* Linden ex E. Forum to high temperatures reduced flowering compared to growing plants at cooler temperatures across irradiance levels (Warner and Erwin, 2001b).

High-temperature tolerance can vary within varieties of a species. Prolonged exposure of *V.* x *wittrockiana* to high temperatures (30°C) compared to a 'non-stress' environment (20°C) reduced flowering of all cultivars; however, the degree of reduction varied with the variety: 23% for 'Crystal Bowl Purple' and 79% for 'Majestic Giants Red and Yellow' (Warner and Erwin, 2003a). Similarly, high-temperature exposure (32/28°C DT/NT) impacted *L. esculentum* flower number per cluster varieties differently (Warner and Erwin, 2003a). Lastly, Strope (1999) showed *I. hawkeri* cultivars varied in their ability to flower at constant 30°C.

High-temperature sensitivity of flowering may be due to reduced carbon assimilation at high temperatures (Ranney and Peet, 1994), increased respiration (Berry and Bjorkman, 1980), or a reduced ability of floral organs to recruit photoassimilates (Dinar and Rudich, 1985). Photosynthetic rates of five birch taxa declined sharply when temperature exceeded 30°C (Ranney and Peet, 1994). This may be in part due to chloroplast biogenesis reduction at temperatures over 32°C in barley (Smillie et al., 1978). Respiration rates increase at temperatures above the optimum temperature for photosynthesis, resulting in a net carbon loss (Berry and Bjorkman, 1980). Lastly, high-temperature impacts on plant growth hormones in reproductive tissues may reduce the ability of those tissues to attract assimilates (Kuo and Tsai, 1984). Additionally, species may have a minimum irradiance required for flower development as was noted with *Rosa* spp. previously (Nell and Rasmussen, 1979).

### *6.8 Gaps in knowledge*

Commercial growers are increasingly called on to produce plants during non-traditional times of a year. In addition, new species are continuously being introduced into the marketplace. Therefore, gaps include:

- Identification of juvenile phase length and specific flower induction requirements on new species.
- Determination of how photoperiodic flower induction interacts with temperature.
- Better understanding of how light quality impacts flower induction of different species.

## 7 Dormancy

In some cases, plant development can be arrested until after some environmental factor, or criteria, is satisfied. For instance, flower initiation is induced and initiated under SD and/or cool NTs in July and August. Following initiation in the beginning stages of flower development, further development is arrested, or flowers go 'dormant'. Flower dormancy in these crops is then overcome by a 6–8-week exposure to 4°C. In some cases, exogenous application of GA3 can overcome dormancy and enhance shoot emergence from tubers (Lurie et al., 1992). The application of GA3 to overcome flower bud dormancy is common commercially in azalea (Rhododendron).

In contrast to woody plant species, bulb dormancy can be more complex. Bulb crops can vary considerably as to when flowers are initiated during the year and during the life cycle of the species. Species have been classified into the following five response groups by Hartsema (1961):

1 Flowers initiated during the spring or early summer of the year, preceding that in which they reach anthesis and before the bulbs are 'lifted' (*Narcissus*, *Galanthus*, *Leucojum*).
2 Flowers initiated following the previous growing period, so that the bulbs have initiated flowers by replanting time in autumn (*Tulipa*, *Hyacinthus*, *Iris reticulata*).
3 Flowers initiated after replanting, at the low temperatures of winter or early spring (bulbous *Iris*, but not *I. reticulata*).
4 Flowers initiated more than a year before anthesis (*Nerine*).
5 Flower initiation alternates with leaf formation through the whole growing period (*Hippeastrum*, *Zephyranthes*).

The physiological basis for flowering in each of these groupings can differ significantly. For instance, there is evidence that group 3 has a vernalization requirement, whereas groups 1 and 2 have a dormancy requirement. By dormancy requirement, flower development can be arrested until environmental or time constraints are fulfilled. For example, ephemerals, that is *Tulipa* x *hybrida* or *Narcissus pseudonarcissus*, are not photoperiodic as the length of a growth cycle is relatively short. Following flower initiation and partial development, plants/bulbs go 'dormant'. Flower development will proceed only after a cool temperature requirement is met in these bulbs.

A cool temperature requirement to 'break' dormancy is common in temperate herbaceous perennial species as well. In contrast to temperate species, some species from dry indigenous climates or deserts can have a moisture requirement to break dormancy. For instance, water stress can induce dormancy in *Achimenes* hybrids (Zimmer and Junker, 1985).

## 8 References

Abe, M., Kaya, H., Watanabe-Taneda, A., Shibuta, M., Yamaguchi, A., Sakamoto, T., Kurata, T., Ausin, I., Araki, T. and Alonso-Blanco, C. 2015. FE, a phloem-specific Myb-related protein, promotes flowering through transcriptional activation of FLOWERING LOCUS T and FLOWERING LOCUS T INTERACTING PROTEIN 1. *Plant J.* 83(6), 1059-68. doi:10.1111/tpj.12951.

Acquaah, G. 2008. *Horticulture: Principles and Practice* (4th edn.). Pearson Publishers, Harlow, UK.

Adams, S. R., Pearson, S., Hadley, P. and Patefield, W. M. 1999. The effects of temperature and light integral on the phases of photoperiod sensitivity in *Petunia x hybrida. Ann. Bot.* 83(3), 263-9. doi:10.1006/anbo.1998.0817.

Agharkar, M., Lomba, P., Altpeter, F., Zhang, H., Kenworthy, K. and Lange, T. 2007. Stable expression of AtGA2ox1 in a low-input turfgrass (*Paspalum notatum* Flugge) reduces bioactive gibberellin levels and improves turf quality under field conditions. *Plant Biotech. J.* 5(6), 791-801. doi:10.1111/j.1467-7652.2007.00284.x.

Aguilar-Martinez, J., Poza-Carrion, C. and Cubas, P. 2007. Arabidopsis BRANCHED1 acts as an integrator of branching signals within axillary buds. Plant Cell 19, 458-72.

Amoo, S. O., Finnie, J. F. and Van Staden, J. 2009. *In vitro* propagation of *Huernia hystrix*; an endangered medicinal and ornamental succulent. *Plant Cell Tiss. Organ Cult.* 96(3), 273-8. doi:10.1007/s11240-008-9484-8.

Anderson, N. (Eds). 2007. *Flower Breeding and Genetics*. Springer Publishers, Heidelberg, Germany.

Aparecida de Oliveira, S., Fátima Pires da Silva Machad, M., José Prioli, A. and Aparecida Mangolin, C. 1995. *In vitro* propagation of *Cereus peruvianus* mill. (cactaceae). *In Vitro Cell Dev. Biol - Plant* 31(1), 47-50. doi:10.1007/BF02632226.

Aragino, M. 1981. Response of macrophytes to temperature. In: Pirson, A. and Zimmerman, M. H. (Eds), *Encyclopedia of Plant Physiology* (vol. 12a), pp. 339-71.

Armitage, A. M. 1996. Forcing perennials in the greenhouse. *GrowerTalks* 60(3), 86, 88, 93, 94, 96, 97.

Armitage, A. M. and Garner, J. M. 1999. Photoperiod and cooling duration influence growth and flowering of six herbaceous perennials. *J. Hortic. Sci. Biotech.* 74(2), 170-4. doi:10.1080/14620316.1999.11511090.

Armitage, A. M. and Tsujita, M. J. 1979. The effect of supplemental light source, illumination and quantum flux density on the flowering of seed-propagated geraniums. *J. Hortic. Sci.* 54(3), 195-8. doi:10.1080/00221589.1979.11514870.

Armitage, A. M., Carlson, W. H. and Flore, J. A. 1981. The effect of temperature and quantum flux density on the morphology, physiology, and flowering of hybrid geraniums. *J. Amer. Soc. Hortic. Sci.* 106(5), 643-7.

Armitage, A. M., Copeland, L., Gross, P. and Green, M. 1996. Cold storage and moisture regime influence flowering of *Oxalis adenophylla* and *Ipheion uniflorum*. *HortScience* 31(7), 1154-5. doi:10.21273/HORTSCI.31.7.1154.

Atherton, J. G., Craigon, J. and Basher, E. A. 1990. Flowering and bolting in carrot. I. Juvenility, cardinal temperatures and thermal times for vernalization. *J. Hortic. Sci.* 65(4), 423-9. doi:10.1080/00221589.1990.11516075.

Bagnall, D. J. 1993. Light quality and vernalization interact in controlling late flowering in *Arabidopsis* ecotypes and mutants. *Ann. Bot.* 71(1), 75-83. doi:10.1006/anbo.1993.1009.

Bangerth, F. 1994. Response of cytokinin concentration in the xylem exudate of bean (*Phaseolus vulgaris* L) plants to decapitation and auxin treatment, and relation to apical dominance. *Planta* 194(3), 439-42. doi:10.1007/BF00197546.

Barendse, G. W. 1964. Vernalization of *Cieranthus allionii* Hort. *Meded. Landbou., Wageningen* 64, 1-64.

Basra, A. 2006. *Handbook of Seed Science and Technology*. CRC Press, Boca Raton, FL.

Berghage, R. 1998. Controlling height with temperature. *HortTechnology* 8(4), 535-9. doi:10.21273/HORTTECH.8.4.535.

Berghage, R. D. and Heins, R. D. 1991. Quantification of temperature effects on stem elongation in poinsettia. *J. Amer. Soc. Hortic. Sci.* 116(1), 14-8. doi:10.21273/JASHS.116.1.14.

Berry, J. and Bjorkman, O. 1980. Photosynthetic response and adaptation to temperature in higher plants. *Annu. Rev. Plant Physiol.* 31(1), 491-543. doi:10.1146/annurev.pp.31.060180.002423.

Beveridge, C., Dun, E. and Rameua, C. 2009. Peas has its tendrils in branching discoveries spanning a century from auxin to strigolactones. Plant Physiol. 151, 985-90.

Bevington, J. M. 1981. Phytochrome action during prechilling induced germination of *Betula papyrifera* Marsh. *Plant Physiol.* 67(4), 705-10. doi:10.1104/pp.67.4.705.

Bewley, J., Bradford, K., Hilhorst, H. and Nonogaki, H. 2013. *Seeds: Physiology of Development, Germination and Dormancy* (3rd edn.). Springer Publishers, Heidelberg, Germany.

Biddington, N. L. 1986. The effects of mechanically-induced stress in plants - a review. *Plant Growth Reg.* 4(2), 103-23. doi:10.1007/BF00025193.

Booiju, R. and Meurs, E. J. J. 1993. Flower induction and initiation in celeriac (*Apium graveolens* L. var. *rapaceum* (Mill.) DC.): effects of temperature and plant age. *Sci. Hortic.* 55(3-4), 227-38. doi:10.1016/0304-4238(93)90034-N.

Borthwick, H. A. and Hendricks, S. B. 1960. Photoperiodism in plants. *Science* 132(3435), 1223-8. doi:10.1126/science.132.3435.1223.

Borthwick, H. A., Hendricks, S. B. and Parker, M. W. 1948. Action spectrum for the photoperiodic control of floral initiation of a long day plant, Wintex Barley (*Hordeum vulgare*). *Bot. Gaz.* 110(1), 103-18. doi:10.1086/335520.

Borthwick, H. A., Hendricks, S. B. and Parker, M. W. 1952a. The reaction controlling floral initiation. *Proc. Natl. Acad. Sci. U.S.A.* 38(11), 929-34. doi:10.1073/pnas.38.11.929.

Borthwick, H. A., Hendricks, S. B., Parker, M. W., Toole, E. H. and Toole, V. K. 1952b. A reversible photoreaction controlling seed germination. *Proc. Natl. Acad. Sci. U.S.A.* 38(8), 662-6. doi:10.1073/pnas.38.8.662.

Brondum, J. J. and Heins, R. D. 1993. Modeling temperature and photoperiod effects on growth and development of Dahlia. *J. Amer. Soc. Hortic. Sci.* 118(1), 36-42. doi:10.21273/JASHS.118.1.36.

Burg, S. P. and Burg, E. A. 1966. Auxin-induced ethylene formation: its relation to flowering in pineapple. *Science* 152(3726), 1269. doi:10.1126/science.152.3726.1269.

Burn, J. E., Bagnall, D. J., Metzger, J. D., Dennis, E. S. and Peacock, W. J. 1993. DNA methylation, vernalization, and the initiation of flowering. *Proc. Natl. Acad. Sci. U.S.A.* 90(1), 287-91. doi:10.1073/pnas.90.1.287.

Butler, K. H., Norris, H., Siegelman, W. and Hendricks, S. B. 1959. Detection assay and purification of the pigment controlling photoresponsive development of plants. *Proc. Natl. Acad. Sci. U.S.A.* 45, 1703-8.

Cabahug, R. A. M., Nam, S. Y., Lim, K., Jeon, J. K. and Hwang, Y. 2018. Propagation techniques for ornamental succulents. *Flower Res. J.* 26(3), 90-101. doi:10.11623/frj.2018.26.3.02.

Calvert, A. 1957. Effect of early environment on the development of flowering in tomato. *J. Hortic. Sci.* 32(1), 9-17. doi:10.1080/00221589.1957.11513884.

Cameron, A., Yuan, M., Heins, R. and Carlson, W. 1996. Juvenility: your perennial crop's age affects flowering. *GrowerTalks* 60(8), 30-2, 34.

Carr, D. J. 1959. Translocation between leaf and meristem in the flowering response of short-day plants. *9th Int. Bot. Congr.* (vol. 2), p. 11.

Carvalho, S., Bakker, M. and Heuvelink, H. 2006. Improving product quality and timing of Kalanchoe. Acta Hortic. 718, 655-63.

Carver, S. A. and Tayama, H. K. 1992. Number of stock plant shoot nodes influences splitting of 'Lilo' poinsettia. *HortTechnology* 2(2), 206-7. doi:10.21273/HORTTECH.2.2.206.

Cendan, C., Sampedro, L. and Zas, R. 2013. The maternal environment determines the timing of germination in *Pinus pinaster*. *Environ. Exp. Bot.* 94, 66-72. doi:10.1016/j.envexpbot.2011.11.022.

Chacko, E. K., Kohli, R. R., Swamy, R. D. and Randhawa, G. S. 1976. Growth regulators and flowering in juvenile mango (*Mangifera indica* L.) seedlings. *Acta Hortic.* 56(56), 173-6. doi:10.17660/ActaHortic.1976.56.17.

Charlton, T. M. 1957. Research on roads in Britain. *Nature* 180(4586), 588-9. doi:10.1038/180588a0.

Chouard, P. 1960. Vernalization and its relations to dormancy. *Annu. Rev. Plant. Physiol.* 11(1), 191-238. doi:10.1146/annurev.pp.11.060160.001203.

Clark, J. R. 1983. Age-related changes in trees. *J. Arboric.* 9, 201-5.

Clark, H. E. and Kerns, K. R. 1942. Control of flowering with phytohormones. *Science* 95(2473), 536-7. doi:10.1126/science.95.2473.536.

Cockshull, K. E. 1984. The photoperiodic induction of flowering in short-day plants. In: Vince-Prue, D., Thomas, B. and Cockschull, K. E. (Eds), *Light and the Flowering Process*. Academic Press, London, UK, pp. 33-50.

Cockshull, K. E. 1985. *Chrysanthemum morifolium*. In: Halevy, A. H. (Ed.), *Handbook of Flowering* (vol. 2). CRC Press, Boca Raton, FL, pp. 238-57.

Cone, J. W. and Kendrick, R. E. 1986. Photocontrol of seed germination. In: Kendrick, R. E. and Kronenberg, G. H. M. (Eds), *Photomorphogenesis in Plants*. Marinus Nijoff Publishing, Dordrecht, the Netherlands, pp. 443-65.

Cooper, W. C. and Peynado, A. 1958. Effect of gibberellic acid on growth and dormancy of Citrus. *Proc. Amer. Soc. Hortic. Sci.* 72, 284-9.

Copeland, L. and McDonald, M. 2001. *Principles of Seed Science and Technology* (4th edn.). Springer Publishers, Heidelberg, Germany.

Corbesier, L., Vincent, C., Jang, S., Fornara, F., Fan, Q., Searle, I., Giakountis, A., Farrona, S., Gissot, L., Turnbull, C. and Coupland, G. 2007. FT protein movement contributes to long- distance signaling in floral induction of Arabidopsis. *Science* 316(5827), 1030-3. doi:10.1126/science.1141752.

Corbineau, F. and Come, S. 1991. Seeds of ornamental plants and their storage. *Acta Hortic.* 298, 313-22.

Corbineau, F., Bagniol, S. and Come, S. 1990. Sunflower (*Helianthus annuus* L.) seed dormancy and its regulation by ethylene. *Isr. J. Bot.* 39, 313-25.

Cumming, B. G. 1969. *Chenopodium rubrum* L. and related species. In: Evans, L. T. (Ed.), *The Induction of Flowering*. Macmillan, Melbourne, Australia and London, UK, pp. 156-85.

Currey, C. J., Hutchinson, V. A. and Lopez, R. G. 2012. Growth morphology, and quality of rooted cuttings of several herbaceous annual bedding plants are influenced by photosynthetic daily light integral during root development. *HortScience* 47(1), 25-30. doi:10.21273/HORTSCI.47.1.25.

Curtis, O. F. and Chang, C. K. 1930. The relative effectiveness of temperature of the crown as contrasted with that of the rest of the plant upon flowering of celery plants. *Amer. J. Bot.* 17, 1047-8.

Davies, F., Geneve, R., Wilson, S., Hartmann, H. and Kester, D. 2017. *Hartmann and Kester's Plant Propagation: Principles and Practice* (9th edn.). Pearson Higher Education, New York, NY.

Deeken, R., Ache, P., Kajahn, I., Klinkenberg, J., Bringmann, G. and Hedrich, R. 2008. Identification of *Arabidopsis thaliana* phloem RNAs provides a search criterion for phloem-based transcripts hidden in complex datasets of microarray experiments. *Plant J.* 55(5), 746-59. doi:10.1111/j.1365-313X.2008.03555.x.

Deitzer, G. F. 1984. Photoperiodic induction in long-day plants. In: Vince-Prue, D., Thomas, B. and Cockschull, K. E. (Eds), *Light and the Flowering Process*. Academic Press, London, UK, pp. 51-64.

De Kerck, G. 2002. Rooting of microcuttings: theory and practice. *In Vitro Cell. Dev. Biol. - Plant* 38(5), 415-22. doi:10.1079/IVP2002335.

de Saint Germain, A., Ligerot, Y., Dun, E. A., Pillot, J. P., Ross, J. J., Breveridge, C. A. and Rameau, C. 2013. Strigolactones stimulate internode elongation independently of gibberellins. *Plant Physiol.* 163(2), 1012-25. doi:10.1104/pp.113.220541.

Dinar, M. and Rudich, J. 1985. Effect of heat stress on assimilate partitioning in tomato. *Ann. Bot.* 56(2), 239-48. doi:10.1093/oxfordjournals.aob.a087008.

Dole, J. M. and Wilkins, H. F. 1999. *Floriculture: Principles and Species*. Prentice Hall, Upper Saddle River, NJ.

Dole, J. M. and Wilkins, H. F. 2004. *Floriculture: Principles and Species* (2nd edn.). Pearson Publishers, Harlow, UK.

Downs, R. J. 1956. Photoreversibility of flower initiation. *Plant Physiol.* 31(4), 279-84. doi:10.1104/pp.31.4.279.

Duke, S. O., Egley, G. H. and Reger, B. J. 1977. Model for variable light sensitivity in imbibed dark-dormant seeds. *Plant Physiol.* 59(2), 244-9. doi:10.1104/pp.59.2.244.

Eltzroth, D. E. and Link, C. B. 1970. The influence of light during vernalization on the flowering response of *Ajuga* and *Dianthus*. *J. Amer. Soc. Hortic. Sci.* 95, 95-8.

Enfield, A. 2011. Influence of the postharvest environment on the storage potential and propagation performance of unrooted cuttings of herbaceous ornamentals. MS Dissertation. Clemson University.

Engelen-Eigles, G. 1996. Irradiance and light quality interact to affect vernalization in *Raphanus sativus* L. M.S. Thesis. Department of Horticultural Science, University of Minnesota, St. Paul, MN, p. 55108.

Engelen-Eigles, G. and Erwin, J. E. 1997. A model plant for vernalization studies. *Sci. Hortic.* 70(2-3), 197-202. doi:10.1016/S0304-4238(97)00037-X.

Erickson, V. L., Armitage, A., Carlson, W. H. and Miranda, R. M. 1980. The effect of cumulative photosynthetically active radiation on the growth and flowering of the seedling geranium, *Pelargonium x hortorum*. *HortScience* 15(6), 815-7.

Erwin, J. E. 1991a. Cool temperatures are still critical on regals. *Minn. Comm. Flow. Grow. Bull.* 40(3), 3-4.

Erwin, J. E. 1991b. Thermomorphogenesis in plants. PhD Thesis. Department of Horticulture, Michigan State University.

Erwin, J. E. 1999. Ivy geranium production. *Ohio Flor. Bull.* 831(1), 15-20.
Erwin, J. 2009. Looking for new ornamentals: flowering studies. *Acta Hortic.* 813, 61-6. doi:10.17660/ActaHortic.2009.813.5.
Erwin, J. E. and Engelen, G. 1992. Regal geranium production. *Minn. Comm. Flow. Grow. Bull.* 41(6), 1-9.
Erwin, J. E. and Engelen-Eigles, G. 1997. Variation in *Lilium* flower induction based on shipping and rooting temperature and year. *J. Amer. Soc. Hortic. Sci.*
Erwin, J. E. and Guenthner, G. 2018. Impact of irradiance abd carbon dioxide concentration on the photosynthesis of five ornamental plant species/hybrids grown in tissue culture containers. *ASHS Annual Conference*, Washington DC.
Erwin, J. E. and Heins, R. D. 1990. Temperature effects on lily development rate and morphology from the visible bud stage until anthesis. *J. Amer. Soc. Hortic. Sci.* 115(4), 644-6. doi:10.21273/JASHS.115.4.644.
Erwin, J. E. and Heins, R. D. 1992. Environmental effects on geranium development. *Minn. Comm. Flow. Grow. Bull.* 41(1), 1-9.
Erwin, J. E. and Heins, R. D. 1995. Thermomorphogenic responses in stem and leaf development. *HortScience* 30(5), 940-9. doi:10.21273/HORTSCI.30.5.940.
Erwin, J. and Kovanda, B. 1990. Fuchsia production. *Minn. Comm. Flow. Grow. Bull.*, 1-3.
Erwin, J. E. and Pierson, G. 1992. Interaction between diurnal temperature fluctuations and gibberellins on *Lycopersicon* stem elongation and chlorophyll content. *HortScience* 27, 657.
Erwin, J. E. and Warner, R. 1999. Temperature. In: Buck, C. A., Carver, S. A., Gaston, M. L., Konjoian, P. S., Kunkle, L. A. and Wilt, M. F. (Eds), *Tips on Growing Bedding Plants* (4th edn.). OFA Services, Columbus, OH, pp. 69-82.
Erwin, J. E. and Warner, R. M. 2002. Determination of photoperiodic response group and effect of supplemental irradiance on flowering of several bedding plant species. *Acta Hortic.* 580(580), 95-9. doi:10.17660/ActaHortic.2002.580.11.
Erwin, J. E., Heins, R. D. and Karlsson, M. G. 1989a. Thermomorphogenesis in *Lilium longiforum* Thunb. *Amer. J. Bot.* 76(1), 47-52. doi:10.1002/j.1537-2197.1989.tb11283.x.
Erwin, J. E., Heins, R. D., Berghage, R. D., Kovanda, B. J., Carlson, W. H. and Biernbaum, J. A. 1989b. Cool mornings can control plant height. *GrowerTalks* 52(9), 73-4.
Erwin, J. E., Heins, R. D., Berghage, R. D. and Kovanda, B. 1990. Temperature effects *Schlumbergera truncata* 'Madisto' flower initiation. *Acta Hortic.* 272(272), 97-102. doi:10.17660/ActaHortic.1990.272.13.
Erwin, J. E., Heins, R. D. and Moe, R. 1991a. Temperature and photoperiod effects on *Fuchsia x hybrida* morphology. *J. Amer. Soc. Hortic. Sci.* 116(6), 955-60. doi:10.21273/JASHS.116.6.955.
Erwin, J. E., Velguth, P. and Heins, R. D. 1991b. Diurnal variations in temperature affect cellular elongation but not division. *HortScience* 26, 105.
Erwin, J. E., Heins, R. D. and Faust, J. E. 1993. Thermomorphogenic and photoperiodic responses of *Nephrolepis exaltata* 'Dallas Jewel'. *HortScience* 28(3), 182-4. doi:10.21273/HORTSCI.28.3.182.
Erwin, J. E., Strefeler, M. and Heins, R. D. 1994a. Diurnal temperature fluctuations affect vegetable development. *Minn Comm Flow Grow. Bull.* 42, 6-10.
Erwin, J. E., Velguth, P. and Heins, R. D. 1994b. Day/night temperature affects cell elongation but not cell division in *Lilium longiflorum* Thunb. *J. Exp. Bot.* 45(7), 1019-25. doi:10.1093/jxb/45.7.1019.

Erwin, J. E., Warner, R., Smith, T. and Wagner, R. 1997. Photoperiod and temperature interact to affect *Petunia x hybrida* Vilm. development. *HortScience* 32, 501.

Erwin, J. E., Warner, R. M. and Smith, A. G. 2002. Vernalization, photoperiod and GA3 interact to affect flowering of Japanese radish (*Raphanus sativus* Chinese Radish Jumbo Scarlet). *Physiol. Plant.* 115(2), 298–302. doi:10.1034/j.1399-3054.2002.1150217.x.

Erwin, J. E., Rohwer, C. and Gesick, E. 2006. Red:far red and photosynthetically active radiation foltering by leaves differs with species. *Acta Hortic.* 711, 195–9.

Erwin, J., Gesick, E., Altman, K. and O'Connell, R. 2015. Plant growth regulator and temperature affect *Echeveria* and *Andromischus* leaf cutting rooting and axillary shoot formation. *HortScience* 50(9), 235.

Erwin, J. E., Alman, K. and Esqueda, F. 2017. Temperature impacts cactus and succulent development rate. *HortTechnology* 27(1), 65–8. doi:10.21273/HORTTECH03515-16.

Evans, L. T. and Wardlaw, I. F. 1966. Independent translocation of $^{14}C$-labelled assimilates and of the floral stimulus in *Lolium temulentum*. *Planta* 68(4), 310–26. doi:10.1007/BF00386331.

Faiss, M., Zalubìlová, J., Strnad, M. and Schmülling, T. 1997. Conditional transgenic expression of the ipt gene indicates a function for cytokinins in paracrine signaling in whole tobacco plants. *Plant J.* 12(2), 401–15. doi:10.1046/j.1365-313x.1997.12020401.x.

Fausey, B. A., Cameron, A. C. and Heins, R. D. 2001. Daily light integral, photoperiod, and vernalization affect flowering of *Digitalis purpurea* L. 'Foxy'. *HortScience* 36(3), 565.

Faust, J. E. and Heins, R. D. 1998. Modeling shoot-tip temperature in the greenhouse environment. *J. Amer. Soc. Hortic. Sci.* 123(2), 208–14. doi:10.21273/JASHS.123.2.208.

Felippe, G. M. 1980. Germination of light-sensitive seeds of *Cucumis anguria* and *Rumex obtusifolius*: effects of temperature. *New Phytol.* 84(3), 439–48. doi:10.1111/j.1469-8137.1980.tb04551.x.

Ferguson, B. J. and Breveridge, C. A. 2009. Roles for auxin, cytokinin, and strigolactone in regulating shoot branching. *Plant Physiol.* 149(4), 1929–44. doi:10.1104/pp.109.135475.

Finch-Savage, W. E. and Bassel, G. W. 2016. Seed vigour and crop establishment: extending performance beyond adaptation. *J. Exp. Bot.* 67(3), 567–91. doi:10.1093/jxb/erv490.

Finnegan, E. J., Genger, R. K., Kovac, K., Peacock, W. J. and Dennis, E. S. 1998. DNA methylation and the promotion of flowering by vernalization. *Proc. Natl. Acad. Sci. U.S.A.* 95(10), 5824–9. doi:10.1073/pnas.95.10.5824.

Fornara, F., Panigrahi, K. C., Gissot, L., Sauerbrunn, N., Ruhl, M., Jarillo, J. A. and Coupland, G. 2009. *Arabidopsis* DOF transcription factors act redundantly to reduce CONSTANS expression and are essential for a photoperiodic flowering response. *Dev. Cell* 17(1), 75–86. doi:10.1016/j.devcel.2009.06.015.

Frattianne, D. G. 1965. The inter relationship between the flowering of dodder and the flowering of some long and short day plants. *Amer. J. Bot.* 52(6 Part 1), 556–62. doi:10.1002/j.1537-2197.1965.tb06822.x.

Friend, D. J. C. 1965. Interaction of red and far red radiations with the vernalization process in winter rye. *Can. J. Bot.* 43(1), 161–70. doi:10.1139/b65-018.

Frydman, V. M. and Wareing, P. F. 1974. Phase change in *Hedera helix* L. III. The effects of gibberellins, abscisic acid and growth retardants on juvenile and adult ivy. *J. Expt. Bot.* 25(2), 420–9. doi:10.1093/jxb/25.2.420.

Fujime, Y. and Hirose, T. 1980. Studies on thermal conditions of curd formation and development in cauliflower and broccoli. *J. Japan Soc. Hortic. Sci.* 49(2), 217-27. doi:10.2503/jjshs.49.217.

Fujiwara, K., Kozai, T. and Watanabe, I. 1987. Fundamental studies on environments in plant tissue culture vessels (3) Measuerments of carbon dioxide gas concentration in closed vessels containing tissue cultured plantlets and estimates of net photosynthetic rates of the plantlets. *J. Agric. Meteorol.* 43(1), 21-30. doi:10.2480/agrmet.43.21.

Fujiwara, K., Kira, S. and Kozai, T. 1992. Time course of $CO_2$ exchange of potato cultures *in vitro* with different sucrose concentrations in the culture medium. *J. Agric. Meteorol.* 48(1), 49-56. doi:10.2480/agrmet.48.49.

Gagnon, S. and Dansereau, B. 1990. Influence of light and photoperiod on growth and development of gerbera. *Acta Hortic.* 272(272), 145-52. doi:10.17660/ActaHortic.1990.272.21.

Garner, W. W. and Allard, H. A. 1920. Effect of relative length of day and night and other factors of the environment on growth and reproduction of plants. *J. Agric. Res.* 18, 553-607.

Geneve, R. 2013. Seed dormancy. *Acta Hortic.* 1014, 243-8.

Giakountis, A. and Coupland, G. 2008. Phloem transport of flowering signals. *Curr. Opin. Plant Biol.* 11(6), 687-94. doi:10.1016/j.pbi.2008.10.003.

Goh, C. J. and Arditti, J. 1985. Orchidaceae. In: Halevy, A. H. (Ed.), *Handbook of Flowering* (vol. I). CRC Press, Boca Raton, FL, pp. 309-36.

Goliber, T. E. and Feldman, L. J. 1989. Osmotic stress, endogenous abscisic acid and the control of leaf morphology in *Hippuris vulgaris* L. *Plant Cell Environ.* 12(2), 163-71. doi:10.1111/j.1365-3040.1989.tb01929.x.

Gomez-Roldan, V., Fermas, S., Brewer, P. B., Puech-Pages, V., Dun, E. A., Pillot, J. P., Letisse, F., Matusova, R., Danoun, S., Portais, J. C., Bouwmeester, H., Bécard, G., Beveridge, C. A., Rameau, C. and Rochange, S. F. 2008. Strigolactones inhibition of shoot branching. *Nature* 455(7210), 189-94. doi:10.1038/nature07271.

Graper, D. F. and Healy, W. 1991. High pressure sodium irradiation and infrared radiation accelerate *Petunia* seedling growth. *J. Amer. Soc. Hortic. Sci.* 116(3), 435-8. doi:10.21273/JASHS.116.3.435.

Gregory, F. G. and de Ropp, R. S. 1938. Vernalization of excised embryos. *Nature* 142(3593), 481-2. doi:10.1038/142481b0.

Guttridge, C. G. 1959. Further evidence for a growth-promoting and flower-inhibiting hormone in strawberry. *Ann. Bot.* 23(4), 612-21. doi:10.1093/oxfordjournals.aob.a083679.

Hackett, W. P. 1980. Control of phase change in woody plants. In: Little, C. H. A. (Ed.), *Control of Shoot Growth in Trees*. Proc. Int. Union For. Res. Organ Working Parties on Xylem Physiology and Shoot Growth Physiology, Fredericton, New Brunswick, NJ, pp. 257-72.

Hackett, W. P. and Srinivasani, C. 1983/1985. *Hedera helix* and *Hedera canatiensis*. In: Halevy, A. H. (Ed.), *Handbook of Flowering* (vol. 3). CRC Press, Boca Raton, FL, pp. 89-97.

Halevy, A. H. 1984. Light and autonomous induction. In: Vince-Prue, D., Thomas, B. and Cockschull, K. E. (Eds), *Light and the Flowering Process*. Academic Press, London, UK, pp. 65-74.

Hamner, K. C. 1940. Interrelation of light and darkness in photoperiodic induction. *Bot. Gaz.* 101(3), 658–87. doi:10.1086/334903.
Hamner, K. C. and Bonner, J. 1938. Photoperiodism in relation to hormones as factors in floral initiation and development. *Bot. Gaz.* 100(2), 388–431. doi:10.1086/334793.
Handro, W. 1977. Photoperiodic induction of flowering on different explanted tissues from *Streptocarpus nobilis* cultured *in vitro*. *Bol. Botan. Univ. Sao Paulo* 5, 21-6.
Harder, R. and Bode, O. 1943. Wirkung von zwischenbelichtungen wahrend der dunkelperiode auf Kalanchoe. *Planta* 33(4), 469–504. doi:10.1007/BF01916538.
Hartmann, H. T. and Kester, D. E. 1959. *Plant Propagation: Principles and Practices*. Prentice-Hall, Inc, Englewood Cliffs, NJ.
Hartsema, A. M. 1961. Influence of temperatures on flower formation and flowering of bulbous and tuberous plants. In: Ruhland, W. (Ed.), *Encyclopedia of Plant Physiology* (vol. 16), pp. 123-61.
Healy, W. E. and wilkins, H. F. 1981. Interaction of soil temperature, air temperature and photoperiod on growth and flowering of *Alstroemeria* 'Regina'. *HortScience* 16, 459.
Healy, W. E. and Wilkins, H. F. 1982. The interaction of temperature on flowering of *Alstroemeria* 'Regina'. *J. Amer. Soc. Hortic. Sci.* 107, 248–51.
Hedley, C. L. 1974. Response to light intensity and daylength of two contrasting flower varieties of *Antirrhinum majus*. *J. Hortic. Sci.* 49(1), 105–12. doi:10.1080/00221589.1974.11514556.
Heide, O. M. 1965. *Campanula isophylla* som langdagsplante. *Gartneryrket (Oslo)* 55, 210–2.
Heide, O. M. and Runger, W. 1985. Begonia. In: Halevy, A. H. (Ed.), *Handbook of Flowering* (vol. II). CRC Press, Boca Raton, FL, pp. 4-23.
Heins, R. D. and Erwin, J. E. 1990. Understanding and applying DIF. *Greehouse Grower* 8(2), 73–8.
Heins, R. D. and Erwin, J. E. 1991. The history of DIF and the use of a morning temperature dip to control height. *Minn. Comm. Flow. Grow. Bull.* 40(6), 1-4.
Heins, R., Lange, N., Wallace Jr., T. F. and Carlson, W. 1994. *Plug Storage*. Greenhouse Grower Meister Pub, Willoughby, OH.
Heinze, P. H., Parker, M. W. and Borthwick, H. A. 1942. Floral initiation in Biloxi soybean as influenced by grafting. *Bot. Gaz.* 103(3), 518–30. doi:10.1086/335066.
Hendricks, L. 1991. Diff bei Beet- und Balkonpflanzen. *Gb+Gw* 39, 1883-9.
Hendricks, S. B. and Taylorson, R. B. 1976. Variation in germination and amino acid leakage of seeds with temperature related to membrane phase change. *Plant Physiol.* 58(1), 7–11. doi:10.1104/pp.58.1.7.
Hendricks, S. B., Borthwick, H. A. and Downs, R. J. 1956. Pigment conversion in the formative responses of plants to radiation. *Proc. Natl. Acad. Sci. U.S.A.* 42(1), 19–26. doi:10.1073/pnas.42.1.19.
Hicks, K. A., Millar, A. J., Carre, I. A., Somers, D. E., Straume, M., Meeks-Wagner, D. R. and Kay, S. A. 1996. Conditional circadian dysfunction of the *Arabidopsis* early-flowering 3 mutant. *Science* 274(5288), 790–2. doi:10.1126/science.274.5288.790.
Hiller, L. K. and Kelly, W. C. 1979. The effect of post-vernalization temperature on seedstalk elongation and flowering in carrots. *J. Amer. Soc. Hortic. Sci.* 104, 253–7.
Hopkins, W. and Huner, N. 2009. *An Introduction to Plant Physiology* (4th edn.). CRC Press, Boca Raton, FL.
Imamura, S. 1953. Photoperiodic initiation of flower primordia in Japanese morning glory, *Pharbitis nil* Chois. *Proc. Japan Acad.* 29(7), 368–73. doi:10.2183/pjab1945.29.368.

Imamura, S. and Takimoto, A. 1955. Transmission rate of photoperiodic stimulus in *Pharbitis nil*. *Bot. Mag.* 68(808), 260-6. doi:10.15281/jplantres1887.68.260.

Iverson, R. R. and Weiler, T. C. 1994. Strategies to force flowering of six herbaceous garden perennials. *HortTechnology* 4(1), 61-5. doi:10.21273/HORTTECH.4.1.61.

Jaffe, M. J. 1973. Thigmomorphogenesis: the response of plant growth and development to mechanical stimulation: with special reference to *Bryonia dioica*. *Planta* 114(2), 143-57. doi:10.1007/BF00387472.

Kaczperski, M. P. 1989. Influence of temperature and irradiance on growth and development of *Petunia* x *hybrida* 'Snow Cloud'. MS Thesis. Department of Horticulture, Michigan State University.

Karlsson, M. G. 1988. Temperature and irradiance effects on *Dendranthema grandiflorum* cv 'Bright Golden Anne' growth and development. PhD Thesis. Department of Horticulture, Michigan State University.

Karlsson, M. G. and Heins, R. D. 1992. Begonias. In: Larson, R. A. (Ed.), *Introduction to Floriculture* (2nd edn.). Academic Press, San Diego, CA, pp. 409-27.

Karlsson, M. G., Heins, R. D. and Erwin, J. E. 1988. Quantifying temperature controlled leaf unfolding rate in 'Nellie White' Easter lily. *J. Amer. Soc. Hortic. Sci.* 113(1), 70-4.

Karlsson, M. G., Heins, R. D., Erwin, J. E. and Berghage, R. D. 1989. Development rate during four phases of chrysanthemum growth as determined by preceding and prevailing temperatures. *J. Amer. Soc. Hortic. Sci.* 114(2), 158-62.

Keatinge, J. D. H., Qi, A., Wheeler, T. R., Ellis, R. H. and Summerfield, R. J. 1998. Effects of temperature and photoperiod on phenology as a guide to the selection of annual legume cover and green manure crops for hillside farming systems. *Field Crops Res.* 57(2), 139-52. doi:10.1016/S0378-4290(97)00122-6.

Keeley, J. E. and Pausas, J. G. 2018. Evolution of 'smoke' induced seed germination in pyroendemic plants. *S. Afr. J. Bot.* 115, 251-5. doi:10.1016/j.sajb.2016.07.012.

Ketellapper, H. J. and Barbaro, A. 1966. The role of photoperiod, vernalization and gibberellic acid in floral induction in *Coreopsis grandiflora* Nutt. *Phyton Buenos Aires* 23, 33-41.

King, R. W., Evans, L. T. and Wardlaw, I. F. 1968. Translocation of the floral stimulus in *Pharbitis nil* in relation to that of assimilates. *Z. Pflanzenphysiol.* 59, 377-88.

Klein, A., Houtin, H., Rond, C., Marget, P., Jacquin, F., Boucherot, K., Huart, M., Rivière, N., Boutet, G., Lejeune-Hénaut, I. and Burstin, J. 2014. QTL analysis of frost damage in pea suggests different mechanisms involved in frost tolerance. *Theor. Appl. Genet.* 127(6), 1319-30. doi:10.1007/s00122-014-2299-6.

Kozai, T., Kushihashi, S., Kubota, C. and Fujiwara, K. 1992. Effect of difference between photoperiod and dark period temperatures, and photosynthetic photon flux density on the shoot length and growth of potato plantlets *in vitro*. *J. Japan Soc. Sci.* 61, 93-8.

Krumweide, D. 1960. Uber de wirkung von starkund schwachlichtkombinationen auf das bluhen von *Kalanchoe blossfeldiana*. *Biol. Zentralbl.* 79, 258-78.

Kuo, C. G. and Tsai, C. T. 1984. Alteration by high temperature of auxin and gibberellin concentrations in the floral buds, flowers and young fruit of tomato. *HortScience* 19, 870-2.

Lang, A. 1952. Physiology of flowering. *Annu. Rev. Plant. Physiol.* 3(1), 265-306. doi:10.1146/annurev.pp.03.060152.001405.

Lang, A. 1957. The effect of gibberellin upon flower formation. *Proc. Natl. Acad. Sci. U.S.A.* 43(8), 709-17. doi:10.1073/pnas.43.8.709.

Lang, A. 1965. Physiology of flower initiation. In: Ruhland, H. (Ed.), *Encyclopedia of Plant Physiology*. Springer-Verlag, Berlin, Germany, pp. 1380–536.

Lang, A. and Melchers, G. 1943. Die photoperiodische Reaktion von Hyoscyamus niger. *Planta* 33(5), 653-702. doi:10.1007/BF01916588.

Lang, A., Chailakhyan, M. K. and Frolova, I. A. 1977. Promotion and inhibition of flower formation in a day-neutral plant in grafts with a short-day plant and a long-day plant. *Proc. Natl. Acad. Sci. U.S.A.* 74(6), 2412-6. doi:10.1073/pnas.74.6.2412.

Langton, F. A., Cockshull, K. E., Cave, C. R. J. and Hemming, E. J. 1992. Temperature regimes to control plant structure: current UK R&D. *Acta Hortic.* 326, 49–59.

Latimer, J. G. 1991. Mechanical conditioning for control of growth and quality of vegetable transplants. *HortScience* 26(12), 1456–61. doi:10.21273/HORTSCI.26.12.1456.

Latimer, J. G. and Thomas, P. A. 1991. Application of brushing for growth control of tomato transplants in a commercial setting. *HortTechnology* 1(1), 109–10. doi:10.21273/HORTTECH.1.1.109.

Leatherwood, W. R. and Dole, J. 2007. Ethylene sensitivity of unrooted cuttings from 28 genera and effectiveness of 1-MCP to prevent ethylene damage. AFE Special Report #433.

Lecharny, A., Scwall, M. and Wagner, E. 1985. Stem extension rate in light-grown plants: effects of photo- and thermoperiodic treatments on the endogenous circadian rhythm in *Chenopodium rubrum*. *Plant Physiol.* 79(3), 625–9. doi:10.1104/pp.79.3.625.

Leduc, N., Roman, H., Barbier, F., Peron, T., Huche-Thelier, L., Lothier, J., Demotes-Mainard, S. and Sakr, S. 2014. Light signaling in bud outgrowth and branching in plants. *Plants* 3(2), 223–50. doi:10.3390/plants3020223.

Lejune-Henaut, I., Hanocq, E., Bethencourt, L., Fontaine, V., Delbreil, B., Morin, J., Petit, A., Devaux, R., Boilleau, M., Stempniak, J. J., Thomas, M., Lainé, A. L., Foucher, F., Baranger, A., Burstin, J., Rameau, C. and Giauffret, C. 2008. The flowering locus Hr colocalizes with a major QTL affecting winter frost tolerance in *Pisum sativum* L. *Theor. Appl. Genet.* 116(8), 1105–16. doi:10.1007/s00122-008-0739-x.

Leyser, O. 2005. The fall and rise of apical dominance. *Curr. Opin. Genet. Dev.* 15(4), 468–71. doi:10.1016/j.gde.2005.06.010.

Leyser, O. and Day, S. 2002. *Mechanisms in Plant Development*. Wiley-Blackwell, Oxford, UK.

Lindow-Lopez, L., Galindez, G., Aparicio-Gonzalez, M., Suhring, S., Rojas-Arechiga, M., Pritchard, H. W. and Ortega-Baes, P. 2018. Effects of alternating temperature on cactus seeds with positive photoblastic response. *J. Arid Environ.* 148, 74-7. doi:10.1016/j.jaridenv.2017.10.006.

Liu, L., Liu, C., Hou, X., Xi, W., Shen, L., Tao, Z., Wang, Y. and Yu, H. 2012. FTIP1 is an essential regulator required for florigen transport. *PLoS Biol.* 10(4), e1001313. doi:10.1371/journal.pbio.1001313.

Ljung, K., Bhalerao, R. P. and Sandberg, G. 2001. Sites and homeostatic control of auxin biosynthesis in *Arabidopsis* during vegetative growth. *Plant J.* 28(4), 465-74. doi:10.1046/j.1365-313x.2001.01173.x.

Lo, S. F., Yang, S. Y., Chen, K. T., Hsing, Y. I., Zeevaart, J. A., Chen, L. J. and Yu, S. M. 2008. Anovel class of gibberellin 2-oxidases control semidwarfism, tillering, and root development in rice. *Plant Cell* 20(10), 2603-18. doi:10.1105/tpc.108.060913.

Lona, F. 1946. Sui fenomeni di induzione, post-effeto e localizzazione fotoperiodica. *Nuovo Giorn. Bot. Ital.* 53, 548–75.

Longman, K. A. 1976. Some experimental approaches to the problem of phase change in forest trees. *Acta Hortic.* 56(56), 81–90. doi:10.17660/ActaHortic.1976.56.6.
Lopes, L. C. and Weiler, T. C. 1977. Light and temperature effects on the growth and flowering of *Dicentra spectabilis* (L.) Lem. *J. Amer. Soc. Hortic. Sci.* 102, 388–90.
Lopez, R. G. and Runkle, E. S. 2008. Photosynthetic daily light integral during propagation influences rooting and growth of cuttings and subsequent development of New Guinea impatiens and Petunia. *HortScience* 43(7), 2052–9. doi:10.21273/HORTSCI.43.7.2052.
Luckwill, L. C. 1970. Control of growth and fruitfulness of apple trees. In: Luckwill, L. C. and Cutting, C. V. (Eds), *Physiology of Tree Crops*, pp. 237–54.
Ludolph, D. 1992. Height control of ornamental plants without chemical growth retardants. *Ohio Florist's Bull.* 748, 1–4.
Lurie, G., Watad, A. A. and Borochov, A. 1992. Aconitum: effect of tuber size, daylength and $GA_3$ on growth, flowering and tuber production. *Acta Hortic.* 325, 113–8.
Lyons, R. E. and Booze-Daniels, J. N. 1986. Characteristics of the photoperiodic response of California poppy. *J. Amer. Soc. Hortic. Sci.* 111(4), 593–6.
Martinez-Zapater, J. M. and Somerville, C. R. 1990. Effect of light quality and vernalization on late-flowering mutants of *Arabidopsis thaliana*. *Plant Physiol.* 92(3), 770–6. doi:10.1104/pp.92.3.770.
Martins, J. P. R., Verdoodt, V., Pasqual, M. and DeProft, M. 2015. Impacts of photoautotrophic and photomixotrophic conditions on *in vitro* propgated *Billbergia zebrina* (Bromeliaceae). *Plant Cell Tiss. Organ Cult.* 123(1), 121–32. doi:10.1007/s11240-015-0820-5.
Mashiguchi, K., Sasaki, E., Shimada, Y., Nagae, M., Ueno, K., Nakano, T., Yoneyama, K., Suzuki, Y. and Asami, T. 2009. Feedback-regulation of strigolactone biosynthetic genes and strigolactone-regulated genes in *Arabidopsis*. *Biosci. Biotechnol. Biochem.* 73(11), 2460–5. doi:10.1271/bbb.90443.
Mason, M. G., Ross, J. J., Babst, B. A., Wienclaw, B. N. and Beveridge, C. A. 2014. Sugar demand, not auxin, is the initial regulator of apical dominance. *Proc. Natl Acad. Sci. U.S.A.* 111(16), 6092–7. doi:10.1073/pnas.1322045111.
Mathieu, J., Warthmann, N., Kuttner, F. and Schmid, M. 2007. Export of FT protein from phloem companion cells is sufficient for floral induction in *Arabidopsis*. *Curr. Biol.* 17(12), 1055–60. doi:10.1016/j.cub.2007.05.009.
Mattson, N. 2002. Photoperiod, Irradiance, and Light Quality affect flowering of herbaceous annuals. M.S. Thesis. Department of Horticultural Science. University of Minnesota, St. Paul, MN.
Mattson, N. and Erwin, J. E. 2003. Impact of photoperiod and irradiance on flowering of several herbaceous ornamentals I. *J. Amer. Soc. Hortic. Sci.*
Mayers, A., Neuman, J., Reid, M. and Dodge, L. 1997. New ethylene inhibitor could extend flower life. Perishables Handling Quarterly Issue No. 92, pp. 9–11.
McDaniel, C. N., Singer, S. R. and Smith, S. M. E. 1992. Developmental stated associated with the floral transmission. *Dev. Biol.* 153(1), 59–69. doi:10.1016/0012-1606(92)90091-t.
McDaniels, C. N. and Harnett, L. K. 1996. Flowering as metamorphosis: two sequential signals regulate floral initiation in *Lolium temulentum*. *Development* 122(11), 3661–8.
McDonald, M. and Kwong, F. 2005. *Flower Seeds Biology and Technology*. CABI Publishers, Wallingford, UK.
Melchers, G. and Lang, A. 1941. Weitere untersuchungen zur frage der bluhhormone. *Biol. Zentralbl.* 61, 16–39.

Metzger, J. D. 1988. Localization of the site of perception of thermo-inductive temperatures in *Thlaspi arvense* L. *Plant Physiol.* 88(2), 424-8. doi:10.1104/pp.88.2.424.
Metzger, J. D. 1996. A physiological comparison of vernalization and dormancy chilling requirement. In: Lang, G. A. (Ed.), *Plant Dormancy: Physiology, Biochemistry and Molecular Biology*. CAB International, Wallingford, UK, pp. 147-56.
Metzger, J. D. and Dusbabek, K. 1991. Determination of the cellular mechanisms regulating thermo-induced stem growth in *Thlaspi arvense* L. *Plant Physiol.* 97(2), 630-7. doi:10.1104/pp.97.2.630.
Metzger, J. D., Dennis, E. S. and Peacock, W. J. 1992. Tissue specificity of thermoinductive processes: *Arabidopsis* roots respond to vernalization. *Plant Physiol.* 99(1 Suppl.), 52.
Millar, A. J., Carre, I. A., Strayer, C. A., Chua, N. H. and Kay, S. A. 1995. Circadian clock mutants in *Arabidopsis* identified by luciferase imaging. *Science* 267(5201), 1161-3. doi:10.1126/science.7855595.
Miller, W. 1991. Yti mechanical stress as a 'natural' growth regulator. *GrowerTalks* 55(7), 81-5.
Miller, R. O. and Kiplinger, D. C. 1966. Reversal of vernalization in northwest Easter lilies. *Proc. Amer. Soc. Hortic. Sci.* 88, 646-50.
Milthorpe, F. L. 1959. Studies on the expansion of the leaf surface. I. The influence of temperature. *J. Exp. Bot.* 10(2), 233-49. doi:10.1093/jxb/10.2.233.
Misic, D., Grubisic, D. and Konjevic, R. 2006. Micropropagation of *Salvia brachyodan* through nodel explants. *Biol. Plant.* 50(3), 473-6. doi:10.1007/s10535-006-0074-5.
Mitchell, C. A., Severson, C. J., Wottand, J. A. and Hammer, P. A. 1975. Seismomorphogenic regulation of plant growth. *J. Amer. Soc. Hortic. Sci.* 100, 161-5.
Moe, R., Heins, R. D. and Erwin, J. E. 1991. Effect of day and night temperature alterations, and plant growth regulators on stem elongation and flowering of the long-day plant *Campanula isophylla* Morettii. *Sci. Hortic.* 48, 141-51.
Mortensen, L. M. and Moe, R. 1992. Effects of various day and night temperature treatments on morphogenesis and growth of some greenhouse and bedding plant species. *Acta Hortic.* 327, 77-86.
Motum, G. J. and Goodwin, P. B. 1987a. Floral initiation in kangaroo paw (*Anigozanthos* spp.): a scanning electron microscope study. *Sci. Hortic.* 32(1-2), 115-22. doi:10.1016/0304-4238(87)90022-7.
Motum, G. J. and Goodwin, P. B. 1987b. The control of flowering in Kangaroo paw (*Anigozanthos* spp.). *Sci. Hortic.* 32(1-2), 123-33. doi:10.1016/0304-4238(87)90023-9.
Murfet, I. C. and Reid, J. B. 1993. Developmental mutants. In: Casey, R. and Davies, D. R. (Eds), *Peas: Genetics, Molecular Biology and Biotechnology*. CAB International, Wallingford, UK, pp. 165-216.
Napp-Zinn, K. 1969. *Arabidopsis thaliana* (L.) Heynh. In: Evans, L. T. (Ed.), *The Induction of Flowering*. Macmillan Press, Melbourne, Australia, pp. 291-304.
Napp-Zinn, K. 1985. *Arabidopsis thaliana*. In: Halevy, A. H. (Ed.), *Handbook of Flowering* (vol. I). CRC Press, Boca Raton, FL, pp. 492-503.
Nell, T. A. 1993. *Flowering Potted Plants, Prolonging Shelf Performance*. Ball Publishing, Batavia, IL.
Nell, T. A. and Rasmussen, H. P. 1979. Blindness in roses: effects of high intensity light and blind shoot prediction severity. *J. Amer. Soc. Hortic. Sci.* 104, 21-5.
Njoku, E. 1958. Effect of gibberellic acid on leaf form. *Nature* 182(4642), 1097-8. doi:10.1038/1821097a0.

Nordstrom, A., Tarkowski, P., Tarkowski, D., Norbaek, R., Astot, C., Dolezal, K. and Sandberg, G. 2004. Auxin regulation of cytokinin biosynthesis in *Arabidopsis thaliana*: a factor of potential importance for auxin-cytokinin-regulated development. *Proc. Natl. Acad. Sci. U.S.A.* 101(21), 8039–44. doi:10.1073/pnas.0402504101.

Nordwig, G. 1999. Evaluation of floral inductive requirements and commercial potential of *Asclepias* species. MS Thesis. Department of Horticultural Science, University of Minnesota.

Oinam, G., et al. 2011. Adventitious root formation: an overview of recent developments. *Propag. Ornam. Plants* 11, 78–90.

Oracz, K. and Karpinski, S. 2016. Phytohormones signaling pathways and ROS involvement in seed germination. *Front. Plant Sci.* 7, 864. doi:10.3389/fpls.2016.00864.

Papenfuss, H. D. and Salisbury, F. B. 1967. Aspects of clock resetting in flowering of *Xanthium*. *Plant Physiol.* 42(11), 1562–8. doi:10.1104/pp.42.11.1562.

Pawar, S. S. and Thompson, H. C. 1950. The effect of age and size of plant at the time of exposure of low temperature on reproductive growth of celery. *Proc. Amer. Soc. Hortic. Sci.* 55, 367–71.

Pearson, S., Hadley, P. and Wheldon, A. E. 1993. A reanalysis of the effects of temperature and irradiance on time to flowering in chrysanthemum (*Dendranthema grandiflora*). *J. Hortic. Sci.* 68(1), 89–97. doi:10.1080/00221589.1993.11516332.

Pharis, R. P. and King, R. W. 1985. Gibberellins and reproductive development in seed plants. *Annu. Rev. Plant Physiol.* 36(1), 517–68. doi:10.1146/annurev.pp.36.060185.002505.

Pietsch, G. M., Carlson, W. H., Heins, R. D. and Faust, J. E. 1995. The effect of day and night temperature and irradiance on development of *Catharanthus* roseus (L.) 'Grape Cooler'. *J. Amer. Soc. Hortic. Sci.* 120(5), 877–81. doi:10.21273/JASHS.120.5.877.

Pimienta-Barrios, E. and del Castillo, R. F. 2002. Reproductive biology. In: Nobel, P. S. (Ed.), *Cacti: Biology and Uses*. University of California Press, Berkeley and Los Angeles, CA, pp. 75–90.

Pinthus, M. J. and Mieri, J. 1979. Effects of the reversal of day and night temperatures on tillering and on the elongation of stem and leaf blades of wheat. *J. Exp. Bot.* 30(2), 319–26. doi:10.1093/jxb/30.2.319.

Poesch, G. H. 1931. Forcing plants with artificial light. *Proc. Amer. Soc. Hortic. Sci.* 28, 402–6.

Poethig, R. S. 1990. Phase change and the regulation of shoot morphogenesis in plants. *Science* 250(4983), 923–30. doi:10.1126/science.250.4983.923.

Purvis, O. N. and Gregory, F. G. 1952. Studies in vernalization of cereals. XII. The reversibility by high temperature of the vernalized condition in Petkus winter rye. *Ann. Bot.* 1, 569–92.

Rameau, C., Bertheloot, J., Leduc, N., Andrieu, B., Foucher, F. and Sakr, S. 2014. Multiple pathways regulate shoot branching. *Front Plant Biol.* 5, 741. doi:10.3389/fpls.2014.00741.

Ramin, A. A. and Atherton, J. G. 1991. Manipulation of bolting and flowering in celery (*Apium graveolens* L. var. dulce) II. Juvenility. *J. Hortic. Sci.* 66(6), 709–17. doi:10.1080/00221589.1991.11516202.

Ramin, A. A. and Atherton, J. G. 1994. Manipulation of bolting and flowering in celery (*Apium graveolens* L. var. dulce). III. Effects of photoperiod and irradiance. *J. Hortic. Sci.* 69(5), 861–8. doi:10.1080/14620316.1994.11516522.

Ranney, T. G. and Peet, M. M. 1994. Heat tolerance of five taxa of birch (*Betula*): physiological responses to supraoptimal leaf temperatures. *J. Amer. Soc. Hortic. Sci.* 119(2), 243–8. doi:10.21273/JASHS.119.2.243.

Roberts, D. R. 1991. DIF. *Greenhouse Manag.* (April), 52–5.

Roberts, E. H. and Sommerfield, R. J. 1987. Measurement and prediction of flowering in annual crops. In: Atherton, J. G. (Ed.), *Manipulation of Flowering*.

Rogler, C. E. and Hackett, W. P. 1975. Phase change in *Hedera helix*: stabilization of the mature form with abscisic acid and growth retardants. *Physiol. Plant.* 34(2), 148–52. doi:10.1111/j.1399-3054.1975.tb03810.x.

Ross, J. J. and Murfet, I. C. 1985. Flowering and branching in *Lathyrus odoratus* L.: environmental and genetic effects. *Ann. Bot.* 55(5), 715–26. doi:10.1093/oxfordjournals.aob.a086950.

Sachs, T. and Thimann, K. V. 1964. Release of lateral buds from apical dominance. *Nature* 201(4922), 939–40. doi:10.1038/201939a0.

Saji, H., Vince-Prue, D. and Furuya, M. 1983. Studies on the photoreceptors for the promotion and inhibition of flowering in dark-grown seedlings of *Pharbitis nil* Choisy. *Plant Cell Physiol.* 67, 1183–9.

Salisbury, F. B. 1963. Biological timing and hormone synthesis in flowering of *Xanthium*. *Planta* 49, 518–24.

Salisbury, F. B. and Bonner, J. 1956. The reactions of the photoinductive dark period. *Plant Physiol.* 31(2), 141–7. doi:10.1104/pp.31.2.141.

Salisbury, F. B. and Ross, C. 1969. *Plant Physiology*. Wadsworth Publishing Co., Belmont, CA, p. 608.

Sandved, G. 1969. Flowering in *Begonia* x *hiemalis* Fotsch. as affected by daylength and temperature. *Acta Hortic.* 14(14), 61–76. doi:10.17660/ActaHortic.1969.14.5.

Schwabe, W. W. 1954. Acceleration of flowering in non-vernalized chrysanthemums by the removal of apical sections of the stem. *Nature* 174(4439), 1022. doi:10.1038/1741022b0.

Seeley, J. G. 1985. Finishing bedding plants, effects of environmental factors - temperature, light, carbon dioxide, growth regulators. In: Mastalerz, J. W. and Holcomb, E. J. (Eds), *Bedding Plants III: A Penn State Manual*. Pennsylvania Flower Growers, University Park, PA, pp. 212–44.

Sheldron, K. G. and Weiler, T. C. 1982a. Regulation of growth and flowering in Basket of Gold, *Autinia saxatilis* (L.) Desv. *HortScience* 17, 338–40.

Sheldron, K. G. and Weiler, T. C. 1982b. Regulation of growth and flowering of *Aquilegia* x *hybrida* Sims. *J. Amer. Soc. Hortic. Sci.* 107, 878–82.

Shillo, R. 1976. Control of flower initiation and development of statice (*Limonium sinuatum*) by temperature and daylength. *Acta Hortic.* 64(64), 197–204. doi:10.17660/ActaHortic.1976.64.25.

Shimizu-Sato, S., Tanaka, M. and Mori, H. 2009. Auxin-cytokinin interactions in the control of shoot branching. *Plant Mol. Biol.* 69(4), 429–35. doi:10.1007/s11103-008-9416-3.

Shi-wei, S., Yu-ling, L., Xin-min, H., Wei, S., Ri-yuan, C. and Yan-wei, H. 2018. Crosstalk of low-temperature and gibberellin treatments on bolting and flowering of Chinese cabbage. *J. Integ. Agric.* 17(0), 60345–7.

Singer, S. R. and McDaniel, C. N. 1986. Floral determination in the terminal and axillary buds of *Nicotiana tabacum* L. *Dev. Biol.* 118(2), 587–92. doi:10.1016/0012-1606(86)90028-X.

Siraj-Ali, Y. S., Tayama, H. K., Prince, T. L. and Carver, S. A. 1990. The relationship between maturity level and splitting of poinsettia. *HortScience* 25(12), 1616–8. doi:10.21273/HORTSCI.25.12.1616.

Smillie, R. M., Critchley, C., Bain, J. M. and Nott, R. 1978. Effect of growth temperature on chloroplast structure and activity in barley. *Plant Physiol.* 62(2), 191-6. doi:10.1104/pp.62.2.191.

Sorefan, K., Booker, J., Haurogne, K., Goussot, M., Bainbridge, K., Foo, E., Chatfield, S., Ward, S., Beveridge, C., Rameau, C. and Leyser, O. 2003. MAX4 and RMS1 are orthologous dioxygenase-like genes that regulate shoot branching in *Arabidopsis* and pea. *Genes Dev.* 17(12), 1469-74. doi:10.1101/gad.256603.

Starling, R. J. and Dodds, J. H. 1983. Tissue-culture propagation of cacti and other succulents. *Bradleya* 1, 84-90. doi:10.25223/brad.n1.1983.a8.

Stokes, P. and Verkerk, K. 1951. Flower formation in Brussels sprouts. *Meded. Landbouwhogesch. Wageningen* 50, 141-60.

Strope, K. 1999. The effects of heat stress on flowering and vegetative growth of New Guinea impatiens (*Impatiens hawkeri* Bull.). MS Thesis. Department of Horticultural Science, University of Minnesota, St. Paul, MN.

Styer, R. C. and Koranski, D. S. 1997. *Plug & Transplant Production: A Grower's Guide*. Ball Publishing, Batavia, IL.

Suge, H. 1984. Re-examination on the role of vernalization and photoperiod in the flowering of *Brassica* crops under controlled environment. *Japan J. Plant Breed.* 34(2), 171-80. doi:10.1270/jsbbs1951.34.171.

Suo, J., Chen, S., Zhao, Q., Shi, L. and Dai, S. 2015. Fern spore germination in response to environmental factors. *Front. Biol.* 10(4), 358-76. doi:10.1007/s11515-015-1342-6.

Tageras, H. 1979. Modifying effects of ancymidol and gibberellins on temperature induced elongation in *Fuchsia x hybrida*. *Acta Hortic.* 91, 411-7.

Taiz, L. and Zeiger, E. (Eds). 1998. *Plant Physiology* (2nd edn.). Sinauer Associates, Inc., Sunderland, MA.

Takaki, M., Kendrick, R. E. and Dietrich, S. M. C. 1981. Interaction of light and temperature on the germination of *Rumex obtusifolium* L. *Planta* 152(3), 209-14. doi:10.1007/BF00385146.

Takimoto, A. and Hamner, K. C. 1964. Effect of temperature and preconditioning on photoperiodic response of Pharbitis. *Plant Physiol.* 39(6), 1024-30. doi:10.1104/pp.39.6.1024.

Tanaka, M., Takei, K., Kojima, M., Sakakibara, H. and Mori, H. 2006. Auxin controls local cytokinin biosynthesis in the nodal stem in apical dominance. *Plant J.* 45(6), 1028-36. doi:10.1111/j.1365-313X.2006.02656.x.

Taylorson, R. B. and Hendricks, S. B. 1976. Interactions of phytochrome and exogenous gibberellic acid on germination of *Lamium amplexicaule* L. seeds. *Planta* 132(1), 65-70. doi:10.1007/BF00390332.

Teichmann, T. and Muhr, M. 2015. Shaping plant architecture. *Front. Plant Sci.* 6(233), 233. doi:10.3389/fpls.2015.00233.

Thimann, K. V. and Skoog, F. 1933. Studies on the growth hormone of plants: III. The inhibiting action of the growth substance on bud development. *Proc. Natl. Acad. Sci. U.S.A.* 19(7), 714-6. doi:10.1073/pnas.19.7.714.

Thomas, B. and Vince-Prue, D. 1984. Juvenility, photoperiodism and vernalization. In: Wilkins, M. B. (Ed.), *Advanced Plant Physiology*. Pitman, London, UK, pp. 408-39.

Thomas, B. and Vince-Prue, D. 1997. *Photoperiodism in Plants* (2nd edn.). Academic Press, New York, NY, pp. 1-26.

Tong, H., Jin, Y., Liu, W., Li, F., Fang, J., Yin, Y., Qian, Q., Zhu, L. and Chu, C. 2009. DWARF and LOW-TILLERING, a new member of the GRAS family, plays

positive roles in brassinosteroid signaling in rice. *Plant J.* 58(5), 803–16. doi:10.1111/j.1365-313X.2009.03825.x.
Toole, E. H., Toole, V. K., Borthwick, H. A. and Hendricks, S. B. 1955. Interaction between temperature and light in germination of seeds. *Plant Physiol.* 30(5), 473–8. doi:10.1104/pp.30.5.473.
Toole, V. K., Toole, E. H., Borthwick, H. A. and Snow, A. G. 1962. Responses of seeds of *Pinus taeda* and *Pinus strobus* to light. *Plant Physiol.* 37, 228–33.
Tudela-Isanta, M., Fernandez-Pascual, E., Wijayasinghe, M., Orsenigo, S., Rossi, G., Pritchard, H. W. and Mondoni, A. 2018. Habitat-related seed germination traits in alpine habitats. *Ecol. Evol.* 8(1), 150–61. doi:10.1002/ece3.3539.
Turck, F., Fornara, F. and Coupland, G. 2008. Regulation and identity of florigen: FLOWERING LOCUS T moves center stage. *Annu. Rev. Plant Biol.* 59, 573–94. doi:10.1146/annurev.arplant.59.032607.092755.
Turnbull, C. G. N., Raymond, M. A. A., Dodd, I. C. and Morris, S. E. 1997. Rapid increases in cytokinin concentration in lateral buds of chickpea (*Cicer arietinum* L.) during release of apical dominance. *Planta* 202(3), 271–6. doi:10.1007/s004250050128.
Umehara, M., Hanada, A., Magome, H., Takeda-Kamiya, N. and Yamaguchi, S. 2010. Contribution of strigolactones to the inhibition of tiller bud outgrowth under phosphate deficiency in rice. *Plant Cell Physiol.* 51(7), 1118–26. doi:10.1093/pcp/pcq084.
Vaid, T. M. and Runkle, E. S. 2013. Developing flowering rate models in response to mean temperature for common annual ornamental crops. *Sci. Hortic.* 161, 15–23. doi:10.1016/j.scienta.2013.06.032.
VanDeWoude, W. J. and Toole, V. K. 1980. Studies of the mechanism of enhancement of phytochrome dependent lettuce seed germination by prechilling. *Plant Physiol.* 66(2), 220–4. doi:10.1104/pp.66.2.220.
Van Gaal, T. and Erwin, J. E. 2005. Diurnal variation in thigmotropic inhibition of stem elongation. *HortTechnology* 15(2), 291–4.
Vince-Prue, D. 1975. *Photoperiodism in Plants*. McGraw-Hill, London, UK.
Visser, T. and DeVries, D. P. 1970. Precocity and productivity of propagated apple and pear seedlings as dependent on the juvenile period. *Euphytica* 19(2), 141–4. doi:10.1007/BF01902937.
Vogelezang, J., Cuijpers, L., deGraaf, M. and Zande, V. D. 1992a. Growth regulation of bedding plants by reversed day/night temperature only? *Acta Hortic.* 305, 37–43.
Vogelezang, J., Moe, R., Schussler, H., Hendricks, L., Cuijpers, L. and Ueber, E. 1992b. Cooperative European research on temperature strategies for bedding plants. *Acta Hortic.* 327(327), 11–6. doi:10.17660/ActaHortic.1992.327.1.
Wahl, V., Ponnu, J., Schlereth, A., Arrivault, S., Langenecker, T., Franke, A., Feil, R., Lunn, J. E., Stitt, M. and Schmid, M. 2013. Regulation of flowering by trehalose-6-phosphate signaling in *Arabidopsis thaliana*. *Science* 339(6120), 704–7. doi:10.1126/science.1230406.
Wang, M., JeMolgne, M. A., Bertheloot, J., Crespel, L., Perez-Garcia, M. D., Oge, L., Demotes-mainard, S., Hamama, L., Daviere, J. M. and Sakr, S. 2019. BRANCHED1: a key hub of shoot branching. *Front. Plant Sci.* 10, 76. doi:10.3389/fpls.2019.00076.
Wareing, P. F. and Saunders, P. F. 1971. Hormones and dormancy. *Annu. Rev. Plant. Physiol.* 22(1), 261–88. doi:10.1146/annurev.pp.22.060171.001401.
Warner, R. M. and Erwin, J. E. 2001a. Effect of high-temperature stress on flower number per inflorescence of 11 *Lycopersicon esculentum* Mill. Genotypes. Poster presented

at the *98th International Conference of the American Society of Horticultural Science*, Sacramento, CA, 22–25 July. *HortScience* 36, 508.
Warner, R. M. and Erwin, J. E. 2001b. Impact of high-temperature and irradiance on development of five bedding plant species. Poster presented at the *98th International Conference of the American Society of Horticultural Science*, Sacramento, CA, 22–25 July. *HortScience* 36, 550.
Warner, R. M. and Erwin, J. E. 2002. Photosynthetic responses of heat-tolerant and heat-sensitive cultivars of *Impatiens hawkeri* and *Viola* x *wittrockiana* to high temperature exposures. *Acta Hortic.* 580(580), 215–9. doi:10.17660/ActaHortic.2002.580.28.
Warner, R. M. and Erwin, J. E. 2003a. *High Temperature Developmentally Reduces Growth and Flowering of Twelve Pansy Cultivars*. IHC Program, Toronto, Canada.
Warner, R. M. and Erwin, J. E. 2003b. Effect of photoperiod and daily light integral on flowering of five *Hibiscus* spp. *Sci. Hortic.* 97, 341–51.
Warner, R., Erwin, J. E. and Wagner, R. 1997. Photoperiod and temperature interact to affect *Gomphrena globosa* L. and *Salvia farinacea* Benth. development. *HortScience* 32, 501.
Wellensiek, S. J. 1958. Vernalization and age in *Lunaria biennis*. *Kon. Ned. Akad. Wet., Proc. Ser. C. (Amsterdam)* 61, 561–71.
Wellensiek, S. J. 1961. Temperature. In: *Encyclopedia of Plant Physiology* (vol. 16). Springer Verlag, Berlin, pp. 1–23.
Wellensiek, S. J. 1962. Phytotronics. *Plant Sci. Symp.* Campbell Soup Co., Camden, NJ, pp. 149–62.
Wellensiek, S. J. 1964. Dividing cells as a pre-requisite for vernalization. *Plant Physiol.* 39(5), 832–5. doi:10.1104/pp.39.5.832.
Wellensiek, S. J. 1965. Recent developments in vernalization. *Acta Bot. Neerl.* 14(3), 308–14. doi:10.1111/j.1438-8677.1965.tb00196.x.
Wellensiek, S. J. 1966. The flower forming stimulus in *Silene armeria*. *Z. Pflanzenphysiol.* 55, 1.
Wellensiek, S. J. and Hakkaart, F. A. 1955. Vernalization and age. *Proc. K. Ned. Akad. Wet. Amst. Proc. Sec. Sci.* C58, 16–21.
Went, F. W. 1944. Plant growth under controlled conditions. II. Thermoperiodicity in growth and fruiting of the tomato. *Amer. J. Bot.* 31(3), 135–50. doi:10.1002/j.1537-2197.1944.tb08011.x.
Went, F. W. 1953. The effect of temperature on plant growth. *Annu. Rev. Plant. Physiol.* 4(1), 347–62. doi:10.1146/annurev.pp.04.060153.002023.
Went, F. W. 1957. *The Experimental Control of Plant Growth*. Chr Bot., The Ronald Press Co., New York, NY.
Werner, T., Motyka, V., Strnad, M. and Schmulling, T. 2001. Regulation of plant growth by cytokinin. *Proc. Natl Acad. Sci. U.S.A.* 98(18), 10487–92. doi:10.1073/pnas.171304098.
Whitman, C. M. 1995. Influence of photoperiod and temperature on flowering of *Campanula carpatica* 'Blue Chips', *Coreopsis grandiflora* 'Early Sunrise', *Coreopsis verticillata* 'Moonbeam', *Rudbeckia fulgida* 'Goldstrum', and *Lavandula angustifolia* 'Munstead'. MS Thesis. Department of Horticulture, Michigan State University, East Lansing, MI.
Whitman, C. M., Heins, R. D., Cameron, A. C. and Carlson, W. H. 1996. Cold treatments, photoperiod, and forcing temperature influence flowering of *Lavandula angustifolia*. *HortScience* 31(7), 1150–3. doi:10.21273/HORTSCI.31.7.1150.

Wiebe, H. J., Habegger, R. and Liebig, H. P. 1992. Quantification of vernalization and devernalization effects for kohlrabi (*Brassica oleracea* convar. *acephala* var. *gonglyodes* L.). *Sci. Hortic.* 50(1–2), 11–20. doi:10.1016/S0304-4238(05)80004-4.

Wigge, P. A., Kim, M. C., Jaeger, K. E., Busch, W., Schmid, M., Lohmann, J. U. and Weigel, D. 2005. Integration of spatial and temporal information during floral induction in *Arabidopsis*. *Science* 309(5737), 1056–9. doi:10.1126/science.1114358.

Wilkins, H. F., Waters, W. E. and Widmer, R. E. 1968. University of Minnesota's Easter Lily Research Report: Paper No. II. An insurance policy: Lighting lilies at shoot emergence will overcome inadequate bulb precooling. *Minn. State Florists' Bull.* (December), 10–12.

Williams, E., Fisher, P. and Han, S. 2006. Post-harvest conditions for rooted cuttings. Young Plant Research Center Report, pp. 1–19.

Wittwer, S. J. and Bukovac, M. J. 1959. Effects of gibberellin on photoperiodic response of some higher plants. In: Withrow, R. B. (Ed.), *Photoperiodism and Related Phenomenon in Plants and Animals*. American Association for the Advancement of Science, Washington DC, pp. 373–80.

Wurr, D. C. E., Fellows, J. R., Phelps, K. and Reader, R. J. 1994. Testing a vernalization model on field-grown crops of four cauliflower cultivars. *J. Hortic. Sci.* 69(2), 251–5. doi:10.1080/14620316.1994.11516452.

Yaping, S. and Heins, R. D. 1996. Influence of day and night temperature on sweet pepper seedling development. *J. Amer. Soc. Hortic. Sci.* 121(4), 699–704.

Yeh, D. M., Atherton, J. G. and Craigon, J. 1997. Manipulation of flowering in cineraria. IV. Devernalization. *J. Hortic. Sci.* 72(4), 545–51. doi:10.1080/14620316.1997.1151 5542.

You, Y., Sawikowska, A., Lee, J. E., Benstein, R. M., Neumann, M., Krajewski, P. and Schmid, M. 2019. Phloem companion cell-specific transcriptomic and epigenomic analysis identify MRF-1, a regulator of flowering. *Plant Cell* 31, 325–45. doi:10.1105/ tpc.17.00714.

Yuan, M. 1995. Effect of juvenility, temperature and cultural practices on flowering of *Coreopsis*, *Gaillardia*, *Heuchera*, *Leucanthemum* and *Rudbeckia*. MS Thesis. Department of Horticulture, Michigan State University, East Lansing, MI.

Yui, S. and Yoshikawa, H. 1991. Bolting resistance breeding of Chinese cabbage. 1. Flower induction of late bolting variety without chilling treatment. *Euphytica* 52(3), 171–6. doi:10.1007/BF00029393.

Zagotta, M. T., Hicks, K. A., Jacobs, C. I., Young, J. C., Hangarter, R. P. and Meeks-Wagner, D. R. 1996. The *Arabidopsis* ELF3 gene regulates vegetative photomorphogenesis and the photoperiodic induction of flowering. *Plant J.* 10(4), 691–702. doi:10.1046/j.1365-313x.1996.10040691.x.

Zanin, P. and Erwin, J. E. 2003. Factors affecting flowering of *Anisodontea* x *hypomandarum* K. Presl. and *Leonotis menthaefolia* (Pers.) R. Br. *HortScience* (in review).

Zeevaart, J. A. D. 1957. Studies on flowering by means of grafting. II: Photoperiodic treatment of detached *Perilla* and *Xanthium* leaves. *Proc. K. Ned. Acad. Wet. Amst.*, 6D, 332–7.

Zeevaart, J. A. D. 1962. Physiology of flowering: flowering is hormonally controlled, but the nature of the hormone remains to be elucidated. *Science* 137(3532), 723–31. doi:10.1126/science.137.3532.723.

Zeevaart, J. A. D. 1969. Perilla. In: Evans, L. T. (Eds), *The Induction of Flowering. Some Case Histories*. Macmillian, Melbourne, Australia, pp. 116–44.

Zeevaart, J. A. D. 1983. Gibberellins and flowering. In: Crozier, A. (Ed.), *The Biochemistry and Physiology of Gibberellins*. Springer-Verlag, New York, NY, pp. 333–74.
Zhang, D., Armitage, A. M., Affolter, J. M. and Dirr, M. A. 1996. Environmental control of flowering and growth of *Achillea millefolium* L. 'Summer Pastels'. *HortScience* 31(3), 364–5. doi:10.21273/HORTSCI.31.3.364.
Zieslin, N. and Tsujita, M. J. 1988. Regulation of stem elongation of lilies by temperature and the effect of gibberellin. *Sci. Hortic.* 37(1–2), 165–9. doi:10.1016/0304-4238(88)90159-8.
Zimmer, K. and Junker, K. 1985. *Achimenes*. In: Halevy, A. H. (Ed.), *Handbook of Flowering* (vol. I). CRC Press, Boca Raton, FL, pp. 391–2.
Zrebiec, V. and Tayama, H. K. 1990. Short-day photoperiod duration influences splitting of poinsettia. *HortScience* 25(12), 1663. doi:10.21273/HORTSCI.25.12.1663.

# Chapter 2

## Exploiting the genetic diversity of ornamentals

*Yoo Gyeong Park, Gyeongsang National University, Republic of Korea; Young Hoon Park, Pusan National University, Republic of Korea; Abinaya Manivannan, National Institute of Horticultural and Herbal Science, Republic of Korea; Prabhakaran Soundararajan, National Institute of Agricultural Science, Republic of Korea; and Byoung Ryong Jeong, Gyeongsang National University, Republic of Korea*



### 1 Introduction

In a rapidly changing world, genetic diversity is crucial to stable agricultural production (Widrlechner, 2009). In improving crops and developing new agricultural products, it is important to account for the genetic variations of economically important plants such as field crops, fruits, vegetables, ornamentals, as well as their weedy and wild relatives. This diversity can be found in traditional agricultural systems as well as in nature. However, both traditional agriculture and natural plant communities are now being threatened

http://dx.doi.org/10.19103/AS.2020.0066.02

to be replaced with extensive habitat degradation, urbanization, and modern agricultural systems with a relatively narrow genetic base.

Plant genebanks have been established throughout the world to sample, conserve, and make readily available the significant components of genetic diversity (Widrlechner, 2009). This is usually done away from traditional agricultural systems or natural plant communities, and under controlled conditions, so that the efficiency of scale can be realized. However, this also presents the challenge of re-creating optimal growing conditions. Genebanks generally focus on major crops, but, collectively, genebanks around the world conserve a wide variety of plants, including ornamental plants.

Global conservation efforts for agricultural plant genetic resources (PGRs) have led to an enormous increase in the number and size of in situ and ex situ germplasm collections over the last three to four decades (IPGRI, 2003). Differing national and international policy frameworks, available expertise, facilities, budget, institutional environments, and the extent of national and international collaboration resulted in a wide variety of conditions that these collections are maintained under (IPGRI, 2003). Additionally, different management regimes are required by a wide variety of germplasm in these collections (IPGRI, 2003). Hence, this chapter discusses the collection, evaluation, propagation, dissemination, and utilization of ornamental germplasm for exploiting genetic diversity.

## 2 Management of ornamental germplasm to exploit genetic diversity: collection strategy

The objective of the collection (e.g. gap filling, reducing the loss of genetic diversity from genetic erosion, and the targeted collection of specific genotypes) and the precise mandate of a specific genebank determine how the germplasm will be sampled. The sampling technique and necessary equipment are determined by what needs to be collected, that is, pollen, vegetative propagules, seeds, or whole plants. However, it should be noted that maintaining the varietal integrity of the population is not a necessary prerequisite to the conservation of genetic diversity per se, which presents the option to maintain the diversity in clonally propagated species as seeds.

Recently, vegetative tissues such as bud wood, apices or shoots, and zygotic embryos have been grown in distant locations after being sampled, treated, and transported. Applying in vitro techniques to germplasm collection has been detailed by Withers (1995). Cases where in vitro collections can be advantageous have been specified by Engelmann (1997), some of which are when the size and weight of the seeds are problematic, when the survival of the vegetative materials and seeds is not guaranteed on long missions to remote areas, when there aren't a sufficient number of seeds to collect, and when soil particles remain on the collected material and pose a high risk of transferring pests and diseases.

When specimens are collected for genebanks or herbaria, DNA-rich materials such as root nodules and leaves can be collected with a little additional effort. The collected material should immediately be immersed in a stabilizing buffer or stored with a desiccant to ensure that the subsequent DNA extraction is successful. Such storage methods are simple and suitable for a long-term storage (Adams, 1997). However, it should be noted that DNA will introduce individual traits only through the application of biotechnological methods. Furthermore, unlike seeds, DNA cannot be regenerated, and DNA stocks will eventually be exhausted. This indicates that DNA storage is never going to be sufficient to replace the storage of living materials, in the form of cryopreserved materials or seeds, or in vitro tissues.

From a genebank management perspective as well as a genetic diversity perspective, an important collection decision is how many seeds to collect for each accession. A significant number of seeds are necessary for each accession when a genebank, such as the Genebank of the Royal Botanic Gardens (Kew, UK), wants to avoid the initial regeneration and use the collected materials to start the conservation effort. However, it must be realized that it may not be easy to collect seeds in their optimal physiological conditions that a curator wants to ensure the seeds' longevity for long-term conservation.

Furthermore, it is important to screen the collected materials carefully, check the phytopathological statuses, characterize the materials, and possibly execute a preliminary evaluation. A genebank manager will be able to make the decisions for effective and efficient management only after completing the proper and thorough screening processes as such. The optimal conditions for collecting a large number of seeds are often not met in practice.

## 3 In situ conservation methods

It is recommended that plant species are grown in their natural habitats, in the dynamic systems that continuously expose the plants to the evolutionary processes (Ogbu et al., 2010). Dispersal of pollens and seeds, mutation, and recombination within and among populations generate genetic variability. New plant types with improved adaptability are developed when selection operates on the genetic variability. In situ conservation needs to be carried out on farms, in addition to natural habitats in protected areas and national reserves, in the areas where farmers cultivate their locally adapted varieties and landraces; this requires farmers to actively participate in conserving the traditional farmers' varieties and landraces. Conservation of the novel genetic resources may even be done in home gardens (Tao, 2001).

An international policy framework is provided by the Convention on Biological Diversity (CBD) for conserving the genetic diversity of plants, which applies particularly to the use of genetic diversity in breeding crops and to utilize it (e.g., to improve crops) (Castañeda-Álvarez et al., 2016; CBD, 1992; Whitlock et al., 2016). The CBD (UNEP, 1992) defines in situ conservation, which applies to domesticated and wild species, as follows: 'The conservation of ecosystems and natural habitats and the maintenance and recovery of viable populations of species in their natural surroundings and, in the case of domesticated or cultivated species, in the surroundings where they have developed their distinctive properties.' Substantial differences may exist in the approaches employed to conserve wild species. A significant impetus has been given to in situ conservation following the conclusion of the 1992 CBD and Agenda 21, and with the participating countries' adoption of the GPA in the Fourth International Technical Conference on Plant Genetic Resources (FAO, 1996). On-farm conservation activities in recent years have become closely tied to development work, which includes farmer empowerment (Jarvis and Hodgkin, 2000).

## 3.1 Protected areas

Protected areas play an instrumental role in conserving wild relatives of plants in situ. Wild relatives of crops and domestic animals can exist in natural and semi-natural ecosystems, beyond the reach of farming influences, and conserving them in the existing nature reserves may work well. Many proposals draw from this approach (Ingram, 1984), but not many were funded until recently. Agrobiodiversity conservation in protected areas is currently unplanned for the most part, and the conservation of wild relatives is usually not specifically addressed. With agrobiodiversity conservation in protected areas, continuous evolutionary processes may change entire populations and sometimes even drive them to become extinct. A disadvantage with agrobiodiversity conservation in protected areas is that the subject of the conservation is not readily available for use in agriculture. Also, because of the limitations in managing the protected areas, there's little opportunity for the germplasm to be characterized and evaluated, which reduces the utility of the germplasm in protected areas as a genetic resource (Maxted et al., 1997b).

## 3.2 On-farm conservation

As long as agriculture has existed, farmers around the world have been conserving their on-farm plant resources as a part of the crop production process. The management practices that combine the greatest food security and the highest yields have been the most effective for farmers. Usually, on-farm conservation practices are based on the diversity within and among the different species, adopted where modern high-input agricultural practices are not in place. Farming gives rise to wild-and-weedy species in addition to the crops. Suggestions have been made to improve the effectiveness of the age-old on-farm conservation processes. Jarvis et al. (2000) provided detailed suggestions and procedures for on-farm resource management in traditional farming systems such that landraces, as well as wild-and-weedy species that rely on traditional agricultural practices for survival, will continue to be maintained and to evolve. A trade-off study between the potential advantages and disadvantages of on-farm conservation practices is necessary for the perspective of how suitable these practices are for conservation and how they impact livelihoods.

### 3.3 Home gardens

A large diversity of crops, vegetables, fruits, domestic livestock, and especially ornamental plants are raised in home gardens. The proximity of home gardens to households allows for the detailed selection of most plants (e.g. color variants) and the generation of the wide range of morphological variations present in many domesticated species. Several authors (Engels, 1995; Damania, 1996) separately address the conservation of genetic diversity for home-garden plants. Home-garden conservation methods are dynamic, as they are for on-farm conservation. The intraspecific diversity of an individual garden is typically limited, whereas it is usually substantial among gardens, such that the inclusion of a community of gardens may be needed for genetic diversity conservation efforts in home gardens (Engels, 2002). Various ideas and recommendations by ethnobotanists have been proposed for the in situ agrobiodiversity conservation, including 'mass reservoirs' (Frankel et al., 1995; Simmonds, 1962; Altieri and Merrick, 1987; Brush, 1999). Others proposed to broaden the genetic base through adapting composite populations in different decentralized sites and contribute to on-farm conservation. Jarvis et al. (2000) provided a good overview of the learned lessons from on-farm conservation practices.

In situ conservation approaches always carry with them the threat of a species becoming extinct or the decline in population due to competition from invasive alien species, demographic variations, environmental changes, genetic drifts, human disturbances, inbreeding, loss of habitat, over-exploitation, and pests (Ogbu et al., 2010). Ex situ maintenance is an alternative to in situ diversity conservation in situations where such threats are substantial.

## 4 Ex situ conservation

Initially, conservation and agricultural biologists captured allelic diversity within ex situ collections, which was derived from the sampling theory for neutral alleles of Marshall and Brown (1975). Recent ex situ guidelines apply to relatively small targets (e.g. collecting seeds from ten individuals in each of the five populations) (Centre for Plant Conservation, CPC, 1991) and also to comprehensive germplasm collections (Brown and Marshall, 1995). These ex situ guidelines are also relevant to in situ conservation of genetic diversity and have thus been extended to encompass it (Dulloo et al., 2008). This development is significant because only 28–38% of threatened plants have ex situ collections with five populations (Godefroid et al., 2011). Furthermore, ex situ populations can genetically diverge rapidly from their source populations (Lauterbach et al., 2012), therefore becoming no longer adaptable to their source environment, and may become inbred (Schoen and Brown, 2001), which emphasizes the need for in situ conservation to complement ex situ conservation.

## 4.1 Seed storage

Genetic diversity storage in the form of seeds is the most widely used, most researched, and the most convenient ex situ conservation method (Widrlechner, 2009). Readers are referred to Harrington (1970) for an early review. Requirements for seed storage include adequate drying, where oily seeds require moisture contents as low as 3% and starchy seeds need 5% or more; proper storage temperature where, for long-term storage, −18°C is recommended; and production of high-quality seeds to warrant the greatest longevity (Rao and Jackson, 1996). Recent research has demonstrated that extremely low moisture content may not be optimal for seeds, and care should be taken to avoid such low moisture levels. For many crop species, especially for tropical shrubs and trees, their seeds will lose viabilities when dried (also known as 'recalcitrant' seeds). Seeds of certain species can withstand some extent of dehydration, but not low-temperature storage, and require more moderate storage characteristics. Moreover, there is no guarantee that seeds of wild relatives behave the same way as those of the domesticated do, which requires individual determination of optimal storage conditions.

## 4.2 Pollen storage

Pollen storage techniques are comparable to those for seed storage, since pollens can also be dried to have less than 5% moisture content of the total dry weight and be stored at or below 0°C (Widrlechner, 2009). The knowledge of the fertilizing and survival capacities of cryopreserved pollens more than 5 years old is limited (Towill, 1985). Pollens may be an interesting alternative in conserving problematic species for a long period (IPGRI, 1996). However, the lifespan of pollens, compared to that of seeds, is relatively short (although this varies significantly among species), and testing pollens for viability can be time consuming and costly. Therefore, the use of pollens for germplasm conservation has not been very common (Hoekstra, 1995). Pollens have the disadvantages of being produced in small quantities by many species, being unable to transfer organelle genomes, losing sex-linked genes in dioecious species, and being generally unable to be regenerated into plants (Hoekstra, 1995). However, pollens rarely transfer pests and diseases (except for some viral diseases), making them useful for germplasm movement and exchange.

## 4.3 Field genebanks

Field genebanks are used to conserve clonal crops, crops whose seeds are recalcitrant, and crops that rarely produce seeds. It is imperative to follow the same propagation techniques the farmer employs in using field genebanks (e.g., not disrupting adapted clones through genetic segregation in a seed cycle). Malaysian oil palm provides an example of the management scale of a field genebank, where genetic resources are planted at 140 palms per hectare, and the collection from Nigeria alone accounts for 200 hectares. Oil palm seeds can be stored up to only 2 years, and pollen for only 3 years, requiring a living collection to be maintained for its conservation despite the high costs. Similarly, over 1600 accessions of coffee trees from the center of the crop diversity comprise the coffee genebank in Jima, Ethiopia.

## 4.4 In vitro conservation

As with field genebanks, conservation methods may be susceptible to inevitable hazards, requiring that an alternative, complementary conservation technique be employed. In vitro conservation maintains explants in a sterile, pathogen-free environment and is in extensive use to conserve and multiply species that produce recalcitrant seeds or no seeds at all (Engelmann, 1997). In vitro techniques have been researched for only about the last 20 years, but they have been used to multiply store and, more recently, collect germplasm of over 1000 species (Ashmore, 1997).

## 4.5 Botanical gardens

Historically, botanical gardens have played an important role in the introduction and exchange of genetic resources of plants. Botanical garden collections typically carry only a few individuals per species (FAO, 1998). In recent years, however, there has been a tendency to establish conservation units in botanical gardens, including seed banks (Laliberté, 1997). However, most botanical gardens, unfortunately, have a limited interest or expertise in the genetic resources of plants, and efforts are being made to change this (Heywood, 1998). The representative botanical gardens in some countries are as shown in Table 1.

**Table 1** The representative botanical gardens in some countries

| Country | The representative botanical gardens |
|---|---|
| Republic of Korea | Hantaek Botanical Garden (http://www.hantaek.co.kr/), Yeomiji Botanical Garden (http://www.yeomiji.or.kr), Seoul Botanic Park (http://botanicpark.seoul.go.kr/), the Garden of Morning Calm (http://www.morningcalm.co.kr/), and Baekdudaegan National Arboretum (http://www.bdna.or.kr) |
| Japan | Kyoto Botanical Garden (http://www.pref.kyoto.jp/plant/), Koishikawa Botanical Garden (https://www.bg.s.u-tokyo.ac.jp/koishikawa/eng/), Hokkaido University Botanical Gardens (https://www.hokudai.ac.jp/fsc/bg/index_e.html), and Higashiyama Zoo and Botanical Gardens (http://www.higashiyama.city.nagoya.jp/) |
| United States | United States Botanic Garden (https://www.usbg.gov/), Missouri Botanical Garden (https://www.missouribotanicalgarden.org/), New York Botanical Garden (https://www.nybg.org/), Desert Botanical Garden (https://www.nybg.org/), Brooklyn Botanic Garden (https://www.bbg.org/), Dallas Arboretum and Botanical Garden (https://www.dallasarboretum.org/), and Atlanta Botanical Garden (https://atlantabg.org/) |
| Canada | Montreal Botanical Garden (https://espacepourlavie.ca/jardin-botanique), the Butchart Gardens (https://www.butchartgardens.com/), VanDusen Botanical Garden (https://vancouver.ca/parks-recreation-culture/vandusen-botanical-garden.aspx), and Allan Gardens (https://www.toronto.ca/data/parks/prd/facilities/complex/41/index.html) |
| The Netherlands | Keukenhof (https://keukenhof.nl/en/), Hortus Botanicus Amsterdam (https://www.dehortus.nl/), Castle Gardens in Arcen (https://kasteeltuinen.nl/nl/), and Hortus Botanicus Leiden (https://www.hortusleiden.nl/) |
| Germany | Botanical Garden and Botanical Museum (https://www.bgbm.org/), Botanical Garden Munich-Nymphenburg (https://www.botmuc.org/), Palmengarten (https://www.palmengarten.de/#/de_DE/index/index), Berggarten Hannover (http://www.berggarten-hannover.de/), and Botanischer Garten Rombergpark (https://www.dortmund.de/de/freizeit_und_kultur/rombergpark/startseite_rp/index.html) |
| United Kingdom | Royal Botanic Garden Edinburgh (https://www.rbge.org.uk/), Royal Botanic Garden Kew (https://www.kew.org/), Cambridge University Botanic Garden (https://www.botanic.cam.ac.uk/), Bodnant Garden (https://www.nationaltrust.org.uk/bodnant-garden), and Birmingham Botanical Gardens (https://www.birminghambotanicalgardens.org.uk/) |
| France | Galleries, Gardens, Zoo (https://www.jardindesplantesdeparis.fr/en), Jardin des Plantes (https://www.umontpellier.fr/patrimoine/jardin-des-plantes), Parc de la Tête d'Or (https://www.lyon.fr/lieu/parcs/parc-de-la-tete-dor), Le jardin exotique d'Eze (https://www.jardinexotique-eze.fr/), and Jardin Botanique de Lyon (http://www.jardin-botanique-lyon.com/jbot/sections/fr/) |

### 4.6 DNA storage

DNA storage is a relatively new technique but has been increasing in importance. The DNA is now routinely extracted from the nuclei, chloroplasts, and mitochondria for storage. For analysis, DNA is often immobilized on nitrocellulose sheets, including cloned genes, to be probed. Specific genes and oligonucleotides can now be routinely amplified, following the development of the polymerase chain reaction (PCR). The efficient use of DNA sequences has been further facilitated by DNA cloning technology. The advances in DNA cloning technology have brought forth an international network of DNA repositories for genomic DNAs (Adams, 1997). The DNA storage has the advantages of being simple, efficient, and capable of overcoming the many constraints and limitations that other forms of storage face. However, the DNA storage does not allow for the regeneration of live organisms, and more disadvantages come with problems in the subsequent gene isolation, cloning, and transfer (Maxted et al., 1997a; refer to www.cgn.wageningen-ur.nl/pgr/for more recent updates).

## 5 Management of plant genetic resources in the Republic of Korea and in Japan

### 5.1 Management of plant genetic resources in Korea

As a result of the distribution of diverse vegetation belts, the coexistence of coastal and forest ecosystems, unique topography, climate, and geographical features, Korea has a relatively diverse array of biotic resources despite its limited land size (100 033 $km^2$). An estimated number of 100 000 native species inhabit Korea. Although not limited to ornamentals, as of 2015, the total number of biological species conserved ex situ by the National Agrobiodiversity Center (NAC), National Institute of Agricultural Sciences (NAS), and Rural Development Administration (RDA) of the Republic of Korea is 2613 species and 229 916 resources (Table 2; Kim et al., 2016). More than 2200 of the 2613 species are known to be endemic, which is a relatively high ratio of endemic species, due to the four distinct seasons resulting from the seasonal winds, the developed coastlines, geological features ranging from high mountains to flat fields, and ecological diversity of inland as well as thousands of islands. That, when combined altogether, provides a highly diverse array of habitat environments. Additionally, the location of the Korean Peninsula in Northeast Asia, bridging the Asian continent and the Pacific Ocean, exposes it to diverse climate patterns, from the cold continental climate to the warm oceanic climate. Biodiversity in Korea has an unlimited potential for future growth, some of which include applications to not only ornamentals but also food industry, novel pharmaceuticals, cosmetics, and eco-tourism.

**Table 2** Plant genetic resources of ex situ conservation in the Republic of Korea (Kim et al., 2016)

| | Seeds | | | | | |
|---|---|---|---|---|---|---|
| | Agricultural crop | Horticultural crop | Specialty crop | Forage/ others | Propagules | Total |
| No. of species | 269 | 533 | 352 | 463 | 996 | 2613 |
| No. of resources (%) | 151 096 (65.7) | 25 359 (11.0) | 21 826 (9.5) | 3608 (1.6) | 28 027 (12.2) | 229 916 (100) |
| No. of native resources (%) | 52 821 (35.0) | 5226 (20.6) | 9686 (44.4) | 1056 (29.3) | - | 68 789 (29.9) |

Tidal flats and wetlands are habitats crucial to forming the diverse and distinct characteristics of the different species in Korea. Wetlands are important to biodiversity because they are the habitat for unique organisms that evolve from an environment that's distinct from that inhabited by terrestrial and aquatic organisms. The recent gradual reduction of wetland areas is threatening the diverse wetland species.

A major threat to the biodiversity in Korea is the loss of habitat resulting from rapid industrialization and urbanization, which has quickly increased the number of endangered flora and fauna. Despite efforts of afforestation, colony restoration, and natural restoration, wetlands, farmland, and forestland have been reduced by approximately 20.4%, 15.9%, and 2.1%, respectively, within the last 2 decades. Climate change and alien species invasion are additional threats to biodiversity. The total number of alien species that were naturally or artificially introduced into Korea is 2167 (333 plants and 1834 animals). Eighteen of these 2167 alien species are designated as invasive alien species and monitored to be kept under control to minimize their disturbance to the ecosystems.

The NAS of the Republic of Korea distributes genetic resources and the relevant information to users in the public and private sectors through collecting, preserving, regenerating, and evaluating agricultural genetic resources (NAC, 2019). The NAC has recently been focusing on collecting genetic resources to secure the diversity of agricultural genetic resources, although not specifically centered on ornamental plants, related to the 'Low Carbon, Green Growth' objectives. The NAC is following up with international issues such as access, benefit sharing, and the protection of intellectual property on agricultural genetic resources. The NAC is especially strengthening the international relations for genetic resources. Four teams currently share the mission of 'establishing the national management system,' 'securing the diversity of agricultural genetic resources,' 'enhancing the utilization of genetic resources,' 'securing sovereignty,' and 'serving as a northeast bank hub for genetic resources.'

The basic components of managing the raw materials for the agricultural sector to meet all future challenges are the collection and conservation of agricultural diversity. Collection of genetic resources for agriculture focuses on broadening them for horticulture to help generate income and for food crops such that the domestic and foreign diversities help ensure food security (NAC, 2019). The conservation activities also focus on developing protocols for long-term preservation, which include using liquid nitrogen for storage; cloning using the dormant buds for apple, pear, and so on; and using short-lived seed germplasm for cloning in the case of perilla, lettuce, and so on. Also, important germplasm from foreign countries and institutes is managed with safe duplication practices for the sustainable use of foreign genetic resources by the recipient party.

Plant genetic resources are essential materials for breeding, agricultural education, and biotechnological research. Characterization, multiplication, and evaluation of PGRs are necessary to enhance their usage (NAC, 2019). As such, the current activities focus on the characterization, regeneration, multiplication, and evaluation of PGRs, in addition to the development of massive evaluation technology using up-to-date instruments and DNA techniques (Fig. 1). All the data produced during characterization, multiplication, and evaluation are provided to the user of the PGRs through the NAC's Genebank Management System (GMS). The GMS, which includes the database on passport, characterization, and distribution, was upgraded and integrated with websites related to genetic resources at http://genebank.rda.go.kr to promote the use of agricultural genetic resources.

## 5.2 Management of plant genetic resources in Japan

The Genetic Resources Center, National Agriculture and Food Research Organization (NARO), is the central coordinating institute in Japan for the conservation of plants, microorganisms, animals, and DNA materials related to agriculture (NARO, 2019). The NIAS Genebank coordinates this activity in collaboration with a network of institutes throughout Japan (NARO, 2019). This genebank project had been carried out by the name of the NIAS Genebank, but for the reorganization of the organizations, it was renamed as the Genebank Project, NARO, or NARO Genebank from April 2016 (NARO, 2019). The center also carries out diversity studies and develops preservation technology to conserve genetic resources and creates new genetic and breeding materials using irradiation mutagenesis and other methods. To date, the conserved genetic resources include about 225 000 plant accessions, about 28 000 microorganism accessions, and about 1000 animal accessions. The NARO is currently focusing on ornamental germplasm development efforts in four priority spices such as chrysanthemum (748 plant accessions), rose (405 plant accessions), azalea (415 plant accessions), and carnation (267 plant accessions) (NARO, 2019).

**Figure 1** The management procedure for agricultural genetic resources in the Republic of Korea (NAC, 2019).

## 6 Management of plant genetic resources in the United States

The National Plant Germplasm System (NPGS) of the U.S. Department of Agriculture-Agricultural Research Service (USDA-ARS) is an organized network of 19 genebanks located throughout the United States that perform research to acquire, characterize, distribute, document, evaluate, maintain, and regenerate PGRs that, through breeding and selection programs, underpin crop improvements and agricultural production (Byrne et al., 2018). The NPGS is administered and coordinated by the USDA's Agricultural Research Service in collaboration with the USDA's Cooperative State Research, Education, and Extension Service and the State Agricultural Experiment Stations and Land Grant Universities (Bretting,

2007). Since its inception in 1980 to 1987, the repositories have collected and identified over 41 500 clonally propagated accessions. Approximately 596 000+ accessions from 159 000+ species have been characterized, maintained, and made available for distribution in these collections [Germplasm Resources Information Network (GRIN); http://www.ars-grin.gov/npgs/collections.html, 2016] (USDA, 2018). The NPGS sites are located to encompass a broad range of growing conditions, which range from arid to very wet, and from high latitude to tropical. These sites represent many environments and conserve widely varying types of plants, including those of interest to IPPS members. The Woody Landscape Plant Germplasm Repository at the National Arboretum and the Ornamental Plant Germplasm Center (OPGC) of the Ohio State University are the two largest collections that focus on ornamental horticulture.

The NPGS has a long-term outlook of preserving the diversity of agricultural plants, in the form of plant collections comprised of distinct accessions that are often managed by the genus and maintain the associated documentation. The overall goal of the NPGS is to support research and education by making these collections and the associated information widely available. Accessions are composed of seed-propagated populations and clonal selections. The NPGS curators employ many of the techniques that commercial propagators use to ensure the health and genetic integrity of plants for clonal selections (Postman et al., 2006). These include grafting, budding, stooling, rooting the cuttings, grafting, layering, virus indexing, using isolation chambers to exclude disease vectors, and preserving tissue cultures in vitro. Clones are often cultivated on-site in long-term field plantings.

A curator's choice of techniques depends strongly on the plant's breeding history (e.g. a uniform inbred line would not need nearly as large a population sample as would a diverse, outcrossed landrace of the same species) and the pollination biology for seed-propagated populations (Sackville Hamilton and Chorlton, 1997; Widrlechner, 2009). Isolation by space and/or time, hand-pollination, and screened cages containing pollinating insects (Widrlechner et al., 1997) are all used to produce seeds. The seed samples are usually maintained on-site, under a temperature of generally at or below 4°C, and controlled humidity.

Collections that are held at only one location are vulnerable to accidental loss or disaster. To reduce such risks, the NPGS places back-up collections at other locations, usually at the National Center for Genetic Resources Preservation (NCGRP) in Fort Collins, Colorado (Widrlechner, 2009). Conventional freezers or cryogenic tanks with liquid nitrogen are used to store most of the seed samples. The NCGRP has also worked to develop cryogenic storage methods for clonal materials with the active sites, for both dormant buds (Towill and Widrlechner, 2004) and vitrified in vitro cultures (Towill and Bonnart, 2003).

Cryogenic preservation techniques employed cooling at a controlled rate that followed immersion in liquid nitrogen and were ultimately developed into protocols that were successfully used in freezing and recovering growing shoot tips (Grout et al., 1978). Planning and implementing cryopreservation storage for clonal materials to safeguard many irreplaceable collections are now possible (Reed, 2001). Numerous genera and species throughout the world employ a range of cryopreservation techniques (Benson, 1999), and such procedures for new genera and species continue to evolve. It is now possible to routinely cryopreserve many species, and the number of species to be able to be cryopreserved is steadily increasing.

The OPGC holds approximately 3200 accessions within about 200 genera of herbaceous ornamental plants that are distributed to investigators, breeders, and educators worldwide upon request (https://opgc.osu.edu/) (Long et al., 2018). The OPGC is currently focusing on ornamental germplasm development efforts in six priority genera such as *Begonia*, *Coreopsis*, *Lilium*, *Phlox*, *Rudbekia*, and *Viola* (OPGC, 2019). These efforts include comprehensive acquisition, characterization, and enhancement (OPGC, 2019). The priority is reviewed periodically as new needs and opportunities arise (OPGC, 2019). Some of the well-known herbaceous ornamental genera included among the more than 200 in the OPGC's collection are *Antirrhinum*, *Aquilegia*, *Asclepias*, *Baptisia*, *Bidens*, *Campanula*, *Chrysanthemum*, *Clematis*, *Cleome*, *Cosmos*, *Delphinium*, *Dianthus*, *Gaillardia*, *Geranium*, *Glandularia*, *Gypsophila*, *Iris*, *Leucanthemum*, *Liatris*, *Ligularia*, *Loberia*, *Monarda*, *Oenothera*, *Pelargonium*, *Penstemon*, *Petunia*, *Portulaca*, *Primula*, *Ratibida*, *Silene*, *Solidago*, *Stokesia*, *Symphiotrichum*, and *Tagetes* (OPGC, 2019). Also, the collection includes many other genera of native North American species that may not yet be important from a commercial standpoint, but that include species with a potential for development and use as ornamental plants (OPGC, 2019).

## 7 Management of plant genetic resources in Canada

Canada's Plant Germplasm System is a network of centers and people dedicated to preserving the genetic diversity of crop plants, their wild relatives, and plants present and unique in the Canadian biodiversity (https://pgrc.agr.gc.ca/about-propos_e.html). The system plays a significant part of Agriculture and Agri-Food Canada's commitment to the Canadian Biodiveristy Strategy in response to the CBD (https://pgrc.agr.gc.ca/about-propos_e.html). Agriculture and Agri-Food Canada appointed the first plant gene resources officer and established Plant Gene Resources of Canada (PGRC) in 1970 (https://pgrc.agr.gc.ca/about-propos_e.html). Until early 1998 it was located on the Central Experimental Farm in Ottawa but moved to a modern facility in Saskatoon (https://pgrc.agr.gc.ca/about-propos_e.html). The PGRC seed genebank, as part of the Saskatoon Research Centre, coordinates Canada's germplasm system and is the main repository for seeds (https://pgrc.agr.gc.ca/about-propos_e.html).

The Canadian Clonal Genebank was designated in 1989 as the primary germplasm repository for fruit trees and small fruit crops (https://pgrc.agr.gc.ca/about-propos_e.html). Originally located near Trenton, Ontario it moved in 1996 to Harrow, Ontario. A multi-nodal system was established in 1992 to respond to recommendations from study committees on the enhancement of germplasm conservation in Canada (https://pgrc.agr.gc.ca/about-propos_e.html). It was initially funded through the Green Plan. This initiative links rejuvenation, evaluation, and documentation to research and plant breeding programs for specific crop plants (https://pgrc.agr.gc.ca/about-propos_e.html). This strategy is consistent with the recommendations of the Food and Agriculture Organization (FAO) of the United Nations (https://pgrc.agr.gc.ca/about-propos_e.html). Canadian agriculture is based on crops that originated from areas outside of Canada. For example, wheat originated in the Near East (in such countries as Iran), corn in Mexico and Guatemala, alfalfa in Turkey, and soybean in China (https://pgrc.agr.gc.ca/about-propos_e.html). Crops of economic importance that are native to Canada are limited and include sunflower, strawberry, raspberry, saskatoon berry, blueberry, currant, and cranberry (https://pgrc.agr.gc.ca/about-propos_e.html). A large number of native forage and grass species have a significant value in parts of Canada, whether for pasture, erosion control, benefit to wildlife, or other usages (https://pgrc.agr.gc.ca/about-propos_e.html). The largest genera are the bluegrasses, brome grass, milk vetch, and wild rye (https://pgrc.agr.gc.ca/about-propos_e.html). Thus, almost all of the germplasm needed to increase the genetic diversity of Canadian agriculture come from foreign locations, used to characterize, rejuvenate, and document the diversity in collections. The Fredericton Research Centre is responsible for potato, while the Saskatoon Research Centre is responsible for crucifers and forage crop germplasm (both legumes and grasses) activities (https://pgrc.agr.gc.ca/about-propos_e.html).

In the multi-nodal system, the two central agencies, the PGRC and the Canadian Clonal Genebank are still the primary contact points for germplasm entering and leaving Canada, and have responsibility for national and international contacts, distribution, rejuvenation, and evaluation of germplasm not assigned to the nodes, seed viability testing, database management, and technical information (https://pgrc.agr.gc.ca/about-propos_e.html).

Canada's food supply is based on intensive agriculture, and this benefits from the genetic uniformity of crops (https://pgrc.agr.gc.ca/about-propos_e.html). However, genetic uniformity increases the potential for crop vulnerability to new pests and stresses (https://pgrc.agr.gc.ca/about-propos_e.html). Genetic diversity gives us more sustained ability to develop new plant cultivars than they can resist these pests, diseases, and environmental stresses (https://pgrc.agr.gc.ca/about-propos_e.html). Wild ancestors and relatives are the keys to genetic diversity. Unfortunately, the land base where wild plants grow continues to shrink, and many plant species and variants are disappearing (https://pgrc.agr.gc.ca/about-propos_e.html). Also, the conservation of Canadian wild plant germplasm is important as considerable material has been identified as unique in the Canadian biodiversity and to the original inhabitants of Canada (https://pgrc.agr.gc.ca/about-propos_e.html). The Canadian Plant Germplasm System exists to conserve, increase utilization, and catalog germplasm of plants that might otherwise be lost (https://pgrc.agr.gc.ca/about-propos_e.html).

Plant genetic resources have been used in Canada for more than a century, and all Canadians have benefited from their use through increases in the quantity and quality of food consumed (https://pgrc.agr.gc.ca/about-propos_e.html). Benefits from their use have accrued through genetic improvements which have contributed to major increases in productivity and resistance to pests, diseases, and adverse growing conditions (https://pgrc.agr.gc.ca/about-propos_e.html). Canola exemplifies Canada's contribution to enriching the world's crop diversity as a result of our use of germplasm acquired in other countries (https://pgrc.agr.gc.ca/about-propos_e.html). In recent times, for example, wide crossing in native plums has been used to improve the adaptability of domesticated plums; *Fragaria chiloensis* is better able to withstand drought; and many species of *Rubus* carry useful traits, including resistance to major fungal pathogens and insect pests (https://pgrc.agr.gc.ca/about-propos_e.html). Shrub roses have been developed which are hardy to zone 3, and these are being widely grown, replacing the tender hybrid tea and floribunda roses (https://pgrc.agr.gc.ca/about-propos_e.html). Many grass genera are adapted to xeric conditions due to drought and/or salt tolerance, while others are more adapted to more mesic situations (https://pgrc.agr.gc.ca/about-propos_e.html). Researchers continue to evaluate germplasm for new crop development and for value in value-added processing (https://pgrc.agr.gc.ca/about-propos_e.html).

Benefits are also derived by conserving Canada's biodiversity of indigenous plant species (https://pgrc.agr.gc.ca/about-propos_e.html). In addition to commonly occurring species, threatened, rare, or endangered species are made available for study or for small habitat restoration projects (https://pgrc.agr.gc.ca/about-propos_e.html). Canada conserves for the world the largest collection of rare and threatened species of native plants in the genus *Lotus* (https://pgrc.agr.gc.ca/about-propos_e.html).

Three tools improve the use of Canada's PGR collection: better knowledge of the characteristics of germplasm accessions, making this knowledge more accessible to germplasm users, and germplasm enhancement programs which integrate new genes and traits into genetic backgrounds more easily used by plant breeders (https://pgrc.agr.gc.ca/about-propos_e.html).

Over 110 000 seed samples are preserved at the Seed Genebank in Saskatoon. PGRC participates in the worldwide network of PGR centers established by the FAO of the United Nations and the International Plant Genetic Resources Institute (IPGRI) (https://pgrc.agr.gc.ca/about-propos_e.html). On Canada's behalf, it has accepted formal responsibility for principal world base collections of barley and oat and duplicate world base collections of pearl millet and oilseed and crucifiers (https://pgrc.agr.gc.ca/about-propos_e.html).

The seed collection includes foreign and indigenous plants, wild-and-weedy relatives of crop species, cultivars and inbred parental lines, elite breeding lines, and some rare and threatened species and genetic stocks (https://pgrc.agr.gc.ca/about-propos_e.html). Genetic stocks include induced and natural mutations; cytological stocks of genetic oddities; and variations on normal chromosomes, marker genes, polyploids, and pest-resistant stocks (https://pgrc.agr.gc.ca/about-propos_e.html).

Approximately 3000 accessions are established at the Canadian Clonal Genebank in Harrow (https://pgrc.agr.gc.ca/about-propos_e.html). Approximately two thirds of the collection consists of indigenous wild relatives of Canadian fruit crops, and the remaining one third is cultivars or breeding selections of Canadian origin or cultivars of interest to Canadian scientists (https://pgrc.agr.gc.ca/about-propos_e.html).

Seed storage facilities at the PGRC consist of long-term, medium-term, and cryopreservation units. For long-term storage a large walk-in vault is available in which seeds are preserved in laminated envelopes at −20°C. For medium-term storage a large walk-in vault stores seeds in paper envelopes at 4°C and 20% relative humidity (https://pgrc.agr.gc.ca/about-propos_e.html). Seeds are evaluated for viability, dried to optimum moisture content of 6–8%, and transferred to either medium- or long-term storage. Cryopreservation in or over liquid nitrogen at −196°C is the most highly developed of these techniques. Depending on species, dry seeds can last from a few years to centuries (https://pgrc.agr.gc.ca/about-propos_e.html).

At the Canadian Clonal Genebank the clones are either stored in the field, screen, or greenhouses. Tissue culture techniques are well advanced and used for many species, and scientists are evaluating the techniques for those species that can't be stored by other means (https://pgrc.agr.gc.ca/about-propos_e.html). Tissue culture is a cloning method that is used extensively at the potato node in Fredericton for the storage and distribution of germplasm (https://pgrc.agr.gc.ca/about-propos_e.html).

Canada's multi-nodal system offers a number of opportunities for Canadian germplasm conservation (https://pgrc.agr.gc.ca/about-propos_e.html). The rejuvenation of the major species is done at the plant breeding locations responsible for the species, allowing the central agencies to concentrate on preservation technology, genetic diversity analysis in the collections, distribution, database development, and rejuvenation and evaluation of species not assigned to nodes (https://pgrc.agr.gc.ca/about-propos_e.html). Characterization is done by plant breeders and specialized technicians who have the expertise in the field (https://pgrc.agr.gc.ca/about-propos_e.html). This permits an accumulation of relevant information on accessions, for which there is currently insufficient information (https://pgrc.agr.gc.ca/about-propos_e.html). Germplasm is evaluated for a number of desirable agronomic traits such as earliness or winter hardiness and is screened for resistance to pests, diseases, and environmental stress and for quality factors such as color and flavor. The results are made available through the national database, GRIN-CA (https://pgrc.agr.gc.ca/about-propos_e.html).

## 8 Management of plant genetic resources in Europe

The European Plant Genetic Resources Search Catalogue (EURISCO, http://eurisco.ipk-gatersleben.de/apex/f?p=103:2:0::NO:::) is a web-based catalog that provides information about ex situ plant collections across Europe. It provides information on more than 2 million accessions of crop plants and their wild relatives, preserved ex situ by about 400 institutes (https://pgrdeu.genres.de/). It is based on a network of national inventories of 43 member countries and represents an important effort for the preservation of the world's agrobiological diversity by providing information on the large genetic diversity kept by the collaborating institutions (https://pgrdeu.genres.de/).

Between 2003 and 2014, the EURISCO was hosted and maintained by Bioversity International, Rome, Italy. Since 2014, the EURISCO is maintained at the Leibniz Institute of Plant Genetics and Crop Plant Research (IPK), Gatersleben, Germany (https://pgrdeu.genres.de/). The EURISCO is maintained on behalf of the Secretariat of the European Cooperative Programme for Plant Genetic Resources (ECPGR), in collaboration with and on behalf of the National Focal Points for the National Inventories (https://pgrdeu.genres.de/). The central goal of the EURISCO, which contains both passport data and phenotypic data, is to provide a one-stop shop for information for the scientific community and for plant breeders (https://pgrdeu.genres.de/).

The intention of the EURISCO is to provide a central interface to information on PGRs and accessions, which are maintained in almost 400 European collections (https://pgrdeu.genres.de/). For legal reasons, the EURISCO does not provide the possibility to order accessions directly (https://pgrdeu.genres.de/). Material requests should be addressed to the holding institutions. The EURISCO catalog contains passport data on about 1.8 million samples of crop diversity representing more than 6200 genera and 41 600 species from 43 countries by January 2016 (https://pgrdeu.genres.de/). These samples of crop diversity represent more than half of the ex situ accessions maintained in Europe and roughly 16% of total worldwide holdings (https://pgrdeu.genres.de/). The EURISCO is a one-stop shop window using international standards for information on ex situ plant collections that enables users to search and access information on food crops, forages, wild-and-weedy species, including cultivars, landraces, farmers' varieties, breeding lines, genetic stocks, and research materials (https://pgrdeu.genres.de/).

Addresses and contact details of the holding institutions are available from the particular accession details pages, the link 'Contact details on FAO-WIEWS website,' which is provided in the section 'Holding institute' of each details page (https://pgrdeu.genres.de/). In addition, for many of the accessions documented in the EURISCO, direct links to the ordering systems of the respective holding institutes are available (https://pgrdeu.genres.de/).

## 8.1 Management of plant genetic resources in the Netherlands

The Center for Genetic Resources, the Netherlands (CGN), botanical gardens, non-governmental organizations, and private companies manage PGR collections of major importance for food and agriculture at a global scale. The CGN is part of Wageningen University and Research. It maintains the Dutch genebank for PGRs for food and agriculture under a mandate of the Netherlands government. It is also involved in on-farm conservation programs. The crop collections of CGN include a total of over 23 000 different genetic samples of more than 30 crops. Gathering information about the material in the collection and making this information available via the internet is an important priority for which appearance and properties such as molecular genetic characteristics are analyzed. The CGN acts under a mandate of the Dutch government in close cooperation with breeding companies and an international network of genebanks. It is currently focusing on development efforts for ornamental germplasm, such as *Lilium* and *Lupin* (CGN, 2019).

The Dutch botanic gardens maintain a very wide array of plant species for many reasons, for example, for research, education, conservation, display, and maintenance. To that end, the Dutch gardens agreed in the past to avoid unnecessary duplication, achieved through specialization. Each participating garden maintains one or more of these specializations, resulting in rich collections, with special emphasis on (mostly) particular plant families or genera. All the specializations were put together to form the Dutch National Plant Collection (www.nationale-plantencollectie.nl/UK/intro.htm). Currently, over 100 specializations such as *Rosa*, *Aechmea*, *Iris*, *Chrysanthemum*, and other plant families or genera are included. The quality of these specializations is maintained through the regular application of quality criteria, and both criteria and assessments are carried out by an independent foundation, the Dutch

Botanic Gardens Collections Foundation, founded in 1988. This cooperative system of specialization and quality assessment is unique in the world.

## 8.2 Management of plant genetic resources in Germany

Germany has ratified the International Treaty on Plant Genetic Resources for Food and Agriculture on March 31, 2004, and is fully supportive of its objectives, which are the conservation and sustainable use of PGRs for food and agriculture and the fair and equitable sharing of the benefits arising from their use, in harmony with the CBD (https://pgrdeu.genres.de/?lang=en).

A core element of the treaty is the multilateral system of access and benefit sharing. The multilateral system (MLS) shall include all PGRs for food and agriculture listed in Annex I of the treaty that are under the management and control of the contracting parties and in the public domain (https://pgrdeu.genres.de/?lang=en). Furthermore, contracting parties shall invite other holders of the PGRs for food and agriculture listed in Annex I to include these in the MLS (https://pgrdeu.genres.de/?lang=en).

Germany notifies the following PGRs (116 394 accessions) for food and agriculture listed in Annex I and maintained in Germany that have been included in the MLS (https://pgrdeu.genres.de/?lang=en). Through the website of the National Inventory for Plant Genetic Resources (PGRDEU, https://pgrdeu.genres.de/) in Germany, detailed data on the composition of the collections and user procedures to order samples are readily available (https://pgrdeu.genres.de/?lang=en).

Facilitated access is provided solely for the purpose of utilization and conservation for research, breeding, and training for food and agriculture. For the implementation of such a facilitated access to PGRs in accordance to the International PGR, a so-called standard material transfer agreement (sMTA) was compiled (https://pgrdeu.genres.de/?lang=en). The sMTA is a standard contract that is used by each transfer of resources that are part of the MLS. It is a private-law contract between the provider and the recipient of the PGRs (https://pgrdeu.genres.de/?lang=en). The contract defines the rights and duties of the involved persons and is at the same time an important instrument for the implementation of the fair and equitable sharing of the benefits arising out of the use of PGRs with the aim to support the conservation and use of PGRs in developing and transition countries (https://pgrdeu.genres.de/?lang=en).

Generally, PGRs listed in Annex I of the MLS can be ordered in institutions in the public domain by signing the sMTA. In Germany the opportunity to receive material under an sMTA has been even extended to non-Annex-I species in gene banks (https://pgrdeu.genres.de/?lang=en).

The PGRDEU is the central documentation of ex situ collections of PGRs in Germany (https://pgrdeu.genres.de/?lang=en). Ex situ conservation is the conservation of genetic resources outside their natural habitats, for example, the maintenance of generative and vegetative reproductive materials such as seeds, embryos, cuttings, and cell cultures in genebanks (https://pgrdeu.genres.de/?lang=en).

The PGRDEU includes data of the following ex situ collections of PGRs for food and agriculture (https://pgrdeu.genres.de/?lang=en). The number of accessions of the respective genebank collection is given in parentheses: Federal Central Genebank of the Leibniz Institute of Plant Genetics and Crop Plant Research (IPK) (150 153 accessions), German Genebank for Ornamentals (14 885 accessions), German Genebank for Fruit Crops (4848 accessions), German Genebank for Grapevine (4094 accessions), German Genebank for Crop Wild Relatives (4603 accessions), Genebank for Tobacco (788 accessions), and conservation cultures of native wild plants in botanic gardens (2580 accessions) (https://pgrdeu.genres.de/?lang=en).

The PGRDEU comprises data of approximately 181 951 accessions of more than 3000 crops. Users can search for taxonomic information and other information such as the country of origin, name, and number of accession and the maintaining institute. In order to guarantee an efficient data exchange, the data fields (descriptors) of the PGRDEU are in accordance with the EURISCO exchange format, which is based on the 'Multi-Crop Passport Descriptor List' of the FAO and Bioversity International (https://pgrdeu.genres.de/?lang=en).

The in situ conservation of PGRs comprises the conservation of ecosystems and natural habitats and the maintenance and recovery of viable populations of species in their natural environments (e.g. protected areas) (https://pgrdeu.genres.de/?lang=en). Searchable data on in situ occurrences of crop wild relatives (CWRs), which are briefly described next, were obtained from completed model and demonstration projects in the area of conservation and innovative sustainable use of biodiversity as well as from publications, from the EURISCO, and from mappings carried out in the context of the Habitats Directive (https://pgrdeu.genres.de/?lang=en).

### *8.2.1 Development of a reporting and monitoring system for the in situ conservation of genetic resources of crop wild relatives in Brandenburg (2007–2010)*

The project aims to contribute to the conservation, support, and sustainable use of in situ PGRs in Brandenburg, Germany, by supplying transparent documentation and information, particularly with regard to CWRs) (https://pgrdeu.genres.de/?lang=en). The model-based reporting infrastructure for Brandenburg is intended to provide further assistance in view of a possible introduction in other German Laender (https://pgrdeu.genres.de/?lang=en).

A database was created that can be used to meet national and international reporting obligations (https://pgrdeu.genres.de/?lang=en). Project partners were the University for Sustainable Development at Eberswalde; Department of Forest and Environment, in cooperation with the Environmental Agency of Brandenburg; and the country's center of excellence, Eberswalde Forestry (https://pgrdeu.genres.de/?lang=en). Data on 1.8 million occurrences of plant species were collected in a database, among them were about 700 000 CWRs with extensive descriptions. These data are now searchable via the PGRDEU (https://pgrdeu.genres.de/?lang=en).

### *8.2.2 Securing the viability of the wild grape* **Vitis vinifera** *L. ssp.* **sylvestris** *C.C. Gmel. in the old Rheinaue wetlands through targeted in situ management (2008–2013)*

The project aims at ensuring the viability of the wild grape and was carried out by the WWF-Auen-Institute and the Botanical Institute at Karlsruhe Institute of Technology (KIT) (https://pgrdeu.genres.de/?lang=en). The remaining diversity of wild grape should be safeguarded and used to recreate a viable, genetically diverse population on the natural sites (https://pgrdeu.genres.de/?lang=en). Within this framework, all the factors relevant for conservation were taken into account on an entire area for the first time. Simultaneously, living collections were comprehensively conserved (https://pgrdeu.genres.de/?lang=en).

## *8.3 Management of plant genetic resources in the United Kingdom*

The United Kingdom (UK) is relatively rich in wild relatives of crops, landraces of cereal, vegetables and fruit crops, and traditional orchard trees (DEFRA, 2010). Maintaining the genetic diversity of plants in the UK is important for the conservation of biological diversity, as well as for economic and cultural reasons (DEFRA, 2010). The UK is a party to the International Treaty on Plant Genetic Resources for Food and Agriculture and has also committed to a five-point action plan in response to the Foresight (2011) report on the future of food and farming (Environmental Audit Committee, 2011), which includes actions on conserving PGRs (DEFRA, 2010).

The indicator presents the genetic diversity of cultivated plants and their wild relatives in the UK, including other socio-economically and culturally valuable plant species, by assessing the genetic diversity of target plants held in UK germplasm collections (DEFRA, 2010). Effectively, it is a measure of the status of ex situ conservation of cultivated plants and their wild relatives in the

UK (DEFRA, 2010). Any plant which grows or is cultivated in the UK, or has grown in or has been cultivated in the UK, could be included in the indicator (DEFRA, 2010). The indicator is based on an Enrichment Index developed by the FAO (2010) of the United Nations to describe the enhancement of ex situ plant germplasm collections (DEFRA, 2010).

There was a 15% increase in the Enrichment Index between 2013 and 2018 (DEFRA, 2010). A rapid rise in the Enrichment Index since 2000 can be attributed to a concerted collection effort by the Millennium Seed Bank (DEFRA, 2010). There is a considerable annual variability in the number of new accessions into UK germplasm collections. The total number of accessions has risen since 1960, totaling 93 786 accessions by June 2018 (DEFRA, 2010).

Plant Heritage's National Plant Collections contain, among other internationally significant socio-economically and culturally valuable resources, 2395 taxa in 7 Annex 1 genera across 14 locations (Plant Heritage, 2018). In the past year these increased by 512 taxa at 3 locations newly registered within the scheme. Unpublished data (2011 to 2018) held within Plant Heritage's Threatened Plants Project (Seymour, 2012) and publicly available data 2 additionally show that 9000 relevant taxa in 54 genera are held across 185 other locations throughout the UK, including significant holdings by the National Trust for Scotland, the Royal Horticultural Society, national, university, and other botanic gardens and arboreta, the Sir Harold Hillier Gardens, and the Eden Project (DEFRA, 2010).

## *8.4 Management of plant genetic resources in France*

Since the early 1990s, a national collection in France has been developed with different species by public-private national networks for the conservation of PGRs (https://www.geves.fr/). These networks were set up by the French Genetic Resources Office (BGR). This national collection is made up of collections from public or private organizations (Act No. 2010-874) that are made available to the state as part of France's international commitments in contributing to the implementation of the International Treaty on Plant Genetic Resources for Food and Agriculture (ITPGRFA) (https://www.geves.fr/).

This collection aims to conserve PGRs in the general interest, with a view to ensure their sustainable use (https://www.geves.fr/). A vital element of the national living agricultural and food heritage, it would be used in particular for scientific research, innovation, and applied plant breeding, with the aim of avoiding the irreversible loss of strategic PGRs (https://www.geves.fr/).

The national collection includes PGRs considered as common heritage (https://www.geves.fr/). it must also meet the previous criteria, and be widely known to be part of the country's agricultural, horticultural, forest, and culinary history. Most often these PGRs reflect this history as they are distributed or present in the area or they are emblematic of a particular region (https://www.geves.fr/).

It is important to organize the conservation of these heritage resources in the general interest and under conditions that facilitate access to samples by citizens, all physical and legal persons, and the international community, given the common interest of these resources for agriculture and food (https://www.geves.fr/).

National collection is included in the MLS of the FAO's ITPGRFA and can be accessed by signing an SMTA established as part of the treaty (https://www.geves.fr/).

Seven species of the national collection have currently been placed in the ITPGRFA multilateral system: aubergine (80 accessions), oat (603 accessions), soft wheat (1775 accessions), fodder (531), maize (533), potato (91), and triticale (36 accessions) (https://www.geves.fr/).

## 9 Utilizing ornamental germplasm to exploit genetic diversity in roses

### 9.1 Taxonomy and classification

Rose, commonly referred to as the 'queen of the flowers,' is a member of the genus *Rosa* that contains approximately 150–200 species. The *Rosa* genus belongs to the Rosaceae family and is divided into four subgenera: *Eurosa*, *Hesperhodos*, *Hulthemia* (synonymous with *Simplicifolia*), and *Platyrhodon* (Rehder, 1949). Examining the classification and taxonomy of the *Rosa* genus can be problematic and confusing due to the continuous variation of characters (Phillips and Rix, 1988). The species freely hybridize and are very variable, making it difficult to delimit the species (Zieliński et al., 2004). The ploidy and chromosome number of roses vary from diploid to octoploid ($2n = 2x = 14$ to $2n = 8x = 56$). Most cultivated modern roses are within the subgenera of *Eurosa* and diploid or tetraploid that can be genetically classified as 'segmental' allopolyploids (a mixture between autopolyploidy and allopolyploidy).

The *Rosa* genus is native to the temperate regions of the northern hemisphere and has been cultivated for 5000 years in China, Europe, India, North America, and Turkey. Western China holds the greatest diversity of rose species (Phillips and Rix, 1988). The earliest records of roses come from the Mediterranean, where the names 'Rhodes' and 'Syria' both translate to rose. The oldest cultivated garden rose was *Rosa* × *richardii*, which was grown in Crete and is found in artworks from the Minoan civilization (Mann, 2002). The flower is now cultivated in gardens alongside roads for decorative purposes, in open fields for the production of rose oil and rose hip, and in greenhouses to produce cut and potted flowers.

Roses not only are the most economically significant woody perennial flowering plant but also represent a model for ornamental plant species that allows traits such as bloom seasonality, flower morphology, and fragrance to be investigated. Owing to the complex evolutionary history and a breeding history that includes inter-specific hybridization and polyploidization, roses display a range of unique characteristics. The unique features of old rose varieties are still maintained in private and public rose gardens and serve as a living historical archive of rose breeding and selection.

*Rosa* × *hybrida* is the scientific name for roses, which is not the name of species in the botanical sense but is rather a description used for most cultivars of different classes, including hybrid tea and floribunda roses. Alba (*R. alba*), Centrifolia (*R. centrifolia*), Damask (*R. damasenca*), Gallica (*R. galllica*), and Moss rose (*R. centrifolia moscosa*) are the five broad classes of roses, which had emerged by the 18th century. These five classes share several features such as double flower; fragrance; muted flower colors; the resistance to black spot, rust, and frost and spring flowering; and susceptibility to mildew in certain climates. Hybrid tea and floribunda roses are cultivated worldwide as potted/field plants or as cut flowers. These cultivars have been derived over centuries through complex crosses that involved many different species from the *Rosa* genus (Phillips and Rix, 1988).

## 9.2 Early genetic improvement

The history of domesticating roses is long and complex, and different rose species have been hybridized across vast geographic areas that include Asia, Europe, and the Middle East. The domestication processes selected several characteristics that affect the quality of the flowers, such as double flowers, fragrance, petal colors, and recurrent flowering. From the early days of breeding roses to 1875, desirable rose species from the wild were collected, out of which several species and cultivars were grown by horticulturists in gardens in China, Egypt, Greece, Italy, Mesopotamia, and Persia. During this period, long before the Common Era, horticulturists around the world created genetic variations and new cultivars exclusively by sowing the seeds obtained from open pollination and possibly inter-specific hybridization. Collectors and travelers have introduced new genotypes and exchanged species and cultivars, which further expanded the diversity of rose cultivars. The majority of those cultivars were initially diploid, but gradually new cultivars of tetraploids emerged.

The advancements in rose breeding technology over the last 200 years have given rise to a wide range of variability in flower type and growth characteristics in the genus *Rosa*. Until the 1980s, however, genetic improvement of roses based on scientific research has been scarce. Difficulties with conventional breeding and genetic analysis were the main causes of high levels of heterozygosity and ploidy that resulted in low sexual reproduction and low hip sets. It is known that only between 8 and 11 wild species have contributed to the breeding and selection of modern commercial roses, which are considered as the present complex hybrid rose cultivars. Crossing between the Chinese and European roses resulted in the development of the modern hybrid teas. Extensive hybridization among the Chinese, European, and Middle Eastern roses formed the genetic basis of the modern rose cultivars. However, only a small percentage of this variability has been used in rose breeding of the present.

## 9.3 Modern hybrids

Many types of modern hybrids of *R.* × *hybrida* have been developed over the centuries through the selection and hybridization of diverse species of *Rosa*. Modern breeding procedures have led to 29 groups or classes of 20 000 rose cultivars, which include hybrid tea and floribunda roses (*R.* × *hybrida*), the most common roses from the 20th century. These rose varieties are highly sought after for their horticultural traits such as flower shape and color, bloom quality, stem length, fragrance, and vase life.

The first success of an aimed species-crossing was with 'Soleil d'Or' in 1900, which had large, recurrently blooming, yellow flowers. A range of miniature cultivars were also developed through a new species, now-named *R. rouletii* Correvon. Breeders were able to develop a large number of miniatures in a relatively short period because dwarfness is controlled by one dominant gene (De Vries and Dubois, 1996).

Amateurs or highly competitive companies were responsible for the majority of rose breeding, making genetic knowledge often proprietary and unpublished (Gudin, 2000). Additionally, because the modern cultivars of *R.* × *hybrida* tend to be triploid or tetraploid, crossing the modern cultivars may not generate viable seeds. Distant crosses are limited due to the incompatibility or differences in ploidy levels among parents, and the gene pool of *R.* × *hybrida* is limited for some traits (Rout et al., 1999). Many cultivars have been born through spontaneous mutation, although there aren't many reports of cultivars that have been given rise through induced mutation breeding (Rout et al., 1999). Nevertheless, there exists sufficient genetic variation in the current progenitors to ensure the vitality and productivity of future rose cultivars in cross-breeding.

## 9.4 Important traits from a breeding aspect

Complex inter-specific crosses among the elemental species have resulted in the present-day improvements in the color, flowering, flower shape, fragrance, growth habit, and size of roses. Indiscriminate hybridization, spontaneous and induced mutations, molecular breeding, and open pollination have become common breeding methods.

A principal focus of breeding roses over the past 250 years has been the aesthetic criteria. The main objective of rose breeding should be to create new cultivars that have greater vigor (maternal inheritance), attractive flower colors, double flower (single dominant gene), and recurrent flowering (a single recessive gene). Thornlessness (single or two complementary recessive genes, introduced from wild species *R. blanda*, *R. Carolina*, and *R. pendulina*) and fragrance (polygenes, *R. damascena* and *R. centifolia*) are important traits for garden roses. Additionally, easy propagation with cuttings, suitability for cultivation in subtropical environments, and high oil contents are also targeted objectives in breeding roses.

Resistance to abiotic or biotic stresses, such as cold (introduced from *R. rugose* and *R. wichuriana*), heat (tropical rose breeding, *R. clinophylla* and *R. gigantean*), and pests and diseases including black spot and powdery mildew (*R. bracteata* and *R. clinophylla*) and rose rosette disease (RRD) are the main targets of rose breeding. Especially, RRD is one of the most destructive diseases of commercial roses that is caused by the Emaravirus, Rose rosette virus (RRV) (Laney et al., 2011). Some wild rose species of the *Rosa* genus (*R. acicularis, R. blanda, R. carolina, R. palustris, R. setigera*, and *R. spinosissim*) have been reported to be resistant to RRD, while no commercial rose cultivars have been confirmed as resistant to either the viral agent or its vector (eriophyid mite) (Epstein and Hill, 1999; Bryne et al., 2015). Recently, however, large field trials for the observation of diverse rose cultivars are under way for screening resistance to RRD, which indicated the existence of resistance among the cultivated rose germplasm (Byrne et al., 2018a).

## 9.5 Limitations of conventional breeding

Classical or selective breeding based on hybridization is still a powerful tool for improving roses. It is essential to understand the genetic background of the variations in the morphological and physiological traits to effectively breed roses. However, the rose is a difficult model system for genetic studies due to some technical factors. The highly heterozygous genetics and complex polyploid nature of rose species result in severe chromosomal disorders and high male/female sterility. Poor seed setting and seed germination hinder the genetic analysis and breeding of roses. Additionally, modern hybrid roses carry the genes inherited from many ancestors, making them practically impossible to predict the outcome of any specific cross and design desired combinations. In general, multiple generations are required for the introgression of a new trait such as disease resistance from a wild species, which would take a minimum of

7 years to develop a new cultivar from one conventional breeding cycle (Byrne et al., 2018b).

In addition to conventional breeding using cross-hybridization, induced mutation with physical and chemical mutagens, irradiation, colchicine, and in vitro mutagenesis are being utilized to expand the genetic variability. Furthermore, the state-of-the-art molecular biology technologies such as molecular marker, next-generation sequencing (NGS), and gene transformation are emerging and being practically incorporated to overcome the current limitations in improving *R.* × *hybrida*.

## 9.6 Molecular marker techniques

The advent of molecular markers makes it possible to detect specific chromosome regions or genes that control important traits and to improve the access and utilization of germplasm resources, genetic analysis of breeding populations, parental selection performance, progeny prediction performance, and MAS and marker-enhanced back-cross-breeding. There have been efforts to identify the most efficient and cost-effective markers to be used in studying roses.

The study of the genetic diversity is of utmost importance to germplasm conservation, accurate cultivar identification, and breeding program enhancement. Molecular techniques provide tools for assessing the evolutional history and genetic relatedness among the rose germplasm. Various molecular marker types including randomly amplified polymorphic DNA (RAPD), restriction fragment length polymorphic (RFLP), and simple sequence repeat (SSR) have been used for genetic relatedness study and phylogenetic analysis within and among the *Rosa* species (Debener and Mattiesch, 1999) as well as for cultivar identification and protection (Hubbard et al., 1992). The ISSR and ScoT markers have also been used for DNA fingerprinting and in assessing the genetic relationship of diverse rose germplasm from different geographical locations, cultivar groups, and species (Agarwal et al., 2019).

Constructing a marker linkage map enables the identification of the structural organization of the rose genome, the genomic regions linked to traits, and the DNA markers that help breeders make trait selection decisions based on the markers. In a genetic map of diploid 94/1 population, a dominant gene for double flower (*Blfo*) was mapped on the linkage group B3, and *Blfa* for pink flowers was mapped on A2 and B2. The dominant gene *Rdr1* for black spot resistance was mapped in three tetraploid populations (Linde et al., 2004), and a sequence-characterized amplified region (SCAR) marker closely related to *Rdr1* was developed. An isolate of nine-specific powdery mildew resistance gene *Rpp1* was mapped using 117 diploid roses of the population 97/9 on the linkage group A3, and the nearest AFLP marker was converted to a SCAR marker. For important horticultural traits, a single recessive gene for the absence of prickles on the petiole was characterized and located at the end of the paternal linkage group 7 in a pseudo-test cross population. Genetic factors for recurrent blooming were also identified. The QTL mapping has been performed for some traits, which include the growth vigor (Yan et al., 2005), thorn quantity (Crespel et al., 2002), flower color, number of petals, and size of leaves and prickles (Zhang, 2003), as well as powdery mildew resistance.

For improved genomics research on roses, access to the whole genome information is necessary. The international consortium of the 'Rose Genome Sequencing Initiative' (RGSI) has recently initiated the genome sequencing for roses. A high-quality reference genome sequence was developed for *R. chinensis* (Hibrand Saint-Oyant et al., 2018) based on NGS, and a genotyping-by-sequencing (GBS) approach was used to develop a high-density SNP-based consensus map that would be useful for the discovery of marker-trait associations in rose through fine mapping. The GBS approach was also used to produce a large number of SNPs from 79 rose cultivars and to design a DNA chip for high-throughput genotyping.

## 9.7 Genetic modification (GM)

The first genetically modified (GM) hybrid of *R.* × *hybrida* was reported by using the *Agrobacterium*-mediated system for the cv. Royalty (Firoozabady et al., 1994) and a biolistic-mediated modification system for the cv. Glad Tidings (Marchant et al., 1998). Since then, gene transfer technologies have been available as a unique opportunity for the fast development of roses with improvement in flower color, disease resistance, fragrance, and senescence.

Production of new flower colors through the development of the transgenic cultivars has been the major commercial benefit of the gene modification technologies thus far. Important cut-flower crops in their genetically modified, color-altered cultivars have now been commercially available for nearly 10 years. Using a gene encoding the chalcone synthase to manipulate the anthocyanin biosynthesis pathway, new flower colors were produced in *R.* × *hybrida* through the development of the transgenic cultivars (Souq et al., 1995).

Consumers desire and breeders select for increased fragrance in cut roses. However, it appears that 'real rose fragrance' is negatively correlated with vase life. Ethylene is a hormonal regulator of senescence in most plant organs, and recent research on the cloning and expression of 1-aminocyclopropane-1-caroxylate synthase from *Rosa* may lead to the regulation of the ethylene response in plants and, therefore, a potentially longer vase life for roses (Wang et al., 2004). Some researchers have developed disease-tolerant transgenic hybrid roses. *Rosa* × *hybrida* cv. Carefree Beauty genetically modified with an antimicrobial protein gene has demonstrated increased resistance to powdery mildew (Li et al., 2003). Inserting various fungal resistance genes has led to the development of some transgenic rose lines but with limited success. The RNAi technology has the potential to provide resistance to viruses.

Other desirable traits of roses, including thornless stems, soil-less cultivation capacity, efficient nutrient usage, and enhanced performance of cultivars on their rootstock, may become harvestable with advances in GM technology (Gudin, 2000).

## 10 Utilizing ornamental germplasm to exploit genetic diversity in cacti

### *10.1 Cacti – a brief introduction*

Cacti are thorny perennial plants that belong to the Cactaceae family with highly xerophytic or drought-tolerant characteristics. All cacti are succulents with distinct morphological features. Cacti generally grow slowly and have meristematic areolar tissues that act as the originating point of thorns, branches, and flowers (Ranjan et al., 2016). The Cactaceae family consists of more than 2000 species of plants divided into the Pereskioideae, Opuntioideae, and Cactoideae subfamilies. Among the subspecies, Opuntioideae includes the majority of cacti (more than 250 species), which have glochids as the distinct morphological trait (Barthlott and Hunt, 1993). The glochids are deciduous spines that are small and barbed, and the seeds of these cacti are completely covered by a hard and bony funicular stalk (Ranjan et al., 2016). Crassulacean acid metabolism (CAM) in cacti gives rise to their adaptation to arid environments and extreme drought tolerance. Opuntioid cacti, such as *Opuntia ficus-indica* (L.) Mill, convert the available water into biomass, which makes them ideal for arid regions (Kluge and Ting, 1978). Moreover, vivipary in cacti enables seedling development even in extreme environments (Cota-Sa´nchez, 2004). Also, several cacti varieties have been used as food, animal fodder, and medicine. Cacti also hold industrial importance (Ranjan et al., 2016) and aid in the conservation of natural resources. The illustrated benefits and resistance to multiple stresses including water scarcity, extreme temperatures, and light intensities make cacti a viable solution to the growing global climate changes. Furthermore, cacti are a genetic resource that contains several vital genes and markers related to stress tolerance that can be incorporated into other commercial crops.

## 10.2 Distribution and diversity

Most cacti are native to the Americas, where several diverse species exist in the eastern Andes, southeast Brazil, and northeast Mexico (Mutke and Barthlott, 2005). Cacti can grow in a wide range of habitats, including deserts, coastal areas, scrublands, deciduous forests, and tropical rainforests (Barthlott and Hunt, 1993). Some cactus species grow in the driest parts of the earth, the Atacama Desert, to name one (Ranjan et al., 2016). Mexico is home to the most diverse population of cacti, followed by Brazil, Argentina, Bolivia, and Peru (Ramawat, 2010). In addition to cacti that inhabit dry land, moisture-affinitive epiphytic and climber cacti species are found in the mountains and forests of Brazil, Bolivia (a diversity center of Rhipsalideae), and Central America (a diversity center for several climbing Hylocereceae) (Anderson, 2001). The *Rhipsalis baccifera* is an exceptional cactus found across the Americas as well as Africa, Madagascar, and Sri Lanka (Ranjan et al., 2016). Birds feeding on the cacti and carrying the seeds in their digestive tracts contributed to the wide dispersion of cacti across the continents. Godínez-Alvarez et al. (2002) attribute the seed dispersal of the gigantic columnar cactus species *Neobuxbaumia tetetzo* to the bat species *Leptonycteris curasoae* Miller that feed on the fruits.

## 10.3 Cacti-renowned ornamentals

Cacti have long been widely cultivated as ornamental crops or plants. The ornamental values of cacti have been recorded since the 16th century, by the establishment of various collection centers by numerous European herbaria and botanical gardens (Ranjan et al., 2016). The ornamental characteristics of cacti are predominantly rendered by beautiful flowers, aesthetic shapes, and spines, as well as their distinct and diverse morphology. Cacti have been grown as potted plants in households and widely planted in xerophytic gardens. Cactus flowers boast a wide array of colors that include white, pink, yellow, violet, and red as well as a mix of different colors in spotted or striped patterns (Barthlott and Hunt, 1993). Furthermore, cactus flowers range from small (1 cm) to very large (35 cm) in size (Anderson, 2001). The *Mammillaria* and *Opuntia* genera are ornamentals that are widely planted in households and gardens in the United States (Ranjan et al., 2016). Apart from these genera, *Astrophytom*, *Cereus*, *Chamaecereus*, *Echinopsis*, *Gymnocalycium*, *Lobivia*, *Melocactus*, *Notocactus*, *Rhipsalis*, *Schlumbergera*, and *Trichocereus* are also widely cultivated across the world as ornamental plants (Ranjan et al., 2016). *Mammillaria* is the most popularly traded cactus genus accounting for 64% of all cactus trades, followed by *Opuntia*, *Coryphanta*, and *Echinocereus* in the global markets (Ranjan et al., 2016). In northern Europe, *Hatiora* and *Schlumbergera* are the most important potted flowering cacti cultivated in glasshouses. The variability in the size and shape of cacti makes them admirably unique plants with different stem shapes. A globular stem, for instance, has a high water-storage capacity and low water loss by transpiration. The tallest cactus species is *Pachycereus pringlei* (19.2 m), and the smallest species is *Blossfeldia liliputiana* (1 cm) (Altesor and Ezcurra, 2003). Cactus prices are highly variable depending upon several characteristics, such as the flowering duration. For instance, the species *Selenicereus grandifloras*, also known as the queen of the night, produce fully opened flowers for only 2 h at night, whereas flowers of other species can last for a whole week (Ranjan et al., 2016). Cacti have remained as important ornamentals to the present day in the horticulture industry due to their myriad of forms and aesthetic morphologies. In addition to the ornamental properties, cacti are also consumed as food in many countries due to their rich nutrient contents. For instance, the juvenile cladodes (nopalitos) of *Opuntia* and *Nopalea* species have been consumed as a vegetable in Mexico and Texas, and, similarly, fruits of several cacti have been consumed in global markets due to the nutritional benefits (Basile, 2001; Inglese et al., 2002).

## 10.4 Utilizing cacti to cope with the global climate change

The recent era of global climate change has been primarily associated with extreme temperatures, aridity, high salinity, and increasing carbon dioxide ($CO_2$) concentration. Therefore, suitable solutions to combat the extreme climate change have to be identified; plants may be considered as a natural solution to the aforementioned problems. Cacti have adaptation mechanisms that enable them to endure adverse environmental conditions, and the associated physiological and molecular adaptation strategies may be a potential solution to the global climate change. Morphologically, the occurrence of spines in cacti is a vital adaptation for the temperature tolerance, since the diminutive surface area of spines as compared to that of leaves drastically reduces the transpiration rate (Rebman and Pinkava, 2001). Besides, the shapes of the spines aid in shading and reduce direct exposure to the sun, as observed in *Ferocactus* species (barrel cacti) (Rebman and Pinkava, 2001).

Being a CAM plant is a vital merit for cacti since the stomata are closed during the day, which prevents latent heat loss, and the vertically oriented cladodes reduce overheating by hindering the interception of light at mid-day (Nobel, 1978). The presence of a thick waxy cuticular region and a minimal number of stomata (about 20–30 per $mm^2$) result in the reduction of water exchange between the cactus tissue and the surrounding environment. Cacti such as *Opuntia* consist of mucilaginous cells that reserve the mucilage, which has a great water retention ability under extreme conditions (Saenz et al., 2004).

Cacti have shallow and extremely fibrous root systems, which allow for effective water uptake from the soil, and temporary rainy roots produced by cacti help absorb rainwater. In *O. ficus*, rainy roots are formed within 1 day of wetting and thus aid in the absorption and conservation of water (Ranjan et al., 2016). The contraction of roots and the pulling of the crown into the ground as a means to escape high temperatures and drought have been identified as an adaptation strategy in *Ariocarpus fissuratus* (living rock) (Garrett et al., 2010). In addition to the mechanisms discussed earlier, cacti possess the ability to rapidly seal tear wounds in the stem caused by external sources (Ranjan et al., 2016).

In addition to the aforementioned morphological adaptations, the CAM photosynthesis is considered as the distinct armor cacti have against extreme conditions. CAM is one of the three major photosynthetic mechanisms adopted by vascular plants, where $CO_2$ fixation occurs at night via PEP carboxylase, which leads to the synthesis of malate (Dodd et al., 2002). The malate is accumulated in the vacuolar region, and the decarboxylation is followed by the fixation of $CO_2$ by rubisco. Thus, the CAM pathway acts as the temporal boundary between the C3 and C4 pathways, giving rise to the plasticity for maximizing the carbon gain and the water usage according to the ambient environmental conditions (Borland et al., 2009). Moreover, the efficient functioning of the CAM pathway under low water conditions is due to the stomatal opening during the night, which prevents water loss and transpiration (Borland et al., 2009; Nobel, 1996). On the other hand, the daytime closure of stomata reduces $CO_2$ leakage and thus synergistically regulates the endogenous $CO_2$ level and internal temperature for effective photosynthesis (Borland et al., 2009; Nobel, 1996).

Overall, the in-depth understanding of the molecular mechanisms and the genomic information of cacti would aid in the supplementation of crops with tailored genes for adaptation to harsh environments. Moreover, the low water requirements and maintenance of cacti, in addition to their aesthetic properties, make them excellent ornamental plants. Furthermore, because cacti are plants that are the least affected by extreme climates, their vital tolerance-related traits have to be investigated and exploited in detail for future genetic resource management.

## 10.5 Cacti in the Republic of Korea

Cacti have been introduced into the Republic of Korea in 1940s by private collectors and commercialization launched in the late 1970s. The total area of cacti cultivation has been steadily increased in Korea since 1985 (Table 3, Jeong et al., 2004). Due to the labor-intensive cultivation approaches of cacti, several programs have been introduced to reduce labor time and cost (Lee and Oda, 2003). Recently, to improve the quality of cacti varieties, various novel approaches and modern technologies such as light quality, night interruption, and in vitro and ex vitro culture techniques have been implemented. Moreover, cultivations of cacti in controlled environment with

**Table 3** Cultivation area and production of total cacti in Korea up to 2002 (Jeong et al., 2004)

| Year | Cultivation area (ha) | Production (1000 plants) | Value (US $1000) |
|---|---|---|---|
| 1985 | 3.5 | 368 | 235 |
| 1990 | 38.3 | 10 819 | 2799 |
| 1995 | 61.6 | 28 747 | 8621 |
| 2000 | 50.2 | 29 160 | 7935 |
| 2001 | 58.4 | 36 267 | 8851 |
| 2002 | 64.2 | 48 970 | 8837 |

defined parameters such as $CO_2$, temperature, irrigation method, nutrient solution, and so on have been reported by several researchers (Song et al., 2006, 2007; Jeong et al., 2006e, 2007b; Lee et al., 2014; Song and Song, 2012; Kim et al., 2019a,b).

Korea exports grafted cacti majorly to the Netherlands, the United States, China, and Canada. Among popularly exported cacti, 70% belong to *G. mihannovichii*, 20% *C. silverstrii*, and 10% *N. capa*, *E. leninghausii*, and *G. baladianum*. In Korea, the Cactus Research Institute (CRI), ARES, Goyang, and the National Horticulture and Herbal Research Institute (NHHRI), RDA, Wanju, are the two important research institutes actively involved in improving cacti breeding to the international standards. A total of 400 species of *Gymnocalycium* and *Lobivia* were collected, and about 80 cultivars have been released by the National Institute of Horticultural and Herbal Science (NIHHS), RDA. In addition, more than 30 cultivars have been released by the CRI, ARES, Goyang (Table 4). Lee et al. (2006) also classified cacti based on their diurnal $CO_2$ assimilation and showed that nine important species have full CAM or weakly CAM characteristics. In addition to their ornamental values, cacti are potentially involved in the detoxification of volatile air pollutants, which increase the demand for cacti in horticulture industries. Future work on classifying cacti germplasm based on their efficiency on repressing the $CO_2$ buildup could be effective in increasing indoor air quality.

## 11 Conclusions

Despite the importance of the genetic diversity of ornamentals, there have been only a few studies on the collection and management of ornamentals. To collect, conserve, and utilize the genetic diversity of ornamentals better and more efficiently, it is necessary to systematize the management of the ornamental germplasm around the world, and international cooperation is important. Scientific societies related to genetic diversity such as 'International Symposium on Germplasm of Ornamentals' and 'International Symposium

**Table 4** Cacti cultivars released by National Horticulture and Herbal Research Institute (NHHRI), RDA and Cactus Research Institute (CRI), Republic of Korea

| Species | Cultivar | Year | Color | Globe shape | Spine | Released | Reference |
|---|---|---|---|---|---|---|---|
| *G .mihanovichii* | Hongil, Hongwoo, Honghwa, Hongjo | 1994 | Data not included | | | NHHRI, RDA | Jeong et al. (2004) |
| | Myeongweol, Seonhong, Hongkwang | 1997 | | | | | |
| | Hongsil, Cheongsil, Yeongji, Konji, Hongweol | 1998 | | | | | |
| | Seolhong, Jinhong, Manhong, Hongsu, Yeonmin Hukwang, Ojak, Saekdong, Sinseong, Binghwa | 1999 | | | | | |
| | Baekya, Mihong, Hongsung, Yeonhwa, Keumhong Keumhwang, Hwangweol, Danhong, Bihong | 2000 | | | | | |
| | Damdam, Yeonhwang, Yeonhong, Hoohwang, Damhong | 2001 | | | | | |
| | Heukhong, Suhwang, Gamhong, Seohong, Seohwang, Noksim | 2002 | | | | | |
| | Hwihwang, Hwisung, Heukseong, | 2003 | | | | | |
| | Hwiseong | 1999 | Orange and yellow | Flat round | Light gray Straight | | Jeong et al. (2006c) |
| | Hwahong | 2002 | Bright red | Flat round | Gray Slant | | Park et al. (2008b) |

(*Continued*)

**Table 4** (*Continued*)

| Species | Cultivar | Year | Color | Globe shape | Spine | Released | Reference |
|---|---|---|---|---|---|---|---|
| | Dahong | 2002 | Bright dark red | Flat round | Gray<br>Slant | | Park et al. (2008a) |
| | Geoseong | 2003 | Orange | Flat round | Brown<br>Straight | | Jeong et al. (2006d) |
| | Suhong | 2003 | Dark red | Flat round | Gray<br>Straight | | Jeong et al. (2006a) |
| | Gohong | 2004 | Bright red | Flat round | Gray<br>Straight | | Park et al. (2009) |
| | Hwangjo | 2005 | Yellow | Flat round | Grayish<br>brown<br>Semierect | | Jeong et al. (2007a) |
| | Simhong | 2005 | Dark red | Flat round | Dark<br>brown<br>Straight | | Jeong et al. (2007c) |
| | Hwangun | 2005 | Bright yellow | Flat round | Brown<br>Straight | | Park et al. (2010a) |
| | Hwangseon | 2005 | Yellow | Flat round | Gray<br>Straight | | Park et al. (2012) |
| | Saekjo | 2005 | Red and yellow | Flat round | Brown<br>Straight | | Park et al. (2013) |
| | Suyeon | 2007 | Pink with black | Flat round | Brown<br>Straight | | Jeong et al. (2007d) |

| | | | | | | | |
|---|---|---|---|---|---|---|---|
| | Ilhong | 2008 | Dark red | Flat round | Gray and brown Straight | | Park et al. (2014) |
| | Isaek | 2012 | Bicolor (red and yellow) | Round oblate | Gray and brown Straight | | Park et al. (2015) |
| | Bulbit | 2015 | Dark red | Flat round | Brown Straight | | Park et al. (2019) |
| | Yeonbit | 2015 | Pink | Flat round | Brown Straight | | Park et al. (2018) |
| | Achi #1, Achi #2, Achi #3, Manchu #1 | 1998 | Data not included | | | CRI | Jeong et al. (2004) |
| | Achi #4, Achi #5, Achi #6, Rainbow 1, Black Pearl 1 | 1999 | | | | | |
| | Heorum, Hedoji, Hemin, Jinsol, Noul | 2000 | | | | | |
| | Saebyul, Ruby, Fire, Gumbich, Gold | 2002 | | | | | |
| | Red Jewelry, Sun Rise, Star Fish | 2003 | | | | | |
| | Black Sun | 2011 | Red patterned | Data not included | | | Soh et al. (2019) |
| *C. silvestrii* | Mangwool, Sumji, Bosong, Doran, Somsom | 1999 | Data not included | | | NHHRI, RDA | Jeong et al. (2004) |
| | Duri, Mongsil, Woori | 2000 | | | | | |
| | Hongwoon, Tosil | 2001 | | | | | |
| | Chongchong | 2002 | | | | | |

*(Continued)*

**Table 4** (*Continued*)

| Species | Cultivar | Year | Color | Globe shape | Spine | Released | Reference |
|---|---|---|---|---|---|---|---|
| | Light, Champion, Chungsol | 2002 | Data not included | | | CRI | Jeong et al. (2004) |
| | Doldol | 2003 | Yellow | Cristate | Data not included | | Jeong et al. (2006b) |
| | Blossom, Aram | 2003 | Data not included | | | | Jeong et al. (2004) |
| | Yellow Tree | 2007 | Yellow | Cylindrical | Data not included | | Cho et al. (2008a) |
| *G. denudatum* | Hwangkeum | 2000 | Data not included | | | NHHRI RDA | Jeong et al. (2004) |
| *Notocactus scopa* | Soonjung | 2002 | | | | CRI | |
| *Lobivia* spp. | Shinbi, Geumhyang | 2003 | | | | | |
| *Copiapo scopa* | Red Stone | 2007 | Red brown | Globose round | Black | | Cho et al. (2008b) |
| *Euphorbia milii* | Panorama | 2009 | Green pink | Data not included | | | Park et al. (2010b) |
| | Dual Pink | 2015 | Pink and yellow | Data not included | | | Lee et al. (2019a) |
| *Rebutia candiae* | Melonia | 2016 | Yellowish green | Cylindrical sphere | White Moderately decline | | Jung et al. (2019) |
| *Astrophytum* spp. | Snow Gem | 2018 | Green | Flat round | White Straight | | Lee et al. (2019b) |

–, Information on color and shape was not available. National Institute of Horticultural and Herbal Science (NIHHS), RDA, Wanju, 55365, Korea, was previously functioning as National Horticultural Research Institute (NHRI), RDA, Suwon 440-310, Korea.

on Woody Ornamentals of the Temperate Zone' have been formed under the International Society for Horticultural Science, and we hope that the activities and roles of such groups become more cohesive and productive in the near future.

## 12 References


Adams, R. P. 1997. Conservation of DNA: DNA banking. In: Callow, J. A., Ford-Lloyd, B. V. and Newbury, H. J. (Eds), *Biotechnology and Plant Genetic Resources Conservation and Use*. CAB International, Wallingford, UK, pp. 163-74.

Agarwal, A., Gupta, V., Ul Haq, S., Jatav, P. K., Kothari, S. L. and Kachhwaha, S. 2019. Assessment of genetic diversity in 29 rose germplasms using SCoT marker. *J. King Saud Univ. Sci.* 31(4), 780-8. doi:10.1016/j.jksus.2018.04.022.

Altesor, A. and Ezcurra, E. 2003. Functional morphology and evolution of stem succulence in cacti. *J. Arid. Environ.* 53(4), 557-67. doi:10.1006/jare.2002.1059.

Altieri, M. A. and Merrick, L. C. 1987. *In situ* conservation of crop genetic resources through maintenance of traditional farming systems. *Econ. Bot.* 41(1), 86-96. doi:10.1007/BF02859354.

Anderson, E. F. 2001. *The Cactus Family*. Timber Press, Portland, OR.

Ashmore, S. 1997. *Status Report on the Development and Application of In Vitro Techniques for the Conservation of Plant Genetic Resources*. IPGRI, Rome, Italy.

Barthlott, W. and Hunt, D. R. 1993. Cactaceae. In: Kubitzki, K., Rohwer, J. G. and Bittrich, V. (Eds), *The Families and Genera of Vascular Plants* (vol. 2). Springer, Berlin, Germany, pp. 161-97.

Basile, F. 2001. Economic aspects of cactus pear production and market. *J. Prof. Assoc. Cactus Dev.* 5, 3-46.

Benson, E. E. 1999. *Plant Conservation Biotechnology*. Taylor and Francis, London, UK.

Borland, A. M., Griffiths, H., Hartwell, J. and Smith, J. A. C. 2009. Exploiting the potential of plants with crassulacean acid metabolism for bioenergy production on marginal lands. *J. Exp. Bot.* 60(10), 2879-96. doi:10.1093/jxb/erp118.

Bretting, P. K. 2007. The U.S. National Plant Germplasm System in an era of shifting international norms for germplasm exchange. *Acta Hortic.* 760(760), 55-60. doi:10.17660/ActaHortic.2007.760.5.

Brown, A. H. D. and Marshall, D. R. 1995. A basic sampling strategy: theory and practice. In: Guarino, L., Rao, V. R. and Reid, R. (Eds), *Collecting Plant Genetic Diversity: Technical Guidelines*. CAB International, Wallingford, UK.

Brush, S. B. 1999. *Genes in the Field: On-Farm Conservation of Crop Diversity*. Lewis Publishers, Boca Raton, FL, IPGRI, Rome, Italy and IDRC, Ottawa, Canada.

Bryne, D. H., Roundey, E., Klein, P. and Yan, M. 2015. Combating rose rosette disease: are there resistant roses? *Amer. Rose* 43(5), 78-83.

Byrne, D. H., Klein, P., Yan, M., Young, E., Lau, J., Ong, K., Shires, M., Olson, J., Windham, M., Evans, T. and Novick, D. 2018a. Challenges of breeding rose rosette-resistant roses. *HortScience* 53(5), 604-8. doi:10.21273/HORTSCI12553-17.

Byrne, P. F., Volk, G. M., Gardner, C., Gore, M. A., Simon, P. W. and Smith, S. 2018b. Sustaining the future of plant breeding: the critical role of the USDA-ARS National Plant Germplasm System. *Crop Sci.* 58(2), 451-68. doi:10.2135/cropsci2017.05.0303.

Castañeda-Álvarez, N. P., Khoury, C. K., Achicanoy, H. A., Bernau, V., Dempewolf, H., Eastwood, R. J., Guarino, L., Harker, R. H., Jarvis, A., Maxted, N., Müller, J. V., RamirezVillegas, J., Sosa, C. C., Struik, P. C., Vincent, H. and Toll, J. 2016. Global conservation priorities for crop wild relatives. *Nat. Plants* 2, 16022. doi:10.1038/nplants.2016.22.

CBD. 1992. *Convention on Biological Diversity*. CBD, Rio de Janeiro, Brazil.

CGN (Center for Genetic Resources, the Netherlands). 2019. Available at: https://www.wur.nl (accessed on 26 April 2019).

Cho, C. H., Nam, S. Y., Leem, S. D. and Park, H. B. 2008a. A new yellow color grafted cactus (*Chamaecereus silvestrii*) cultivar 'Yellow Tree'. *Korean J. Breed. Sci.* 40, 521–4.

Cho, G. H., Nam, S. Y., Lee, S. D. and Park, Y. S. 2008b. A new color grafted cactus (*Copiapoa scopa*) cultivar 'Red Stone'. *Korean J. Breed. Sci.* 40, 525–8.

Cota-Sánchez, J. H. 2004. Vivipary in the Cactaceae: its taxonomic occurrence and biological significance. *Flora* 199(6), 481–90. doi:10.1078/0367-2530-00175.

CPC. 1991. Genetic sampling guidelines for conservation collections of endangered plants. In: Falk, D. A. and Holsinger, K. E. (Eds), *Genetics and Conservation of Rare Plants*. Oxford University Press, Oxford, UK, pp. 225–38.

Crespel, L., Chirollet, M., Durel, E., Zhang, D., Meynet, J. and Gudin, S. 2002. Mapping of qualitative and quantitative phenotypic traits in Rosa using AFLP markers. *Theor. Appl. Genet.* 105(8), 1207–14. doi:10.1007/s00122-002-1102-2.

Damania, A. B. 1996. Biodiversity conservation: a review of options complementary to standard *ex situ* methods. *Plant Genet. Resour. Newsl.* 107, 1–18.

Debener, T. and Mattiesch, L. 1999. Construction of a genetic linkage map for roses using RAPD and AFLP markers. *Theor. Appl. Genet.* 99(5), 891–9. doi:10.1007/s001220051310.

DEFRA. 2010. United Kingdom: country report on plant genetic resources for food and agriculture. DEFRA, London, UK, 40p. Available at: http://www.fao.org/fileadmin/templates/agphome/documents/PGR/SoW2/country_reports/europe/UK_genetic_resources_country_report.pdf.

De Vries, D. P. and Dubois, L. A. M. 1996. Rose breeding: past, present, prospects. *Acta Hortic.* 424(424), 241–8. doi:10.17660/ActaHortic.1996.424.43.

Dodd, A. N., Borland, A. M., Haslam, R. P., Griffiths, H. and Maxwell, K. 2002. Crassulacean acid metabolism: plastic, fantastic. *J. Exp. Bot.* 53(369), 569–80. doi:10.1093/jexbot/53.369.569.

Dulloo, M. E., Labokas, J., Iriondo, J. M., Maxted, N., Lane, A., Laguna, E., Jarvis, A. and Kell, S. P. 2008. Genetic reserve location and design. In: Iriondo, J. M., Maxted, N. and Dulloo, M. E. (Eds), *Conserving Plant Genetic Diversity in Protected Areas: Population Management of Crop Wild Relatives*. CAB International, Wallingford, UK, pp. 23–64.

Engelmann, F. 1997. *In vitro* conservation methods. In: Ford-Lloyd, B. V., Newbury, J. H. and Callow, J. A. (Eds), *Biotechnology and Plant Genetic Resources: Conservation and Use*. CAB International, Wallingford, UK, pp. 119–62.

Engels, J. M. M. 1995. *In situ* conservation and sustainable use of plant genetic resources for food and agriculture in developing countries. Report of a *DSE/ATSAF/IPGRI Workshop*, 2–4 May 1995, Bonn-Röttgen, Germany. IPGRI, Rome, Italy and DSE, Feldafing, Germany.

Engels, J. M. M. 2002. Home gardens-a genetic resources perspective. In: Watson, J. W. and Eyzaguirre, P. B. (Eds), *Home Gardens and In Situ Conservation of Plant Genetic*

*Resources in Farming Systems. Proceedings of the Second International Home Gardens Workshop*, 17–19 July 2001, Witzenhausen, Germany. IPGRI, Rome, Italy.

Environmental Audit Committee. 2011. Supplementary written evidence submitted by Professor Sandy Thomas, Head of Foresight, Government Office for Science. Department for Business Innovation and Skills. Available at: http://www.publications.parliament.uk/pa/cm201012/cmselect/cmenvaud/879/879we20.htm.

Epstein, A. H. and Hill, J. H. 1999. Status of rose rosette disease as a biological control for multiflora rose. *Plant Dis.* 83(2), 92–101. doi:10.1094/PDIS.1999.83.2.92.

FAO. 1996. *Global Plan of Action for the Conservation and Sustainable Utilization of Plant Genetic Resources for Food and Agriculture*. FAO, Rome, Italy.

FAO. 1998. *The State of the World's Plant Genetic Resources for Food and Agriculture*. FAO, Rome, Italy.

FAO. 2010. Enrichment Index: an indicator of enhancement for *ex situ* plant germplasm collections. FAO Plant Production and Protection Division. Unpublished Report, 5p.

Firoozabady, E., Moy, Y., Courtney-Gutterson, N. and Robinson, K. 1994. Regeneration of transgenic rose (*Rosa hybrida*) plants from embryogenic tissue. *Biotechnology* 12(6), 609–13. doi:10.1038/nbt0694-609.

Foresight. 2011. The future of food and farming. Final Project Report. The Government Officer for Science, London, UK. Available at: https://www.gov.uk/government/uploads/system/uploads/attachment_data/file/288329/11-546-future-of-food-and-farming-report.pdf.

Frankel, O. H., Brown, A. H. D. and Burdon, J. J. 1995. *The Conservation of Plant Biodiversity*. Cambridge University Press, Cambridge, UK.

Garrett, T. Y., Huynh, C. V. and North, G. B. 2010. Root contraction helps protect the 'living rock' cactus from lethal high temperatures when growing in rocky soil. *Am. J. Bot.* 97(12), 1951–60. doi:10.3732/ajb.1000286.

Godefroid, S., Rivière, S., Waldren, S., Boretos, N., Eastwood, R. and Vanderborght, T. 2011. To what extent are threatened European plant species conserved in seed banks? *Biol. Conserv.* 144(5), 1494–8. doi:10.1016/j.biocon.2011.01.018.

Godínez-Alvarez, H., Valiente-Banuet, A. and Rojas-Martínez, A. 2002. The role of seed dispersers in the population dynamics of the columnar cactus *Neobuxbaumia tetetzo*. *Ecology* 83(9), 2617–29. doi:10.1890/0012-9658(2002)083[2617:TROSDI]2.0.CO;2.

Grout, B. W. W., Westcott, R. J. and Henshaw, G. G. 1978. Survival of shoot meristems of tomato seedlings frozen in liquid nitrogen. *Cryobiology* 15(4), 478–83. doi:10.1016/0011-2240(78)90068-8.

Gudin, S. 2000. Rose: genetics and breeding. *Plant Breed. Rev.* 17, 159–90.

Harrington, J. F. 1970. Seed and pollen storage for conservation of plant gene resources. In: Frankel, O. H. and Bennett, E. (Eds), *Genetic Resources in Plants–Their Exploration and Conservation*. IBP Handbook No. 11. Blackwell Scientific Publications, Oxford, UK, pp. 501–21.

Heywood, V. H. 1998. The role of botanic gardens in *ex situ* conservation of agrobiodiversity. In: Gass, T., Begemann, F. and Frese, L. (Eds), *Implementation of the Global Plan of Action in Europe–Conservation and Sustainable Utilization of Plant Genetic Resources for Food and Agriculture. Proceedings of the European Symposium*, 30 June–3 July 1998, Braunschweig, Germany. IPGRI, Rome, Italy.

Hibrand Saint-Oyant, L., Ruttink, T., Hamama, L., Kirov, I., Lakhwani, D., Zhou, N. N., Bourke, P. M., Daccord, N., Leus, L., Schulz, D., Van de Geest, H., Hesselink, T., Van Laere, K., Debray, K., Balzergue, S., Thouroude, T., Chastellier, A., Jeauffre, J., Voisine, L., Gaillard,

S., Borm, T. J. A., Arens, P., Voorrips, R. E., Maliepaard, C., Neu, E., Linde, M., Le Paslier, M. C., Bérard, A., Bounon, R., Clotault, J., Choisne, N., Quesneville, H., Kawamura, K., Aubourg, S., Sakr, S., Smulders, M. J. M., Schijlen, E., Bucher, E., Debener, T., De Riek, J. and Foucher, F. 2018. A high-quality genome sequence of *Rosa chinensis* to elucidate ornamental traits. *Nat. Plants* 4(7), 473–84. doi:10.1038/s41477-018-0166-1.

Hoekstra, F. A. 1995. Collecting pollen for genetic resources conservation. In: Guarino, L., Rao, V. R. and Reid, R. (Eds), *Collecting Plant Genetic Diversity: Technical Guidelines*. CAB International, Wallingford, UK, pp. 527–50.

Hubbard, M., Kelly, J., Rajapakse, S., Abbott, A. and Ballard, R. 1992. Restriction fragment length polymorphisms in rose and their use for cultivar identification. *HortScience* 27(2), 172–3. doi:10.21273/HORTSCI.27.2.172.

Inglese, P., Basile, F. and Schirra, M. 2002. Cactus pear fruit production. In: Nobel, P. S. (Ed), *Cacti: Biology and Uses*. University of California, Berkeley, CA, pp. 163–83.

Ingram, G. B. 1984. *In situ* conservation of the genetic resources of plants: the scientific and technical base. Forest Resources Division, FAO, Rome, Italy.

IPGRI (The International Plant Genetic Resources Institute). 1996. Programme activities, germplasm maintenance and use. Annual Report. IPGRI, Rome, Italy, pp. 56–65.

IPGRI (The International Plant Genetic Resources Institute). 2003. A guide to effective management of germplasm collections germplasm. NBPGR, New Delhi, India and IPGRI and FAO, Rome, Italy, p. 51.

Jarvis, D. and Hodgkin, T. 2000. Farmer decision making and genetic diversity: linking multidisciplinary research to implementation on-farm. In: Brush, S. B. (Ed.), *Genes in the Field. On Farm Conservation of Crop Diversity*. Lewis Publishers, Boca Raton, FL, IPGRI, Rome, Italy, and IDRC, Ottawa, Canada, pp. 261–78.

Jarvis, D. I., Myer, L., Klemick, H., Guarino, L., Smale, M., Brown, A. H. D., Sadiki, M., Sthapit, B. and Hodgkin, T. 2000. *A Training Guide for In Situ Conservation On-Farm*, Version 1. IPGRI, Rome, Italy.

Jeong, M. I., Cho, C. H. and Lee, J. M. 2004. Production and breeding of Cacti for grafting Korea. *Chron. Hortic.* 44, 7–10.

Jeong, M. I., Chung, B. N., Kim, M. S., Song, J. S., Kim, J. Y. and Park, Y. J. 2006a. A new dark red graft cactus (*Gymnocalycium mihanovichii*) cultivar, 'Suhong'. *Korean J. Breed.* 38, 127–8.

Jeong, M. I., Chung, B. N., Lee, Y. R., Bang, C. S., Kim, J. Y. and Bae, J. H. 2006b. A new yellow graft cactus (*Chamaecereus silvestrii*) cultivar, 'Doldol' with cristate shape. *Korean J. Breed.* 38, 131–2.

Jeong, M. I., Chung, B. N., Kim, M. S., Song, J. S., Kim, J. Y. and Park, Y. J. 2006c. A new orange and yellow graft cactus (Gymnocalycium mihanovichii) cultivars, "Hwiseong". Korean J. Breed. 38, 125–6.

Jeong, M. I., Chung, B. N., Lee, Y. R., Song, J. S., Kim, J. Y. and Park, Y. J. 2006d. A new orange graft cactus (*Gymnocalycium schickendantziix × G. mihanovichii*) cultivar, 'Geoseong'. *Korean J. Breed.* 38, 129–30.

Jeong, S. J., Lee, J. S., Lee, S. C. and Kim, W. S. 2006e. Effects of vascular bundle sizes of scions and stocks on graft-take and growth of grafted cactus ruby ball. *Hortic. Environ. Biotechnol.* 74, 144–9.

Jeong, M. I., Chung, B. N., Park, P. M., Kim, W. H., Lee, E. K. and Kim, J. Y. 2007a. A new yellow graft cactus (*Gymnocalycium marsoneri × mihanovichii× G. mihanovichii*) Cultivar, 'Hwangjo'. *Korean J. Breed. Sci.* 39, 568–9.

Jeong, S. J., Kim, W. S. and Lee, J. S. 2007b. Effect of stationary room temperature on graft-take and post-graft growth of grafted cactus ruby ball. *Hortic. Environ. Biotechnol.* 48, 393–6.

Jeong, M. I., Chung, B. N., Park, P. M., Kim, W. H., Lee, E. K. and Kim, J. Y. 2007c. A new dark red graft cactus (Gymnocalycium mihanovichii) cultivar, "Simhong". J. Korean Breed. Sci. 39, 566-7.

Jeong, M. I., Chung, B. N., Park, P. M., Kim, W. H., Lee, E. K. and Kim, J. Y. 2007d. A new pink and black graft cactus (Gymnocalycium mihanovichii) cultivar "Suyeon". Korean J. Breed. Sci. 39, 391-2.

Jung, J. H., Lee, G. Y., Kim, Y. H., Soh, H. S., Lee, S. D. and Shin, B. E. 2019. A new cultivar of *Rebutia* spp. 'Melonia'. *Hortic. Sci. Technol.* 37(Suppl. I), 147.

Kim, S. S., Heo, J. H. and Lee, H. J. 2016. A study on conservation and utilization of agriculture and forest genetic resources in response to the Nagoya Protocol. Korea Rural Economic Institute.

Kim, Y. H., Lee, G. Y., Kim, H. H., Lee, J. H., Jung, J. H. and Lee, S. D. 2019a. Establishment of tissue culture and acclimatization method for *in vitro* mass propagation of *Echeveria lauci* and *Echeveria elegan*. *J. Plant Biotechnol.* 46(1), 22–31. doi:10.5010/JPB.2019.46.1.022.

Kim, Y. H., Lee, G. Y., Kim, H. H., Lee, J. H., Jung, J. H. and Lee, S. D. 2019b. *In vitro* mass propagation and acclimatization of *Haworthia truncate*. *J. Plant Biotechnol.* 46(2), 127–35. doi:10.5010/JPB.2019.46.2.127.

Kluge, M. and Ting, I. P. 1978. *Crassulacean Acid Metabolism: an Ecological Analysis*. Ecological Studies Series vol. 30. Springer, Berlin.

Laliberté, B. 1997. Botanic garden seed banks/genebanks worldwide, their facilities, collections and network. *Bot. Gard. Conserv. News* 2(9), 18–23.

Laney, A. G., Keller, K. E., Martine, R. R. and Tzanetakis, I. E. 2011. A discovery 70 years in th emaking: characterization of the Rose rosette virus. *J. Gen. Virol.* 92(7), 1727-32. doi:10.1099/vir.0.031146-0.

Lauterbach, D., Burkart, M. and Gemeinholzer, B. 2012. Rapid genetic differentiation between *ex situ* and their *in situ* source populations: an example of the endangered *Silene otitis* (Caryophyllaceae). *Bot. J. Linn. Soc.* 168(1), 64–75. doi:10.1111/j.1095-8339.2011.01185.x.

Lee, S. D., Kim, S. J., Jung, S. I., Son, K. C. and Kays, S. J. 2006. Diurnal $CO_2$ assimilation patterns in nine species of CAM-type succulent plants. *HortScience* 41(6), 1373–6. doi:10.21273/HORTSCI.41.6.1373.

Lee, J. H., Lee, H. K., Hong, S. M. and Song, C. Y. 2014. Effect of tray height and irrigation hole on the growth of grafted cactus (*Gymnocalycium mihannovichii* var. *friedrichii*) in hydroponic culture. *Flower Res. J.* 22(1), 1–6. doi:10.11623/frj.2014.22.1.1.

Lee, G. Y., Lee, J. H. and Kim, H. H. 2019a. A new cultivar of *Euphorbia milii* 'Dual Pink' with pink and yellow Cyanthophyll. *Hortic. Sci. Technol.* 37(Suppl. I).

Lee, J. H., Lee, J. H., Soh, H. S., Lee, S. D. and Hong, S. M. 2019b. A new grafted cactus *Astrophytum spp.* 'Snow Gem' with white hairy scales. *Hortic. Sci. Technol.* 37(Suppl. I), 203–4.

Li, X., Gasic, K., Cammue, B., Broekaert, W. and Korban, S. S. 2003. Transgenic rose lines harboring an antimicrobial protein gene, Ace-AMP1, demonstrate enhanced resistance to powdery mildew (*Sphaerotheca pannosa*). *Planta* 218(2), 226–32. doi:10.1007/s00425-003-1093-5.

Linde, M., Mattiesch, L. and Debener, T. 2004. *Rpp1*, a dominant gene providing race-specific resistance to rose powdery mildew (*Podosphaera pannosa*): molecular mapping, SCAR development and confirmation of disease resistance data. *Theor. Appl. Genet.* 109(6), 1261–6. doi:10.1007/s00122-004-1735-4.

Long, C., Chen, Z., Zhou, Y. and Long, B. 2018. The role of biodiversity and plant conservation for ornamental breeding. In: Van Huylenbroech, J. (Ed.), *Ornamental Crops*. Springer, Berlin, Germany, p. 9.

Mann, R. 2002. *Roses: a Practical Guide to Over 30 Roses for Australia and New Zealand*. HarperCollins, Pymble, NSW, p. 464.

Marchant, R., Power, J. B., Lucas, J. A. and Davey, M. R. 1998. Biolistic transformation of rose (*Rosa hybrida* L.). *Ann. Bot.* 81(1), 109–14. doi:10.1006/anbo.1997.0538.

Marshall, D. R. and Brown, A. H. D. 1975. Optimum sampling strategies in genetic conservation. In: Frankel, O. H. and Hawkes, J. H. (Eds), *Crop Genetic Resources for Today and Tomorrow*. Cambridge University Press, Cambridge, UK, pp. 53–80.

Maxted, N., Ford-Lloyd, B. V. and Hawkes, J. G. 1997a. Complementary conservation strategies. In: Maxted, N., Ford- Lloyd, B. V. and Hawkes, J. G. (Eds), *Plant Genetic Conservation: the In Situ Approach*. Chapman and Hall, London, UK, pp. 15–39.

Maxted, N., Hawkes, J. G., Ford-Lloyd, B. V. and Williams, J. T. 1997b. A practical model for *in situ* genetic conservation. In: Maxted, N., Ford-Lloyd, B. V. and Hawkes, J. G. (Eds), *Plant Genetic Conservation: the In Situ Approach*. Chapman and Hall, London, UK, pp. 339–63.

Mutke, J. and Barthlott, W. 2005. Patterns of vascular plant diversity at continental to global scales. *Biol. Skr.* 55, 521–31.

NAC (National Agrobiodiversity Center). 2019. Available at: https://www.naas.go.kr:2361/english/research/dep10.do (accessed on 17 April 2019).

NARO (National Agriculture and Food Research Organization). 2019. Available at: http://www.naro.affrc.go.jp/english/index.html (accessed on 26 April 2019).

Nobel, P. S. 1978. Surface temperatures of cacti: influences of environmental and morphological factors. *Ecology* 59(5), 986–95. doi:10.2307/1938550.

Nobel, P. S. 1996. Response of soma North American CAM plants to freezing temperature and doubled $CO_2$ concentrations: implications of global climate change for extending cultivation. *J. Arid. Environ.* 34(2), 187–96. doi:10.1006/jare.1996.0100.

Oda, J. L. M. and Lee, M. 2003. Grafting of herbaceous vegetable and ornamental crops. In: Janick, J. (Eds), Horticultural Reviews. John Wiley and Sons, New Jersey, USA, pp. 61–124.

Ogbu, J. U., Essien, B. A., Essien, J. B. and Anaele, M. U. 2010. Conservation and management of genetic resources of horticultural crops in Nigeria: issues and biotechnological strategies. *J. Hortic. For.* 2, 214–22.

OPGC (The Ornamental Plant Germplasm Center). 2019. Available at: https://opgc.osu.edu/(accessed on 25 April 2019).

Park, P. M., Jeong, M. I., Lee, D. W., Kim, Y. J., Kim, W. H., Lee, E. K. and Kim, S. T. 2008a. A new grafted cactus 'Dahong' with bright dark red color. *Flower Res. J.* 16, 196–9.

Park, P. M., Jeong, M. I., Lee, D. W., Kim, Y. J., Kim, W. H., Lee, E. K. and Kim, S. T. 2008b. A new grafted cactus 'Hwahong' with bright red color. *Korean J. Hortic. Sci. Tech.* 26, 48–50.

Park, P. M., Jeong, M. I., Kim, J. K., Kim, W. H., Lee, E. K., Kim, S. T. and Yoo, B. S. 2009. A new *Gymnocalycium mihanovichii* 'Gohong' with bright red color. *Flower Res. J.* 17, 197–200.

Park, P. M., Kim, W. H., Lee, E. K., Kim, S. T., Jeong, M. I. and Yoo, B. S. 2010a. A new *Gymnocalyciummi hanovichii* 'Hwangun' with bright yellow color. *Flower Res. J.* 18, 291–4.

Park, H. B., Nam, S. Y., Chung, J. W., Lee, J. J. and Park, I. T. 2010b. A new *Euphobia* cultivar 'Panorama' with green and pink colored bract. *Korean J. Breed. Sci.* 42, 717–20.

Park, P. M., Jeong, M. I., Yae, B. W., Kim, M. S., Lee, Y. R., Park, P. H. and Yoo, B. S. 2012. A new grafted cactus with bright yellow color, 'Hwangseon'. *Korean J. Hortic. Sci. Technol.* 30, 342–4.

Park, P. M., Kim, M. S., Lee, Y. R., Park, P. H., Yoo, B. S. and Lee, J. S. 2014a. Breeding of Moon cactus 'Ilhong' with dark red and good propagation. *Flower Res. J.* 22(4), 274–7. doi:10.11623/frj.2014.22.4.5.

Park, P. M., Yae, B. W., Kim, M. S., Lee, Y. R., Park, P. H., Jeong, M. I. and Yoo, B. S. 2013. Breeding of bicolor cultivar cv. 'Saekjo' in Moon cactus. *Flower Res. J.* 21(1), 35–7, doi:10.11623/frj.2013.21.1.4.

Park, P. M., Kim, M. S., Lee, Y. R., Park, P. H., Yoo, B. S. and Lee, J. S. 2015. Breeding of bicolor moon cactus 'Isaek' with red and yellow colors. *Flower Res. J.* 23(3), 208–11. doi:10.11623/frj.2015.23.3.34.

Park, P. M., Kwon, O. K., Park, P. H., An, H. R. and Lee, S. Y. 2018. Breeding of moon cactus 'Yeonbit' with pink color. *Flower Res. J.* 26(3), 161-5. doi:10.11623/frj.2018.26.3.12.

Park, P. M., Kwon, O. K., Park, P. H., An, H. R. and Lee, S. Y. 2019. Breeding of moon cactus 'Bulbit' with dark red color and excellent propagation. *Curr. Germplasm Release*. doi:10.7235/HORT.20190066.

Phillips, R. and Rix, M. 1988. *Roses*. Random House, p. 224.

Plant Heritage. 2018. The national plant collections 2018 directory. *Plant Heritage*, March 2018, 162p.

Postman, J., Hummer, K., Stover, E., Krueger, R., Forsline, P., Grauke, L. J., Zee, F., Ayala-Silva, T. and Irish, B. 2006. Fruit and nut genebanks in the U.S. National Plant Germplasm System. *HortScience* 41(5), 1188–94. doi:10.21273/HORTSCI.41.5.1188.

Ramawat, K. G. 2010. *Desert Plants.* Springer, Heidelberg, Germany and Dordrecht, the Netherlands.

Ranjan, P., Ranjan, J. K., Misra, R. L., Dutta, M. and Singh, B. 2016. Cacti: notes on their uses and potential for climate change mitigation. *Genet. Resour. Crop Evol.* 63(5), 901–17. doi:10.1007/s10722-016-0394-z.

Rao, N. K. and Jackson, M. T. 1996. Seed production environment and storage longevity of japonica rices (*Oryza sativa* L.). *Seed Sci. Res.* 6(1), 17–21. doi:10.1017/S0960258500002956.

Rebman, J. P. and Pinkava, D. J. 2001. *Opuntia* cacti of North America: an overview. *Fla Entomol.* 84(4), 475–83. doi:10.2307/3496374.

Reed, B. M. 2001. Implementing cryogenic storage of clonally propagated plants. *Cryo Lett.* 22(2), 97–104.

Rehder, A. 1949. *Bibliography of Cultivated Trees and Shrubs in the Cooler Temperate Regions of the Northern Hemisphere*. Arnold Arboretum.

Rout, G. R., Samantaray, S., Mottley, J. and Das, P. 1999. Biotechnology of the rose: a review of recent progress. *Sci. Hortic.* 81(3), 201–28. doi:10.1016/S0304-4238(99)00025-4.

Sackville Hamilton, N. R. and Chorlton, K. H. 1997. *Regeneration of Accessions in Seed Collections: A Decision Guide*. Handbook for Genebanks 5. IPGRI, Rome, Italy.
Saenz, C., Sepulveda, E. and Matsuhiro, B. 2004. *Opuntia* spp. mucilage's: a functional component with industrial perspectives. *J. Arid. Environ.* 57(3), 275–90. doi:10.1016/S0140-1963(03)00106-X.
Schoen, D. J. and Brown, A. H. D. 2001. The conservation of wild plant species in seed banks. *BioScience* 51(11), 960–6. doi:10.1641/0006-3568(2001)051[0960:TCOWPS]2.0.CO;2.
Seymour, K. 2012. Conserving cultivars. *Plantsman* 11(3), 154–9.
Simmonds, N. W. 1962. Variability in crop plants, its use and conservation. *Biol. Rev.* 37(3), 422–65. doi:10.1111/j.1469-185X.1962.tb01620.x.
Soh, H. S., Lee, J. H., Lee, J. H., Lee, S. D. and Hong, S. M. 2019. A new variety of grafted cactus (*Gymnocalycium mihanovichii*) 'Black Sun' with red patterned color. *Hortic. Sci. Technol.* 37(Suppl. I).
Song, C. Y. and Song, E. K. 2012. Effects of indoor position and LED on growth of two grafted cacti and one succulent planted in a pot. *Flower Res. J.* 20, 64–70.
Song, C. Y., Lee, S. D., Kim, S. J. and Lee, J. S. 2006. Effects of shading and watering intervals on survivals and growth during acclimatization of cacti and succulents mingled planting in a pot. *Korean J. Hortic. Sci. Technol.* 42, 81–9.
Song, C. Y., Lee, S. D., Park, I. T. and Cho, C. H. 2007. Effect of media and planting depth on growth of cacti and succulents in a pot. *Korean J. Hortic. Sci. Technol.* 25, 429–35.
Souq, F., Coutos-Thevenot, P., Yean, H., Delbard, G., Maziere, Y., Barbe, J. P. and Boulay, M. 1995. Genetic transformation of roses, 2 examples: one on morphogenesis, the other on anthocyanin biosynthetic pathway. In: *II International Rose Symposium 424*, pp. 381–8. doi:10.17660/ActaHortic.1996.424.73.
Tao, K. L. 2001. Complementary conservation strategy for plant genetic resources. In: Said Saad, M. and Ramanatha Rao, V. (Eds), *Establishment and Management of Field Genebank: a Training Manual*. IPGRI-APO, Serdang, Malaysia, p. 46.
Tay, D., Widrlechner, M. P. and Corfield, J. L. 2004. Establishment of a new genebank for herbaceous ornamental plants. *FAO/IPRGI Plant Genet. Resour. Newsl.* 137, 26–33.
Towill, L. E. 1985. Low temperature and freeze-vacuum-drying preservation of pollen. In: Kartha, K. K. (Ed.), *Cryopreservation of Plant Cells and Organs*. CRC Press, Boca Raton, FL, pp. 171–98.
Towill, L. E. and Bonnart, R. 2003. Cracking in a vitrification solution during cooling or warming does not affect growth of cryopreserved mint shoot tips. *Cryo Lett.* 24(6), 341–6.
Towill, L. E. and Widrlechner, M. 2004. Cryopreservation of *Salix* species using sections from winter vegetative scions. *Cryo Lett.* 25(1), 71–80.
UNEP. 1992. Convention on biological diversity. UNEP, Nairobi, Kenya.
USDA. 2018. National genetic resources program. Germplasm Resources Information Network-Global (GRIN-Global). USDA-ARS National Germplasm Resources Laboratory, Beltsville, MD. Available at: https://www.ars-grin.gov/npgs/collections.html (accessed on 17 April 2019).
Wang, D., Fan, J. and Ranu, R. S. 2004. Cloning and expression of 1-aminocyclopropane-1-carboxylate synthase cDNA from rosa (*Rosa* × *hybrida*). *Plant Cell Rep.* 22(6), 422–9. doi:10.1007/s00299-003-0721-7.

Whitlock, R., Hipperson, H., Thompson, D. B. A., Butlin, R. K. and Burke, T. 2016. Consequences of in-situ strategies for the conservation of plant genetic diversity. *Biol. Conserv.* 203, 134–42. doi:10.1016/j.biocon.2016.08.006.
Widrlechner, M. P. 2009. Ornamental plants and the U.S. National Plant Germplasm System: conserving, evaluating, seeking, and sharing. *Combined Proceedings International Plant Propagators' Society* (vol. 58).
Widrlechner, M. P., Abel, C. A. and Wilson, R. L. 1997. Ornamental seed production in field cages with insect pollination. *Comb. Proc. Intl. Plant Prop. Soc.* 46, 512–6.
Withers, L. A. 1995. Collecting in vitro for genetic resources conservation. In: Guarino, L., Rao, V. R. and Reid, R. (Eds), *Collecting Plant Genetic Diversity. Technical Guidelines.* CAB International, Wallingford, UK, pp. 511–25.
Yan, Z., Denneboom, C., Hattendorf, A., Dolstra, O., Debener, T., Stam, P. and Visser, P. B. 2005. Construction of an integrated map of rose with AFLP, SSR, PK, RGA, RFLP, SCAR and morphological markers. *Theor. Appl. Genet.* 110(4), 766–77. doi:10.1007/s00122-004-1903-6.
Zhang, L. 2003. Genetic linkage mapping in tetraploid and diploid rose. PhD Thesis. Clemson University, pp. 1–160.
Zieliński, J., Petrova, A. and Tan, K. 2004. Taxonomic status of the roses (*Rosa*) described by S. G. Dimitrov from Bulgaria. *Ann. Bot. Fenn.* 41, 449–51.

# Chapter 3

# Advances in conventional breeding techniques for ornamentals

*Traud Winkelmann and Philipp Braun, Leibniz Universität Hannover, Germany; and Emmy Dhooghe and Johan van Huylenbroeck, Flanders Research Institute for Agriculture, Fisheries and Food, Belgium*



## 1 Introduction

Sustainability in the floricultural value chain comprises a multitude of aspects, several of which can be addressed in breeding programs. Sustainability covers the economical use of resources, mainly energy, water, nutrients, and substrate ingredients such as peat, as well as the avoidance of environmental pollution by plastic wastes, nutrients, and pesticides (Mariotti and Roccotiello, 2013; Bermejo Dominguez et al., 2017); in its social dimension, sustainability also addresses the well-being of employees, who are becoming increasingly important in times of skilled labor shortage (Meyerding, 2017). Many of these topics can be and have been approached by technological innovations during production, transportation, and storage (Bermejo Dominguez et al., 2017), but the breeding of improved cultivars seems to be a very efficient strategy for quite a number of these goals as discussed in the subsequent section.

For the majority of ornamental plants, breeding still mainly relies on the conventional methods of cross breeding and mutation breeding (Dohm, 2017). The germplasm of many species and their relatives has not been fully explored, and consequently breeders currently also succeed in creating innovations with these classical breeding methods. However, since many ornamentals are complex hybrids with polyploidy and a highly heterozygous character, some goals may be difficult to achieve by classical cross breeding, and modern approaches may be helpful.

http://dx.doi.org/10.19103/AS.2020.0066.03

Within this chapter, we will briefly review advanced conventional breeding techniques, namely cross breeding and selection, interspecific and intergeneric hybridization, and somatic and gametic polyploidization, before discussing their value in reaching breeding goals in the context of sustainable floricultural production.

## 2 Current common breeding techniques in ornamentals

### 2.1 Conventional cross breeding and selection

Ornamental plant breeding became an objective of private seed companies in the second half of the nineteenth century (Horn, 2002). The selection of novel phenotypes in progenies from controlled crosses or even open pollination has been and still is a successful approach in many ornamental crops due to the high levels of heterozygosity observed in the majority of species. Even for the economically most important crop in floriculture worldwide, the rose, cross breeding is the prevalent breeding technology. The high variability in the germplasm of multispecies hybrid modern roses (Gudin, 2003; Chaanin, 2003) has resulted in thousands of released varieties. Conversely, breeding has resulted in a narrowing of the gene pool of cultivated roses, especially cut roses, which may limit the future potential of conventional breeding (Leus et al., 2018).

Breeding strategies and technologies differ depending on the techniques used for commercial propagation of cultivars. Breeders have become specialized either toward vegetatively propagated ornamentals or seed propagated ornamental crops (Horn, 2002), with the former having higher commercial relevance. If selected elite plants matching the breeding goals can be propagated by cuttings, grafting, or in vitro culture, maintenance breeding mainly includes measures to prevent the spread of disease and mutations. In the case of seed-propagated plants, the reproductive system must be considered; most ornamentals are outbreeding species (Horn, 2002). Predominant breeding methods for seed-propagated ornamentals are mass selection and $F_1$ hybrid breeding (Horn, 2002).

Parental plants for crosses are selected based on their phenotypes and according to the breeding goals, which can be divided into external characters, such as flower color, shape, and size; growth habit; leaf color, shape, and size; and internal characters, such as pest and disease resistance, abiotic stress tolerance, fragrance, longevity, sterility, and repeated, continued flowering, among others (Kuligowska et al., 2016). Many traits of interest when considering sustainability are internal ones that are controlled by more than one gene. For several of these, selection could ideally be supported by the use of molecular markers to choose parents (Debener, 2002; Smulders and Arens, 2018).

The execution of crossings in breeding programs typically involves the selection of maternal flowers in an early developmental stage and hand emasculation (excluding self-incompatible species) before pollen shedding as first steps. The pollen of the paternal crossing partner may be collected, tested for viability, dried and stored if necessary, and is transferred to the stigma of the pollen recipient at full stigma receptivity. In some species, such as *Cyclamen persicum*, repeated pollination may be recommended due to successive ripening of the ovules. To prevent pollination by other pollens, isolation bags are used in several species. For seed harvest, processing, and sowing, a species-dependent methodology is applied.

Breeders are interested in reducing the number of seedlings, typically several tens to hundreds of thousands raised in a breeding program as early as possible. However, early selection is possible only for a minority of traits that can be visualized in early seedling stages. Many important breeding goals, for example, those concerning flower characters, require adult plants to select the desired phenotypes and others even demand sophisticated testing setups, such as tests for pest and disease resistance or low temperature tolerance (see Section 3). Selection for these traits is typically carried out in later stages with only a limited number of genotypes or lines. Again, molecular markers would be extremely helpful to facilitate selection (Debener, 2002; Smulders and Arens, 2018), but in many cases, they are either not yet available for the trait of interest or not implemented by ornamental plant breeders.

## 2.2 Interspecific and intergeneric hybridization

Interspecific and intergeneric hybridization are major drivers toward creating breakthrough innovations in a specific plant genus or species. They enable the introgression of novel genes (e.g. biotic or abiotic stress resistance, novel colors, and different plant architecture) that are absent in the target species. In many ornamentals, several species form the genetic background of the actual breeding gene pool. The breeding history of economically important ornamental crops, such as rose, chrysanthemum, orchids, lily, tulip, and many others, confirms that interspecific hybridization has commonly served as the basis for most of today's commercially available assortment (Marasek-Ciolakowska et al., 2018). However, in modern advanced breeding programs, such as those for rose or chrysanthemum, the use of wild relatives to introgress new traits is very difficult, requiring great efforts and time because several backcrosses are necessary after obtaining the first hybrid generation. Therefore, commercial companies sometimes do not invest enough in prebreeding work. Public-private partnerships in which the initial prebreeding work is performed in collaboration with a public research institute may help tackle this problem. Although such long-term partnerships for prebreeding work have been realized in agricultural crops, on the one hand, they are rather rare in ornamentals due to the high diversity and limited economic revenue of such specimens and, on the other hand, due to a decreasing number of research institutions with programs in ornamental horticulture.

Interspecific hybridization can occur in nature. However, artificial pollination between species is typically necessary. As a general rule, hybridization among species belonging to the same taxonomic section or closely related ones can be achieved by normal pollination. With greater genetic distances between species or genera, success rates will drop quickly or hybridization will become impossible (Table 1). Current molecular techniques enable the calculation of genetic distances between different species, which is useful information for predicting the chances of successful pollination, as shown in *Helleborus*

**Table 1** Different steps in an interspecific breeding program with the bottlenecks that might occur or require examination and the respective techniques that can be used

| Steps in an interspecific breeding program | Which problem needs to be solved/studied? | Which (scientific) technique can be used to investigate or to solve the occurring problem? |
|---|---|---|
| 1. Selection of breeding population/ germplasm | • Characterization of genetic distances | • Molecular marker technology<br>• Chromosome counts<br>• Karyotyping<br>• Genome size measurements<br>• Ploidy level measurements |
| 2. Flowering | • Differences in flowering period<br>• Flower induction<br>• Flower biology | • Controlled growth conditions<br>• Pollen storage<br>• Dissection of meristem<br>• Literature/knowledge on flower structure and natural pollination |
| 3. Pollination | • Pollen quality<br>• Female receptivity | • Pollen germination (in vitro)<br>• Pollen viability tests<br>• Controlled emasculation<br>• Categorization in flower stages for female receptivity and male maturation |
| 4. Pollen germination, pollen tube growth, fertilization | • Prezygotic barriers<br>• Incompatibility | • Aniline blue staining<br>• Mentor or mixed pollen<br>• Style cutting and bud or stump pollination<br>• Stigma treatment (e.g. with plant growth regulators)<br>• Crossings at higher temperatures<br>• Crossings at different physiological ages<br>• Reciprocal crosses |
| 5. Seed development | • Postzygotic barriers | • Immature seed harvest and in vitro maturation<br>• Embryo rescue<br>• Ovary slice culture |
| 6. Germination and growth | • Hybridization success<br>• Hybrid breakdown<br>• Hybrid vigor | • Molecular marker technology<br>• Morphological screening<br>• Genomic In Situ Hybridization (GISH)<br>• Reciprocal crosses |
| 7. Hybrid selection | • Fertility restoration | • Polyploidization |

by Meiners and Winkelmann (2012) or in *Hydrangea* by Granados Mendoza et al. (2013). Breeding barriers can be situated at different levels. In nature, the spatial distribution of different species can form a physical barrier. In addition, differences in flowering period avoid pollination and prevent interspecies crosses. In practical breeding, these barriers can easily be circumvented by growing plants under controlled environmental conditions. Crossing failures can be due to existing pre- or postzygotic barriers (for reviews see Eeckhaut et al., 2006; Kato and Mii, 2012). Prezygotic barriers include differences in chromosome numbers, ploidy levels, and features that hamper successful pollen growth through the style of a different species or genus. To overcome this stigmatic incompatibility, a number of different techniques such as cut-style pollination, intrastylar pollination, use of mentor pollen or hormones, and in vitro pollination can be applied (Van Tuyl and De Jeu, 1997). A textbook example of the successful application of all these methods is observed in lily breeding, resulting in a huge new palette of lily hybrids in the market (Van Tuyl et al., 2018).

Even when fertilization is successful, postzygotic barriers might appear due to lack of endosperm development, embryo abortion, albinism, sterility, or lack of growth vigor. In such cases, embryo rescue, ovary culture, or ovule culture can be useful tools (Winkelmann et al., 2010). In hybrid plants, fertility problems frequently limit further use in breeding. This last hurdle can be overcome by chromosome doubling of the interspecific hybrid, allowing the restoration of fertility. If no hybrid plants can be generated via sexual crossing, then protoplast fusion can be a final solution. However, the number of examples of successful protoplast fusion products that are further used in breeding is very limited (Eeckhaut et al., 2013; Bridgen et al., 2018).

Therefore, when starting a new breeding program in a new genus, it is recommended to collect basic information on genetic relationships, chromosome numbers, and DNA content, among other features, in addition to essential morphological information on pollination and seed development.

## 2.3 Polyploidization

Polyploidization can be divided into (1) somatic or mitotic and (2) gametic or meiotic polyploidization (Ramsey and Schemske, 1998). The goals for ploidy breeding are often similar - to induce new variation or to facilitate crosses. For mitotic polyploidization, ploidy of somatic tissue is doubled, while meiotic polyploidization results in the production of 2n gametes. By manipulating the ploidy level in plants, changes in plant and flower morphology are often observed (Dhooghe et al., 2011). Therefore, ploidy breeding is of special interest for ornamentals (Eeckhaut et al., 2018).

Somatic polyploidization often uses plantlets from tissue culture; however, this approach is not compulsory. Different explant types can be used; the most popular are shoots or shoot meristems (Dhooghe et al., 2011; Eeckhaut et al., 2018). These explants are treated with mitosis-arresting chemicals - colchicine is the most commonly used, followed by oryzalin. Mitotic polyploidization has demonstrated several positive effects on plant morphology, with the most noted effects being increased leaf thickness, increased width/length ratio of the leaves, and a more intense green color of the leaves (Dhooghe et al., 2011; Eeckhaut et al., 2018). In *Petunia hybrida*, polyploidization causes dwarfing and stunted growth (Ning et al., 2019), and similar results for plant compactness have been observed in polyploidization experiments for *Escallonia* (Denaeghel et al., 2018). Additionally, alterations in flower morphology are frequently detected. The most frequently observed morphological changes are larger flowers, a greater number of petals per flower, and deeper-coloration of the flowers (Dhooghe et al., 2011; Eeckhaut et al., 2018). An important quality criterion for cut flowers is the strength of the stems. Polyploidization in lily results in stronger stems (Van Tuyl and Van Holsteijn, 1996). Apart from a direct morphological effect, which can be of interest in breeding, polyploidization can also result in invisible but important selection criteria, such as increased abiotic stress tolerance. This phenomenon was observed for *Spathiphyllum,* in which polyploid plants showed increased drought tolerance (Eeckhaut et al., 2004), and for *Dendranthema,* where polyploidization improved tolerance to cold, drought, and salinity (Liu et al., 2011).

All these examples demonstrate that polyploidization can induce variation. However, polyploidization can also be used as a breeding tool to extend hybridization. It can be used to overcome interspecific or intergeneric barriers or facilitate crosses, by bringing crossing partners to an equal ploidy level, or to overcome (or induce) sterility in triploid plants. For example, *Hibiscus syriacus* 'DVPAzurri' is a commercial, sterile cultivar that was obtained after a cross of *Hibiscus syriacus* 'Oiseau Bleu' and a chromosome doubled 'Oiseau Bleu' (Van Laere et al., 2006). *Lavandula × intermedia,* commonly known as the lavandins, are naturally occurring hybrids of *Lavandula angustifolia* and *Lavandula latifolia* and are sterile. Urwin (2014) restored fertility to lavandins by polyploidization with colchicine, although all other plant characteristics remained similar to the parents. This fertility creates the opportunity to broaden the hybridization possibilities within this genus. Similarly, fertility could be restored for wide interspecific lily hybrids by somatic chromosome doubling techniques (Van Tuyl et al., 2002). Chromosome doubling can facilitate interspecific breeding when parent species differ in ploidy level. For *Buddleja globosa* (2n = 2x = 38), chromosome doubling was undertaken to allow interspecific hybridization with *Buddleja davidii* (2n = 4x = 76) (Van Laere et al., 2011b).

Undeniably, somatic polyploidization approaches have contributed substantially to the development of plant cultivars. However, the formation of naturally occurring polyploids, which plays a crucial role in plant diversification (Schatlowski and Köhler, 2012), originates mostly from unreduced gametes

(Ramsey and Schemske, 1998). The spontaneous occurrence of male or female gametes that possess the somatic chromosome number (or higher) is caused not only by aberrations of meiosis I (First Division Restitution, FDR) or II (Second Division Restitution, SDR) but also by premeiotic and postmeiotic doubling (Lelley et al., 1987; Falistocco et al., 1995; De Storme and Geelen, 2013a,b). The utilization of unreduced gametes has great potential for plant breeding (Dewitte et al., 2011), and several experimental studies have already been conducted in this context (Younis et al., 2014). Unreduced gametes have the potential to overcome ploidy-based hybridization barriers (Schatlowski and Köhler, 2012). Another advantage of gametic polyploidization approaches is that meiotic recombination results in greater heterozygosity and allelic diversity among the generated progeny (Bretagnolle and Thompson, 1995). This aspect is of particular importance in ornamental plant breeding, which is reliant on genetic variability that becomes increasingly difficult to access as a result of narrowing gene pools in a target species and due to restricted access to germplasm due to the Nagoya protocol (Dohm, 2017).

Spontaneous formation of unreduced gametes has been detected in many plant species, but their frequency is often too low to use them efficiently in cross pollinations. In experimental studies, which have mostly focused on male gametes because of their generally greater numbers and easier accessibility, the formation of unreduced gametes was stimulated by heat (Negri and Lemmi, 1998; Pécrix et al., 2011), cold (Mason et al., 2011), colchicine, oryzalin, trifluralin, and $N_2O$ (Younis et al., 2014). In this context, screening for unreduced male gametes and determination of their percentages among the total pollen volume are a challenging necessity. Various methods have been applied for such screenings, each with advantages and disadvantages (Dewitte et al., 2011).

As unreduced pollen grains are typically larger than those possessing the reduced chromosome number (Ramsey and Schemske, 1998), morphological screenings using a microscope are a simple approach to detect 2n pollen (Bretagnolle and Thompson, 1995). However, such screenings are time consuming and do not always obtain clear-cut results due to the occurrence of naturally varying pollen diameters (Bell, 1959; Jansen and Den Nijs, 1993; Aguilar-García et al., 2012; Braun and Winkelmann, 2015). Microscopic analysis of microsporogenesis provides insights into the formation of unreduced gametes (Crespel et al., 2006; Dewitte et al., 2009; Pécrix et al., 2011; De Storme and Geelen, 2011), but it is not applicable for the determination of their frequency. The DNA content of sperm nuclei in pollen tubes can be measured and compared by microfluorometry (Friedman, 1999), but the number of samples that can potentially be analyzed in this way is limited. Flow cytometry (FCM) is a tool for the automation of this technique, and it was therefore suggested as an efficient method to detect unreduced male gametes (Van Tuyl et al., 1989; Bino et al., 1990). Unfortunately, recent studies have shown that persistent junctions between sperm cells and vegetative pollen nuclei (referred

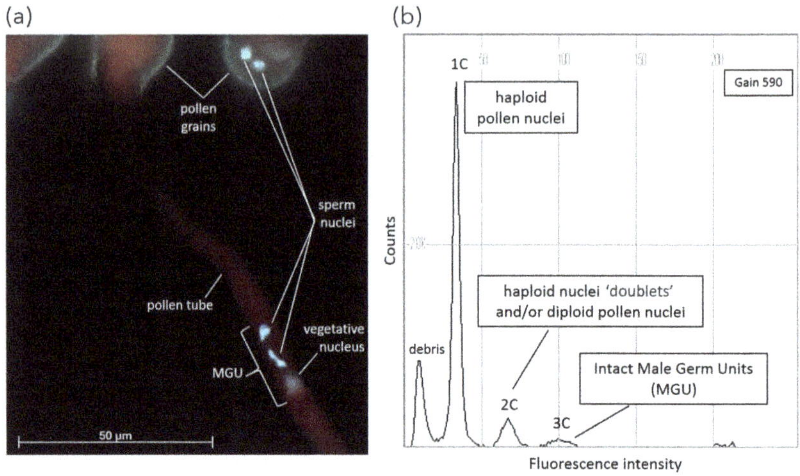


**Figure 1** (a) DAPI-stained *Delosperma* (Aizoaceae) pollen nuclei in an in vitro-grown pollen tube. The persistent connections between the nuclei that form the MGU are not visible in this micrograph. (b) Detection of pollen nuclei 'doublets' and MGU in a flow cytometric screening for unreduced *Lampranthus* (Aizoaceae) gametes.

to as Male Germ Units, MGU) can lead to false-positive screening results in flow cytometric measurements (Hirano and Hoshino, 2009; Kron and Husband, 2015; Braun and Winkelmann, 2016; Fig. 1). For this reason, the dissociation of MGUs prior to FCM measurements is required to avoid recordings of 2C- and 3C signals caused by associated pollen nuclei. A further method to verify the presence of unreduced gametes is the interploidy cross followed by progeny analysis (Bretagnolle and Thompson, 1995). Depending on the plant species of interest, this can be a time-consuming process and provides only limited information about the frequency of 2n gamete formation.

## 3 Conventional breeding for sustainable production of ornamentals

Sustainability is 'becoming an increasingly important breeding target' by ornamental plant breeders (Dohm, 2017) and encompasses the resources needed for production and the reduced application of pesticides and chemicals for postharvest treatments. In the following section, we select a number of breeding goals that will pave the way for more sustainable production of ornamental plants.

### *3.1 Drought tolerance*

Most modern experts accept anthropogenic climate change and the accompanying increasing probability of drought events on our planet as a fact (Li et al., 2009; Dai, 2013; Cook et al., 2016). Prospectively, water conservation will therefore also play a more prominent role in the ornamental plant industry. To facilitate the sustainable handling of water resources and to retain consumer acceptance for ornamentals, the efforts of all value chain participants are required. Initially, commercial plant breeders are challenged to develop drought-tolerant assortments and cultivars, which can be produced by water-saving practices and withstand weather extremes at the consumer end.

The widely used concept of drought adaptation mechanisms in plants comprises three categories, namely drought escape, drought avoidance (or dehydration avoidance), and drought tolerance, sensu stricto (Kumar et al., 2012; Rauf et al., 2016). Drought escape is defined as the ability of a plant to complete its life cycle before the onset of drought stress (Rauf et al., 2016). In natural populations of herbaceous plants, drought escape is mainly restricted to annuals and flowering time is often the primary associated trait (Kooyers, 2015). Most genetic mechanisms underlying drought escape remain uncharacterized (Kooyers, 2015), but experiments in *Arabidopsis* have shown that reduced water availability influences the signaling of abscisic acid (ABA) and several genes involved in flower induction (Riboni et al., 2013; Han et al., 2013). In contrast to vegetable and agricultural crop breeding, in which early fruit maturation and harvest are sometimes targeted traits (Kumar et al., 2012), ornamental breeding usually aims at developing cultivars with long life spans and flowering periods. Strategies of plants to escape from unfavorable environmental conditions therefore mainly manifest in seasonal production and sales, for example, in spring flowering ornamentals such as *Primula sp.* and *Viola sp.*

Drought avoidance is described as the ability of plants to maintain tissue-water potential despite a shortage of soil moisture (Dhakar et al., 2017), and it occurs in annuals, biennials, and perennials. This strategy is expressed in morpho-physiological traits such as limited vegetative growth, higher root-to-shoot growth ratios prior to drought conditions, reduced stomatal and epidermal transpiration, early vigor, and osmotic adjustments (Kooyers, 2015; Dhakar et al., 2017; Rauf et al., 2016). Plant characteristics that enable the reduction of transpirational water losses are a smaller leaf area, greater succulence, increased leaf reflectance, accentuated leaf lobing, and an altered stomatal size and density (Kooyers, 2015; Eggli and Nyffeler, 2009). As for drought escape, multiple quantitative trait loci (QTL) associated with drought avoidance responses have been identified in model plants, but their specific phenotypic effects remain largely unknown (Kooyers, 2015).

The third category of drought response mechanisms, namely drought tolerance, sensu stricto, is defined as the relative capacity to sustain or conserve plant function in a dehydrated state (Blum, 2005). It comprises increased cell elasticity, decreased cell size, the accumulation of osmoprotectants, and again osmotic adjustment (Kumar et al., 2012). Several genes with upregulated expression in response to dehydration have been identified in genetic model systems and crop plants (Ingram and Bartels, 1996; Hu and Xiong, 2014).

One option for ornamental breeders to develop sustainable cultivars could be a stronger focus on plant families and genera that express one or more drought avoidance traits and have high aesthetic value. The plant family Aizoaceae (midday flowers), for instance, which is mainly distributed in arid and semi-arid regions of southern Africa (Hartmann, 2001), includes many species with attractive flowers and morpho-physiological adaptations that reduce transpirational water losses (Fig. 2). Many midday flowers can switch their mode

**Figure 2** Attractive flowers and morpho-physiological adaptations that reduce transpirational water losses in midday flowers (Aizoacaea): Flowers of two *Lampranthus* genotypes ((a), (b)), Succulent *Drosanthemum* young plant (c) covered with reflective bladder cells (d); Ombrohydrochoric seed dispersal (storage in dry capsules, capsules open after moistening, seed expulsion by raindrops) in *Delosperma*: Closed dry capsule (e) and open capsule after moistening (f).

of $CO_2$ fixation from $C_3$ to Crassulacean acid metabolism (CAM); Winter, 1973; Adams et al., 1998; Matimati et al., 2012), and various types of leaf succulence can be found within the family (Ihlenfeldt, 1960). Succulence is regarded as a drought avoidance strategy that enables plants to temporarily survive arid conditions. Succulent tissues store water to maintain active metabolism and to delay desiccation during periods without rainfall (Eggli and Nyffeler, 2009). Members of the Aizoaceae have relatively small genome sizes (De Rocher et al., 1990; Bennett and Leitch, 1995; Braun and Winkelmann, 2016). Because nuclear DNA contents and cell volumes are positively correlated in plants (Jovtchev et al., 2006), midday flowers and other succulent plants with small genome sizes utilize a deviant mode of the normal cell cycle (referred to as endoreduplication) to develop water-storing tissues (De Rocher et al., 1990; Del Angel et al., 2006; Braun and Winkelmann, 2015, 2016). During endoreduplication, nuclear DNA contents are doubled repeatedly in the absence of chromosomal separation and cytokinesis, leading to increased cell sizes (D'Amato, 1984; Breuer et al., 2014; Fig. 3).

Growns and Webb (2012) reported on an Australian long-term breeding program with native plants aimed at the identification of new ornamentals with climate adaptability and low water requirements. This program also included the development of intraspecific, interspecific, and intergeneric hybrids.

Polyploidization is another promising tool to reduce susceptibility to drought in ornamentals. Van Laere et al. (2011a) compared different morphological and physiological characteristics of diploid *Spathiphyllum* genotypes and their tetraploid derivatives. After the induction of drought stress, tetraploid plants showed fewer wilting symptoms, almost no changes in leaf water potential and relative water content (RWC), and a reduced increase in proline contents and

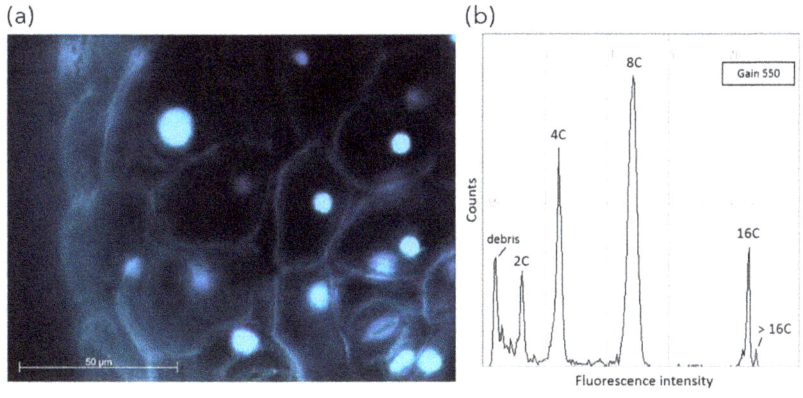


**Figure 3** (a) DAPI-stained epidermal cells of a *Lampranthus* (Aizoaceae) genotype showing differently sized cell nuclei. (b) Flow cytometric analysis of succulent *Lampranthus* leaf tissue revealing endoreduplicative cell nuclei with at least five different C-values.

stomatal resistance. The stomatal density of the tetraploids was significantly lower than in the diploid genotypes. Improved drought resistance has also been described in polyploid *Lonicera japonica,* where it was associated with a thicker leaf epidermis and palisade tissue, as well as a denser pubescence (Li et al., 2011).

The application of realistic drought conditions in experimental setups and selection screens is difficult because drought can combine the effects of simultaneously occurring environmental stressors such as water deficit, heat stress (Kooyers, 2015), and ozone stress. Considering the variety of plant drought responses, the selection of appropriate parameters to assess drought stress is not a simple task. Koźmińska et al. (2019) analyzed drought responses of ornamental *Sedum* species and revealed that a higher tolerance to water deficit is associated with (a) reduced degradation of chlorophyll and carotenoids, (b) constant malondialdehyde (MDA) levels, and (c) higher proline contents. They suggested that these parameters are appropriate biochemical markers when screening for drought tolerance. Franco et al. (2006) reviewed selection (and nursery production) methods for ornamental plants for landscaping or xerophytic gardening in semi-arid environments. They identified physiological characteristics that are applicable in seedling screening, such as low stomatal conductance, leaf water potential, leaf turgor potential, and RWC. Additional drought stress parameters are chlorophyll fluorescence, leaf gas exchange, and reactive oxygen species (ROS; Toscano et al., 2019).

### 3.2 Low temperature tolerance

Greenhouse cultivation of tropical and subtropical ornamental plants requires large amounts of heating energy, which is typically provided by fossil fuels and elevated greenhouse gas emission. Reduction of the cultivation temperature during greenhouse tomato production by only 2 K resulted in an energy savings of 16% (Elings et al., 2005). However, the growth of plants at temperatures below the optimal range results in chilling stress. Usadel et al. (2008) showed that *Arabidopsis thaliana* plants express an increasing cold stress response with decreasing temperatures from 17 to 8°C, but with the same principal changes in gene expression and metabolism. Low temperature stress ultimately causes reduced growth and prolongs the cultivation time due to disrupted photosynthesis, alterations in carbohydrate metabolism and phytohormone homeostasis, and accumulation of proteins, among others dehydrins, other late embryogenesis abundant proteins, heat shock proteins, and the accumulation of osmolytes (Ruelland and Collin, 2012; Bauerfeind, 2016). It is important to differentiate chilling injury, sensu stricto, that severely affects tropical and subtropical plants such as African Violet or orchids and causes symptoms like loss of water, discoloration up to necrotic lesions (Ruelland and Collin, 2012) from low temperature tolerance, that is, the maintenance of growth, and major physiological processes under suboptimal temperatures. The following considerations will focus on the latter case of low temperature tolerance in ornamental crops.

Attempts to select for chilling-tolerant ornamental plants were first published in the 1980s during the first oil crisis in Europe. For ornamentals that required high cultivation temperatures, research aimed to select tolerant genotypes and simultaneously to identify physiological markers for chilling tolerance. Enhanced proline contents at lower cultivation temperatures were observed in tolerant varieties of African violets (*Saintpaulia ionantha*) (Amberger et al., 1984). When comparing two poinsettia (*Euphorbia pulcherrima*) cultivars that differed in their low temperature tolerance, Tantau and Dörffling (1991) found that the tolerant genotype maintained its cytokinin level and more rapidly accumulated ABA.

Another background to assess low temperature tolerance in roses was the damage to outdoor plants in early spring or late autumn. Chlorophyll fluorescence dropped significantly in more sensitive genotypes than in the low-temperature-tolerant ones (Hakam et al., 2000). For ornamental pepper, Gajanayake et al. (2011) distinguished between cold tolerance during the vegetative growth phase and during reproduction. They suggested to test in vitro pollen germination at a low temperature to characterize low-temperature-tolerant genotypes in the generative phase, and calculated an index from a range of physiological parameters to evaluate the chilling tolerance during the vegetative phase.

In recent years, transcriptomic and metabolomic studies were undertaken to characterize the low temperature response, for example, in petunia (Bauerfeind et al., 2015) and chrysanthemum (Wang et al., 2018a). Based on these data, candidate genes can be deduced, such as those involved in oxidative stress protection, cold response signaling, osmotic adjustment, or membrane protection, which contribute to low temperature tolerance. Molecular markers linked to these candidate genes can then be used for marker-assisted selection. For cucumber, QTLs have been identified for tolerance to low temperatures during the germination phase (Song et al., 2018), but to the best of our knowledge, there are no comparable reports available thus far for ornamental crops.

Since polyploid plants often express a higher tolerance to abiotic stressors, polyploidization might also be an approach to improve low temperature tolerance, as reported, for example, for tetraploid *Citrus* rootstocks (Oustric et al., 2017); however, as polyploidy also affects several other traits and may result in slower growth, this remains to be investigated.

Thus, to date, genotypes of ornamental species have only been selected for good performance under low temperatures, whereas breeding for this trait involving controlled crosses, interspecific hybridization, or polyploidization have only scarcely been reported, so far. In contrast, breeding efforts to achieve chilling tolerance have been summarized by Patterson and Reid (1990) with examples given for chilling-tolerant intergeneric hybrids of orchids. Like many other stress tolerance traits, low temperature tolerance can be expected to be a highly complex trait - affecting different developmental phases in different ways (Ruelland and Collin, 2012) - that is controlled by a fine-tuned network of several pathways, and targeted breeding will not be an easy task. Furthermore, as also observed for drought tolerance, the establishment of a screening system requires the consideration of many factors to identify genotypes with better vegetative growth at low temperatures but with a similar total cultivation time.

### 3.3 Elevated pH tolerance

Soil pH plays a role in nutrient uptake and is an important factor for the successful growth of plants in gardens and landscaping. Most plants grow well in a pH range from 5.5 to 7.0. Some plant families, such as the Ericaceae (including some economically important ornamental genera such as *Rhododendron*, *Erica*, and *Calluna*), prefer acidic and nutrient-poor growing conditions, limiting the use of and market for these plants. In addition, these calcifuge plants must be grown in peat substrates. However, worldwide peat resources are limited, and natural peat reserves are important for carbon storage. Plants that are adapted to a higher pH can be grown in substrates with little or no peat amendment, leading to more sustainable production. In the future, the need to produce ornamental plants in peat-free substrates will become more pressing.

Despite the frequent occurrence of soils with an elevated pH and the adverse reactions of plants to such growing conditions, relatively little breeding work has been directed toward selecting for an elevated pH tolerance in woody plant species (Van Laere et al., 2018). However, the breeding and selection of new cultivars with a higher pH tolerance and 'lime tolerance' could significantly increase the use of specific genera. In *Rhododendron,* extensive work has been conducted to identify species that grow well under higher pH conditions and to understand the underlying physiological mechanisms of tolerance. McAleese and Rankin (2000) and Kaisheva (2006) reviewed different hypotheses regarding lime tolerance. Genotype data at the species level were collected from natural growing areas or resulted from screening studies based on manipulation of the pH of soilless media for container-grown plants. Krebs (2018) listed a number of *Rhododendron* species that grow in near neutral or higher pH soils. Recently, Wang et al. (2018b) published an interesting approach to identify valuable genetic resources using a large amount of data from specimens present in the Chinese herbarium. The geochemical data for the collection site of more than 30 000 *Rhododendron* specimens were associated with information for the soil pH and $CaCO_3$ concentration of the habitat soil as designated in the world soil database. This allowed the authors to identify valuable resources at the taxon level for tolerance against lime soils characterized by a high pH and high $CaCO_3$ concentrations. The predicted tolerant taxa have paved the way in the long term for using this genetic material in breeding for abiotic stress tolerance.

However, in the practical breeding of *Rhododendron*, to date, a rather limited number of these species are used. Breeding for tolerance is also challenging, as the genetic basis and the heritability of lime tolerance are unknown. Effective and reliable screening methods are necessary to identify seedlings with improved tolerance at a young stage. Preil and Ebbinghaus (1994) used an in vitro screen to identify lime-tolerant seedlings in *Rhododendron* breeding populations. After a thorough further *in vivo* screening, this strategy resulted in market introduction of the so-called INKARHO® rootstocks with better tolerance to soils in the pH range from 6.5 to 7.0 and bicarbonate concentrations of approximately 400 mg $kg^{-1}$ soil (Chaanin, 1998). Susko et al. (2018) utilized an in vitro colorimetric assay to measure the capacity for rhizosphere acidification in deciduous azalea. Significant variation among families in terms of the capacity to acidify the rhizosphere were noted. A high-throughput system to screen seedlings in a very young stage might be based on chlorophyll fluorescence data. This technique allowed us to establish an efficient screen in very early seedling stages (Fig. 4). Clear differences between individual seedlings with respect to tolerance were observed. This technique seemed to be appropriate for selection of those genotypes that might be more tolerant to increased pH values as early as the early seedling stage, providing a potentially valuable instrument for breeders interested in pH tolerance.

### 3.4 Resistance and tolerance to biotic stressors

In the past, pest and disease resistance were considered as breeding goals only after addressing the major ornamental traits such as flower size, color, shape, compactness, and fragrance, among others (Debener, 2009; Leus, 2018). Perception of the awareness of disease resistance as an important breeding goal will most likely increase in the future due to (i) an increasing interest of consumers and the trading chain in ornamental plants produced with limited or no chemical plant protection and (ii) the lack of registered and effective chemical plant protection measures. Furthermore, many ornamentals are propagated vegetatively, which increases the risk of spreading diseases, especially viral and bacterial diseases, compared with seed propagation. For sustainable production of ornamental plants, disease resistance is of utmost importance, because it will enable a reduction or even eliminate the use of plant protection chemicals, many of which have a negative impact on the environment, on beneficial organisms including pollinators and on consumers.

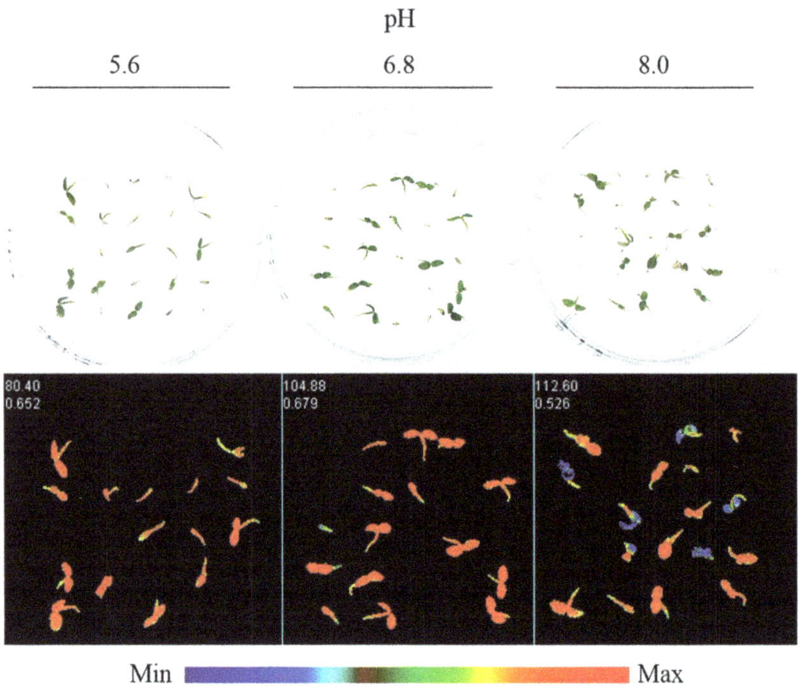


**Figure 4** Example of chlorophyll fluorescence (Fv/Fm) images of 40-day-old *R. fortunei* seedlings grown at different pH levels. (Red color: healthy seedlings, blue color: stressed seedlings).

Ornamentals comprise an enormous range of different species and varieties, each of which is affected by a multitude of pathogens and pests with different life cycles and epidemiological traits. The identification of disease resistance donors, the establishment of a screening test, and unraveling of the genetic basis of resistance are complex tasks requiring enormous long-term efforts. Consequently, breeding for disease resistance in ornamentals is not as advanced as in agronomic crops (Leus, 2018) and, thus far, has mainly focused on economically most relevant host-pathogen interactions.

Conventional cross breeding and interspecific hybridization can be used to introduce resistance to biotic stressors into the existing gene pool or cultivars of interest. Recurrent selection after backcrossing is often implemented in breeding programs (Leus, 2018) to reduce unwanted traits of the resistance donor. In the progeny, screening for resistant genotypes depends on a bio-assay involving inoculations that account for the genetic diversity of pathogens.

Thus far, breeding for disease resistance in ornamentals is frequently based on selection for resistance genes (R-genes). An alternative approach is to obtain more tolerant plants based on the loss-of-function of a susceptibility gene (S-gene) (Eckardt, 2002; Van Schie and Takken, 2014). In rose, increased resistance to powdery mildew was obtained after inactivation of the mildew resistance locus o (MLO) S-gene (Qiu et al., 2015). Additionally, fundamental insights into host-pathogen interactions and the role of plant hormones as a signaling and regulatory network following hostile biotic attacks can open new horizons for plant breeding. In general, the jasmonic acid (JA) pathway is known to play a key role in defense against herbivores (Howe and Jander, 2008) and necrotrophic pathogens (Glazebrook, 2005), while the salicylic acid (SA) pathway acts in plant defense against biotrophic pathogens (Vlot et al., 2009). Ethylene modulates these two phytohormones (Diezel et al., 2009), and extensive crosstalk exists between SA, JA, and ET signaling (Pieterse et al., 2009).

Future approaches will most likely be based on molecular markers that are linked to the resistance locus or loci. Whereas numerous publications deal with the identification of donors for disease resistance, reports on conventional or interspecific breeding approaches for disease resistance in ornamental plants are relatively scarce. They include, for instance, the fungal pathosystems black spot and mildew and rose (reviewed by Debener and Byrne, 2014), *Fusarium* and lily (Shahin et al., 2009), and *Calonectria pseudonaviculata* and boxwood (Van Laere et al., 2019). De Backer (2012) studied the possibility of marker-assisted selection for chrysanthemum cultivars resistant to chrysanthemum white rust (*Puccinia horiana*), a major threat in the cultivation of this species.

## 3.5 Nutrient use efficiency

Reducing fertilizer application, particularly of nitrogen and phosphorus, is important to sustain ornamental plant production. Nitrogen fertilizer that is not completely used by the crop may either leach and contaminate the ground water and eutrophy ecosystems or be converted by bacteria into nitrogen-containing gases; $N_2O$ is a very potent greenhouse gas (Alvarez et al., 2012). Natural phosphorus reserves are limited (Gilbert, 2009), so phosphorus should be applied parsimoniously. The nitrogen use efficiency of plants is a complex trait that can be divided into nitrogen-uptake efficiency and nitrogen-use efficiency (Hirel et al., 2007), and it involves diverse morphological and physiological traits (Ranjan and Yadav, 2019). Breeding for nitrogen use efficient crops is thus extremely difficult, although the first QTL for nitrogen use efficiency have been identified in major cereal crops (Ranjan and Yadav, 2019). The same holds true for phosphorus use efficiency, so research on this complex trait is urgently needed (Veneklaas et al., 2012).

For ornamentals, genetic and breeding aspects of nutrient use efficiency have not yet been studied, but insights into the physiological and molecular mechanisms underlying these traits in model plants will eventually be transferred to ornamental crops.

### 3.6 Compact growing plants

Compactness is an important selection criterion in the ornamental production chain. The increasing population living in apartments with limited or almost no space for plants and the smaller size of gardens within the suburban area necessitate compact, easy to grow plants. In addition, cost reduction during production and the transport of ornamental plants make compactness an elementary quality criterion in the ornamental production chain. In horticultural practice, compact growth is often achieved by the application of chemical growth retardants. However, environmental concerns and the potential future ban of the use of these chemicals necessitates the search for plants with naturally compact growth habit (Van Huylenbroeck et al., 2019). In many plants species, the plant habit, architecture, and compact growth are major traits of importance to breeders. Understanding the genetic and physiological processes underlying plant habit enables the selection of appropriate genotypes. Cross breeding and subsequent selection have recently led to many new introductions of compact growing plants. Alternative approaches to obtaining more compact growing plants are polyploidization, as shown in petunia (Regalado et al., 2017) and *Escallonia* (Denaeghel et al., 2018), and the introduction of *rol*-genes from *Rhizobium rhizogenes* (Christensen and Müller, 2009). In the latter case, wild-type strains of *Rhizobium rhizogenes* were co-cultivated in vitro with plant explants, which can result in hairy roots. Plant regeneration from these hairy roots can provide plants with a distinct phenotype, often with shortened internodes and reduced apical dominance (Tepfer, 1984). By using these plants as prebreeding material, compactness can be introduced into the next generation. In *Kalanchoe blossfeldiana*, this process resulted in commercial genotypes that could be grown in a more sustainable manner without the use of plant growth regulators (Mackenzie et al., 2018). Figure 5 shows an application of this approach resulting in more compact *Osteospermum rol*-transgenic plants.

### 3.7 Postharvest life

A long vase or shelf life is an important breeding goal for cut flowers and for potted plants. Because global production and trading may include a long transportation, postharvest quality is of the outmost importance for most ornamental crops. In addition to temperature, the gas atmosphere and prevention of pathogens, as well as protection from ethylene damage, are major measures in postharvest treatments (Mibus, 2018). Most chemicals that are used to preserve flowers are environmentally harmful, have been banned in several countries, and are costly. Thus, an additional interest

**Figure 5** Compact growth induced by transformation with *Rhizobium rhizogenes* in *Osteospermum fruticosum*. Left: control plant; Right: the same cultivar with *rol*-genes introduced.

in genetic approaches to improve longevity can be observed from the perspective of sustainability. Depending on the ornamental sector (i.e. cut flowers vs potted plants) and even the species, the molecular and physiological processes underlying vase and shelf life are diverse. The prevention of leaf yellowing and leaf and flower abscission are important traits for potted plant postharvest quality, whereas the senescence of floral organs or bent neck must be considered first in cut flowers. Therefore, for each ornamental crop, the testing conditions and parameters for evaluation must first be established (Mibus, 2018). In the past, postharvest physiology studies had a strong focus on ethylene-induced processes, but future work is needed to unravel other pathways, for example, stomata management, contributing to complex traits.

Genetic variation for longevity has been reported in some ornamentals, with a strong focus on commercially important groups (reviewed in Mibus, 2018), such as the rose (Ahmadi et al., 2009; Carvalho et al., 2015), carnation (Onozaki et al., 2011), lily (van der Meulen-Muisers et al., 1999), and *Phalaenopsis* orchids (Vo et al., 2015), underscoring the possibility of, for example, the selection of desired types and of breeding for long-lasting flowers and plants (Mibus, 2018). Marker-assisted selection integrated into conventional intraspecific breeding programs and applied after interspecific hybridization and backcrossing may ideally support breeding for improved shelf- and vase life in the future.

## 4 Conclusion and future trends

The main goals in ornamental breeding have remained unchanged in a changing world: to develop new cultivars with improved floral attributes (color, shape, perfume, and enhanced vase life), leaf characteristics, or plant habits. Breeding for sustainability, in the form of enhanced disease and pest resistance, increased tolerance against abiotic stresses, and a reduced input of chemicals during the production phase is a challenge, but it has great potential for future markets. Conventional cross breeding, ploidy breeding, and interspecific hybridization are currently, and will be, important in the future to contribute to major novel breeding goals. However, breeding will be more efficient if the genetic base of the novel traits is resolved, for which research efforts are needed in model and crop plants. Breeding depends heavily on genetic resources that must be preserved but are also amenable for ornamental plant breeders. An alternative approach to further improve established ornamental crops may be the introduction and domestication of new ornamental species that can be produced with fewer resources because they are native to nutrient-poor and arid environments.

It can be expected that marker-assisted breeding and novel breeding techniques, especially gene editing, will allow progress for many breeding aims that are important for a sustainable production of ornamental plants. However, a profound knowledge on classical breeding techniques in horticultural crops will always be the prerequisite of successful breeding strategies. Thus, theorectical but also research-based education of students in these topics is needed in future, although horticultural faculties are endangered at many universties worldwide. Furthermore, public research institutes working on ornamentals are essential partners of small and medium private-breeding enterprises and should be conserved and should collaborate in international joint projects.

## 5 References

Adams, P., Nelson, D. E., Yamada, S., Chmara, W., Jensen, R. G., Bohnert, H. J. and Griffiths, H. 1998. Growth and development of *Mesembryanthemum crystallinum* (Aizoaceae). *New Phytol.* 138(2), 171-90. doi:10.1046/j.1469-8137.1998.00111.x.

Aguilar-García, S. A., Figueroa-Castro, D. M. and Castañeda-Posadas, C. 2012. Pollen morphology of *Pachycereus weberi* (Cactaceae): an evaluation of variation in pollen size. *Plant Syst. Evol.* 298(10), 1845-50. doi:10.1007/s00606-012-0685-6.

Ahmadi, N., Mibus, H. and Serek, M. 2009. Characterization of ethylene-induced organ abscission in F1 breeding lines of miniature roses (*Rosa hybrida* L.). *Postharvest Biol. Technol.* 52(3), 260-6. doi:10.1016/j.postharvbio.2008.12.010.

Alvarez, J. M., Vidal, E. A. and Gutiérrez, R. A. 2012. Integration of local and systemic signalling pathways for plant N responses. *Curr. Opin. Plant Biol.* 15(2), 185-91. doi:10.1016/j.pbi.2012.03.009.

Amberger, S., Schlegel, G. and Horn, W. 1984. Selection for chilling tolerance in *Saintpaulia ionantha*. *Report of 12th Meeting, Eucarpia Section Ornamentals*, Arslev, Denmark, pp. 27–42.

Bauerfeind, M. A. 2016. Molecular physiology of chilling tolerance of *Petunia hybrida*. Dissertation, Leibniz University Hannover, Hannover, Germany.

Bauerfeind, M. A., Winkelmann, T., Franken, P. and Druege, U. 2015. Transcriptome, carbohydrate, and phytohormone analysis of *Petunia hybrida* reveals a complex disturbance of plant functional integrity under mild chilling stress. *Front. Plant Sci.* 6, 583. doi:10.3389/fpls.2015.00583.

Bell, C. R. 1959. Mineral nutrition and flower to flower pollen size variation. *Am. J. Bot.* 46(9), 621–4. doi:10.1002/j.1537-2197.1959.tb07062.x.

Bennett, M. D. and Leitch, I. J. 1995. Nuclear DNA amounts in angiosperms. *Ann. Bot.* 76(2), 113–76. doi:10.1006/anbo.1995.1085.

Bermejo Dominguez, G., Mibus-Schoppe, H. and Sparke, K. 2017. Evaluation of existing research concerning sustainability in the value chain of ornamental plants. *Eur. J. Sustain. Devel.* 6, 11-9. doi:10.14207/ejsd.2017.v6n3p11.

Bino, R. J., Van Tuyl, J. M. V. and De Vries, J. N. 1990. Flow cytometric determination of relative nuclear DNA contents in bicellulate and tricellulate pollen. *Ann. Bot.* 65(1), 3–8. doi:10.1093/oxfordjournals.aob.a087904.

Blum, A. 2005. Drought resistance, water-use efficiency, and yield potential-are they compatible, dissonant, or mutually exclusive? *Aust. J. Agric. Res.* 56(11), 1159–68. doi:10.1071/AR05069.

Braun, P. and Winkelmann, T. 2015. Cytological investigations in midday flowers (Aizoaceae) reveal high DNA contents in different somatic tissues and potential occurrence of unreduced male gametes. *Acta Hortic.* 1087(1087), 437–44. doi:10.17660/ActaHortic.2015.1087.59.

Braun, P. and Winkelmann, T. 2016. Flow cytometric analyses of somatic and pollen nuclei in midday flowers (Aizoaceae). *Caryologia* 69, 1–12. doi:10.1080/00087114.2016.1188359.

Bretagnolle, F. and Thompson, J. D. 1995. Gametes with the somatic chromosome number: mechanisms of their formation and role in the evolution of autopolyploid plants. *New Phytol.* 129,1–22. doi:10.1111/j.1469-8137.1995.tb03005.x.

Breuer, C., Braidwood, L. and Sugimoto, K. 2014. Endocycling in the path of plant development. *Curr. Opin. Plant Biol.* 17, 78–85. doi:10.1016/j.pbi.2013.11.007.

Bridgen, M. P., Van Houtven, W. and Eeckhaut, T. 2018. Plant tissue culture techniques for breeding. In: Van Huylenbroeck, J. (Ed.), *Ornamental Crops, Handbook of Plant Breeding 11*. Springer, Cham, pp. 127–44. doi:10.1007/978-3-319-90698-0_6.

Carvalho, D. R. A., Koning-Boucoiran, C. F. S., Fanourakis, D., Vasconcelos, M. W., Carvalho, S. M. P., Heuvelink, E., Krens, F. A. and Maliepaard, C. 2015. QTL analysis for stomatal functioning in tetraploid *Rosa × hybrida* grown at high relative air humidity and its implications on postharvest longevity. *Mol. Breed.* 35(8), 172. doi:10.1007/s11032-015-0365-7.

Chaanin, A. 1998. Lime tolerance in rhododendron. *Comb. Proc. Int. Plant Prop. Soc.* 48, 180–2.

Chaanin, A. 2003. Selection strategies for cut roses. In: Roberts, A. V., Debener, T. and Gudin, S. (Eds), *Encyclopedia of Rose Science*. Elsevier Academic Press, Oxford, pp. 33–41.

Christensen, B. and Müller, R. 2009. The use of *Agrobacterium rhizogenes* and its *rol*-genes for quality improvement in ornamentals. *Eur. J. Hort. Sci.* 74, 275–87.
Cook, J., Oreskes, N., Doran, P. T., Anderegg, W. R. L., Verheggen, B., Maibach, E. W., Carlton, J. S., Lewandowsky, S., Skuce, A. G., Green, S. A., Nuccitelli, D., Jacobs, P., Richardson, M., Winkler, B., Painting, R. and Rice, K. 2016. Consensus on consensus: a synthesis of consensus estimates on human-caused global warming. *Environ. Res. Lett.* 11(4). doi:10.1088/1748-9326/11/4/048002.
Crespel, L., Ricci, S. C. and Gudin, S. 2006. The production of 2n pollen in rose. *Euphytica* 151(2), 155–64. doi:10.1007/s10681-006-9136-1.
Dai, A. 2013. Increasing drought under global warming in observations and models. *Nat. Clim. Change* 3(1), 52–8. doi:10.1038/nclimate1633.
D'Amato, F. 1984. Role of polyploidy in reproductive organs and tissues. In: Johri, B. M. (Ed.), *Embryology of Angiosperms*. Springer-Verlag, pp. 519–66.
De Backer, M. 2012. Characterization and detection of *Puccinia horiana* on chrysanthemum for resistance breeding and sustainable control. Dissertation, Ghent University, Faculty of Bioscience Engineering, Ghent, Belgium.
Debener, T. 2002. Molecular markers as a tool for analyses of genetic relatedness and selection in ornamentals. In: Vainstein, A. (Ed.), *Breeding for Ornamentals: Classical and Molecular Approaches*. Kluwer Academic Publishers, Dordrecht, pp. 329–46.
Debener, T. 2009. Current strategies and future prospects of resistance breeding in ornamentals. *Acta Hortic.* 836(836), 125–30. doi:10.17660/ActaHortic.2009.836.17.
Debener, T. and Byrne, D. H. 2014. Disease resistance breeding in rose: current status and potential of biotechnological tools. *Plant Sci.* 228, 107–17. doi:10.1016/j.plantsci.2014.04.005.
Del Angel, C., Guadalupe, P., García, A. and Méndez, I. 2006. Nuclear genome size and karyotype analysis in *Mammillaria* species (Cactaceae). *Caryologia* 59(2), 177–86. doi:10.1080/00087114.2006.10797914.
Denaeghel, H. E. R., Van Laere, K., Leus, L., Lootens, P., Van Huylenbroeck, J. and Van Labeke, M. C. 2018. The variable effect of polyploidization on the phenotype in Escallonia. *Front. Plant Sci.* 9, 354. doi:10.3389/fpls.2018.00354.
De Rocher, E. J., Harkins, K. R., Galbraith, D. W. and Bohnert, H. J. 1990. Developmentally regulated systemic endopolyploidy in succulents with small genomes. *Science* 250(4977), 99–101. doi:10.1126/science.250.4977.99.
De Storme, N. and Geelen, D. 2011. The *Arabidopsis* mutant *jason* produces unreduced first division restitution male gametes through a parallel/fused spindle mechanism in meiosis II. *Plant Physiol.* 155, 1403–15. doi:10.1104/pp.110.170415.
De Storme, N. and Geelen, D. 2013a. Tansley review: sexual polyploidization in plants – cytological mechanisms and molecular regulation. *New Phytol.* 198(3), 670–84. doi:10.1111/nph.12184.
De Storme, N. and Geelen, D. 2013b. Pre-meiotic endomitosis in the cytokinesis-defective tomato mutant *pmcd1* generates tetraploid meiocytes and diploid gametes. *J. Exp. Bot.* 64(8), 2345–58. doi:10.1093/jxb/ert091.
Dewitte, A., Eeckhaut, T., Van Huylenbroeck, J. and Van Bockstaele, E. 2009. Occurrence of viable unreduced pollen in a *Begonia* collection. *Euphytica* 168(1), 81–94. doi:10.1007/s10681-009-9891-x.
Dewitte, A., Van Laere, K. and Van Huylenbroeck, J. 2011. Use of 2n gametes in plant breeding. In: Abdurakhmonov, I. Y. (Ed.), *Plant Breeding*. IntechOpen, pp. 59–86.

Available at: https://www.intechopen.com/books/plant-breeding/use-of-2n-gametes-in-plant-breeding.

Dhakar, S., Soni, A. and Kumari, P. 2017. Breeding for abiotic stress tolerance in ornamental crops: a review. *Chem. Sci. Rev. Lett.* 6, 1549–54.

Dhooghe, E., Van Laere, K., Eeckhaut, T., Leus, L. and Van Huylenbroeck, J. 2011. Mitotic chromosome doubling of plant tissues *in vitro*. *Plant Cell Tiss. Org. Cult.* 104(3), 359–73. doi:10.1007/s11240-010-9786-5.

Diezel, C., von Dahl, C. C., Gaquerel, E. and Baldwin, I. T. 2009. Different Lepidopteran elicitors account for cross-talk in herbivory-induced phytohormone signaling. *Plant Physiol.* 150(3), 1576–86. doi:10.1104/pp.109.139550.

Dohm, A. 2017. Ornamental plant breeding in a changing competitive environment. *Julius Kühn-Archiv* 457, 61–3. doi:10.5073/jka.2017.457.014.

Eckardt, N. A. 2002. Plant disease susceptibility genes? *Plant Cell* 14(9), 1983–6. doi:10.1105/tpc.140910.

Eeckhaut, T., Werbrouck, S., Leus, L., Van Bockstaele, E. and Debergh, P. 2004. Chemically induced polyploidization of *Spathiphyllum wallisii* Regel through somatic embryogenesis. *Plant Cell Tiss. Org. Cult.* 78, 241–6.

Eeckhaut, T., Van Laere, K., De Riek, J. and Van Huylenbroeck, J. 2006. Overcoming interspecific barriers in plant breeding. In: Teixeira De Silva, J. A. (Ed.), *Floriculture, Ornamental and Plant Biotechnology: Advances and Topical Issues* (1st edn.). Global Science Publications Books, London, UK, pp. 540–51.

Eeckhaut, T., Lakshmanan, P. S., Deryckere, D., Van Bockstaele, E. and Van Huylenbroeck, J. 2013. Progress in plant protoplast research. *Planta* 238(6), 991–1003. doi:10.1007/s00425-013-1936-7.

Eeckhaut, T., Van der Veken, J., Dhooghe, E., Leus, L., Van Laere, K. and Van Huylenbroeck, J. 2018. Ploidy breeding in ornamentals. In: Van Huylenbroeck, J. (Ed.), *Ornamental Crops, Handbook of Plant Breeding 11*. Springer, Cham, pp. 145–73. doi:10.1007/978-3-319-90698-0_7.

Eggli, U. and Nyffeler, R. 2009. Living under temporarily arid conditions - succulence as an adaptive strategy. *Bradleya* 27(27), 13–36. doi:10.25223/brad.n27.2009.a10.

Elings, A., Kempkes, F. L. K., Kaarsemaker, R. C., Ruijs, M. N. A., van de Braak, N. J. and Dueck, T. A. 2005. The energy balance and energy-saving measures in greenhouse tomato cultivation. *Acta Hortic.* 691(691), 67–74. doi:10.17660/ActaHortic.2005.691.5.

Falistocco, E., Tosti, N. and Falcinelli, M. 1995. Cytomixis in pollen mother cells of diploid *Dactylis*, one of the origins of 2n gametes. *J. Hered.* 86(6), 448–53. doi:10.1093/oxfordjournals.jhered.a111619.

Franco, J. A., Martínez-Sánchez, J. J., Fernández, J. A. and Bañon, S. 2006. Selection and nursery production of ornamental plants for landscaping and xerogardening in semi-arid environments. *J. Hortic. Biotechnol.* 81(1), 3–17. doi:10.1080/14620316.2006.11512022.

Friedman, W. E. 1999. Expression of the cell cycle in sperm of *Arabidopsis*: implications for understanding patterns of gametogenesis and fertilization in plants and other eukaryotes. *Development* 126(5), 1065–75.

Gajanayake, B., Trader, B. W., Reddy, K. R. and Harkess, R. L. 2011 Screening ornamental pepper cultivars for temperature tolerance using pollen and physiological parameters. *HortScience* 46(6), 878–84. doi:10.21273/HORTSCI.46.6.878.

Gilbert, N. 2009. Environment: The disappearing nutrient. *Nature* 461(7265), 716–8. doi:10.1038/461716a.

Glazebrook, J. 2005. Contrasting mechanisms of defence against biotrophic and necrotrophic pathogens. *Annu. Rev. Phytopathol.* 43, 205–27. doi:10.1146/annurev.phyto.43.040204.135923.

Granados Mendoza, C., Wanke, S., Goetghebeur, P. and Samain, M. S. 2013. Facilitating wide hybridization in *Hydrangea* S. L. cultivars: a phylogenetic and marker-assisted breeding approach. *Mol. Breed.* 32(1), 233–9. doi:10.1007/s11032-012-9822-8.

Growns, D. and Webb, M. 2012. Breeding Australian native plants for a changing climate. *Acta Hortic.* 937(937), 1093–6. doi:10.17660/ActaHortic.2012.937.136.

Gudin, S. 2003. Breeding/overview. In: Roberts, A. V., Debener, T. and Gudin, S. (Eds), *Encyclopedia of Rose Science*. Elsevier Academic Press, Oxford, pp. 25–30.

Hakam, N., DeEll, J. R., Khanizadeh, S. and Richer, C. 2000. Assessing chilling tolerance in roses using chlorophyll fluorescence. *HortScience* 35(2), 184–6. doi:10.21273/HORTSCI.35.2.184.

Han, Y., Zhang, X., Wang, Y. and Ming, F. 2013. The suppresion of WRKY44 by GIGANTEA-MiR172 pathway is involved in drought response in *Arabidopsis thaliana*. *PLoS ONE* 8(11), e73541.

Hartmann, H. E. K. 2001. *Illustrated Handbook of Succulent Plants: Aizoaceae A-E; F-Z.* Springer, Berlin.

Hirano, T. and Hoshino, Y. 2009. Detection of changes in the nuclear phase and evaluation of male germ units by flow cytometry during in vitro pollen tube growth in *Alstroemeria aurea*. *J. Plant Res.* 122(2), 225–34. doi:10.1007/s10265-008-0208-2.

Hirel, B., Le Gouis, J., Ney, B. and Gallais, A. 2007. The challenge of improving nitrogen use efficiency in crop plants: towards a more central role for genetic variability and quantitative genetics within integrated approaches. *J. Exp. Bot.* 58(9), 2369–87. doi:10.1093/jxb/erm097.

Horn, W. 2002. Breeding methods and breeding research. In: Vainstein, A. (Ed.), *Breeding for Ornamentals: Classical and Molecular Approaches*. Kluwer Academic Publishers, Dordrecht, pp. 47–83.

Howe, G. A. and Jander, G. 2008. Plant immunity to insect herbivores. *Annu. Rev. Plant Biol.* 59, 41–66. doi:10.1146/annurev.arplant.59.032607.092825.

Hu, H. and Xiong, L. 2014. Genetic engineering and breeding of drought-resistant crops. *Annu. Rev. Plant Biol.* 65, 715–41. doi:10.1146/annurev-arplant-050213-040000.

Ihlenfeldt, H. D. 1960. Entwicklungsgeschichtliche, morphologische und systematische Untersuchungen an Mesembryanthemen. *Feddes Rep.* 63, 1–104.

Ingram, J. and Bartels, D. 1996. The molecular basis of dehydration tolerance in plants. *Annu. Rev. Plant Physiol. Mol. Biol.* 47, 377–403.

Jansen, R. C. and Den Nijs, A. P. M. 1993. A statistical mixture model for estimating the proportion of unreduced pollen grains in perennial ryegrass (*Lolium perenne* L.) via the size of pollen grains. *Euphytica* 70(3), 205–15. doi:10.1007/BF00023761.

Jovtchev, G., Schubert, V., Meister, A., Barow, M. and Schubert, I. 2006. Nuclear DNA content and nuclear cell volume are positively correlated in angiosperms. *Cytogenet. Genome Res.* 114(1), 77–82. doi:10.1159/000091932.

Kaisheva, M. V. 2006. The effect of metals and soil pH on the growth of Rhododendron and other alpine plants in limestone soil. Dissertation, Doctor of Philosophy, University of Edinburgh, Edinburgh, UK.

Kato, J. and Mii, M. 2012. Production of interspecific hybrids in ornamental plants. In: Loyola-Vargas, V. and Ochoa-Alejo, N. (Eds), *Plant Cell Culture Protocols. Methods in Molecular Biology (Methods and Protocols)* (vol. 877). Humana Press, Totowa, NJ, pp. 233–45.

Kooyers, N. J. 2015. The evolution of drought escape and avoidance in natural herbaceous populations. *Plant Sci.* 234, 155–62. doi:10.1016/j.plantsci.2015.02.012.

Koźmińska, A., Al Hassan, M., Wiszniewska, A., Hanus-Fajerska, E., Boscaiu, M. and Vicente, O. 2019. Responses of succulents to drought: comparative analysis of four Sedum (Crassulaceae) species. *Scientia Horticulturae* 243, 235–42. doi:10.1016/j.scienta.2018.08.028.

Krebs, S. L. 2018. Rhododendron. In: Van Huylenbroeck, J. (Ed.), *Ornamental Crops, Handbook of Plant Breeding 11*. Springer, Cham, pp. 673–718. doi:10.1007/978-3-319-90698-0_26.

Kron, P. and Husband, B. C. 2015. Distinguishing 2n gamete nuclei from doublets in pollen using flow cytometry and pulse analysis. *Cytom. A* 87(10), 943–57. doi:10.1002/cyto.a.22696.

Kuligowska, K., Lütken, H. and Müller, R. 2016. Towards development of new ornamental plants: status and progress in wide hybridization. *Planta* 244(1), 1–17. doi:10.1007/s00425-016-2493-7.

Kumar, R., Shekhar, S. S. and Singh, M. 2012. Breeding for drought tolerance in vegetables. *Veg. Sci.* 39, 1–15.

Lelley, T., Mahmoud, A. A. and Lein, V. 1987. Genetics and cytology of unreduced gametes in cultivated rye (*Secale cereale* L.). *Genome* 29(4), 635–8. doi:10.1139/g87-106.

Leus, L. 2018. Breeding for disease resistance in ornamentals. In: Van Huylenbroeck, J. (Ed.), *Ornamental Crops, Handbook of Plant Breeding 11*. Springer, Cham, pp. 97–125. doi:10.1007/978-3-319-90698-0_5.

Leus, L., Van Laere, K., De Riek, J. and Van Huylenbroeck, J. 2018. Rose. In: Van Huylenbroeck, J. (Ed.), *Ornamental Crops, Handbook of Plant Breeding 11*. Springer, Cham, pp. 719–67. doi:10.1007/978-3-319-90698-0_27.

Li, Y., Ye, W., Wang, M. and Yan, X. 2009. Climate change and drought: a risk assessment of crop-yield impacts. *Clim. Res.* 39, 31–46. doi:10.3354/cr00797.

Li, W. D., Hu, X., Liu, J. K., Jiang, G. M., Li, O. and Xing, D. 2011. Chromosome doubling can increase heat tolerance in *Lonicera japonica* as indicated by chlorophyll fluorescence imaging. *Biol. Plant.* 55(2), 279–84. doi:10.1007/s10535-011-0039-1.

Liu, S., Chen, S., Chen, Y., Guan, Z., Yin, D. and Chen, F. 2011. *In vitro* induced tetraploid of *Dendranthema nankingense* (Nakai) Tzvel. shows an improved level of abiotic stress tolerance. *Sci. Hortic.* 127(3), 411–9. doi:10.1016/j.scienta.2010.10.012.

Mackenzie, K. K., Lütken, H., Coelho, L. L., Kaaber, M. D., Hegelund, J. N. and Müller, R. 2018. Kalanchoë. In: Van Huylenbroeck, J. (Ed.), *Ornamental Crops, Handbook of Plant Breeding 11*. Springer, Cham, pp. 453–79. doi:10.1007/978-3-319-90698-0_19.

Marasek-Ciolakowska, A., Nishikawa, T., Shea, D. J. and Okazaki, K. 2018. Breeding of lilies and tulips–interspecific hybridization and genetic background. *Breed. Sci.* 68(1), 35-52. doi:10.1270/jsbbs.17097.

Mariotti, M. G. and Roccotiello, E. 2013. *Sustainable Floriculture: Handbook and Guidelines*. LIFE+ 09 ENV/IT/067', Del Gallo editori s.r.l. Green Printing. Available at: http://ec.europa.eu/environment/life/about/index.htm.

Mason, A. S., Nelson, M. N., Yan, G. J. and Cowling, W. A. 2011. Production of viable unreduced gametes in *Brassica* interspecific hybrids is genotype specific and stimulated by cold temperatures. *BMC Plant Biol.* 11, 103. doi:10.1186/1471-2229-11-103.

Matimati, I., Musil, C. F., Raitt, L. and February, E. C. 2012. Diurnal stem diameter variations show CAM and $C_3$ photosynthetic modes and CAM-$C_3$ switches in arid

South African succulent shrubs. *Agric. Forest Meteorol.* 161, 72–9. doi:10.1016/j.agrformet.2012.03.020.

McAleese, A. J. and Rankin, D. W. H. 2000. Growing rhododendrons on limestone soils: is it really possible? *J. Am. Rhododendron Soc.* 54, 126–34.

Meiners, J. and Winkelmann, T. 2012. Evaluation of reproductive barriers and realisation of interspecific hybridisations depending on the genetic distances between species in the genus *Helleborus*. *Plant Biol.* 14(4), 576–85. doi:10.1111/j.1438-8677.2011.00542.x.

Meyerding, S. G. H. 2017. Analyzing job satisfaction and preferences of employees: the case of horticultural companies in Germany. *Int. Food Agribusiness Manag. Rev.* 20(5), 765–88. doi:10.22434/IFAMR2016.0158.

Mibus, H. 2018. Breeding and genetics for shelf and vase life. In: Van Huylenbroeck, J. (Ed.), *Ornamental Crops, Handbook of Plant Breeding 11*. Springer, Cham, pp. 63–95. doi:10.1007/978-3-319-90698-0_4.

Negri, V. and Lemmi, G. 1998. Effect of selection and temperature stress on the production of 2n gametes in *Lotus tenuis*. *Plant Breed.* 117(4), 345–9. doi:10.1111/j.1439-0523.1998.tb01952.x.

Ning, G.-G., Shi, X.-P., Hu, H.-R., Yan, Y. and Bao, M.-Z. 2019. Development of a range of polyploid lines in *Petunia hybrida* and the relationship of ploidy with the single-/double-flower trait. *HortScience* 44(2), 250–5. doi:10.21273/HORTSCI.44.2.250.

Onozaki, T., Yagi, M., Tanase, K. and Shibata, M. 2011. Crossings and selections for six generations based on flower vase life to create lines with ethylene resistance or ultra-long vase life in carnations (*Dianthus caryophyllus* L.). *J. Japan. Soc. Hortic. Sci.* 80(4), 486–98. doi:10.2503/jjshs1.80.486.

Oustric, J., Morillon, R., Luro, F., Herbette, S., Lourkisti, R., Giannettini, J., Berti, L. and Santini, J. 2017. Tetraploid Carrizo citrange rootstock (*Citrus sinensis* Osb. × *Poncirus trifoliata* L. Raf.) enhances natural chilling stress tolerance of common clementine (*Citrus clementina* Hort. ex Tan). *J. Plant Physiol.* 214, 108–15. doi:10.1016/j.jplph.2017.04.014.

Patterson, B. D. and Reid, M. S. 1990. Genetic and environmental influences on the expression of chilling injury. In: Wang, C. Y. (Ed.), *Chilling Injury of Horticultural Crops*. CRC Press, Boca Raton, FL, pp. 87–112.

Pécrix, Y., Rallo, G., Folzer, H., Cigna, M., Gudin, S. and Le Bris, M. 2011. Polyploidization mechanisms: temperature environment can induce diploid gamete formation in *Rosa* sp. *J. Exp. Bot.* 62(10), 3587–97. doi:10.1093/jxb/err052.

Pieterse, C. M. J., Leon-Reyes, A., Van der Ent, S. and Van Wees, S. C. M. 2009. Networking by small-molecule hormones in plant immunity. *Nat. Chem. Biol.* 5(5), 308–16. doi:10.1038/nchembio.164.

Preil, W. and Ebbinghaus, R. 1994. Breeding of lime tolerant Rhododendron rootstocks. *Acta Hortic.* 364, 61–70.

Qiu, X., Wang, Q., Zhang, H., Jian, H., Zhou, N., Ji, C., Yan, H., Bao, M. and Tang, K. 2015. Antisense RhMLO1 gene transformation enhances resistance to the powdery mildew pathogen in *Rosa multiflora*. *Biol. Rep.* 33(6), 1659–65. doi:10.1007/s11105-015-0862-1.

Ramsey, J. and Schemske, D. W. 1998. Pathways, mechanisms, and rates of polyploidy formation in flowering plants. *Annu. Rev. Ecol. Syst.* 29(1), 467–501. doi:10.1146/annurev.ecolsys.29.1.467.

Ranjan, R. and Yadav, R. 2019. Targeting nitrogen use efficiency for sustained production of cereal crops. *J. Plant Nutr.* 42(9), 1086–113. doi:10.1080/01904167.2019.1589497.

Rauf, S., Al-Khayri, J. M., Zaharieva, M., Monneveux, P. and Khalil, F. 2016. Breeding strategies to enhance drought tolerance in crops. In: Al-Khayri, J. M., Jain, S. M. and Johnson, D. V. (Eds), *Advances in Plant Breeding Strategies. Volume 2: Agronomic, Abiotic, and Biotic Stress Traits*. Springer, Cham. doi:10.1007/978-3-319-22518-0_11.

Regalado, J. J., Carmona-Martin, E., Querol, V., Velez, C. G., Encina, C. L. and Pitta-Alvarez, S. I. 2017. Production of compact petunias through polyploidization. *Plant Cell Tiss. Org. Cult.* 129(1), 61–71. doi:10.1007/s11240-016-1156-5.

Riboni, M., Galbiati, M., Tonelli, C. and Conti, L. 2013. GIGANTEA enables drought escape response via abscisic acid-dependent activation of the florigens and SUPPRESSOR OF OVEREXPRESSION OF CONSTANS. *Plant Physiol.* 162(3), 1706–19. doi:10.1104/pp.113.217729.

Ruelland, E. and Collin, S. 2012. Chilling stress. In: Shabala, S. (Ed.), *Plant Stress Physiology*. CAB International, Wallingford, UK, pp. 94–117.

Schatlowski, N. and Köhler, C. 2012. Tearing down barriers: understanding the molecular mechanisms of interploidy hybridizations. *J. Exp. Bot.* 63(17), 6059–67. doi:10.1093/jxb/ers288.

Shahin, A., Arens, P., van Heusden, S. and van Tuyl, J. M. 2009. Conversion of molecular markers linked to Fusarium and virus resistance in Asiatic lily hybrids. *Acta Hortic.* 836(836), 131–6. doi:10.17660/ActaHortic.2009.836.18.

Smulders, M. J. M. and Arens, P. 2018. New developments in molecular techniques for breeding in ornamentals. In: Van Huylenbroeck, J. (Ed.), *Ornamental Crops, Handbook of Plant Breeding 11*. Springer, Cham, pp. 213–30. doi:10.1007/978-3-319-90698-0_9.

Song, Z., Wang, W., Shi, L., Zhang, S., Xie, Q., Wei, S., Wang, Y., Bo, K., Miao, H., Zhang, S. and Gu, X. 2018. Identification of QTLs controlling low-temperature tolerance during the germination stage in cucumber (*Cucumis sativus* L.). *Plant Breed.* 137(4), 629–37. doi:10.1111/pbr.12601.

Susko, A. Q., Rinehart, T. A., Bradeen, J. M. and Hokanson, S. C. 2018. An evaluation of two seedling phenotyping protocols to assess pH adaptability in deciduous azalea (*Rhododendron* sect. *Pentanthera* G. Don). *Hortscience* 53(3), 268–74. doi:10.21273/HORTSCI12520-17.

Tantau, H. and Dörffling, K. 1991. Effects of chilling on physiological responses and changes in hormone levels in two *Euphorbia pulcherrima* varieties with different chilling tolerance. *J. Plant Physiol.* 138(6), 734–40. doi:10.1016/S0176-1617(11)81324-6.

Tepfer, D. 1984. Transformation of several species of higher plants by *Agrobacterium rhizogenes*: sexual transmission of the transformed genotype and phenotype. *Cell* 37(3), 959–67. doi:10.1016/0092-8674(84)90430-6.

Toscano, S., Ferrante, A. and Romano, D. 2019. Response of Mediterranean ornamental plants to drought stress. *Horticulturae* 5(1), 6. doi:10.3390/horticulturae5010006.

Urwin, N. A. R. 2014. Generation and characterisation of colchicine-induced polyploid *Lavandula × intermedia*. *Euphytica* 197(3), 331–9. doi:10.1007/s10681-014-1069-5.

Usadel, B., Bläsing, O. E., Gibon, Y., Poree, F., Höhne, M., Günter, M., Trethewey, R., Kamlage, B., Poorter, H. and Stitt, M. 2008. Multilevel genomic analysis of the response of transcripts, enzyme activities and metabolites in *Arabidopsis rosettes* to a progressive decrease of temperature in the non-freezing range. *Plant Cell Environ.* 31(4), 518–47. doi:10.1111/j.1365-3040.2007.01763.x.

Van der Meulen-Muisers, J. J. M., van Oeveren, J. C., Jansen, J. and van Tuyl, J. M. 1999. Genetic analysis of postharvest flower longevity in Asiatic hybrid lilies. *Euphytica* 107(2), 149–57. doi:10.1023/A:1026493902668.

Van Huylenbroeck, J., Desmet, S., Dhooghe, E., De Keyser, E. and Geelen, D. 2019. Breeding for compact growing ornamentals. *Acta Hortic.* 1237(1237), 1–6. doi:10.17660/ActaHortic.2019.1237.1.

Van Laere, K., Van Huylenbroeck, J. and Van BockStaele, E. 2006. Breeding strategies to increase genetic variability within *Hibiscus syriacus*. *Acta Hortic.* 714(714), 75–82. doi:10.17660/ActaHortic.2006.714.9.

Van Laere, K., França, S. C., Vansteenkiste, H., Van Huylenbroeck, J., Steppe, K. and Van Labeke, M.-L. 2011a. Influence of ploidy level on morphology, growth and drought susceptibility in *Spathiphyllum wallisii*. *Acta Physiol. Plant.* 33(4), 1149–56. doi:10.1007/s11738-010-0643-2.

Van Laere, K., Van Huylenbroeck, J. and Van BockStaele, E. 2011b. Introgression of yellow flower colour in *Buddleja davidii* by means of polyploidisation and interspecific hybridisation. *HortScience (Prague)* 38(3), 96–103. doi:10.17221/61/2011-HORTSCI.

Van Laere, K., Hokanson, S. C., Contreras, R. and Van Huylenbroeck, J. 2018. Woody ornamentals of the temperate zone. In: Van Huylenbroeck, J. (Ed.), *Ornamental Crops, Handbook of Plant Breeding 11*. Springer, Cham, pp. 803–87. doi:10.1007/978-3-319-90698-0_29.

Van Laere, K., Heungens, K., Gehesquière, B., Leus, L., Hermans, D. and Van Huylenbroeck, J. 2019. Breeding and selection of *Buxus* for resistance to *Calonectria pseudonaviculata*. *J. Phytopathol.* 167(6), 363–70. doi:10.1111/jph.12807.

Van Schie, C. C. N. and Takken, F. L. W. 2014. Susceptibility genes 101: how to be a good host. *Annu. Rev. Phytopathol.* 52, 551–81. doi:10.1146/annurev-phyto-102313-045854.

Van Tuyl, J. M. and De Jeu, M. J. 1997. Methods for overcoming interspecific crossing barriers. In: Sawhney, V. K. and Shivanna, K. R. (Eds), *Pollen Biotechnology for Crop Production and Improvement*. Cambridge University Press, Cambridge UK, pp. 273–93.

Van Tuyl, J. M. and Van Holsteijn, H. C. M. 1996. Lily breeding research in the Netherlands. *Acta Hortic.* 414(414), 35–46. doi:10.17660/ActaHortic.1996.414.3.

Van Tuyl, J. M., De Vries, J. N., Bino, R. J. and Kwakkenbos, T. A. M. 1989. Identification of 2n-pollen producing interspecific hybrids of *Lilium* using flow cytometry. *Cytologia* 54(4), 737–45. doi:10.1508/cytologia.54.737.

Van Tuyl, J. M., Maas, I. W. G. M. and Lim, K.-B. 2002. Introgression in interspecific hybrids of lily. *Acta Hortic.* 570(570), 213–8. doi:10.17660/ActaHortic.2002.570.25.

Van Tuyl, J. M., Arens, P., Shahin, A., Marasek-Ciolakowska, A., Barba-Gonzalez, R., Kim, H. T. and Lim, K.-B. 2018. *Lilium*. In: Van Huylenbroeck, J. (Ed.), *Ornamental Crops, Handbook of Plant Breeding 11*. Springer, Cham, pp. 481–512. doi:10.1007/978-3-319-90698-0_20.

Veneklaas, E. J., Lambers, H., Bragg, J., Finnegan, P. M., Lovelock, C. E., Plaxton, W. C., Price, C. A., Scheible, W. R., Shane, M. W., White, P. J. and Raven, J. A. 2012. Opportunities for improving phosphorus-use efficiency in crop plants. *New Phytol.* 195(2), 306–20. doi:10.1111/j.1469-8137.2012.04190.x.

Vlot, A. C., Dempsey, D. A. and Klessig, D. F. 2009. Salicylic acid, a multifaceted hormone to combat disease. *Annu. Rev. Phytopathol.* 47, 177–206. doi:10.1146/annurev.phyto.050908.135202.

Vo, T. C., Mun, J. H., Yu, H. J., Hwang, Y. J., Chung, M. Y., Kim, C. K., Kim, H. Y. and Lim, K. B. 2015. Phenotypic analysis of parents and their reciprocal F-1 hybrids in *Phalaenopsis*. *Hortic. Environ. Biotechnol.* 56(5), 612–7. doi:10.1007/s13580-015-0063-8.

Wang, K., Bai, Z. Y., Liang, Q. Y., Liu, Q. L., Zhang, L., Pan, Y. Z., Liu, G. L., Jiang, B. B., Zhang, F. and Jia, Y. 2018a. Transcriptome analysis of chrysanthemum (*Dendranthema grandiflorum*) in response to low temperature stress. *BMC Genomics* 19(1), 319. doi:10.1186/s12864-018-4706-x.

Wang, S., Leus, L., Van Labeke, M. C. and Van Huylenbroeck, J. 2018b. Prediction of lime tolerance in *Rhododendron* based on herbarium specimen and geochemical data. *Front. Plant Sci.* 9, 1538. doi:10.3389/fpls.2018.01538.

Winkelmann, T., Doil, A., Reinhardt, S. and Ewald, A. 2010. Embryo rescue. In: Davey, M. and Anthony, P. (Eds), *Plant Cell Culture – Essential Methods*. John Wiley & Sons, Chichester, UK, pp. 79–95.

Winter, K. 1973. NaCl-induzierter Crassulaceensäurestoffwechsel bei einer weiteren *Aizoaceae*: *Carpobrotus edulis*. *Planta* 115(2), 187–8. doi:10.1007/BF00387783.

Younis, A., Hwang, Y. J. and Lim, K. B. 2014. Exploitation of induced 2n-gametes for plant breeding. *Plant Cell Rep.* 33(2), 215–23. doi:10.1007/s00299-013-1534-y.

# Chapter 4

# Advances in tissue culture techniques for ornamental plant propagation

*G. R. Rout, Odisha University of Agriculture and Technology, India; and S. Mohan Jain, University of Helsinki, Finland*



## 1 Introduction

Ornamental horticulture is an important branch of horticulture. A wide diversity of ornamental plants are used in various areas such as home gardening, professional landscaping, and as cut flowers (Jain, 2002; Dobres, 2011). Thousands of varieties of cutflowers, pot plants, hanging plants, bedding plants, shrubs, lawn and turf, ornamental tree and aquatic plants are propagated commercially (Rout et al., 2006). In addition to their traditional uses in horticulture, ornamental plants have a new potential role to play in carbon sequestration to mitigate climate change (Brethour et al., 2007; Hall and Dickson, 2011). In addition, planting trees and other ornamental plants around a building can significantly reduce extreme temperatures in the ambient environment, thus lowering the energy cost of heating and cooling and, in turn, reducing its environmental impact (Bowler et al., 2010). Interior plants can also help regulate the temperature and humidity of an office environment to promote human health and comfort (Lohr and Pearson-Mims, 1996). Ornamental plants have also been associated with improved mental well-being (Dunnett and Qasim, 2000; Chebet et al., 2003; Younis et al., 2008).

To meet these needs the floriculture industry requires new varieties with traits such as improved plant architecture, vase life and keeping quality, colour and disease resistance (Chandler and Sanchez, 2012; Azadi et al., 2016). To complement traditional breeding techniques, a range of biotechnological techniques have been developed (Chebet et al., 2003; Noman et al., 2016). This chapter reviews the application of *in vitro* tissue culture techniques for

http://dx.doi.org/10.19103/AS.2020.0066.04

**Table 1** *In vitro* propagation via meristem culture

| Species/cultivars | Explant used | Response | References |
|---|---|---|---|
| *Alocasia micholitziana* 'Green Velvet' | axm | ms, r, pf | Thao et al. (2003) |
| *Anthurium andraeanum* | axm, apm | ms, r, pf | Pierik et al. (1975), Pierik (1976) |
| *Anthurium scherzerianum* (flamingo flower) | axm, apm | ms, r, pf | Liu and Xu (1992) |
| *Anthurium* spp. | apm | ms, r, pf | Matsumoto and Kuehnle (1997) |
| *Anthurium parvispathum* | axm | ms, r, pf | Atta-Alla et al. (1998) |
| *Begonia × elatior* cvs. Aphrodite Rosa, Claudis Mayer, Mayers Rote, Mayers Rosa | axm, apm | ms, r, pf | Reuter and Bhandari (1981) |
| *B. × hiemalis* cv. Schwabenland Red | apm, le, pt | ms, r, pf | Simmonds (1984) |
| *B. × elatior* | axm | ms, r, pf | Jain (1997) |
| *Chrysanthemum morifolim* cvs. Blue Bird, Montana, Meladion, Dalaware | axm | ms, r, pf | Wang and Ma (1978) |
| *C. morifolium* cvs. Winter westland, Yellow westland, Dark westland, Snowdon, Yellow Snowdon, Altis, Blanche | apm, axm | ms, r, pf | Ahmed and Sagi (1983) |
| *C. morifolium* | apm, axm | ms, r, pf | Kaul et al. (1990) |
| *C. morifolium* | apm, axm | ms, r, pf | Mandal et al. (2000); Belarmino and Gabon (1999) |
| *Dracaena deremensis* cv. Warneckii | axm | ms, r | Debergh (1975) |
| *Dracaena marginata* Tricolour | axm | ms, r, pf | Chua et al. (1981) |
| *Euphorbia fulgans* | axm | ms, r, pf | Jianhua et al. (1997) |
| *Ficus lyrata* | axm | ms, r | Debergh and DeWael (1977) |
| *Ficus benjamina* cv. Golden King | apm, axm | ms, r | Gabryszewska and Rudnicki (1997) |
| *Ficus carica* var. Bursa siyahi | axm | ms, r, pf | Demiralay et al. (1998) |
| *Ficus carica* cvs. Berbera, Lampa | axm | ms, r, pf | Nobre and Romano (1998) |
| *Ficus religiosa* | apm, axm | ms, r, pf | Deshpande et al. (1998) |
| *Ficus religiosa* | apm, axm | ms, r, pf | Nagaraju et al. (1998) |
| *Ficus carica* cv. Gular | apm, axm | ms, r, pf | Kumar et al. (1998) |
| *Ficus benjamina* cvs. Natasja, Starlight | apm, axm | ms, r, pf | Rzepka-Plevnes and Kurek (2001) |
| *Petunia hybrida* | apm | ms, r, pf | Sharma and Mitra (1976) |
| *Pelargonium × hortorum* | apm, axm | ms, r | Horst et al. (1976); Skirvin and Janick (1976) |
| *Pelargonium* spp. | apm, axm | ms, r, pf | AboEl-Nil (1983) |
| *Pelargonium* spp. | apm, axm | ms, r, pf | Theiler (1977) |

| Species/cultivars | Explant used | Response | References |
|---|---|---|---|
| *Pelargonium zonale* hybrid | apm, axm | ms, r, pf | Jalaska and Jelencic (1980) |
| *Rhododendron* spp. | apm, axm | ms, r | Economou and Read (1984) |
| *Rhododendron* spp. | apm, axm | ms, r, pf | Anderson (1984) |
| *Rhododendron* spp. | apm, axm | ms, r, pf | Norton and Norton (1985) |
| *Rhododendron* P.J.M. hybrid | apm, axm | ms, r | Ettinger and Preece (1985) |
| *Rhododendron 'Montego'* | | ms, r, pf | Brand and Kiyomoto (1997) |
| *Rosa hybrida* cvs. Crimson Glory, Glenfiditch | apm, axm | ms, r, pf | Barve et al. (1984) |
| *Rosa hybrida* cv. Amanda | axm | ms, r | De Vries and Dubois (1988) |
| *Rosa hybrida* cvs. Landora, Virgo, Happiness, Sea Pearl, Super Star, Queen-Elizabeth | apm, axm | ms, r, pf | Rout et al. (1990) |
| *Rosa chinensis* var. minima. cvs. Debut, Ginny | apm, axm | ms, r, pf | Rogers and Smith (1992) |
| *Rosa chinensis* var. minima (cvs. Baby Katie, Lavender Jewel, Red Sunblaze, Royal Sunblaze) | apm, axm | ms, r, pf | Chu et al. (1993) |
| *Hybrid tea* 'Dr. Verhage' | apm, axm | ms, r, pf | Voyiatzi et al. (1995) |
| Hybrid tea rose cv. Peace | apm, axm | ms, r, pf | Ara et al. (1997) |
| *Saintpaulia ionantha* | apm, axm | ms, r, pf | Molgaard et al. (1991) |
| *Saintpaulia ionantha* | apm, axm | ms, r, pf | Hoshino et al. (1995) |
| *Saintpaulia ionantha* | apm, axm | ms, r, pf | Jain (1997) |
| *Saintpaulia ionantha* cvs. Benjamin, William | apm, axm | ms, r, pf | Mithila et al. (2003) |
| *Spathiphyllum cv.* Clevelandii | apm, axm | ms, r, pf | Fonnesbech and Fonnesbech (1979) |
| *Spathiphyllum* | apm, axm | ms, r, pf | Orlikowska et al. (1995); Warbrouck and Debergh (1995) |
| *Spathiphyllum* 'Petite' | apm, axm | ms, r, pf | Watad et al. (1997) |
| *Yucca aloifolia* | axm | ms, r, pf | Atta-Alla and Van Staden (1997) |

Abbreviation: axm - axillary meristem, apm - apical meristem, ms - multiple shoot, r - rooting, Pf - plantlet formation, le - leaf explants, pt - petiole.

the propagation of ornamental plants. The *in vitro* plant cell culture technique is the culture of plant cells, tissues or organs grown in a synthetic medium in controlled conditions to produce large number of true-to-type plants. It is also called as the cell, tissue and organ culture (Debergh and Read, 1991). There are different variants of *in vitro* culture such as shoot tip culture, callus culture, organ culture, suspension culture, single cell culture, embryo culture, anther culture, root culture, ovule culture, pollen culture, protoplast culture, somatic embryogenesis and thin layer cell culture (Tables 1 and 2).

**Table 2** *In vitro* propagation of ornamental horticultural plants via organogenesis

| Species/cultivars | Explant used | Response | References |
|---|---|---|---|
| *Anthurium patulum* | lf, | ads, r | Eapen and Rao (1985) |
| *Anthurium andraeanum* cvs. Tinora Red, Senator | lf, pt | ads, r, pd | Martin et al. (2003) |
| *Anthurium andraeanum* Cvs. 'Arizona' and 'Sumi' | lf | c, sbr, r | Çimen and Özge (2009) |
| *Begonia × elatior* cvs. Aphrodite Rose, Aphrodite Rose Pale, Nixe, Schwabenland Orange, Tacora | lf, st | ads, r, pd | Bigot (1981a,b) |
| *Begonia × elatior* cvs. Krefeld Orange, Schwabenland Orange, Schwabenland Pink, Schwabenland Red | lf, st | sbr, r, pd | Takayama and Misawa (1982) |
| *B. tuberhybrida* | st | sbr, r | Peak and Cumming (1984) |
| *C. morifolium* cv. Super Yellow | st, lf | ads, r, pd | Lazar and Cachita (1983) |
| *C. hortorum* cvs. Pink Camino, Super Yellow, Spider | St | sbr, r, pd | Gertsson and Andersson (1985) |
| *C. coccineum* | St | sbr, r, pd | Fujii and Shimzu (1990) |
| *C. morifolium* cv. Royal Purple | st | sbr, r, pd | Lu et al. (1990) |
| *C. morifolium* | st, lf | ads, r, pd | Bhattacharya et al. (1990) |
| *C. maximum* | | ads, r, pd | Kumar and Kumar (1995) |
| *C. morifolium* cv. Deep Pink | st, pt | sbr, r, pd | Rout et al. (1996) |
| *Cyclamen persicum* | st, pt | ads, r, pt | Geier (1977); Bouman et al. (2001) |
| *Cyclamen persicum* | st, pt | ads, r, pd | Ando and Murasaki (1983) |
| *Cyclamen persicum* | st, pt | ads, r, pd | Wainwright and Harwood (1985) |
| *Cyclamen persicum* | st, pt, lf | ads, r, pd | Hawkes and Wainwright (1987) |
| *Euphorbia pulcherrima* | st, ca | ads, r, pd | Langhe et al. (1974) |
| *Euphorbia pulcherrima* cv. Angelika | st, ca | ads, r, pd | Osternack et al. (1999) |
| *Ficus religiosa* | lf, st | sbr, r | Narayan and Jaiswal (1986) |
| *Petunia hybrida, Petunia inflate* | lf, st, ca | sbr, r, pd | Rao et al. (1973) |
| *Rosa hybrida* cv. Bridal Pink | lf, st, ca | ads, r | Burger et al. (1990) |
| *Rosa damascena* | lf, st | ads, r, pd | Ishiooka and Tanimoto (1990) |
| *Rosa hybrida* cv. Landora | lf, st | ads, r, pd | Rout et al. (1992) |
| *Rosa multiflora* | st, lf, ca | ads, r, pd | Rosu et al. (1995) |
| *Rosa hybrida* | st, ca | ads, r, pd | Van der Salm et al. (1996); Lloyd et al. (1988) |
| *Saintpaulia ionantha* | st, pt | sbr, r, pd | Starts and Cumming (1976), Grunewaldt (1977), Vasquez et al. (1977) |

| Species/cultivars | Explant used | Response | References |
|---|---|---|---|
| *Saintpaulia ionantha* 2 confusa hybrids | pt, lf | sbr, r, pd | Lo et al. (1997) |
| *Spathiphyllum floribundum* cv. Petite | pt, lf | sbr, r | Werbrouck and Debergh (1995) |

Abbreviation: lf - leaf, ca- callus, st - stem, pt - prtiole, sbr - shoot bud regeneration, r- rooting, pd - plantlet development.

About 156 ornamental genera are propagated through tissue culture techniques in different commercial laboratories in the world. Major producers are the Netherlands (33%), Japan (24%), Italy (11%), the United States (12%), Thailand (10%) and others (14%) (O'Riordain, 1999; Schiva, 2000; Anon., 2003). The major exporting countries are the Netherlands (59%), United Kingdom (8%), Colombia (10%), Italy (16%), Israel (4%), Spain (2%) and Kenya (1%). The five leading exporters (the Netherlands, United Kingdom, Colombia, Italy and Israel) constitute about 80% of the world market. The share of developing countries in Africa, Asia and Latin America (including countries such as Thailand, Ecuador and India) is less than 20% (Rajagopalan, 2000).

## 2 Factors affecting tissue culture techniques

A key foundation of tissue culture is somoclonal variation, that is, variation among regenerated plants derived from tissue culture. Factors affecting the control of variability (e.g. to improve traits such as flowering have been reviewed by many authors (Skirvin, 1978; Larkin and Scowcroft, 1981; Scowcroft and Larkin, 1988; Jain et al., 1998a,b,c; Skirvin et al., 1993; Bouman and DeKlerk, 1996; Kawata and Oono, 1997; Jain, 1997; Jain and DeKlerk, 1998; Jain, 2001)).

The success of any of these mass propagation methods depends on various factors such as (Kim et al., 2003a):

- genotype
- type of explants
- nutrient culture media

Protocols for mass propagation need to be tailored to species/genotypes as differences exist in organogenesis potential and morpho-physiological response across plant families, genera and species; even genotypes of a species can show different responses to culture conditions (George, 1993, 1996; Cassells and Curry, 2001; Lee, 2004; Rout et al., 2006).

Selected explants need physiological adjustment to culture conditions to achieve an enhanced clonal propagation (Lee, 2004; Kozai and Xiao, 2006;

Yıldız, 2011). Response also varies with the type of donor explants, particularly the age of selected explants (Hagege, 1995; Brar and Jain, 1998; Bairu and Kane, 2011). Successful regeneration/propagation also depends on the size of the explant. Larger explants such as tubers and bulbs with more nutritional reserves can easily regenerate *in vitro* and are less dependent on nutrients and hormones in the growth medium (Pierik, 1987). The explant position also has an impact on mass propagation with, for example, explants excised from the top of the seedling more successful in propagation (Evers, 1984; Pierik, 1987). The explants subsequently grown under greenhouse conditions give rise to better results than the ones grown in field conditions though it is important to manage the transition to greenhouse conditions (Pierik, 1987; Pospisilova et al., 1999; Gürel and Türker, 2001; Hazarika, 2003, 2006). As an example, carbon dioxide enrichment in the greenhouse has a positive impact on the production, reducing water stress of micro-cuttings by closing the stomata (Matysiak and Nowak, 1995).

The most important treatment prior to culture initiation is surface sterilization of the explant. Surface sterilization aims to eliminate all microorganisms so that explants will easily grow under *in vitro* conditions with a high regeneration capacity. A wide range of surface disinfectants, such as ethanol, hydrogen peroxide, bromine water, mercuric chloride, silver nitrate and antibiotics, are used for surface sterilization. Sodium hypochlorite (NaOCl) has been the most widely used disinfectant. NaOCl is highly effective against all kinds of bacteria, fungi and viruses (Mercer and Somers, 1957; Smith, 1968; Spaulding, 1968). Moreover, NaOCl has a strong oxidizing property which makes it highly reactive with amino acids (Bietz and Sandford, 1971; Kantouch and Ardel-Fattah, 1971), nucleic acids (Hayatsu et al., 1971), amines and amides (Sandford et al., 1971a,b). The general reaction between amino acids and NaOCl produces the aldehyde, $NH_4Cl$, and $CO_2$ (Kantouch and Ardel-Fattah, 1971).

A nutrient culture medium for tissue culture is typically composed of inorganic salts, a carbon source, vitamins, growth regulators and a gelling agent. Carbohydrate is one of the major components for nutrient media as a carbon source. Concentration of the carbon source and other nutrients, light intensity and $CO_2$ levels in the culture vessel are the limiting factors that can induce physiological and structural modifications of *in vitro* culture plantlets (Van Huyelenbroeck and Debergh, 1996; Pospisilova et al., 1999; Ticha et al., 1998; Isah, 2015).

*In vitro* culture also depends on the balanced use of plant growth regulators like auxins and cytokinins (Rout et al., 2000; Kane, 2005; Preil, 2003; Rout and Jain, 2004; van Staden et al., 2006; Smulders and de Klerk, 2011). Addition of cytokinins into the culture medium induces morphogenesis or rejuvenation of explants while auxins are necessary for cellular division and differentiation, callus induction and somatic embryogenesis (Takayama and Misawa 1981, 1982; Meins, 1989; Mendoza and Kaeppler, 2002; Preil, 2003). Low concentrations of cytokinins promote a high rate of shoot bud regeneration in different ornamental horticultural plants (Murashige and Skoog, 1962; Reuter and Bhandari, 1981; Bigot, 1981a,b; Roest et al., 1981; Mikkelson and Sink, 1978a,b; Welander, 1977, 1979, 1981; Simmonds, 1984; Appelgren, 1976; Rout et al., 1992; Pati et al., 2006; Jain et al., 2006). Examples of successful use of cytokinins include *Begonia* (Jain, 1997), *Ficus* species (Gabryszewska and Rudnicki, 1997; Debergh and DeWael, 1977; Nobre and Romano, 1998) and *Euphorbia pulcherrima* (Osternack et al., 1999).

Activated charcoal adsorbs growth regulators, prompting a better rooting response. As an example, the addition of 1-2 g/L activated charcoal in the culture medium increased rooting efficiency from excised shoots of *Begonia × hiemalis* (Bigot, 1981a,b). The use of half-strength MS salts has also been found to promote a higher rate of shoot bud regeneration (Beyramizade et al., 2008; Çimen and Özge, 2009).

## 3 Tissue culture techniques

### 3.1 Clonal propagation and organogenesis

Established tissue culture techniques include clonal propagation: the multiplication of genetically identical individuals by asexual reproduction (Steward et al., 1958). It involves the stimulation of axillary meristems which are usually present in the axil of each leaf to develop into a shoot. In some species, removal of the apical meristem is essential to break the apical dominance and stimulate the bud to grow into shoot. Clonal propagation has been reported for a range of species, including:

- *Dracaena deremensis* (Debergh, 1975, 1976; Miller and Murashige, 1976; Chua et al., 1981).
- *Rosa multiflora* (Elliott, 1970; Hasegawa, 1979, 1980; Rout et al., 1989, 1999; Skirvin and Chu, 1979; Khosh-Khui and Sink, 1982a,b,c; Bressan et al., 1982; Chu et al., 1993; Kintzios et al., 1999; Singh and Syamal, 2000; Chakrabarty et al., 2000; Kim et al., 2003c; Rout and Jain, 2004).
- Rhododendron (Ettinger and Preece, 1985; McCown and Lloyd, 1982; Brand and Kiyamoto, 1994, 1997).
- *Spathiphyllum* (Fonnesbech and Fonnesbech, 1979; Orlikowska et al., 1995; Wated et al., 1997).
- *Ficus benjamina* (Gabryszewska and Rudnicki, 1997; Rzepka-Plevnes and Kurek, 2001); *Ficus carica* (Demiralay et al., 1998).

*In vitro* organogenesis is another established micropropagation technique (Atta-Alla et al., 1998; Martin et al., 2003; Joseph et al., 2003). It has been used for cyclamen (Geier, 1977; Geier et al., 1983; Schwenkel, 1991; Dillen et al., 1996), *Heuchera sanguinea* (Hosoki and Kajino, 2003), *Begonia* (Takayama, 1983), *Saintpaulia ionantha* (Jain, 1997) and *Alocasia micholitziana* (Thao et al., 2003).

### 3.2 Embryo rescue

Embryo rescue is one of the earliest tissue culture techniques. It has been used successfully where there are problems of low sexual reproduction and poor germination, for example, in the case of roses (Maluszynski et al., 1995; Mohapatra and Rout, 2005) (Fig. 1).

### 3.3 Thin cell layer (TCL)

Thin cell layer (TCL) technology is a simple and effective tool for *in vitro* regeneration and micropropagation. It is based on small explants derived from a limited number of cells from homogenous tissue (Teixeira da Silva, 2003a,b). Its efficiency is higher compared to conventional tissue culture techniques. The TCL method is also very useful in virus elimination in combination with antiviral compounds. The technology has also been effectively used in the micropropagation of various floricultural crops (Tran Thanh Van and Bui, 2000;

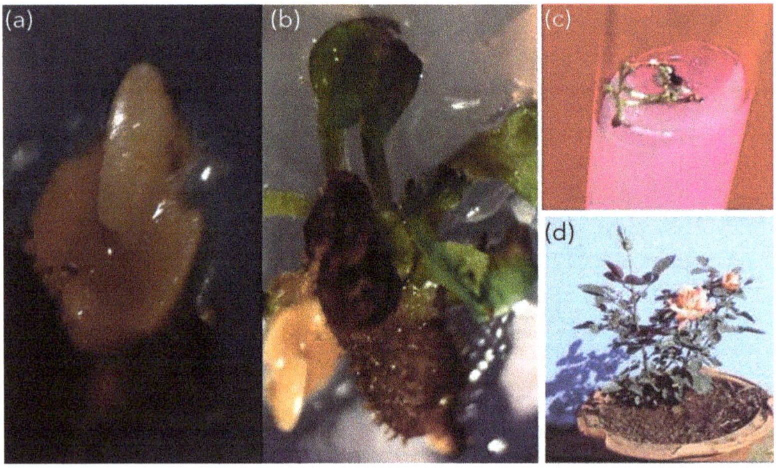


**Figure 1** Protocol on embryo rescue of floribunda rose, 'Arunima'. (a) Culture of immature embryo on MS basal medium supplemented with 2.5 mg/L BAP, 0.5 mg/L $GA_3$ and 3% sucrose under dark condition. (b) and (c) Embryos proliferated on MS medium and 3% sucrose under 16 h photoperiod. (d) Plant grown in the pot bearing flower under greenhouse condition (Mohapatra and Rout, 2005).

Fiore et al., 2002; Nhut et al., 2003a,b; Teixeira da Silva and Nhut, 2003). Examples of the use of TCL technique include: *P. hybrida* (Mulin and Tran Thanh Van, 1989), *S. ionantha* (Ohki, 1994), geranium (Gill et al., 1993) and *R. hybrida* (Hsia and Korban, 1996).

### 3.4 Somatic embryogenesis

*In vitro* somatic embryogenesis is an alternative method in achieving fast and large-scale plant propagation (Haccius, 1978). In somatic embryogenesis a plant or embryo is derived from a single cell. Cells sourced from the tissue are cultured using plant growth regulators. Key steps include: initiation and proliferation of an embryonic culture, prematuration and maturation of somatic embryos, and plant development (Fig. 2a–d). The embryos are developed either directly from somatic cells (direct embryogenesis) or from embryogenic structures (indirect embryogenesis). The major advantage of the process is the potential to propagate plants on a large scale. The disadvantages include genotypic dependence of somatic embryo production, malformation, asynchronous development, deficiencies in maturation and a low germination rate of somatic embryos (Winkelmann et al., 2015). Reviews of somatic embryogenesis in major ornamental plants are summarized in Table 3. Hohe

**Figure 2** Somatic embryogenesis in *Hybrid rose* cv. Landora. (a) Embryogenic callus from stem tissues on MS medium +2.2 uM BA,0.05 uM NAA,0.3uM GA3. (b) Development of group of somatic embryos. (c) Germination of somatic embryos. (d) Somatic embryo-derived plantlets established in growth chamber. (e) Established in the field.

**Table 3** *In vitro* propagation of ornamental plants via Somatic embryogenesis

| Species/cultivars | Culture response | References |
|---|---|---|
| *Begonia gracilis* | emc, gse, pt | Castillo and Smith (1997) |
| *Chrysanthemum morifolium* cv. Yellow Spider | emc, gse, pt | Sauvadet et al. (1990) |
| *C. morifolium* | emc, gse, pt | May and Trigiano (1991) |
| *C. morifolium* cv. Yellow Spider | emc, gse, pt | Pavingerova et al. (1994) |
| *C. morifolium* | emc, gse, pt | Tanaka et al. (2000) |
| *Cyclamen persicum* | ecs, gse, pt | Hohe et al. (2001), Schwenkel (1991) |
| *Cyclamen persicum* | emc, gse, pt | Pueschel et al. (2003); Hohe et al. (1999) |
| *Cyclam enpersicum* | emc, se, pt | Winkelmann et al. (2015) |
| *Clematis tangutica* | emc, pt | (Weber et al., 1994) |
| *Euphorbia pulcherrima* cv. Angelika | emc, gse, pt | Osternack et al. (1999) |
| *Rosa hybrida* cvs. Domingo, Vickey Brown, Tanja, Azteca | emc, gse | de Wit et al. (1990) |
| *Rosa hybrida* cv. Landora | emc, gse | Rout et al. (1991) |
| *Rosa rugosa* | emc, gse, pt | Kunitake et al. (1993) |
| *Rosa* sp. Cvs. Baccara, Mercedes, Ronto, Soray | emc, gse, pt | Kintzios et al. (1999) |
| *Rosa hybrida, Rosa chinensis minima* | emc, gse, pt | Li et al. (2002a) |
| *Rosa hybrida* cv. Sumpath | emc, gse, pt | Kim et al. (2003c) |
| *Saintpaulia ionantha* cvs. Benjamin, William | emc, gse, pt | Mithila et al. (2003) |
| *Saintpaulia ionantha* cv. Benjamin | emc, gse, pt | Murch et al. (2003) |

Abbreviation: emc - embryogenic callus, ecs - embryogenic cell suspension, gse - germination of somatic embryos, pt - plantlet development.

(2010) has reviewed developments in somatic embryogenesis in ornamental plants.

Considerable success has been achieved in using somatic embryogenesis in ornamental plants such as:

- Rose (*R. hybrida*) (de Wit et al., 1990; Short and Roberts, 1991; Rout et al., 1991; Kunitake et al., 1993; Roberts et al., 1995; Rout and Jain, 2004)
- Chrysanthemum (*Dendrathema grandiflorum*) (May and Trigiano, 1991; Tanaka et al., 2000)
- *Cyclamen persicum* (Wicart et al., 1984; Winkelmann et al., 1998; Hohe et al., 2001; Pueschel et al., 2003)
- *Begonia gracilis* (Castillo and Smith, 1997)
- *Saintpaulia ionantha* cv. Benjamin (Murch et al., 2003)
- *E. Pulcherrima* (Osternack et al., 1999)
- *Gladiolus* and *Caladium* (Mujib et al., 2016)

These studies suggest that the factors affecting success include: the orientation of the petiole explants and auxin transport and control of the number of subcultures from embryogenic cell suspension cultures (Hohe et al., 2001; Bouman et al., 2001; Pueschel et al., 2003; Murch et al., 2003).

Direct somatic embryogenesis has also recently been successfully used in various orchids including:

- *Phalaenopsis* (Chen and Chang, 2006; Gow et al., 2009, 2010)
- *Dendrobium* (Chung et al., 2005, 2007)
- *Epipactis* (Moradi et al., 2017)
- *Malaxis* (Mahendran and Bai, 2016)
- *Rhynchostylis* (Islam and Bhattacharjee, 2015)
- *Tolumnia*-related genus, *Oncidium* (Chen and Chang, 2002; Su et al., 2006; Hong et al., 2008a,b; Shen et al., 2018)
- *Anthurium* (Bhavana et al., 2018)

As with other studies, research suggests that plant growth regulators, such as auxins, cytokinins, ethylene and abscisic acid (ABA), have a major role in triggering the signalling response to initiate somatic embryos (Méndez-Hernández et al., 2019).

## 3.5 The use of tissue culture in mutagenesis and mutation breeding

In recent decades induced mutations have been at the forefront of breeding improved varieties with desirable traits such as improved yield, plant architecture, disease resistance and abiotic stress tolerance (Jain and Suprasanna, 2011; Lusser et al., 2012; Hallerman and Grabau, 2016; Penna and Jain, 2017). High-throughput DNA sequencing methods such as next-generation sequencing have made mutation detection more efficient and cost-effective (King et al., 2015). New techniques include artificially engineered hybrid enzymes, zinc-finger nucleases (ZFNs), transcription activator-like effector nucleases (TALENs) and the CRISPR (clustered regularly interspaced short palindromic repeats)-Cas system for genome editing of plants (Kumar and Jain, 2015), though there are, as yet, few reports of their use in ornamentals.

The heterozygous nature of many ornamentals results in a high mutation frequency rate. The resulting success of mutation breeding of ornamental plants in improving phenotypic characteristics such as colour, shape or size of flower and chlorophyll variegation in leaves has been reviewed by a number of authors (Jain, 2006; Datta, 2009; Suprasanna et al., 2017). As a result, chrysanthemum has more than 250 varieties as listed mutants followed by rose with 67 (FAO, IAEA and MVDB, 2016). Some notable mutant cultivars in ornamentals include

*Amaryllis*, *Bougainvillea*, *Canna*, *Chrysanthemum*, *Dahlia*, *Gerbera*, *Gladiolus*, *Hibiscus*, hybrid lily, *Tagetes* spp., rose, tuberose and *Narcissus* spp. (Datta, 2009).

*In vitro* culture methods have facilitated the use of mutation techniques for improvement of both the seed and the vegetativelypropagated plants. In many vegetativelypropagated crops, mutation induction in combination with *in vitro* culture techniques may be the only effective method for plant improvement (Jain, 2002). There has been considerable work done on the induced mutations in ornamentals such as roses using techniques like ethylmethanesulphonate (Kaicker, 1982) and ionizing radiation (Preil et al., 1983, 1991; Benetka, 1985; Smilansky et al., 1986; Walther and Sauer, 1986; Huttema et al., 1989, 1991). Recent examples of the use of radiation to induce mutations in ornamentals include chrysanthemum (Lamseejan et al., 2003), rose (Senapati and Rout, 2011; Baig et al., 2012), *D. grandiflora* (Mandal et al., 2000; Rout and Das, 1997) and *Dendranthema* (Misra et al., 2004). The improved characteristics have included colour, scent and stature.

## 3.6 Slow-growth conservation techniques

Short- and medium-term conservation is routinely used in many laboratories in order to increase the intervals between subcultures required for multiplication in tissue culture. At the end of a storage period, cultures are transferred onto a fresh medium and usually placed for a short period in optimal conditions to stimulate re-growth before entering the next storage cycle (Engelmann, 2012; Rakosy-Tican et al., 2012; Alberto Cruz-Cruz et al., 2013). In the case of ornamentals, storage in dark conditions is commonly used for conservation, using shorter photoperiods (8–12 h), instead of the 16 h commonly used as standard in *in vitro* plant cultivation (Reed et al., 2003; Pedroso et al., 2010; Carvalho et al., 2014; Arda et al., 2016). The use of low-light intensities (3–8 μmol $m^{-2s-1}$) has also been shown to be effective in reducing plant metabolism (Reed et al., 2003). Different carbohydrates like sucrose, mannitol and sorbitol have also been used for *in vitro* slow-growth conservation to reduce the osmotic potential of the culture medium (Grout, 1991, 1995). Another strategy is the use of a low salt concentration in the culture medium to reduce plant growth and development (Ozudogru et al., 2013)

Cryopreservation of tissue and cells has been utilized for long-term storage of elite genetic material, including ornamentals such as chrysanthemum (Sakai, 1960; Fukai et al., 1988; Fukai and Oe, 1990; Chartier-Hollis et al., 1996). Prior to freezing, the cells must be treated with a cryoprotectant solution such as glycerol, dimethyl sulfoxide (DMSO) or ethylene glycol (Quatrano, 1968). These substances protect the cells and their membranes from damage during the freezing process. After the cells have been exposed to the freezing medium containing the cryoprotectant, they must be dehydrated so that the water inside the cells will not form ice crystals that can damage cell tissues. The cells/tissues are dehydrated and cooled slowly prior to plunging them into liquid nitrogen. Hitmi et al. (2000) have used a simple method for efficient cryopreservation of *Chrysanthemum cinerariaefolium*. The shoot tip explants were treated with 0.55 M sucrose and 4 mM ABA for 3 days and immersed in liquid nitrogen and subsequently rapid warming was done at 4°C. The treated shoot tips were cultured on a solid nutrient medium containing 11 μM NAA and 4.5 mM BA, with about 75% shoot tips proliferated into plantlets. This technique helps in *in vitro* conservation of germplasm on a long-term basis.

## 3.7 Bioreactor/liquid culture

Bioreactors provide a fast and efficient plant propagation system for many horticultural species, using a liquid medium to reduce production costs and allow scaling up (Takayama and Misawa, 1981; Preil, 1991; Preil and Beck, 1991; Preil et al., 1991; Ziv, 2005a; Yesil-Celiktas et al., 2010). Bioreactors with computer control systems offer various advantages over conventional culture techniques including more precise process control, automation and reduced labour costs (Aitkens-Christie, 1991; Preil, 1991; Ziv, 1991, 1992, 1995; Aitkens-Christie et al., 1995; Paek et al., 2001; Curtis, 2002; Eide et al., 2003).

Eide et al. (2003) and Welander et al. (2014) have assessed two liquid culture systems for plant propagation: temporary immersion systems and permanent submersion of the plant cells/tissue, with the former better for shoot proliferation. Bioreactor systems have been combined with meristem, embryogenic and organogenic cell culture techniques (Preil et al., 1988; Takayama and Akita, 1994; Takayama, 2000, 2002; Eide et al., 2003). Problems in using bioreactors include contamination, which can be addressed by the use of a disposable pre-sterilized plastic bioreactor (Ziv, 2005a,b), and hyperhydricity (morphogenic malformation), which can be controlled by adding growth retardants to decrease rapid proliferation. This technology has been successfully used in micropropagation of plants, including ornamentals such as Chrysanthemum, Lilium, Phalaenopsis and Cyclamen (Takayama and Akita, 1994, 1998; Ziv, 2000; Paek et al., 2001). Examples include: *Begonia* (Takayama et al., 1986), *C. persicum* (Hvoslof-Eide and Munster, 2001), chrysanthemum (Kim et al., 2003b) and *Clematis tangutica* Preil (2003)

## 3.8 Plant improvement through genetic engineering

The use of biotechnology to improve traits in ornamental plants suchas colour, fragrance, shape and vase life has been reviewed by Casanova et al. (2005) and Noman et al. (2017). Table 4 summarizes examples of the use of genetic engineering in ornamentals breeding. Particular targets include fragrance

**Table 4** Genetic improvement of ornamental plants via genetic transformation

| Species/Cultivars | Foreign genes | References |
|---|---|---|
| *Begonia tuberhybrida* | rol A, B and C | Kiyokawa et al. (1996) |
| Carnation (*Dianthus caryophyllus L.*) | GUS, NPT II | Lu et al. (1991) |
| *Cyclamen persicum* | GUS, NPT II, HPT (hygromycin phosphotransferase) | Aida et al. (1999) |
| *Chrysanthemum morifolium* | NPT II, GUS | Van Wordragen et al. (1991) |
| *Chrysanthemum* (*Dendranthema grandiflorum* cv. Kitamura) | NPT II, GUS | Seiichi and DeJong, (1995) |
| *Dendranthema grandiflorum* cvs. Polaris, Hekla, Iridon | GUS, NPT II | Sherman et al. (1998) |
| *Chrysanthemum morifolium* cv. Peach Margaret | NPT II | Boase et al. (1998a,b) |
| *Dendranthema grandiflorum* | GUS | Seo et al. (2003) |
| *Petunia hybrida* var. Ultra Blue | Delta-9, fatty acid desaturase | Choudhury et al. (1994) |
| *Petunia hybrida* | Tryptophan decarboxylase [aromatic L-aminoacid decarboxylase] | Thomas et al. (1999) |
| (*Petunia axillaries*× (*P. axillaris* × *P. hybrida*)) | rol C | Winefield et al. (2000) |
| *Petunia hybrida* | [naringenin-]chalcone synthase-A (CHS A) | Shao et al. (1996) |
| *Petunia hybrida* | GUS (uid A) naringenin-chalcone synthase (GTCHSI) | Kobayashi et al. (1998) |
| *Pelargonium* (*Pelargonium* × *domesticum*) 'Dubonnet' | NPT II [Neo], GUS A, als, dfr | Boase et al. (1996) |
| *Pelargonium* × *hortorum* | NPT II, hygromycin B phosphotransferase, GUS | Robichon et al. (1995) |
| *Pelargonium* cv. Frensham | Antimicrobial protein (*Ace-AMP1*) | Bi et al. (1999) |
| *Rhododendron* cultivars ('America', 'Catawbiense grandiflorum roseum', 'Madame Carvalho', 'Mars' and 'Nova Zembla') *Rhododendron* species | GUS, NTP II GUS (uid A), NPT II | Pavingerova et al. (1995, 1997) |
| *Rhododendron yakushimanum* cv. Percy Wiseman | NPT II, GUS | Ueno et al. (1996) |
| *Rhododendron* hybrids cvs. Hino-Crimson, Fuchsia | uid A, HPT | Hsia et al. (1998) |

| Species/Cultivars | Foreign genes | References |
|---|---|---|
| *Rosa hybrida* cv. Moneyway | NPT II, rol A, B and C | Vander Salm et al. (1997) |
| *Rosa hybrida* cv. Glad Tidings | GUS (uid A), chitinase | Marchant et al. (1998a,b) |
| Floribunda roses cvs. Heckenzauber, Pariser charme | Chitinase, GUS | Dohm et al. (2002) |
| *Rosa hybrida* cv. Carefree Beauty | uid A<br>*Ace-AMP1*, NPT II | Li et al. (2002b, 2003) |
| *Saintpaulia ionantha* | uid A, NPT II | Mercuri et al. (2000) |

(Muhlemann et al., 2012) and leaf and flower longevity (Teixeira da Silva et al., 2013; Kamthan et al., 2016).

Genetic transformation of ornamentals, such as chrysanthemum, by using *Agrobacterium*-based gene vectors is a well-established technique (Ledger et al., 1991; Firoozabady et al., 1991a,b; Urban et al., 1992; Hutchinson et al., 1992; Van Wordragen et al., 1992; Lowe et al., 1993; Chandler and Sanchez, 2012). This technique has been extensively used in the genetic transformation of rose to improve traits such as disease resistance (Marchant et al., 1998a,b; Matthews et al., 1991; Firoozabady et al., 1991b, 1994; Robinson and Firoozabady, 1993; Van der Salm et al., 1997; Dohm et al., 2002; Li et al., 2002b, 2003; Condliffe et al., 2003). Other examples of genetic transformation include Carnation (*Dianthus caryophyllus L.*) (Lu et al., 1991), *D. grandiflora* (Teixeira da Silva and Fukai, 2002a,b; Seo et al., 2003), rhododendron (Hsia and Korban, 1998; Pavingerova et al., 1995, 1997), *Cyclamen* (Aida et al., 1999), *S. ionantha* (Mercuri et al., 2000) and *Petunia hybrida* (Winefield et al., 2000).

## 3.9 Use of molecular markers in tissue culture

Molecular markers associated with important traits could significantly reduce the time and cost involved in developing new genotypes because they help in selecting the best parents and accelerating the rate of genetic gain in the breeding programme (Table 5) (Ben-Meir et al., 1997; Li and Quiros, 2001; Dandekar, 2003; Perera et al., 2016). The types of markers used for molecular characterization of tissue-cultured plants which include: restriction fragment length polymorphisms (RFLPs), random amplified polymorphic DNA (RAPD), arbitrarily primed PCR (AP-PCR), sequence-tagged sites (STSs), inter-simple sequence repeat (ISSR), DNA amplification fingerprinting (DAF) and amplified fragment length polymorphism (AFLP) markers (Rout and Mohapatra, 2006).

**Table 5** Genetic purity assessment of micropropagated plants using molecular markers

| Name of plant | Molecular markers used | Results | References |
|---|---|---|---|
| *Alstromeria* spp. | RAPD | Genetic variation | Anastassopoulos and Keil (1996) |
| *Ananas comosus variegatum* | RAPD | Identification and analysis of somaclonal variation | El-Tarras et al. (1999) |
| *Chrysanthemum cultivars* | RAPD | Identification of cultivars find out of sporting and chimerism | Wolff (1996) |
| *Dendranthema grandiflora* | RAPD | Genetic variability | Wolff and Peters-vanrijn (1993) |
| *Dendranthema morifolium* | RAPD, ISSR, RFLP | Cultivar identification | Wolff et al. (1995) |
| *Eustoma grandiflorum* | RAPD | Identification of Somaclonal variant | Cheong et al. (1999) |
| *Gymnnadenia conopsea* | Microsatellite | Genetic variation | Gustafsson (2000) |
| *Iris ensata, I. germanica* | RAPD | Analysis of somatic hybrids | Shimizu et al. (1999) |
| *Pelargonium cultivars* | RAPD, RFLP, ISSR and AFLP | Identification of cultivars, Genetic variation | Lesur et al. (2001); Barcaccia et al. (1999) |
| *Pelargonium hortorum cv. Rio* | Flow cytometry, RAPD | Identification and assessment of somaclonal variants | Cassells and Plunkett (1984) |
| *Petunia hybrida, Cylamen persicum,* | RAPD | Genetic purity analysis | Jianhua et al. (1997) |
| *Phalaenopsis* cv. True Lady 'B79-19' | RAPD marker | Somaclonal variation | Chen et al. (1998) |
| *Rhododendron arborescence,* | RAPD | Clonal stability and genetic variability | Iqbal et al. (1995) |
| *Helianthus* sps | RAPD | Stability of somatic embryos | Binsfeld et al. (1999) |
| *Helianthus anus* | RAPD, ISSR | Transgenic plants | Binsfeld et al. (2000) |
| *Phalaenopsis* sps | RAPD | Somaclonal variant analysis | Chen et al. (1998) |
| *Ananas comous* variegatum | RAPD | Somaclonal variant analysis | El-tarras et al. (1999) |
| Hybrid rose | RAPD | Identification of mutant | Senapati and Rout (2011) |
| *Rosa hybrida* cvs First Red, Cri Cri, Pusa Gaurav | RAPD, ISSR | Genetic fidelity analysis | Senapati et al. (2012) |

Abbreviation: RAPD - randomly amplified polymorphic DNA; ISSR - inter-simple sequence repeats; RFLP - restriction fragment length polymorphism; AFLP - amplified fragment length polymorphism.

These markers have been used to characterize a range of ornamental plants (Williams et al., 1990; Rajapakse et al., 1992; Rayburn et al., 1993; Fu et al., 1994; Benner et al., 1995; Wolff et al, 1995; Paden et al., 1996; Benedetti et al., 2000). Specific examples include: chrysanthemum (Huang et al., 2000), rose (Debener et al., 1997, 2001; Ben-Meir et al., 1997; Benedetti et al., 2001), rhododendron (Kobayashi et al., 2000) and *Anthurium* (Ranamukhaarachchi et al., 2001). The role of molecular markers in micropropagation and *in vitro* culture of ornamental plants has been reviewed by Rout and Mohapatra (2006) and Teixeira da Silva et al. (2007). Roles include monitoring somaclonal variation and identifying genetic fidelity of micropropagated plants and desired genotype response to *in vitro* culture conditions for plants such as rose (Senapati and Rout, 2011; Senapati et al., 2012) and *Hibiscus* (Prasad, 2014).

## 4 Future trends

*In vitro* propagation of ornamental plants is widely used commercially for rapid plant multiplication and germplasm conservation. A key challenge is to reduce costs, for example, by using bioreactor technology, provided proper precautions are taken to prevent contamination. Tissue culture techniques will benefit from modern biotechnological tools, such as genome editing and gene silencing, in enhancing the quality and yield of ornamental plants.

## 5 References

AboEl-Nil, M. M. 1983. Geranium (*Pelargonium*). In: Ammirato, P. V., Evans, D. R., Sharp, W. R. and Bajaj, Y. P. S. (Eds), *Handbook of Plant Cell Culture* (vol. 5). McGraw Hill Publishing Co., New York, NY, pp. 439–60.

Ahmed, K. Z. and Sagi, F. 1983. Use of somaclonal variation and *in vitro* selection for induction of plant disease resistance: prospects and limitations. *Acta Phytopathol. Entomol. Hung.* 28, 143–59.

Aida, R., Hirose, Y., Kishimoto, S. and Shibata, M. 1999. *Agrobacterium tumefaciens*-mediated transformation of *Cyclamen persicum* Mill. *Plant Sci.* 148(1), 1–7. doi:10.1016/S0168-9452(99)00072-2.

Aitkens-Christie, J. 1991. Automation. In: Debergh, P. C. and Zimmerman, R. H. (Eds), *Micropropagation Technology and Application*. Kluwer Academic Publishers, the Netherlands, pp. 363–88.

Aitkens-Christie, J., Kozai, T. and Takayama, S. 1995. Automation in plant tissue culture: general introduction and overview. In: Aitken-Christie, J., Kozai, T. and Smith, M. A. L. (Eds), *Automation and Environmental Control in Plant Tissue Culture*. Kluwer Academic Publishers, the Netherlands, pp. 1–18.

Alberto Cruz-Cruz, C., González-Arnao, M. T. and Engelmann, F. 2013. Biotechnology and conservation of plant biodiversity. *Resources* 2(2),73–95. doi:10.3390/resources2020073.

Anastassopoulos, E. and Keil, M. 1996. Assessment of natural and induced genetic variation in *Alstromeria* using random amplified polymorphic DNA (RAPD) markers. *Euphytica* 90(2), 235–44. doi:10.1007/BF00023864.

Anderson, W. C. 1984. A revised tissue culture medium for shoot multiplication of Rhododendron. *J. Amer. Soc. Hortic. Sci.* 109, 343–7.

Ando, T. and Murasaki, K. 1983. *In vitro* propagation of *Cyclamen* by the use of etiolated petioles. *Tech. Bull. Fac. Hortic. Chiba Univ.* 32, 1–5.

Anon. 2003. Omzettabel Kamerplanten. *Vakblad Voorde Bloemisterij*. 21a, 136–7.

Appelgren, M. 1976. Regeneration of *Begonia hiemalis in vitro*. *Acta Hortic.* 64(64), 31–8. doi:10.17660/ActaHortic.1976.64.4.

Ara, K. A., Hossain, M. M., Quasem, M. A., Ali, M. and Ahmed, J. U. 1997. Micropropagation of rose: *Rosa* sp. cv. Peace. *Plant Tissue Cult.* 7(2), 135–42.

Arda, H., Dayan, S., Kartal, C. and Güler, N. 2016. *In vitro* conservation of critically endangered *Dianthus ingoldbyi* Turrill under slow growth conditions. *Trakya Univ. J. Nat. Sci.* 17, 47–54.

Atta-Alla, H. and Van Staden, J. 1997. Micropropagation and establishment of *Yucca aloifolia*. *Plant Cell Tiss. Org. Cult.* 48, 209–212.

Atta-Alla, H., McAlister, B. G. and Van Staden, J. 1998. *In vitro* culture and establishment of *Anthurium parvispathum*. *S. Afr. J. Bot.* 64(5), 296–8. doi:10.1016/S0254-6299(15)30904-2.

Azadi, P., Bagheri, H., Nalousi, A. M., Nazari, F. and Chandler, S. F. 2016. Current status and biotechnological advances in genetic engineering of ornamental plants. *Biotechnol. Adv.* 34(6), 1073–90. doi:10.1016/j.biotechadv.2016.06.006.

Baig, M. M. Q., Hafiz, I. A., Abbasi, N. A., Yaseen, M., Akram, Z. and Donnelly, D. J. 2012. Reduced-stature *Rosa* species through *in vitro* mutagenesis. *Can. J. Plant Sci.* 92(6), 1049–55. doi:10.4141/cjps2011-199.

Bairu, M. W. and Kane, M. E. 2011. Physiological and developmental problems encountered by *in vitro* cultured plants. *Plant Growth Regul.* 63(2), 101–3. doi:10.1007/s10725-011-9565-2.

Barcaccia, G., Albertini, E. and Falcinelli, M. 1999. AFLP fingerprinting in *Pelargonium peltatum*: its development and potential in cultivar identification. *J. Hort. Sci. Biotechnol.* 74(2), 243–50. doi:10.1080/14620316.1999.11511102.

Barve, D. M., Iyer, R. S., Kendurkar, S. and Mascarenhas, A. F. 1984. An efficient method for rapid propagation of some budded rose varieties. *Indian J. Hortic.* 41, 1–7.

Belarmino, M. M. and Gabon, C. F. 1999. Low cost micropropagation of chrysanthemum (*Chrysanthemum morifolium*) through tissue culture. *Philippine J. Sci.* 128(2), 125–43.

Benedetti, L. D., Burchi, G., Mercuri, A., Schiva, T., De-Benedetti, L. and Cadic, A. 2000. Use of RAPD analysis for genotype identification in *Alstroemeria*. *Acta Hortic.* 508, 277–9.

Benedetti, L. D., Mercuri, A., Bruna, S., Burchi, G., Schiva, T., De-Benedetti, L., Dore, C., Dosba, F. and Baril, C. 2001. Genotype identification of ornamental species by RAPD analysis. *Acta Hortic.* 546(546), 391–4. doi:10.17660/ActaHortic.2001.546.49.

Benetka, V. 1985. Some experience of methodology with the isolation of somatic mutants in the rose cultivar 'Sonia'. *Acta Pruboniciana* 50,9–25.

Ben-Meir, H., Scovel, G., Ovadis, M. and Vainstein, A. 1997. Molecular markers in the breeding of ornamentals. *Acta Hortic.* 447, 599–601.

Benner, M. S., Braunstein, M. D. and Weisberg, M. U. 1995. Detection of DNA polymorphisms within the genus *Cattleya* (Orchidaceae). *Plant Mol. Biol. Rep.* 13(20), 147–55.

Beyramizade, E., Azadi, P. and Mii, M. 2008. Optimization of factors affecting organogenesis and somatic embryogenesis of *Anthuriumandreanum* lind. 'Tera'. *Propag. Ornamen. Plants* 8(4), 198-203.

Bhattacharya, P., Dey, S., Das, N. and Bhattacharya, B. C. 1990. Rapid masspropagation of *Chrysanthemum morifolium* by callus derived from stem and leaf explants. *Plant Cell Rep.* 9(8),439–42. doi:10.1007/BF00232268.

Bhavana, G. P., Satyan, K. B. and Aswath, C. 2018. A regenerative protocol and SEM study for *in vitro* propagation of *Anthurium* crossed lines via indirect somatic embryogenesis. *Biosci. Biotech. Res. Comm.* 11(1), 31–40. doi:10.21786/bbrc/11.1/5.

Bi, Y. M., Cammue, B. P. A., Goodwin, P. H., KrishnaRaj, S. and Saxena, P. K. 1999. Resistance to *Botrytis cinerea* in scented Geranium transformed with a gene encoding the antimicrobial protein Ace-AMP1. *Plant Cell Rep.* 18(10), 835–40. doi:10.1007/s002990050670.

Bietz, J. A. and Sandford, P. A. 1971. Reaction of sodium hypochlorite with amines and amides: automation of the method. *Anal. Biochem.* 44(1), 122–33. doi:10.1016/0003-2697(71)90353-8.

Bigot, C. 1981a. Multiplication vegetative *in vitro* de *Begonia* × *hiemalis* (Rieger et Schwabenland). I. *Methodol. Agron.* 1, 433–40.

Bigot, C. 1981b. Multiplication vegetative *in vitro* de *Begonia* × *hiemalis* (Rieger et Schwabenland). II. *Conformite des Plantlets Elevees en Serre. Agron.* 1,441–7.

Binsfeld, P. C., Wingender, R., Wunder, J. and Schnabl, H. 1999. Direct embryogenesis in the genus *Helianthus* and RAPD analysis of obtained clones. *J. Appl. Bot.* 73(3–4), 63–8.

Binsfeld, P. C., Wingender, R. and Schnabl, H. 2000. Characterization and molecular analysis of transgenic plants obtained by microprotoplast fusion in sunflower. *Theor. Appl. Genet.* 101(8), 1250–8. doi:10.1007/s001220051604.

Boase, M. R., Deroles, S. C., Winefield, C. S., Butcher,S. M., Borst, N. K. and Butler,R. C. 1996. Genetic transformation of regal pelargonium (*Pelargonium* × *domesticum* "Dubonet") by *Agrobacterium tumefaciens*. *Plant Sci.* 121(1), 47-61. doi:10.1016/S0168-9452(96)04508-6.

Boase, M. R., Bradley, J. M. and Borst, N. K. 1998a. Genetic transformation mediated by *Agrobacterium tumefaciens* of florists' chrysanthemum (*Dendranthema* × *grandiflorum*) cultivar 'Peach Margaret'. *In Vitro Cell. Dev. Biol. Plant* 34(1), 46–51. doi:10.1007/BF02823122.

Boase, M. R., Butler, R. C. and Borst, N. K. 1998b. Chrysanthemum cultivar-Agrobacterium interactions revealed by GUS expression time course experiments. *Sci. Hortic.* 77(1–2), 89–107. doi:10.1016/S0304-4238(98)00142-3.

Bouman, H. and DeKlerk, G. J. 1996. Somaclonal variation in biotechnology of ornamental plants. In: Geneve, R., Preece, J. and Merkle, S. (Eds), *Biotechnology of Ornamental Plants*. CAB International, UK, pp. 165–83.

Bouman, H., Kuijpers, A. and DeKlerk, G. J. 1995. Measurement of somaclonal variation in *Beginoa*. *Acta Hortic.* 420, 98–100.

Bouman, H., Kuijpers, A. M., Hol, T. and Schoo, W. 2001. Differences in growth and regeneration of suspension cultures derived from various callus lines of *Cyclamen persicum* cv. Purple Flamed. In: Schwenkel, H. G. (Ed.), *Reproduction of Cyclamen persicum Mill. through Somatic Embryogenesis Using Suspension Culture Systems.* COST Action 822. European Commission, Brussels, Belgium, pp. 46–54.

Bowler, D. E., Buyung-Ali, L., Knight, T. M. and Pullin, A. S. 2010. Urban greening to cool towns and cities: a systematic review of the empirical evidence. *Landsc. Urban Plan.* 97(3), 147–55. doi:10.1016/j.landurbplan.2010.05.006.

Brand, M. H. and Kiyamoto, R. 1994. Behaviour of *Rhododendron* tissue affected by tissue proliferation. *In Vitro Cell. Dev. Biol.* 30A, 72.

Brand, M. H. and Kiyomoto, R. 1997. The induction of tissue proliferation-like characteristics in *in vitro* cultures of *Rhododendron* "Montego". *HortScience* 32(6), 989–94. doi:10.21273/HORTSCI.32.6.989.

Brar, D. S. and Jain, S. M. 1998. Somaclonal variation: mechanism and applications in crop improvement. In: Jain, S. M., Brar, D. S. and Ahloowalia, B. S. (Eds), *Somaclonal Variation and Induced Mutations in Crop Improvement*. Kluwer Academic Publishers, Dordrecht, the Netherlands, pp. 15-37.

Bressan, P. H., Kim, Y. J., Hyndman, S. E., Hasegawa, P. M. and Bressan, R. A. 1982. Factors affecting *in vitro* propagation of rose. *J. Am. Soc. Hortic. Sci.* 107(6), 979–90.

Brethour, C., Watson, G., Sparling, B., Bucknell, D. and Moore, T. 2007. Literature review of documented health and environmental benefits derived from ornamental horticulture products. Final Report. Agriculture and Agri-Food Canada Markets and Trade, Ottawa, ON.

Burger, D. W., Liu, L., Zary, K. W. and Lee, C. I. 1990. Organogenesis and plant regeneration from immature embryos of *Rosa hybrida* L. *Plant Cell Tiss. Org. Cult.* 21(2), 147–52. doi:10.1007/BF00033434.

Carvalho, V., Santos, D. S. and NIevola, C. C. 2014. *In vitro* storage under slow growth and *ex vitro* acclimatization ofthe ornamental bromeliad *Acanthostachys strobilacea*. *S. Afr. J. Bot.* 92, 39–43.

Casanova, E., Trillas, M. I., Moysset, L. R. and Vainstein, A. 2005. Influence of rol genes in floriculture. *Biotechnol. Adv.* 23(1), 3–39. doi:10.1016/j.biotechadv.2004.06.002.

Cassells, A. C. and Curry, R. F. 2001. Oxidative stress and physiological, epigenetic and genetic variability in plant tissue culture:implications for micropropagators and genetic engineers. *Plant Cell Tiss. Org. Cult.* 64(2/3), 145–57. doi:10.1023/A:1010692104861.

Cassells, A. C. and Plunkett, A. 1984. Production and growth analysis of plants from leaf cuttings, and from tissue cultures of disks from mature leaves and young axenic leaves of African violet (*Saintpaulia ionantha* Wendl.). *Sci. Hortic.* 23(4), 361–9. doi:10.1016/0304-4238(84)90033-5.

Castillo, B. and Smith, M. A. L. 1997. Direct somatic embryogenesis from *Begonia gracilis* explants. *Plant Cell Rep.* 16(6), 385–8. doi:10.1007/BF01146778.

Chakrabarty, D., Mandal, A. K. and Datta, S. K. 2000. *In vitro* propagation of rose cultivars. *Ind. J. Plant Physiol.* 5(2), 189–92.

Chandler, S. F. and Sanchez, C. 2012. Genetic modification; the development of transgenic ornamental plant varieties. *Plant Biotechnol. J.* 10(8), 891–903. doi:10.1111/j.1467-7652.2012.00693.x.

Chartier-Hollis, J. M., Harris, W., Lynch, P. T., Morisot, A. and Ricci, P. 1996. Cryopreservation of shoot tips of *Rosa multiflora*. *Acta Hortic.* 424, 367–8. doi:10.17660/ActaHortic.1996.424.70.

Chebet, D. K., Okeno, J. A. and Mathenge, P. 2003. Biotechnological approaches to improve horticultural crop production. *Acta Hortic.* 625, 473-7. doi:10.17660/ActaHortic.2003.625.56.

Chen, J. T. and Chang, W. C. 2002. Effects of tissue culture conditions and explants characteristics on direct somatic embryogenesis in *Oncidium* 'Gower Ramsey'. *Plant Cell Tiss. Org. Cult.* 69, 41-4.

Chen, J. T. and Chang, W. C. 2006. Direct somatic embryogenesis and plant regeneration from leaf explants of *Phalaenopsis amabilis*. *Biol. Plant.* 50(2), 169-73. doi:10.1007/s10535-006-0002-8.

Chen, W. H., Chen, T. M., Fu, Y. M., Hsieh, R. M. and Chen, W. S. 1998. Studies on somaclonal variation in *Phalaenopsis*. *Plant Cell Rep.* 18(1-2), 7-13. doi:10.1007/s002990050523.

Cheong, C., Yu, K. W. and Paek, K. Y. 1999. Randomly amplified polymorphic DNA (RAPD) analysis of the *Lisianthus* (*Eustoma grandiflorum* Shinn.) variants obtained during tissue culture. *Korean J. Hortic. Sci. Tech.* 17(3), 352-4.

Choudhury, M. L., Chin, C. K., Polashock, J. J. and Martin, C. E. 1994. Agrobacterium mediated transformation of *Petunia hybrida* with yeast Δ-9 fatty acid desaturase. *Plant Growth Regul.* 15(2), 113-6. doi:10.1007/BF00024099.

Chu, C. Y., Knight, S. L. and Smith, M. A. L. 1993. Effect of liquid culture on the growth and development of miniature rose (*Rosa chinensis* Jacq. 'Minima'). *Plant Cell Tiss. Org. Cult.* 32(3), 329-34. doi:10.1007/BF00042296.

Chua, B. U., Kunisaki, J. T. and Sagawa, Y. 1981. *In vitro* propagation of *Dracaena marginata* tricolour. *HortScience* 16, 494.

Chung, H. H., Chen, J. T. and Chang, W. C. 2005. Cytokinins induce direct somatic embryogenesis of *Dendrobium* Chiengmai Pink and subsequent plant regeneration. *In Vitro Cell. Dev. Biol. Plant* 41(6), 765-9. doi:10.1079/IVP2005702.

Chung, H. H., Chen, J. T. and Chang, W. C. 2007. Plant regeneration through direct somatic embryogenesis from leaf explants of *Dendrobium*. *Biol. Plant.* 51(2), 346-50. doi:10.1007/s10535-007-0069-x.

Çimen, A. and Özge, Ç. 2009. Micropropagation of *Anthurium andraeanum* from leaf explants. *Pak. J. Bot.* 41(3), 1155-61.

Condliffe, P. C., Davey, M. R., Power, J. B., Koehorst-van Putten, H. and Visser, P. B. 2003. An optimised protocol for rose transformation applicable to different cultivars. *Acta Hortic.* 612, 115-20. doi:10.17660/ActaHortic.2003.612.14.

Curtis, W. R. 2002. Application of bioreactor design principles to plant micropropagation. *Ist InternationalSymposium 'Liquid Systems for In Vitro Micropropagation of Plants'*, 29th May - 2nd June 2002, Norway, pp. 58-9.

Dandekar, A. M. 2003. Techniques for manipulating quality and productivity traits in horticultural crops. *Acta Hortic.* 625, 293-305.

Datta, S. K. 2009. A report on 36 years of practical work on crop improvement through induced mutagenesis. In: Shu, Q. Y. (Ed.), *Induced Plant Mutations in the Genomics Era*. Food and Agriculture Organization of the United Nations, Rome, Italy, pp. 253-6.

Debener, T., Bartels, C. and Spethmann, W. 1997. Parentage analysis in interspecific crosses between rose species with RAPD markers. *Gartenbauwissenschaft* 62(4), 180-4.

Debener, T., Mattiesch, L., Vosman, B., Zieslin, N. and Agbaria, H. 2001. A molecular marker map for roses. *Acta Hortic.* 547, 283-7. doi:10.17660/ActaHortic.2001.547.33.

Debergh, P. 1975. Intensified vegetative multiplication of *Dracaena deremensis*. *Acta Hortic.* 54, 83-92. doi:10.17660/ActaHortic.1975.54.10.

Debergh, P. 1976. An *in vitro* technique for the vegetative multiplication of chimeral plants of *Dracaena* and Cordyline. *Acta Hortic.* 64, 17–9.

Debergh, P. and DeWael, J. 1977. Mass propagation of *Ficus lyrata*. *Acta Hortic.* 78, 361–4. doi:10.17660/ActaHortic.1977.78.45.

Debergh, P. C. and Read, P. E. 1991. Micropropagation. In: Debergh, P. C. and Zimmerman, R. H. (Eds), *Micropropagation*. Kluwer Academic Publishers, Dordrecht, the Netherlands, pp. 1–13.

Demiralay, A., Yalcin-Mendi, Y., Aka-Kacar, Y. and Cetiner, S. 1998. *In vitro* propagation of *Ficus carica* L. var. Bursa siyahi through meristem culture. *Acta Hortic.* 480, 165–8. doi:10.17660/ActaHortic.1998.480.28.

Deshpande, S. R., Josekutty, P. C. and Prathapasenan, G. 1998. Plant regeneration from axillary buds of a mature tree of *Ficus religiosa*. *Plant Cell Rep.* 17(6–7), 571–3. doi:10.1007/s002990050444.

De Vries, D. P. and Dubois, L. A. M. 1988. The effect of BAP and IBA on sprouting and adventitious root formation of 'Amanda' rose single-node softwood cuttings. *Sci. Hortic.* 34, 115–21.

de Wit, J. C., Esendam, H. F., Horkanen, J. J. and Tuominen, U. 1990. Somatic embryogenesis and regeneration of flowering plants in rose. *Plant Cell Rep.* 9(8), 456–8. doi:10.1007/BF00232272.

Dillen, W., Dijkstra, I. and Oud, J. 1996. Shoot regeneration in long-term callus cultures derived from mature flowring plants of *Cyclamen persicum* Mill. *Plant Cell Rep.* 15(7), 545–8. doi:10.1007/BF00232991.

Dobres, M. S. 2011. Prospects for commercialisation of transgenic ornamentals. In: Mou, B. and Scorza, R. (Eds), *Transgenic Horticultural Crops: Challenges and Opportunities*. CRC press, Boca Raton, FL, pp. 305–16.

Dohm, A., Ludwig, C., Schilling, D. and Debener, T. 2002. Transformation of roses with genes for antifungal proteins to reduce their susceptibility to fungal diseases. *Acta Hortic.* 572, 105–11. doi:10.17660/ActaHortic.2002.572.12.

Dunnett, N. and Qasim, M. 2000. Perceived benefits to human well-being of urban gardens. *HortTechnology* 10(1), 40–5. doi:10.21273/HORTTECH.10.1.40.

Eapen, S. and Rao, P. S. 1985. Plant regeneration from immature inflorescence callus cultures of wheat, rye and triticale. *Euphytica* 34(1), 153–9. doi:10.1007/BF00022875.

Economou, A. S. and Read, P. E. 1984. *In vitro* shoot proliferation of *Minnesota deciduous* azaleas. *HortScience* 19(1), 60–1.

Eide, A. K. H., Munster, C., Heyerdahl, P. H., Lyngved, R. and Olsen, O. A. S. 2003. Liquid culture systems for plant propagation. *Acta Hortic.* 625, 173–85. doi:10.17660/ActaHortic.2003.625.18.

Elliott, R. F. 1970. Axenic culture of meristem tips of *Rosa multiflora*. *Planta* 95(2), 183–6. doi:10.1007/BF00387250.

El-Tarras, A., Hossni, Y. A. and Elbanna, A. A. 1999. Application of RAPD analysis for identification and genetic analysis of somaclonal variation in *Ananascomous* variegatum. *Egypt. J. Hortic.* 26(1), 97–108.

Engelmann, F. 2012. Germplasm collection, storage and conservation. In: Altman, A. and Hasegawa, P. M. (Eds), *Plant Biotechnology and Agriculture*. Academic Press, Oxford, UK, pp. 255–68.

Ettinger, T. L. and Preece, J. E. 1985. Aseptic micropropagation of *Rhododendron*. P.J.M. hybrids. *J. Hortic. Sci.* 60(2), 269–74. doi:10.1080/14620316.1985.11515628.

FAO, IAEA and MVDB. 2016. Available at: http://www-mvd.iaea.org.

Fiore, S., Pasquale, F. de, Carimi, F. and Sajeva, M. 2002. Effect of 2,4-D and 4-CPPU on somatic embryogenesis from stigma and style transverse thin cell layers of *Citrus*. *Plant Cell Tiss. Org. Cult.* 68(1), 57–63. doi:10.1023/A:1012944100210.

Firoozabady, E., Noriega, C., Sondahl, M. R. and Robinson, K. E. P. 1991a. Genetic transformation of rose (*Rosa hybrida* cv. Royalty) via *Agrobacterium tumefaciens*. *In Vitro* 27, 154A.

Firoozabady, E., Lemieux, C. S., Moy, Y. S., Moll, B., Nicholas, J. A. and Robinson, K. E. P. 1991b. Genetic engineering of ornamental crops. *In Vitro* 27, 96A.

Firoozabady, E., Moy, Y., Courtney-Gutterson, N. and Robinson, K. 1994. Regeneration of transgenic rose (*Rosa hybrida*) plants from embryogenic tissue. *Biotechnology* 12(6), 609–13. doi:10.1038/nbt0694-609.

Fonnesbech, M. and Fonnesbech, A. 1979. *In vitro* propagation of *Spathiphyllum*. *Sci. Hortic.* 10(1), 21–5. doi:10.1016/0304-4238(79)90065-7.

Fu, Y. M., Chen, W. H., Hsieh, R. M., Tsai, W. T., Wu, C. C., Chyou, M. S. and Lin, Y. S. 1994. Studies on DNA amplification fingerprinting techniques of *Phalaenopsis* oechid. Report of the Taiwan Sugar Research Institute No. 146, pp. 9–22.

Fujii, Y. and Shimizu, K. 1990. Regeneration of plants from achenes and petals of *Chrysanthemum coccineum*. *Plant Cell Rep.* 8(10), 625–7. doi:10.1007/BF00270069.

Fukai, S. and Oe, M. 1990. Morphological observations of *Chrysanthemum* shoot tips cultured after cryoprotection and freezing. *J. Jpn. Soc. Hortic. Sci.* 59(2), 383–7. doi:10.2503/jjshs.59.383.

Fukai, S., Morii, M. and Oe, M. 1988. Storage of chrysanthemum (*Dendrathema grandiflorum* Ramat.) plantlets *in vitro*. *Plant Cell. Tiss. Org. Cult.* 5, 20–5.

Gabryszewska, E. and Rudnicki, R. M. 1997. The effects of light quality on the growth and development of shoots and roots of *Ficus benjaminain vitro*. *Acta Hortic.* 418(418), 163–8. doi:10.17660/ActaHortic.1997.418.22.

Geier, T. 1977. Morphogenesis and plant regeneration from cultured organ fragments of *Cyclamen persicum*. *Acta Hortic.* 78, 167–74. doi:10.17660/ActaHortic.1977.78.20.

Geier, T., Kohlenbach, H. W. and Reuther, G. 1983. Cyclamen. In: Ammirato, P. V., Evans, D. R., Sharp, W. R.and Bajaj, Y. P. S. (Eds), *Hand Book of Plant Cell Culture* (vol. 5). McGraw-Hill Publishing Co., New York, NY, pp. 352–74.

George, E. F. 1993. *Plant Propagation by Tissue Culture: Part 1 – In Practice*. Exegetics, Basingstoke, UK.

Gertsson, U. E. and Andersson, E. 1985. Propagation of *Chrysanthemum × hortorum* and *Philodendron scandens* by tissue culture. Rapport-Sveriges Lantbruks universitet (agris.fao.org). ISSN: 0348-4157.

Gill, R., Gerrath, J. M. and Saxena, P. K. 1993. High-frequency direct somatic embryogenesis in thin layer cultures of hybrid seed geranium (*Pelargonium × hortorum*). *Can. J. Bot.* 71(3), 408–13. doi:10.1139/b93-045.

Gow, W. P., Chen, J. T. and Chang, W. C. 2009. Effects of genotype, light regime, explants position and orientation on direct embryo formation from leaf explants of *Phalaenopsis* orchids. *Acta Physiol. Plant.* 31(2), 363–9. doi:10.1007/s11738-008-0243-6.

Gow, W. P., Chen, J. T. and Chang, W. C. 2010. Enhancement of direct somatic embryogenesis and plantlet growth from leaf explants of *Phalaenopsis* by adjusting culture period and explant length. *Acta Physiol. Plant.* 32(4), 621–7. doi:10.1007/s11738-009-0438-5.

Grout, B. W. V. 1991. Conservation *in vitro*. *Acta Hortic.* 289, 171-8. doi:10.17660/ActaHortic.1991.289.45.
Grout, B. W. W. 1995. Introduction to the *in vitro* preservation of plant cells, tissue and organs. In: Grout, B. W. W. (Ed.), *Genetic Preservation of Plant Cells In Vitro*. Springer-Verlag, Berlin Heidelberg and New York, pp. 1-20.
Grunewaldt, J. 1977. Adventivknospenbildung und pflanzenregeneration bei Gesneriaceae in vitro. *Gartenbauwissenschaft* 42, 171-5.
Gürel, E. and Türker, A. U. 2001. Organogenesis. In: Babaoğlu, M., Gürel, E. and Özcan, S. (Eds), *Plant Biotechnology, Tissue Culture and Applications*. Selcuk University Publications, Konya, Turkey, pp. 36-70.
Gustafsson, S. 2000. Patterns of genetic variation in *Gymnadenia conopsea*, the fragrant orchid. *Mol. Ecol.* 9(11), 1863-72. doi:10.1046/j.1365-294x.2000.01086.x.
Haccius, B. 1978. Question of unicellular origin of non-zygote embryos in callus cultures. *Phytomorphology* 28, 74-81.
Hagege, D. 1995. Habituation in plant cell cultures: adaptation to the free radical attacks. *C. R. Seances Soc. Biol. Fil.* 189, 1183-90.
Hall, C. R. and Dickson, M. W. 2011. Economic, environmental, and health/well-being benefits associated with green industry products and services: a review. *J. Environ. Hortic.* 29, 96-103.
Hallerman, E. and Grabau, E. 2016. Crop biotechnology: a pivotal moment for global acceptance. *Food Energy Secur.* 5(1), 3-17. doi:10.1002/fes3.76.
Hasegawa, P. M. 1979. *In vitro* propagation of rose. *HortScience* 14, 610-2.
Hasegawa, P. M. 1980. Factors affecting shoot and root initiation from cultured rose shoot tips. *J. Am. Soc. Hortic. Sci.* 105(2), 216-20.
Hawkes, H. Y. and Wainwright, H. 1987. *In vitro* organogenesis of *Cyclamen persicum* mill. seedling tissue. *Acta Hortic.* 212(212), 711-4. doi:10.17660/ActaHortic.1987.212.122.
Hayatsu, H., Pan, S. and Ukita, T. 1971. Reaction of sodium hypochlorite with nucleic acids and their constituents. *Chem. Pharm. Bull.* 19(10), 2189-92. doi:10.1248/cpb.19.2189.
Hazarika, B. N. 2003. Acclimatization of tissue-cultured plants. *Curr. Sci.* 85(12), 1704-12.
Hazarika, B. N. 2006. Morpho-physiological disorders in *in vitro* cultured plants. *Sci. Hortic.* 108(2), 105-20. doi:10.1016/j.scienta.2006.01.038.
Hitmi, A., Barthomeuf, C. and Sallanon, H. 2000. Cryopreservation of *Chrysanthemumcinerariaefolium* shoot tips. *J. Plant Physiol.* 156(3), 408-12. doi:10.1016/S0176-1617(00)80081-4.
Hohe, A. 2010. Ambitious expectations: somatic embryogenesis in ornamental plant propagation. *Propag. Ornament. Plants* 10(4), 176-82.
Hohe, A., Winkelmann, T. and Schwenkel, H. G. 1999. The effect of oxygen partial pressure in bioreactors on cell proliferation and subsequent differentiation of somatic embryos of *Cyclamen persicum*. *Plant Cell Tiss. Org. Cult.* 59(1), 39-45. doi:10.1023/A:1006323009860.
Hohe, A., Winkelmann, T. and Schwenkel, H. G. 2001. Development of somatic embryos of *Cyclamen persicum* Mill. in liquid culture. *Gartenbauwissenschaft* 66, 219-24.
Hong, P. I., Chen, J. T. and Chang, W. C. 2008a. Effects of salicylic and acetylsalicylic acid on direct embryogenesis in *Oncidium*. *J. Plant Biochem. Biotechnol.* 17(2), 149-53. doi:10.1007/BF03263276.

Hong, P. I., Chen, J. T. and Chang, W. C. 2008b. Promotion of direct somatic embryogenesis of *Oncidium* by adjusting carbon sources. *Biol. Plant.* 52(3), 597–600. doi:10.1007/s10535-008-0119-z.

Horst, R. K., Smith, S. H., Horst, H. T. and Oglevee, W. A. 1976. *In vitro* regeneration of shoot and root growth from meristematic tips of *Pelargonium × hortorum* Bailey. *Acta Hortic.* 59, 131–42.

Hoshino, Y., Nakano, M. and Mii, M. 1995. Plant regeneration from cell suspension-derived protoplasts of *Saintpaulia ionantha* Wendl. *Plant Cell Rep.* 14(6), 341–4. doi:10.1007/BF00238593.

Hosoki, T. and Kajino, E. 2003. Shoot regeneration from petioles of coral bells (*Heuchera sanguinea* Engelm.) cultured *in vitro*, and subsequent planting and flowering *ex vitro*. *In Vitro Cell Dev. Biol. Plant* 39(2), 135–8. doi:10.1079/IVP2002363.

Hsia, C. and Korban, S. S. 1996. Factors affecting*in vitro* establishment and shoot proliferation of *Rosa hybrida* L. and *Rosa chinensis minima*. *In Vitro Cell. Dev. Biol. Plant* 32(4), 217–22. doi:10.1007/BF02822690.

Hsia, C. N. and Korban, S. S. 1998. Microprojectile mediated genetic transformation of *Rhododendron* hybrids. *J. Am. Rhododendron Soc.* 52(4), 187–91.

Huang, S. C., Tsai, C. C. and Sheu, C. S. 2000. Genetic analysis of chrysanthemum hybrids based on RAPD molecular markers. *Bot. Bull. Acad. Sin.* 41(4), 257–62.

Hutchinson, J. F., Kaul, V., Maheswaran, G., Moran, J. R., Graham, M. W. and Richard, D. 1992. Genetic improvement of floricultural crops using biotechnology. *Aust. J. Bot.* 40(6), 765–87. doi:10.1071/BT9920765.

Huttema, J. B. M., Preil, W., Gussenhoven, G. C. and Schneidereit, M. 1989. Methods for the selection of low-temperature tolerant mutants of *Chrysanthemum morifolium* Ramat. by using irradiated cell suspension cultures. I. Selection of regenerants *in vitro* under subtropical temperature conditions. *Plant Breed.* 102(2), 140–7. doi:10.1111/j.1439-0523.1989.tb00327.x.

Huttema, J. B. M., Preil, W. and DeJong, J. 1991. Methods for selection of low-temperature tolerant mutants of *Chrysanthemum morifolium* Ramat. using irradiated cell suspension cells III. Comparison of mutants selected with or without preselection *in vitro* at low temperature. *Plant Breed.* 107, 135–40.

Hvoslof-Eide, A. K. and Munster, C. 2001. Light quality effects on somatic embryogenesis of *Cyclamen persicum* Mill. in bioreactors. In: Schwenkel, H. G. (Ed.), *Reproduction of Cyclamen persicum Mill. through Somatic Embryogenesis Using Suspensor Culture Systems*. Report of Working Group 2. COSTEuropean Commission, pp. 79–84.

IAEA-TECDOC-1384. 2004. *Low Cost Options for Tissue Culture Technology for Developing Countries*. International Arts and Entertainment Alliance, Vienna, Austria.

Iqbal, M. J., Panden, D. W. and Rayburn, A. L. 1995. Clonal stability of RAPD markers in three *Rhododendron* species. *News Lett.* 13(1), 43–6.

Isah, T. 2015. Adjustments to *in vitro* culture conditions and associated anomalies in plants. *Acta Biol. Cracoviensia S. Bot.* 57(2), 9–28.

Ishiooka, N. and Tanimoto, S. 1990. Plant regeneration from Bulgarian rose callus. *Plant Cell Tiss. Org. Cult.* 22(3), 197–9. doi:10.1007/BF00033636.

Islam, S. M. S. and Bhattacharjee, B. 2015. Plant regeneration through somatic embryogenesis from leaf and root explants of *Rhynchostylis retusa* (L.) Blume. *Appl. Biol. Res.* 17(2), 158–65. doi:10.5958/0974-4517.2015.00025.7.

Jain, S. M. 1997. Micropropagation of selected somaclones of *Begonia* and *Saintpaulia*. *J. Biosci.* 22(5), 585–92. doi:10.1007/BF02703396.

Jain, S. M. 2001. Tissue culture-derived variation in crop improvement. *Euphytica* 118(2), 153-66. doi:10.1023/A:1004124519479.

Jain, S. M. 2002. Feeding the world with induced mutations and biotechnology. *Proceedingsof the International Nuclear Conference 2002 - Global Trends and Perspectives. Seminar 1: Agriculture and Bioscience*. MINT, Bangi, Malaysia, pp. 1-14.

Jain, S. M. 2006. Mutation-assisted breeding for improving ornamental plants. *Acta Hortic.* 714, 85-98.

Jain, S. M. and DeKlerk, G. J. 1998. Somaclonal variation in breeding and propagation of ornamental crops. *Plant Tiss. Cult. Biotechnol.* 4(2), 63-75.

Jain, S. M. and Suprasanna, P. 2011. Induced mutations for enhancing nutrition and food production. *Gene Conserve* 40, 201-15.

Jain, S. M., Brar, D. S. and Ahloowalia, B. S. (Eds). 1998a. *Somaclonal Variation and Induced Mutations in Crop Improvement*. Kluwer Academic Publishers, Dordrecht, the Netherlands, p. 603.

Jain, S. M., Ahloowalia, B. S. and Veilleux, R. E. 1998b. Somaclonal variation in crop plants. In: Jain, S. M. and Brar, D. S. and Ahloowalia, B. S. (Eds), *Somaclonal Variation and Induced Mutations in Crop Improvement*. Kluwer Academic Publishers, Dordrecht, the Netherlands, pp. 203-18.

Jain, S. M., Buiatti, M., Gimelli, F. and Saccardo, F. 1998c. Somaclonal variation in improving ornamental plants. In: Jain, S. M., Brar, D. S. and Ahloowalia, B. S. (Eds), *Somaclonal Variation and Induced Mutation in Crop Improvement*. Kluwer Academic Publishers, the Netherlands, pp. 81-104.

Jain, S. M., Jenks, M. A., Rout, G. R. and Radojevic, L. 2006. Micropropagation of ornamental potted plants. *Propag. Ornament. Plants* 6(2), 67-82.

Jelaska, S. and Jelencic, B. 1980. Plantlet regeneration from shoot tip culture of *Pelargonium zonale* hybrid. *Acta Bot. Croat.* 39, 59-63.

Jianhua, Z., Mcdonald, M. B., Sweeney, P. M. and Zhang, J. H. 1997. Testing for genetic purity in *Petunia* and *Cyclamen* seed using random amplified polymorphic DNA markers. *HortScience* 32(2), 246-7. doi:10.21273/HORTSCI.32.2.246.

Joseph, D., Martin, K. P., Madassery, J. and Philip, V. J. 2003. *In vitro* propagation of three commercial cut flower cultivars of *Anthurium andraeanum* Hort. *Indian J. Exp. Biol.* 41(2), 154-9.

Kaicker, U. S. 1982. Mutation breeding in roses. *India Rose. Ann. Rep.* 2, 35-42.

Kamthan, A., Chaudhuri, A., Kamthan, M. and Datta, A. 2016. Genetically modified (GM) crops: milestones and new advances in crop improvement. *Theor. Appl. Genet.* 129(9), 1639-55. doi:10.1007/s00122-016-2747-6.

Kane, M. E. 2005. Shoot culture procedures. In: Trigiano, R. N. and Gray, D. J. (Eds), *Plant Development and Biotechnology*. CRC Press, Washington DC.

Kantouch, A. and Ardel-Fattah, S. H. 1971. Action of sodium hypochlorite on a-amino acids. *Chem. Zvesti* 25, 222-30.

Kaul, V., Miller, R. M., Hutchinson, J. F. and Richards, D. 1990. Shoot regeneration from stem and leaf explants of *Dendrathema grandiflora* Tzvelev (Syn. *Chrysanthemum morifolium* Ramat.). *Plant Cell Tiss. Org. Cult.* 21, 21-30.

Kawata, M. and Oono, K. 1997. Protoclonal variation in crop improvement. In: Jain, S. M., Brar, D. S. and Ahloowalia, B. S. (Eds), *Somaclonal Variation and Induced Mutation for Crop Improvement*. Kluwer Academic Publishers, the Netherlands, pp. 135-48.

Khosh-Khui, M. and Sink, K. C. 1982a. Micropropagation of new and old world rose species. *J. Hortic. Sci.* 57(3), 315-9. doi:10.1080/00221589.1982.11515058.

Khosh-Khui, M. and Sink, K. C. 1982b. Callus induction and culture of *Rosa*. *Sci. Hortic.* 17(4), 361–70. doi:10.1016/0304-4238(82)90117-0.
Khosh-Khui, M. and Sink, K. C. 1982c. Rooting enhancement of *Rosa hybrida* for tissue culture propagation. *Sci. Hortic.* 17(4), 371–6. doi:10.1016/0304-4238(82)90118-2.
Kim, C. K., Chung, J. D., Jee, S. O. and Oh, J. Y. 2003a. Somatic embryogenesis from *in vitro* grown leaf explants of *Rosa hybrida* L. *Afr. J. Plant Biotechnol.* 5(3), 169–72.
Kim, S. J., Hahn, E. J., Paek, K. Y. and Murthy, H. N. 2003b. Application of bioreactor culture for large scale production of Chrysanthemum transplants. *Acta Hortic.*625(625), 187–91. doi:10.17660/ActaHortic.2003.625.19.
Kim, S. W., Oh, S. C., In, D. S. and Liu, J. R. 2003c. Plant regeneration of rose (*Rosa hybrida*) from embryogenic cell derived protoplasts. *Plant Cell Tiss. Org. Cult.* 73(1), 15–9.
King, R., Bird, N., Ramirez-Gonzalez, R., Coghill, J. A., Patil, A., Hassani-Pak, K., Uauy, C. and Phillips, A. L. 2015. Mutation scanning in wheat by exon capture and next-generation sequencing. *PLoS ONE* 10(9), e0137549. doi:10.1371/journal.pone.0137549.
Kintzios, S., Manos, C. and Makri, O. 1999. Somatic embryogenesis from mature leaves of rose (*Rosasps*). *Plant Cell Rep.* 18(6), 467–72. doi:10.1007/s002990050605.
Kiyokawa, S., Kikuchi, Y., Kamada, H. and Harada, H. 1996. Genetic transformation of *Begonia tuberhybrida* by Ri rol genes. *Plant Cell Rep.* 15(8), 606–9. doi:10.1007/BF00232462.
Kobayashi, H., Oikawa, Y., Koiwa, H. and Yamamura, S. 1998. Flower-specific gene expression directed by the promoter of a chalcone synthase gene from *Gentiana triflora* in *Petunia hybrida*. *Plant Sci.* 131(2), 173–80. doi:10.1016/S0168-9452(97)00236-7.
Kobayashi, N., Handa, T., Takayanagi, K., Arisumi, K., Plas, L. H. W., Dons, J. J. M., Vanderleyden, J. and Loose, M. 2000. Chloroplast DNA polymorphisms and morphological variation in Japanese wild azaleas, the origin of evergreen azalea cultivars. *Acta Hortic.* 521, 173–8. doi:10.17660/ActaHortic.2000.521.18.
Kozai, T. and Xiao, Y. 2006. A commercialized photoautotrophic micropropagation system. In: Gupta, S. D. and Ibaraki, Y. (Eds), *Plant Tissue Culture Engineering*. Springer, the Netherlands, pp. 355–71.
Kumar, V. and Jain, M. 2015. The CRISPR–Cas system for plant genome editing: advances and opportunities. *J. Exp. Bot.* 66(1), 47–57. doi:10.1093/jxb/eru429.
Kumar, A. and Kumar, V. A. 1995. High-frequency *in vitro* propagation in *Chrysanthemum maseimum*. *Indian Hortic.*, 37–8.
Kumar, V., Radha, A. and Kumar Chitta, S. 1998. *In vitro* plant regeneration of fig (*Ficus carica* L. cv. Gular) using apical buds from mature trees. *Plant Cell Rep.* 17(9), 717–20. doi:10.1007/s002990050471.
Kunitake, H., Imamizo, H. and Mii, M. 1993. Somatic embryogenesis and plant regeneration from immature seed-derived calli of Rugosa rose (*Rosa rugosa* Thurb.). *Plant Sci.* 90(2), 187–94. doi:10.1016/0168-9452(93)90239-V.
Lamseejan, S., Jompuk, P. and Deeseepan, S. 2003. Improvement of *Chrysanthemum* var. 'Taihei' through *in vitro* induced mutation with chronic and acute gamma rays. *J. Nucl. Soc. Thai.* 4, 2–13.
Langhe, E. D., Debergh, P. and van Rijk, R. 1974. *In vitro* culture as a method for vegetative propagation of *Euphorbia pulcherrima*. *Z. Pflanzenphysiol.* 71(3), 271–4. doi:10.1016/S0044-328X(74)80032-2.
Larkin, P. J. and Scowcroft, W. R. 1981. Somaclonal variation-a novel source of variability from cell cultures for plant improvement. *Theor. Appl. Genet.* 60(4), 197–214. doi:10.1007/BF02342540.

Lazar, M. and Cachita, C. 1983. Micropropagation of *Chrysanthemum*: III. Chrysanthemum multiplication *in vitro* from capitulum explants. *Prod. Veg. Hortic.* 32, 44–7.

Ledger, S. E., Deroles, S. C. and Given, N. K. 1991. Regeneration and *Agrobacterium*-mediated transformation of chrysanthemum. *Plant Cell Rep.* 10(4), 195–9. doi:10.1007/BF00234294.

Lee, J. M. 2004. Plant cell culture and its applications. In: Goodman, R. M. (Ed.), *Encyclopaedia of Plant and Crop Sciences*. Marcel Dekker, New York, NY, pp. 931–3.

Lesur, C., Becher, A., Wolff, K., Weising, K., Steinmetz, K., Peltier, D., Boury, S., Dore, C., Dosba, F. and Baril, C. 2001. DNA fingerprints for *Pelargonium* cultivar identification. *Acta Hortic.* 546, 325–30. doi:10.17660/ActaHortic.2001.546.40.

Lloyd, D., Roberts, A. V. and Short, K. C. 1988. The induction of *in vitro* of adventitious shoots in *Rosa*. *Euphytica* 37(1), 31–6. doi:10.1007/BF00037220.

Li, G. and Quiros, C. F. 2001. Sequence related amplified polymorphism (SRAP), a new marker system based on a simple PCR reaction: its application to mapping and gene tagging in Brassica. *Theor. Appl. Genet.* 103(2–3), 455–61. doi:10.1007/s001220100570.

Li, X. Q., Krasnyanski, S. F. and Korban, S. S. 2002a. Somatic embryogenesis, secondary somatic embryogenesis, and shoot organogenesis in *Rosa*. *J. Plant Physiol.* 159(3), 313–9. doi:10.1078/0176-1617-00688.

Li, X. Q., Krasnyanski, S. F. and Korban, S. S. 2002b. Optimization of the uidA gene transfer into somatic embryos of rose via *Agrobacterium tumefaciens*. *Plant Physiol. Biochem.* 40(5), 453–9. doi:10.1016/S0981-9428(02)01394-3.

Li, X. Q., Gasic, K., Cammue, B., Broekaert, W. and Korban, S. S. 2003. Transgenic rose lines harbouring an antimicrobial protein gene, Ace-AMP1, demonstrate enhanced resistance to powdery mildew (*Sphaerotheca pannosa*). *Planta* 218(2), 226–32. doi:10.1007/s00425-003-1093-5.

Lim, S. H., Teng, P. C. P., Lee, Y. H. and Goh, C. J. 1999. RAPD analysis of some species in the genus *Vanda* (Orchidaceae). *Ann. Bot.* 83(2), 193–6. doi:10.1006/anbo.1998.0801.

Liu, C. M. and Xu, Z. H. 1992. An efficient procedure for micropropagation of *Anthurium scherzerianum* Schott (flamingo flower). *Chin. J. Bot* 4, 49–55.

Lohr, V. I. and Pearson-Mims, C. H. 1996. Particulate matter accumulation on horizontal surfaces in interiors: influence of foliage plants. *Atmos. Environ.* 30(14), 2565–8. doi:10.1016/1352-2310(95)00465-3.

Lowe, J. M., Davey, M. R., Power, J. B. and Blundy, K. S. 1993. A study of some factors affecting Agrobacterium transformation and plant regeneration of *Dendranthema grandiflora* Tzvelev (syn. *Chrysanthemum morifolium* Ramat.). *Plant Cell Tiss. Org. Cult.* 33(2), 171–80. doi:10.1007/BF01983231.

Lu, C. Y., Nugent, G. and Wardley, T. 1990. Efficient, direct plant regeneration from stem segments of chrysanthemum (*Chrysanthemum morifolium* Ramat cv. Royal Purple. *Plant Cell Rep.* 8(12), 733–6. doi:10.1007/BF00272106.

Lu, C. Y., Nugent, G., Wardley-Richardson, T., Chandler, S. F., Young, R. and Dalling, M. J. 1991. *Agrobacterium*-mediated transformation of carnation (*Dianthus caryophyllus* L.). *Biotechnology* 9, 864–8. doi:10.1038/nbt0991-864.

Lusser, M., Parisi, C., Plan, D. and Rodríguez-Cerezo, E. 2012. Deployment of new biotechnologies in plant breeding. *Nat. Biotechnol.* 30(3), 231–9. doi:10.1038/nbt.2142.

Mahendran, G. and Bai, V. N. 2016. Direct somatic embryogenesis of *Malaxis densiflora* (A. Rich.) Kuntze. *J. Genet. Eng. Biotechnol.* 14, 77–81.

Maluszynski, M., Ahloowalia, B. S. and Sigurbjornsson, B. 1995. Application of *in vitro* and *in vivo* mutation techniques for crop improvement. *Euphytica* 85(1-3), 303-15. doi:10.1007/BF00023960.

Mandal, A. K. A., Chakrabarty, D. and Datta, S. K. 2000. *In vitro* isolation of solid novel flower colour mutants from induced chimeric ray florets of chrysanthemum. *Euphytica* 114(1), 9-12. doi:10.1023/A:1003960906646.

Marchant, R., Power, J. B., Lucas, J. A. and Davey, M. R. 1998a. Biolistic transformation of rose (*Rosa hybrida* L.). *Ann. Bot.* 81(1), 109-14. doi:10.1006/anbo.1997.0538.

Marchant, R., Davey, M. R., Lucas, J. A., Lamb, C. J., Dixon, R. A. and Power, J. B. 1998b. Expression of a chitinase transgene in rose (*Rosa hybrida* L.) reduces development of blackspot disease (*Diplocarpon rosae* Wolf). *Mol. Breed.* 4(3), 187-94. doi:10.1023/A:1009642707505.

Martin, K. P., Joseph, D., Madassery, J. and Phillip, V. J. 2003. Direct shoot regeneration from lamina explants of two commercial cut flower cultivars of *Anthurium andraeanum* Hort. *In Vitro Cell. Dev. Biol. Plant* 39(5), 500-4. doi:10.1079/IVP2003460.

Matsumoto, T. K. and Kuehnle, A. R. 1997. Micropropagation of *Anthurium*. In: Bajaj, Y. P. S. (Ed.), *Biotechnology in Agriculture and Forestry 40: High-Tech and Micropropagation* (vol. VI). SpringerVerlag, New York, NY, pp. 15-29.

Matthews, D., Mottley, J., Horan, I. and Roberts, A. V. 1991. A protoplast to plant system in roses. *Plant Cell Tiss. Org. Cult.* 24(3), 173-80. doi:10.1007/BF00033473.

Matysiak, B. and Nowak, J. 1995. Acclimatization of *ex vitro Homalomena* 'Emerald Gem' as affected by nutrient solution concentration and $CO_2$ enrichment. *Acta Hortic.* 390, 157-60. doi:10.17660/ActaHortic.1995.399.17.

May, R. A. and Trigiano, R. N. 1991. Somatic embryogenesis and plant regeneration from leaves of *Dendrathema grandiflora*. *J. Am. Soc. Hort. Sci.* 116(2), 366-71. doi:10.21273/JASHS.116.2.366.

McCown, B. H. and Lloyd, G. B. 1982. A survey of the response of Rhododendron to *in vitro* culture. *Plant Cell Tiss. Org. Cult.* 2(1), 77-85. doi:10.1007/BF00033555.

Meins, F. 1989. Habituation: heritable variation in the requirement of cultured plant cells for hormones. *Annu. Rev. Genet.* 23(1), 395-408. doi:10.1146/annurev.ge.23.120189.002143.

Méndez-Hernández, H. A., Ledezma-Rodríguez, M., Avilez-Montalvo, R. N., Juárez-Gómez, Y. L., Skeete, A., Avilez-Montalvo, J., De-la-Peña, C. and Loyola-Vargas, V. M. 2019. Signaling overview of plant somatic embryogenesis. *Front. Plant Sci.* 10, 77. doi:10.3389/fpls.2019.00077.

Mendoza, M. G. and Kaeppler, H. F. 2002. Auxin and sugar effects on callus induction and plant regeneration frequencies from mature embryos of wheat (*Triticum aestivum* L.). *In Vitro Cell. Dev. Biol. Plant* 38(1), 39-45. doi:10.1079/IVP2001250.

Mercer, W. A. and Somers, I. I. 1957. Chlorine in food plant sanitation. *Adv. Food Res.* 7, 129-69. doi:10.1016/S0065-2628(08)60248-7.

Mercuri, A., De Benedetti, L., Burchi, G. and Schiva, T. 2000. *Agrobacterium*-mediated transformation of African violet. *Plant Cell Tiss. Org. Cult.* 60(1), 39-46. doi:10.1023/A:1006457716959.

Mikkelson, E. P. and Sink Jr., K. C. 1978a. *In vitro* propagation of Rieger *Elatior begonias*. *HortScience* 13, 242-4.

Mikkelson, E. P. and Sink Jr., K. C. 1978b. Histology of adventitious shoot and root formation on leaf-petiole cuttings of *Begonia × hiemalis* Fotsch 'Aphrodite Peach'. *Sci. Hortic.* 8, 179-92.

Miller, L. R. and Murashige, T. 1976. Tissue culture propagation of tropical foliage plants. *In Vitro* 12(12), 797–813. doi:10.1007/bf02796365.

Misra, P., Datta, S. K. and Chakrabarty, D. 2004. Mutation in flower colour and shape of *Chrysanthemum morifolium* induced by γ-radiation. *Biol. Plant.* 47(1), 153–6. doi:10.1023/A:1027365822769.

Mithila, J., Hall, J. C., Victor, J. M. R. and Saxena, P. K. 2003. Thidiazuron induces shoot organogenesis at low concentrations and somatic embryogenesis at high concentrations on leaf and petiole explants of African violet (*Saintpaulia ionantha* Wendl.). *Plant Cell Rep.* 21(5), 408–14. doi:10.1007/s00299-002-0544-y.

Mohapatra, A. and Rout, G. R. 2005. Study of embryo rescue in floribunda rose. *Plant Cell Tiss. Org. Cult.* 81(1), 113–7. doi:10.1007/s11240-004-3128-4.

Molgaard, J. P., Roulund, N., Deichmann, V., Irgens-Moller, L., Andersen, S. B. and Farestveit, B. 1991. *In vitro* multiplication of *Saintpaulia ionantha* Wendl. by homogenization of tissue cultures. *Sci. Hortic.* 48(3–4), 285–92. doi:10.1016/0304-4238(91)90137-N.

Moradi, S., Dianati Daylami, S., Arab, M. and Vahdati, K. 2017. Direct somatic embryogenesis in *Epipactis veratrifolia*, a temperate terrestrial orchid. *J. Hortic. Sci. Biotechnol.* 92, 88–97.

Muhlemann, J. K., Maeda, H., Chang, C. Y., San Miguel, P., Baxter, I., Cooper, B., Perera, M. A., Nikolau, B. J., Vitek, O., Morgan, J. A. and Dudareva, N. 2012. Developmental changes in the metabolic network of snapdragon flowers. *PLoS ONE* 7(7), e40381. doi:10.1371/journal.pone.0040381.

Mujib, A., Ali, M., Tonk, D., Isah, T. and Zafar, N. 2016. Embryogenesis in ornamentalmonocots: plant growth regulators as signalling element. In: Mujib, A. (Ed.), *Somatic Embryogenesis in Ornamentals and Its Applications*. Springer India, India, pp. 187–200. Chapter 12. doi:10.1007/978-81-322-2683-3_12.

Mulin, M. and Tran Thanh Van, K. 1989. Obtention of *in vitro* flowers from thin epidermal cell layers of *Petunia hybrida* (Hort.). *Plant Sci.* 62(1), 113–21. doi:10.1016/0168-9452(89)90195-7.

Murashige, T. and Skoog, F. 1962. A revised medium for rapid growth and bioassays with tobacco tissue cultures. *Plant Physiol.* 15(3), 473–97. doi:10.1111/j.1399-3054.1962.tb08052.x.

Murch, S. J., Victor, J. M. R. and Saxena, P. K. 2003. Auxin, calcium and sodium in somatic embryogenesis of African violet (*Saintpaulia ionantha* Wendl.) cv. Benjamin. *Acta Hortic.* 625, 201–9. doi:10.17660/ActaHortic.2003.625.21.

Nagaraju, S., Reddy, S. K. and Farook, S. A. 1998. Propagation of *Ficus reliosa* L. from maxillary buds and shoot tips. *Adv. Plant Sci.* 11(2), 287–90.

Narayan, P. and Jaiswal, V. S. 1986. Differentiation of plantlets from leaf callus of *Ficus religiosa*. *Indian J. Exp. Biol.* 24, 193–4.

Nhut, D. T., Teixeira da Silva, J. A., Bui, V. L., Thorpe, T. and Tran Thanh Van, K. 2003a. Woody plant micropropagation and morphogenesis by thin cell layers. In: Nhut, D. T., Van Le, B., Tran Thanh Van, K. and Thorpe, T. (Eds), *Thin Cell Layer Culture System: Regeneration and Transformation applications*. Kluwer Academic Publishers, Dordrecht, the Netherlands, pp. 473–93.

Nhut, D. T., Teixeira da Silva, J. A., Bui, V. L. and Tran Thanh Van, K. 2003b. Thin cell layer (TCL) morphogenesis as a powerful tool in woody plant and fruit crop micropropagation and biotechnology, floral genetics and genetic transformation.

In: Jain, S. M. and Ishii, K. (Eds), *Micropropagation of Woody Trees and Fruits*. Kluwer Academic Publishers, Dordrecht, the Netherlands, pp. 783–814.

Nobre, J. and Romano, A. 1998. *In vitro* cloning of *Ficus carica* L. adult trees. *Acta Hortic.* 480, 161–4. doi:10.17660/ActaHortic.1998.480.27.

Noman, A., Aqeel, M. and He, S. 2016. CRISPR-Cas9: tool for qualitative and quantitative plant genome editing. *Front. Plant Sci.* 7, 1740. doi:10.3389/fpls.2016.01740.

Noman, A., Aqeel, M., Deng, J., Khalid, N., Sanaullah, T. and Shuilin, H. 2017. Biotechnological advancements for improving floral attributes in ornamental plants. *Front. Plant Sci.* 8, 530. doi:10.3389/fpls.2017.00530.

Norton, M. E. and Norton, C. R. 1985. *In vitro* propagation of Ericaceae: a comparison of the activity of the cytokinin N6-benzyladenine and N6 -isopentenyladenine in shoot proliferation. *Sci. Hortic.* 27(3–4), 335–40. doi:10.1016/0304-4238(85)90038-X.

Ohki, S. 1994. Scanning electron microscopy of shoot differentiation *in vitro* from leaf explants of the African violet. *Plant Cell Tiss. Org. Cult.* 36(2), 157–62. doi:10.1007/BF00037714.

O'Riordain, F. 1999. Directory of European plant tissue culture laboratories, 1996-97. COST Action 822. Commission of the European Communities, Brussels, Belgium.

Orlikowska, T., Sabala, I. and Nowak, E. 1995. Adventitious shoot regeneration on explants of *Anthurium*, *Codiaeum*, *Dieffenbachia*, *Gerbera*, *Rosa* and *Spathiphyllum* for breeding purposes. *Acta Hortic.* 420, 115–7. doi:10.17660/ActaHortic.1995.420.32.

Osternack, N., Saare-Surminski, K., Preil, W. and Lieberei, R. 1999. Induction of somatic embryos, adventitious shoots and roots in hypocotyls tissue of *Euphorbia pulcherrima* Willd. Ex Klotzsch: comparative studies on embryogenic and organogenic competence. *J. Appl. Bot.* 73, 197–201.

Ozudogru, A., Da Silva, D. P. C., Kaya, E., Dradi, G., Paiva, R. and Lambardi, M. 2013. *In vitro* conservation and cryopreservation of *Nandina domestica*, an outdoor ornamental shrub. *Not. Bot. Horti Agrobot. Cluj-Napoca* 41(2). doi:10.15835/nbha4129335.

Paden, D. W., Iqbal, M. J. and Rayburn, A. L. 1996. DNA fingerprinting of *Rhododendron*s: phase II. *Am. Rhododendron Soc. J.* 50(2), 97–9.

Paek, K. Y., Hahn, E. J. and Son, S. H. 2001. Application of bioreactors for large scale micropropagation systems of plants. *In Vitro Cell Dev. Biol. Plant* 37(2), 149–57. doi:10.1007/s11627-001-0027-9.

Pati, P. K., Rath, S. P., Sharma, M., Sood, A. and Ahuja, P. S. 2006. *In vitro* propagation of rose - a review. *Biotechnol. Adv.* 24(1), 94–114. doi:10.1016/j.biotechadv.2005.07.001.

Pavingerova, D., Dostal, J., Biskova, R. and Benetka, V. 1994. Somatic embryogenesis and *Agrobacterium*-mediated transformation of *Chrysanthemum*. *Plant Sci.* 97(1), 95–101. doi:10.1016/0168-9452(94)90111-2.

Pavingerova, D., Biskova, R., Niedermeierova, H. and Kodytek, K. 1995. *Agrobacterium*-mediated transformation of *Rhododendrons*. *Acta Hortic.* 420, 89-91. doi:10.17660/ActaHortic.1995.420.23.

Pavingerova, D., Briza, J., Kodytek, K. and Niedermeierova, H. 1997. Transformation of *Rhododendron* spp. using *Agrobacterium tumefaciens* with a GUS-intron chimeric gene. *Plant Sci.* 122(2), 165-71. doi:10.1016/S0168-9452(96)04544-X.

Peak, D. E. and Cumming, B. G. 1984. *In vitro* propagation of *Begonia* × *tuberhybrida* from leaf sections. *HortScience* 19, 395–7.

Pedroso, A. N. V., Lazarini, R. A. M., Tamaki, V. and Nievola, C. C. 2010. *In vitro* culture at low temperature and *ex vitro* acclimatization of *Vriesea inflata* an ornamental bromeliad. *Braz. J. Bot.* 33,407–14.

Penna, S. and Jain, S. M. 2017. Mutant Resources and Mutagenomics in crop plants. *Emirates J. Food Agric.* 29(9), 651–7.

Perera, M. F., Racedo, J., García, M. G., Pardo, E. M. and Rocha, C. M. L. 2016. Use of molecular markers to improve the agro-industrial productivity in the North West of Argentina. *Mol. Biol.* 5, 153.

Pierik, R. L. M. 1975. Vegetative vermeerdering van Cyclamen Vakbl. *Bloemisterij* 30, 13.

Pierik, R. L. M. 1976. *Anthurium andraeanum* plantlets produced from callus tissues cultivated *in vitro*. *Physiol. Plant.* 37(1), 80–2. doi:10.1111/j.1399-3054.1976.tb01876.x.

Pierik, R. L. M. 1987. *In vitro Culture of Higher Plants*. Martinus Nijhoff Publishers, Dordrecth, the Netherlands.

Preil, W. 1991. Application of bioreactors in plant propagation. In: Debergh, P. C. and Zimmerman, R. H. (Eds), *Micropropagation Technology and Application*. Kluwer Academic Publishers, the Netherlands, pp. 425–45.

Preil, W. 2003. Micropropagation of ornamental plants. In: Laimer, M. and Rucker, W. (Eds), *Plant Tissue Culture 100 Years Since Gottlieb Haberlandt*. Springer-Verlag, New York, NY, pp. 115–33.

Preil, W. and Beck, A. 1991. Somatic embryogenesis in bioreactor culture. *Acta Hortic.* 289, 179–92. doi:10.17660/ActaHortic.1991.289.46.

Preil, W., Engelhardt, M. and Walther, F. 1983. Breeding of low temperature tolerant Poinsettia (*Euphorbia pulcherrima*) and *Chrysanthemum* by means of mutation induction in *in vitro* culture. *Acta Hortic.* 131, 345–55. doi:10.17660/ActaHortic.1983.131.41.

Preil, W., Florek, P., Wix, U. and Beck, A. 1988. Towards mass propagation by use of bioreactors. *Acta Hortic.* 226(226), 99–106. doi:10.17660/ActaHortic.1988.226.9.

Preil, W., Huitema, J. B. M. and Jong, J. 1991. Methods for selection of low-temperature tolerant mutants of *Chrysanthemum morifolium* Ramat. using irradiated cell suspension cultures II. Preselection *in vitro* under low-temperature stress. *Plant Breed.* 107(2), 131–4. doi:10.1111/j.1439-0523.1991.tb00540.x.

Pospisilova, J., Ticha, I., Kadlecek, S., Haisel, D. and Plzáková, Š. 1999. Acclimatization of micropropagated plants in *ex vitro* conditions. *Biol. Plant.* 42(4), 481–97. doi:10.1023/A:1002688208758.

Prasad, M. P. 2014. Molecular characterization and genetic diversity determination of Hibiscus species using RAPD molecular markers. *Asian J. Plant Sci. Res.* 4(3), 50–6.

Pueschel, A. K., Schwenkel, H. G. and Winkelmann, T. 2003. Inheritance of the ability for regeneration via somatic embryogenesis in *Cyclamen persicum*. *Plant Cell Tiss. Org. Cult.* 72, 43–51.

Quatrano, R. S. 1968. Freeze preservation of cultured flax cells utilizing DMSO. *Plant Physiol.* 43(12), 2057–61. doi:10.1104/pp.43.12.2057.

Rajagopalan, C. 2000. Export potential of Indian floriculture and need of policy environment. *Floriculture Today* 9, 29–33.

Rajapakse, S., Hubbard, M., Kelly, J. W., Abbott, A. G. and Ballard, R. E. 1992. Identification of rose cultivars by restriction fragment length polymorphism. *Sci. Hortic.* 52(3), 237–45. doi:10.1016/0304-4238(92)90024-7.

Rakosy-Tican, E., Bors, B. and Szatmari, A. M. 2012. *In vitro* culture and medium-term conservation of the rare wild species *Gladiolus imbricatus*. *Afr. J. Biotech.* 11, 14703–12.

Ranamukhaarachchi, D. G., Henny, R. J., Guy, C. L. and Li, Q. B. 2001. DNA fingerprinting to identify nine *Anthurium* pot plant cultivars and examine their genetic relationship. *HortScience* 36(4), 758–60. doi:10.21273/HORTSCI.36.4.758.

Rao, P. S., Handro,W. and Harada, H. 1973. Hormonal control of differentiation of shoots, roots and embryos in leaf and stem cultures of *Petunia inflata* and *Petunia hybrida*. *Physiol. Plant.* 28(3), 458–63. doi:10.1111/j.1399-3054.1973.tb08589.x.

Rayburn, A. L., Iqbal, M. J. and Panden, D. W. 1993. Positive identification of *Rhododendron* species through DNA fingerprinting. *Am. Rhododendron Soc. J.* 47(3), 137–8.

Reed, B. M., Okut, N., Achino, J. D., Narver, L. and Denoma, J. 2003. Cold storage and cryopreservation of hops (*Humulus* L.) shoot cultures through application of standard protocols. *CryoLett.* 24(6), 389–96.

Reuter, G. and Bhandari, N. N. 1981. Organogenesis and histogenesis of adventitious organs induced on leaf blade segments of *Begonia*-elatior hybrids (*Begonia* × *hiemalis*) in tissue culture. *Gartenbauwissenschaft* 46, 241–9.

Roberts, A. V., Yokoya, K., Walker, S. and Mottley, J. 1995. Somatic embryogenesis in *Rosa* spp. In: Jain, S. M., Gupta, P. K. and Newton, R. J. (Eds), *Somatic Embryogenesis in Woody Plants* (vol. 2). Kluwer Academic Publishers, the Netherlands, pp. 277–89.

Robichon, M. P., Renou, J. P. and Jalouzot, R. 1995. Genetic transformation of *Pelargonium* × *hortorum*. *Plant Cell Rep.* 15(1–2), 63–7. doi:10.1007/BF01690255.

Robinson, K. E. P. and Firoozabady, E. 1993. Transformation of floriculture crops. *Sci. Hortic.* 55(1–2), 83–99. doi:10.1016/0304-4238(93)90026-M.

Roest, S., Van Bakel, M. A. E., Bokelmann, G. S. and Broertjes, C. 1981. The use of an *in vitro* adventitious bud technique for mutation breeding of *Begonia* × *hiemalis*. *Euphytica* 30(2), 381–8. doi:10.1007/BF00034000.

Rogers, R. B. and Smith, M. A. L. 1992. Consequences of *in vitro* and *ex vitro* root initiation for miniature rose production. *J. Hortic. Sci.* 67(4), 535–40. doi:10.1080/00221589.1992.11516280.

Rosu, A., Skirvin, R. M., Bein, A., Norton, M. A., Kushad, M. and Otterbacher, A. G. 1995. The development of putative adventitious shoots from a chimeral thornless rose (*Rosa multiflora* Thurb. ex J. Murr.) *in vitro*. *J. Hortic. Sci.* 70(6), 901–7. doi:10.1080/14620316.1995.11515365.

Rout, G. R. and Das, P. 1997. Recent trends in the biotechnology of *Chrysanthemum*: a critical review. *Sci. Hortic.* 81, 201–28.

Rout, G. R., Debata, B. K. and Das, P. 1989. *In vitro* mass scale propagation of *Rosahybrida* cv. Landora. *Curr. Sci.* 58(15), 876–8.

Rout, G. R. and Jain, S. M. 2004. Micropropagation of ornamental plants–cut flowers. *Propag. Ornam. Plants* 4(2), 3–28.

Rout, G. R. and Mohapatra, A. 2006. Use of molecular markers in ornamental plants-a critical reappraisal. *Eur. J. Hortic. Sci.* 71(2),53–68.

Rout, G. R., Debata, B. K. and Das, P. 1990. *In vitro* clonal multiplication of roses. *Proc. Natl. Acad. Sci. (India)* 60(3), 311–8.

Rout, G. R., Debata, B. K. and Das, P. 1991. Somatic embryogenesis in callus cultures of *Rosa hybrida* L. cv. Landora. *Plant Cell. Tiss. Org. Cult.* 27(1), 65–9. doi:10.1007/BF00048208.

Rout, G. R., Debata, B. K. and Das, P. 1992. *In vitro* regeneration of shoots from callus cultures of *Rosa hybrida* cv. *Landora Ind. J. Exp. Biol.* 30, 15–8.

Rout, G. R., Palai, S. K., Panday, P. and Das, P. 1996. Direct plant regeneration of *Chrysanthemum morifolium* Ramat. cv. Deep Pink: influence of explant source, age of explants, culture environment, carbohydrates, nutritional factors and hormone regime. *Proc. Natl. Acad. Sci. (India)* 67, 57–66.

Rout, G. R., Samantaray, S., Mottley, J. and Das, P. 1999. Biotechnology of the rose: a review of recent progress. *Sci. Hortic.* 81(3), 201–28. doi:10.1016/S0304-4238(99)00025-4.

Rout, G. R., Samantaray, S. and Das, P. 2000. *In vitro* manipulation and propagation of medicinal plants. *Biotechnol. Adv.* 18(2), 91–120. doi:10.1016/s0734-9750(99)00026-9.

Rout, G. R., Mohapatra, A. and Jain, S. M. 2006. Tissue culture of ornamental pot plant: a critical review on present scenario and future prospects. *Biotechnol. Adv.* 24(6), 531–60. doi:10.1016/j.biotechadv.2006.05.001.

Rzepka-Plevnes, D. and Kurek, J. 2001. The influence of media composition on the proliferation and morphology of *Ficus benjamina* plantlets. *Acta Hortic.* 560, 473–6. doi:10.17660/ActaHortic.2001.560.94.

Sakai, A. 1960. Survival of the twigs of woody plants at −196°C. *Nature* 185, 392–4.

Sandford, P. A., Nafziger, A. J. and Jeanes, A. 1971a. Reaction of sodium hypochlorite with amines and amides: a new method for quantitating amino sugars in manomeric form. *Anal. Biochem.* 42(2), 422–36. doi:10.1016/0003-2697(71)90056-x.

Sandford, P. A., Nafziger, A. J. and Jeanes, A. 1971b. Reaction of sodium hypochlorite with amines and amides: a new method for quantitating polysaccharides containing hexosamines. *Anal. Biochem.* 44(1), 111–21. doi:10.1016/0003-2697(71)90352-6.

Sauvadet, M. A., Brochard, P. and Boccon-Gibod,J. 1990. A protoplast-to-plant system in *Chrysanthemum*: differential responses among several commercial clones. *Plant Cell Rep.* 8(11), 692–5. doi:10.1007/BF00269995.

Schiva, T. 2000. Strategies for development of commercial floriculture in Asia and the Pacific. Report of the *APO seminar*, 2–6 May 2000, New Delhi, India, pp. 27–38.

Schwenkel, H. G. 1991. Regeneration von *Cyclamen persicum* Mill. *in vitro* und deren integration in die Cyclamenzuchtung. PhD thesis. University of Hannover, Germany.

Scowcroft, W. R. and Larkin, P. J. 1988. Somaclonal variation. In: Bock, G. and March, L. (Eds), *Applications of Plant Cell and Tissue Culture*. CIBA Foundation Symposium 137. Wiley, Chichester, UK, pp. 21–35. doi:10.1002/9780470513651.ch3.

Seiichi, F., DeJong,J. and Rademaker, W. 1995. Efficient genetic transformation of Chrysanthemum (*Dendranthema grandiflorum* Ramat. Kitamura) using stem segments. *Breed. Sci.* 45(2), 179–84.

Senapati, S. K. and Rout, G. R. 2011. *In vitro* mutagenesis in *Rosa hybrida* using oryzalin as a mutagen and screening of mutants by randomly amplified polymorphic DNA (RAPD) marker. *Afr. J. Biotechnol.* 10, 5705–12.

Senapati, S. K., Aparajita, S. and Rout, G. R. 2012. An assessment of genetic fidelity of *in vitro* grown plantlets of rose (*Rosa hybrida*) through molecular markers. *Afr. J. Biotechnol.* 11, 16532–8.

Seo, S. Y., Choi, D. C., Kim, J. M., Lim, H. C., Kim, H. J., Choi, J. S. and Choi, Y. G. 2003. Plant regeneration from leaf explants and efficient *Agrobacterium*-mediated transformation system of *Chrysanthemum* (*Dendranthema grandiflorum*). *Acta Hortic.* 625, 403–9.

Shao, L., Li, Y., Yang, M. Z., Song, Y. and Chen, Z. L. 1996. Gene expression of chalcone synthase-A (CHSA) in flower colour alterations and male sterility of transgenic Petunia. *Acta Bot. Sin.* 38(7), 517–24.

Sharma, A. K. and Mitra, G. C. 1976. *In vitro* culture of shoot apical meristem of *Petunia hybrida* for mass production of plants. *Indian J. Exp. Biol.* 14, 348–50.

Shen, H. J., Chen, J. T., Chung, H. H. and Chang, W. C. 2018. Plant regeneration via direct somatic embryogenesis from leaf explants of Tolumnia Louise Elmore 'Elsa'. *Bot. Stud.* 59(4), 1-7.

Sherman, J. M., Moyer, J. W. and Daub, M. E. 1998. A regeneration and *Agrobacterium* mediated transformation system for genetically diverse *Chrysanthemum* cultivars. *J. Am. Soc. Hortic. Sci.* 123(2), 189-94. doi:10.21273/JASHS.123.2.189.

Shimizu, K., Miyabe, Y., Nagaike, H., Yabuya, T. and Adachi, T. 1999. Production of somatic hybrid plants between *Iris ensata* and *I. germanica*. *Euphytica* 107(2), 105-13. doi:10.1023/A:1026431800693.

Short, K. C. and Roberts, A. V. 1991. *Rosa* spp. (Roses): *in vitro* culture, micropropagation and production of secondary products. In: Bajaj, Y. P. S. (Ed.), *Medicinal and Aromatic Plants III. Biotechnology in Agriculture and Forestry* (vol. 15). Springer-Verlag, Berlin, Heidelberg, pp. 376-97.

Simmonds, J. 1984. Induction, growth and direct rooting of adventitious shoots of *Begonia × hiemalis*. *Plant Cell Tiss. Org. Cult.* 3(4), 283-9. doi:10.1007/BF00043079.

Simmonds, J. and Werry, T. 1987. Liquid shake cultures for improved micropropagation of *Begonia × hiemalis*. *HortScience* 22, 122-4.

Singh, S. K. and Syamal, M. M. 2000. Anti-auxin enhance *Rosa hybrida* L. micropropagation. *Biol. Plant.* 43(2), 279-81. doi:10.1023/A:1002768613872.

Skirvin, R. M. 1978. Natural and induced variation in tissue culture. *Euphytica* 27(1), 241-66. doi:10.1007/BF00039141.

Skirvin, R. M. and Chu, M. C. 1979. *In vitro* propagation of "Forever Yours" rose. *HortScience* 14(5), 608-10.

Skirvin, R. M. and Janick, J. 1976. Tissue culture induced variation in scented *Pelargonium* spp. *J. Am. Soc. Hortic. Sci.* 101, 281-90.

Skirvin, R. M., Norton, M. and McPheeters, K. D. 1993. Somaclonal variation: has it proved useful for plant improvement. *Acta Hortic.* 336, 333-40.

Smilansky, Z., Uniel, N. and Zieslin, N. 1986. Mutagenesis in roses (cv. Mercedes). *Environ. Expt. Bot.* 26(3), 279-83. doi:10.1016/0098-8472(86)90040-7.

Smith, C. R. 1968. Mycobactericidal agents. In: Lawrence, C. A. and Block, S. S. (Eds), *Disinfection, Sterilization, and Preservation*. Lea and Febiger, pp. 504-14.

Smulders, M. J. M. and De Klerk, G. J. 2011. Epigenetics in plant tissue culture. *Plant Growth Regul.* 63(2), 137-46. doi:10.1007/s10725-010-9531-4.

Spaulding, E. H. 1968. Chemical disinfection of medical and surgical materials. In: Lawrence, C. A. and Block, S. S. (Eds), *Disinfection, Sterilization, and Preservation*. Lea and Febiger, pp. 517-31.

Starts, N. D. and Cummings, B. G. 1976. *In vitro* propagation of *Saintpaulia ionantha* Wendl. *HortScience* 11, 204-6.

Steward, F. C., Mapes, M. O. and Mears, K. 1958. Growth and organised development of cultured cells. II. Organisation in cultured grown from freely suspended cells. *Am. J. Bot.* 45(10), 705-8. doi:10.1002/j.1537-2197.1958.tb10599.x.

Su, Y. J., Chen, J. T. and Chang, W. C. 2006. Efficient and repetitive production of leaf-derivedsomatic embryos of *Oncidium*. *Biol. Plant.* 50, 107-10.

Suprasanna, P., Ganapathi, T. R., Ghag, S. B. and Jain, S. M. 2017. Genetic modifications of horticultural plants by induced mutations and transgenic approach. In: Abul-Soad, A. A., et al. (Eds), *Proceedings of the IX International Symposium on In Vitro Culture and Horticultural Breeding*. *Acta Hortic.* 1187, 219-31.

Takayama, S. 1983. Begonia. In: Ammirato, P. V., Evans, D. A., Sharp, W. R. and Bajaj, Y. P. S. (Eds), *Handbook of Plant Cell Culture* (vol. 5). McGraw-Hill Publishing Co., New York, NY, pp. 253–83.

Takayama, S. 2000. Bioreactor, airlift. In: Spier, R. E., Griffiths, B. and Scragg, A. H. (Eds), *The Encyclopedia of Cell Technology* (vol. I & II). John Wiley & Son Inc, pp. 201–18.

Takayama, S. 2002. Practical aspects of bioreactor application in mass propagation. *Ist International Symposium "Liquid Systems for In Vitro Mass Propagation of Plants"*, 29 May – 2 June 2002. Norway, pp. 60–2.

Takayama, S. and Akita, M. 1994. The types bioreactors used for shoots and embryos. *Plant Cell Tiss. Org. Cult.* 39(2), 147–56. doi:10.1007/BF00033922.

Takayama, S. and Akita, M. 1998. Bioreactor techniques for large scale culture of plant propagules. *Adv. Hortic. Sci.* 12, 93–100.

Takayama, S. and Misawa, M. 1981. Mass propagation of *Begonia* × *hiemalis* plantlets by shake culture. *Plant Cell Physiol.* 22, 461–7.

Takayama, S. and Misawa, M. 1982. Factors affecting differentiation *in vitro* and a mass-propagation scheme for *Begonia* × *hiemalis*. *Sci. Hortic.* 16, 65–75.

Takayama, S., Arima, Y. and Akita, M. 1986. Mass propagation of plants by fermenter culture techniques. In: Somers, D. A., Gengenbach, B. G., Biesboer, D. D., Hackett, W. P. and Green, C. E. (Eds), *Book of Abstracts: VI International Congress of Plant Tissue and Cell Culture*. University of Minnesota, p. 449.

Tanaka, K., Kanno, Y., Kudo, S. and Suzuki, M. 2000. Somatic embryogenesis and plant regeneration in *Chrysanthemum* (*Dendranthema grandiflorum* Ramat.) Kitamura. *Plant Cell Rep.* 19(10), 946–53. doi:10.1007/s002990000225.

Teixeira da Silva, J. A. 2003a. Thin Cell Layer technology in ornamental plant micropropagation and biotechnology. *Afr. J. Biotech.* 2(12), 683–91.

Teixeira da Silva, J. A. 2003b. Tissue culture and cryopreservation of chrysanthemum: a review. *Biotech. Adv.* 21, 715–66.

Teixeira da Silva, J. A. and Fukai, S. 2002a. Change in transgene expression following transformation of chrysanthemum by four gene introduction methods. *Propag. Ornament. Plants* 2, 28–37.

Teixeira da Silva, J. A. and Fukai, S. 2002b. Increasing transient and subsequent stable transgene expression in chrysanthemum (*Dendranthema* × *grandiflora* (Ramat.) Kitamura) following optimization of particle bombardment and agroinfection parameters. *Plant Biotechnol.* 19(4), 229–40. doi:10.5511/plantbiotechnology.19.229.

Teixeira da Silva, J. A. and Fukai, S. 2003. *Chrysanthemum* organogenesis through thin cell layer technology and Plant Growth Regulator control. *Asian J. Plant Sci.* 2, 505–14.

Teixeira da Silva, J. A. and Nhut, D. T. 2003a. Cells: functional units of TCLs. In: Nhut, D. T., Van Le, B., Tran Thanh Van, K. and Thorpe, T. (Eds), *Thin Cell Layer Culture System: Regeneration and Transformation Applications*. Kluwer Academic Publishers, Dordrecht, the Netherlands, pp. 65–134.

Teixeira da Silva, J. A. and Nhut, D. T. 2003b. Thin cell layers and floral morpho-genesis, floral genetics and in vitro flowering. In: Nhut, D. T., Van Le, B., Tran Thanh Van, K. and Thorpe, T. (Eds), *Thin Cell Layer Culture System: Regeneration and Transformation Applications*. Kluwer Academic Publishers, Dordrecht, the Netherlands, pp. 65–134.

Teixeira da Silva, J. A., Bolibok, H. and Rakoczy-Trojanowska, M. 2007. Molecular markers micropropagation, tissue culture and *in vitro* plant research. *Genes Genom. Genet.* 1(1), 66–72.

Teixeira da Silva, J. A., Shinoyama, H., Aida, R., Matsushita, Y., Raj, S. K. and Chen, F. 2013. Chrysanthemum biotechnology: quovadis? *Crit. Rev. Plant Sci.* 32(1), 21-52. doi:10.1080/07352689.2012.696461.

Thao, N. T. P., Ozaki, Y. and Okubo, H. 2003. Callus induction and plantlet regeneration in ornamental *Alocasia micholitziana*. *Plant Cell. Tiss. Org. Cult.* 73(3), 285-9. doi:10.1023/A:1023025717271.

Theiler, R. 1977. *In vitro* culture of shoot tips of *Pelargonium* species. *Acta Hortic.* 78, 403-14. doi:10.17660/ActaHortic.1977.78.51.

Thomas, J. C., Akroush, A. M. and Adamus, G. 1999. The indole alkaloid tryptamine produced in transgenic *Petunia hybrida*. *Plant Physiol. Biochem.* 37(9), 665-70. doi:10.1016/S0981-9428(00)80096-0.

Ticha, I., Cap, F., Pacovska, D., Hofman, P., Haisel, D., Capkova, V. and Schafer, C. 1998. Cultures on sugar medium enhance photosynthetic capacity and high light resistance of plantlets grown *in vitro*. *Physiol. Plant.* 102(2), 155-62. doi:10.1034/j.1399-3054.1998.1020201.x.

Tran Thanh Van, K. and Bui, V. L. 2000. Current status of thin cell layer method for the induction of organogenesis or somatic embryogenesis. In: Jain, S. M. and Gupta, P. K., Newton, R. J. (Eds), *Somatic Embryogenesis in Woody Plants* (vol. 6). Kluwer Academic Publishers, Dordrecht, the Netherlands, pp. 51-92.

Ueno, K., Fukunaga, Y. and Arisumi, K. 1996. Genetic transformation of Rhododendron by *Agrobacterium tumefaciens*. *Plant Cell Rep.* 16(1-2), 38-41. doi:10.1007/BF01275445.

Urban, L. A., Sherman, J. M., Moyer, J. W. and Daub, M. E. 1992. Regeneration and *Agrobacterium*-mediated transformation of *Chrysanthemum*. *In Vitro Culture and Horticulture Breeding*, 28 June - 2 July 1992, Baltimore, MD, p. 49 (Abstract).

Van der Salm, T. P. M., Van der Toorn, C. J. G., Hanisch-ten Cate, C. H. and Dons, H. J. M. 1996. Somatic embryogenesis and shoot regeneration from excised adventitious roots of the root stock *Rosa hybrida* cv. Money Way. *Plant Cell Rep.* 15, 522-6.

Van der Salm, T. P. M., Van der Toorn, C. J. G., Bouwer, R., Hanisch-ten Cate, C. H. and Dons, H. J. M. 1997. Production of rol gene transformed plants of *Rosa hybrida* L. and characterization of their rooting ability. *Mol. Breed.* 3(1), 39-47. doi:10.1023/A:1009617704014.

Van Huylenbroeck, J. M. and Debergh, P. C. 1996. Physiological aspects in acclimatization of micropropagated plantlets. *Plant Tissue Cult. Biotechnol.* 2, 137-41.

Van Staden, J., Fennell, C. W. and Taylor, N. J. 2006. Plant stress *in vitro*: the role of phytohormones. *Acta Hortic.* 725(725), 55-62. doi:10.17660/ActaHortic.2006.725.2.

Van Wordragen, M. F., Dejong, J., Huitema, H. B. N. and Dons, H. J. M. 1991. Genetic transformation of *Chrysanthemum* using wild-type *Agrobacterium* strain and cultivar specificity. *Plant Cell Rep.* 9(9), 505-8. doi:10.1007/BF00232106.

Van Wordragen, M. F., Ouwerkerk, P. B. F. and Dons, H. J. M. 1992. *Agrobacteriumrhizogenesis* mediated induction of apparently untransformed roots and callus in *Chrysanthemum*. *Plant Cell. Tiss. Org. Cult.* 30, 149-57.

Vazquez, A. M., Davey, M. R. and Short, K. C. 1977. Organogenesis in cultures of *Saintpaulia ionantha*. *Acta Hortic.* 78(78), 249-58. doi:10.17660/ActaHortic.1977.78.32.

Voyiatzi, C., Voyiatzi, D. G. and Tsiakmaki, V. 1995. *In vitro* shoot proliferation rates of the rose cv. (Hybrid tea) 'Dr. Verhage', as affected by apical dominance regulating substances. *Sci. Hortic.* 61(3-4), 241-9. doi:10.1016/0304-4238(94)00750-A.

Wainwright, H. and Harwood, A. C. 1985. *In vitro* organogenesis and plant regeneration of *Cyclamen persicum* Mill. using seedling tissue. *J. Hortic. Sci.* 60(3), 397–403. doi:10.1080/14620316.1985.11515644.

Walther, F. andSauer, A. 1986. *In vitro* mutagenesis in roses. *Acta Hortic.* 189, 37–46. doi:10.17660/ActaHortic.1986.189.4.

Wang, S. O. and Ma, S. S. 1978. Clonal multiplication of *Chrysanthemum in vitro*. *J. Agric. Assoc. China* 32, 64–73.

Wated, A. A., Raghothama, K. G., Kochba, M., Nissim, A. and Gaba, V. 1997. Micropropagation of *Spathiphyllum* and *Syngonium* is facilitated by use of Interfacial membrane rafts. *HortScience* 32(2), 307–8. doi:10.21273/HORTSCI.32.2.307.

Weber, J., Preil, W. and Lieberei, R. 1994. Somatic embryogenesis in bioreactor culture of Clematis tangutica. *VIIIth International Congress for Plant Tissue and Cell Culture*, Firenze, Italy, 1994, p. 202 (Abstract).

Welander, T. 1977. *In vitro* organogenesis in explants from different cultivars of *Begonia × hiemalis*. *Physiol. Plant.* 41(2), 142–5. doi:10.1111/j.1399-3054.1977.tb05546.x.

Welander, T. 1979. Influence of medium composition on organ formation in explants of *Begonia × hiemalisin vitro*. *Swed. J. Agric. Res.* 9, 163–8.

Welander, T. 1981. Effect of polarity on and origin of *in vitro* formed organs in explant of *Begonia elatior* hybrid. *Swed. J. Agric. Res.* 11, 77–83.

Welander, M., Persson, J., Asp, H. and Zhu, L. H. 2014. Evaluation of a new vessel system based on temporary immersion system for micropropagation. *Sci. Hortic.* 179, 227–32. doi:10.1016/j.scienta.2014.09.035.

Werbrouck, S. P. O. and Debergh, P. C. 1995. Imazalil enhances the shoot-inducing effect of benzyladenine in *Spathiphyllum floribundum* Schott. *J. Plant Growth Regul.* 14(2), 105–7. doi:10.1007/BF00203121.

Wicart, G., Mouras, A. and Lutz, A. 1984. Histological study of organogenesis and embryogenesis in *Cyclamen persicum* Mill. tissue cultures: evidence for a single organogenetic pattern. *Protoplasma* 119(3), 159–67. doi:10.1007/BF01288870.

Williams, J. G. K., Kubelik, A. R., Livak, K. J., Rafiski, J. A. and Tingey, S. V. 1990. DNA polymorphisms amplified by arbitrary primers are useful as genetic markers. *Nucleic Acids Res.* 18(22), 6531–5. doi:10.1093/nar/18.22.6531.

Winefield, C., Lewis, D., Arathoon, S. and Deroles, S. 2000. Alteration of *Petunia* plant form through the introduction of the rol C gene from *Agrobacterium rhizogenes*. *Mol. Breed.* 5(6), 543–51.

Winkelmann, T., Hohe, A. and Schwenkel, H. G. 1998. Establishing embryogenic suspension cultures in *Cyclamen persicum* "Purple Flamed". *Adv. Hort. Sci.* 12, 25–30.

Winkelmann, T., Ratjens, S., Bartsch, M., Rode, C., Niehaus, K. and Bednarz, H. 2015. Metabolite profiling of somatic embryos of *Cyclamenpersicum* in comparison to zygotic embryos, endosperm, and testa. *Front. Plant Sci.* 6, 597.

Wolff, K. 1996. RAPD analysis of sporting and chimerism *Crysanthemum*. *Euphytica* 89(2), 159–64. doi:10.1007/BF00034601.

Wolff, K. and Peters-van Rijn, J. 1993. Rapid detection of genetic variability in chrysanthemum (*Dendranthema grandiflora* Tzvelev) using random primers. *Heredity (Edinb)* 71(4), 335–41. doi:10.1038/hdy.1993.147.

Wolff, K., Zietkiewicz, E. and Hofstra, H. 1995. Identification of chrysanthemum cultivars and stability of DNA fingerprint patterns. *Theor. Appl. Genet.* 91(3), 439–47. doi:10.1007/BF00222971.

www.iaea.org. Web site of International Atomic Energy Agency Organisation, Vinnea, Austria.

Yıldız, M. 2011. Evaluation of the effect of *in vitro* stress and competition on tissue culture response of flax. *Biol. Plant.* 55(3),541–4. doi:10.1007/s10535-011-0121-8.

Younis, A., Qasim, M. and Riaz, A. 2008. Case study: impact of a well-planned landscape in delivering quality of life to city dwellers. *Acta Hortic.* 775, 147–54. doi:10.17660/ActaHortic.2008.775.17.

Ziv, M. 1991. Morphogenic patterns of plants micropropagated in liquid medium in shaken flasks or large-scale bioreactor cultures. *Isr. J. Bot.* 40, 145–53.

Ziv, M. 1992. Morphogenetic control of plants micropropagated in bioreactor cultures and its possible impact on acclimatization. *Acta Hortic.* 319, 119–24.

Ziv, M. 1995. The control of bioreactor environment for plant propagation in liquid culture. *Acta Hortic.* 393, 25–38. doi:10.17660/ActaHortic.1995.393.3.

Ziv, M. 2000. Bioreactor technology for plant micropropagation. *Hortic. Rev.* 24, 1–30.

Ziv, M. 2005a. Simple bioreactors for mass propagation of plants. In: Hvoslef-Eide, A. K. and Preil, W. (Eds), *Liquid Culture Systems for In Vitro Plant Propagation*. Springer Netherlands, Dordrecht, the Netherlands, pp. 79-93.

Ziv, M. 2005b. Simple bioreactors for mass propagation of plants. *Plant Cell Tiss. Org. Cult.* 81(3), 277–85. doi:10.1007/s11240-004-6649-y.

# Chapter 5

# Advances in molecular breeding of ornamentals

*Neil O. Anderson, University of Minnesota, USA*



## 1 Introduction

For both classical and molecular flower breeders/geneticists, 'flower power' and convenience - flower color, flower size, and/or doubleness (extra petalage) - have remained the most important traits of interest. Secondary traits include fragrance, resistance to abiotic stress, diseases, and pests, as well as extending postharvest vase life. The immense number of floricultural crops on the market provide a challenging range of opportunities for continued breeding and research to achieve the crop ideotypes. Despite the extensive amount of classical breeding and selection in this wide array of crops, very few species have classical genetic or linkage maps. Today's needs for fingerprinting clones and related cultivars, marker-assisted selection, and molecular maps necessitate complementing classical breeding with molecular technologies. The use of molecular markers, whole genome sequencing, genome-wide-associated sequencing, and marker-assisted selection will hasten the rate and efficiency

http://dx.doi.org/10.19103/AS.2020.0066.06

of cultivar development. Regeneration and transformation systems have been and are being developed for floricultural crops to aid in transformation. Gene editing holds promise as a mechanism to evoke greater change in floricultural crops.

## 2 Flowers as study organisms for early genetic research

Integrated archeological genetics continue to demonstrate domestication's power of selection to change a wild species into a crop (Purugganan and Fuller, 2009), proving that the use of flowers, their propagation, selection, and breeding has been incorporated in most cultures for millennia as an integral component of human life. For example, the ancient Neanderthals used flowers for medicinal purposes and sympathy arrangements (Solecki, 1975), in ~900 BC, cultivated *Chrysanthemum* were transported from King Solomon's gardens to Egypt by Pharaoh Thutmose III (Schweinfurth, 1919), while the Koreans gifted cultivated chrysanthemums to Japan in AD 385 (Ackerson, 1967). A comprehensive history of the importance of flowers throughout all cultures has been reviewed elsewhere (Van Tuyl et al., 2014). For centuries, pollination has been a known technique to produce seeds and fruit, and propagate crops, as noted by ancient Egyptian reliefs of pollinating date palms. Thus, the propagation, selection, and breeding of flowering crops is not a recent development even though the publication of Gregor Mendel's discovery of genetic segregation, *Versuche über Pflanzen-Hybriden*, in 1866 (Mendel, 1866) might lead a novice to assume that this was the first time flower breeding had been done.

Prior to Mendel, there were numerous floriculturists and horticulturists who engaged in plant breeding, for example, Drs. Gaertner, Herbert, Koelreuter, Lecoq, and Michura (de Beer, 1966). Their work, however, was not based on the science of genetics and segregation, which produced inconsistent results. While Gregor Mendel acknowledged their contributions, they paled in comparison to Mendel's own research, which is of lasting significance to our understanding of the breeding of ornamental or floricultural crops.

The origin of modern genetics did not commence until Sir Gavin de Beer rediscovered Mendel's extensive research in 1900 and republished it in English (de Beer, 1966). This discovery, coupled with the publication of the joint work of Charles Darwin and Alfred Wallace on evolutionary selection (Wallace, 1858), produced the discipline of modern-day genetics. Darwin studied many floricultural crops, providing extensive research on reproductive biology, for example, trisylous *Lythrum salicaria* and other plants with heterostyly and/or self incompatibility (Darwin, 1865, 1868). Population genetics also commenced during this time period with W. E. Castle/J. C. Phillips (Castle and Phillips, 1914) and G. H. Hardy/W. Weinberg independently discovering what came to be known as the 'Hardy-Weinberg equilibrium' (Hardy, 1908; Weinberg, 1908) as well as work by subsequent population geneticists (S. Wright, R. Fisher, and J. B. S. Haldane).

Early plant geneticists during this time period worked primarily with floricultural crops to validate Mendel's laws of heredity. For instance, C.E. Correns, who discovered Mendel's work along with E. Tschermak von Seysenegg, Hugo de Vries, and W. J. Spillman, studied four o'clocks (*Mirabilis jalapa*) to publish the first paper on cytoplasmic inheritance and a non-Mendelian gene in 1908-9 (Correns, 1908). Correns also, unknowingly, confirmed Mendel's findings in peas, *Pisum sativum*; although he did not find Mendel's papers until later. This gave rise to the concepts of linkage of traits inherited together, as first proposed by Thomas Hunt Morgan. Hugo Marie de Vries discovered Mendel's works in concert with Correns and Erich Tschermak von Seysenegg. de Vries studied another ornamental, the evening primrose (*Oenothera lamarckiana*), discovering the phenomenon of mutations (albinos, dwarfs, and gigantism), which appeared separately from Darwinian natural selection (de Vries, 1901-1903). de Vries encouraged the famous US ornamental breeder at that time, Luther Burbank, to explore the phenomenon of mutations, but Burbank had nothing to do with it (Anderson and Olsen, 2015). Other researchers, particularly E. M. East and A. J. Mangelsdorf (Turelli, 2016), soon followed to expand the science of genetics, heredity, self incompatibility (termed self sterility at that time), and flower breeding, researching diverse ornamental and fruit crops such as *Nicotiana alata*, *N. forgetiana* (East, 1934, 1935), *Linaria* spp., *Lythrum salicaria* (East, 1932), and *Fragaria* (Mangelsdorf and East, 1927).

Subsequent breeding and genetic research commenced on a wide variety of crops with particular foci on edible grain crops, for example, *Zea mays*, *Triticum vulgare*, *Glycine max*, *Hordeum vulgare*, *Oryza sativa*, and so on. Many historical plant breeding treatises were devoted solely to grain crops as well as other edible horticultural crops, for example, *Solanum tuberosum*, *Daucus carota*, *Beta vulgaris*, *Brassica* spp., *Solanum lycopersicon*, and fruits (*Malus domestica*, *Pyrus pyrifolia*, *Musa* spp., *Olea europaea*, *Ficus carica*, and *Fragaria* spp.) (Allard, 1960; Schlegel, 2018). Fortunately, more recent monographs for plant breeding students have devoted some attention to floricultural crops (Bernardo, 2014).

## 3 Flower breeding in the nineteenth and twentieth centuries

While breeding in the early twentieth century proceeded with ornamental crops, they no longer held the primary focus of genetic research and resource allocation with worldwide governmental or agency funding being primarily devoted to edible crops. This noticeable shift caused much of the floricultural breeding to proceed predominantly funded by the private (companies) and public (universities) sector breeding programs. Classical breeding has held sway although, to a lesser extent, some programs have been fortunate enough to integrate molecular tools.

Numerous amateur as well as professional flower breeders operated during the mid to late nineteenth and early twentieth centuries with notables such as Ernst Benary (whose first product was *Lychnis haageana*; the company also created the first $F_1$ hybrid flower from seed, *Begonia semperflorens*, Fibrous Begonia Primadonna®; https://www.benary.com/en/node/1304), George Russell (Russell Lupines derived from *Lupinus polyphyllus*, https://www.hortmag.com/archive/return_of_the_russell_lupines), George Ball (famous for developing cut snapdragons, *Antirrhinum majus)*, Luther Burbank (bred or selected numerous crops, particularly the Shasta Daisy, *Leucanthemum ×superbum*; Anderson and Olsen, 2015), and W. Atlee Burpee (famous for breeding many floricultural crops as well as the still-famous head lettuce, *Lactuca sativa* 'Iceberg').

Luther Burbank's breeding and his art of selection were summarily documented in the monographic set titled 'Luther Burbank, his methods and discoveries and their practical application. Prepared from his original field notes covering more than 100 000 experiments made during forty years devoted to plant improvement' (Burbank, 1914–1915), which covers many of the 91 floricultural genera he bred or selected as well as the 1000 ornamental cultivars released to the industry (Anderson and Olsen, 2015). He is known as the 'Father of American Ornamental Breeding' (Anderson and Olsen, 2015) and has set up various breeding or selection practices that continue through much of the floricultural breeding programs to this day. Numerous amateur breeders associated with horticultural societies in both the United States and Europe named for each genus also worked on particular floral species, such as *Lilium*, *Hemerocallis*, *Chrysanthemum*, *Dahlia*, and *Hosta*, producing a wide array of cultivars for the floricultural market.

Following the publication of Burbank's field notes, additional monographs devoted to breeding and genetics of floricultural crops were published covering specific taxa or areas of scientific study. No single work is exhaustive, due to the immense number of floricultural crops on the market (Anderson, 2006a). Crane and Lawrence (1947) of the John Innes Horticultural Institute were students of William Bateson and introduced flower breeders to cytology, sterility, self incompatibility, interspecific hybridization, and genetics, as well as provided coverage of delphinium (*Delphinium* spp.), iris (*Iris* spp.), sweet pea (*Lathyrus odoratus*), stock (*Matthiola incana*), and primrose (*Primula sinensis*). Likewise, reporting on important vegetable crops and heterosis was provided for tomato (*Solanum lycopersicon*), lettuce (*Lactuca sativa*), and so on. The causes and methods of deriving sports or mutations to create variegated plants were also provided for vines or foliage plants (*Cytisus adami*, *Solenostemon* spp.), cut flowers (*Bouvardia* spp.; *Rosa* ×*hybrida*), and bedding plants (*Pelargonium* spp.). The newly described phenomenon of self incompatibility was examined in *Nicotiana*, *Veronica*, and *Verbascum* (Crane and Lawrence, 1947). They also introduced the chemistry and genetics of flower color in *Callistephus*, *Dahlia*, *Papaver*, and *Streptocarpus*, the foundation of which was expanded in later years across a wide range of floricultural crops. Understandably, flower color has remained one of the primary traits of interest to flower breeders and geneticists.

Gorer (1970) provided a treatise on selection for double flowers and gigantism as well as extensive historical coverage of the breeding of a wide range of floricultural crops. Chinese plant breeding on a range of native crops included *Callistephus*, *Camellia*, *Chrysanthemum*, *Paeonia*, *Primula*, and *Rhododendron* (Gorer, 1970). Geophytic species bred by Turkish flower breeders included *Anemone*, *Hyacinthus*, *Ranunculus*, *Syringa*, and *Tulipa*. Separate chapters are also devoted to specific crops such as *Begonia*, *Dianthus*, *Fuchsia*, *Iris*, *Narcissus*, *Pelargonium*, and *Viola*.

Continuous interest in the creation of double-flowered (extra petalage) crops culminated in Reynolds and Tampion's (1983) monograph on the topic. Their work provided the physiological and genetic basis for the creation of extra petalage both by flower breeders and those naturally occurring in the wild. Five floricultural crops are studied in depth, that is, *Aquilegia vulgaris*, *Calendula* 'Orange Cockade', *Narcissus* 'Cheerfulness', *Petunia*, and *Rosa*. They also provide a nearly exhaustive listing of genera with reported double-flowered mutants (Reynolds and Tampion, 1983).

More recent monographs on flower breeding and genetics have focused on procedural mechanics and plant breeding suppliers (Callaway and Callaway, 2000) as well as marketing trends, annualization of perennials, cultivar trialing, germplasm banks, prevention of invasiveness, and intellectual property protection throughout the world (Anderson, 2006a). Both these monographs also provide historical and methodological techniques for specific crops, which are authored by the world experts on each taxa. Callaway and Callaway (2000) harness the experts in breeding the Araceae, Gesneriaceae, *Camellia*, *Hemerocallis*, *Hippeastrum*, *Hosta*, *Iris*, *Kalmia*, *Magnolia*, *Narcissus*, *Penstemon*, *Quercus*, *Rhododendron*, *Saintpaulia*, and *Syringa*. Anderson (2006a) provides coverage of bedding plant crops (*Ageratum houstonianum*, *Anagallis monellii*, *Begonia*, *Impatiens wallerana*, *Petunia* ×*hybrida*, *Zinnia*), flowering potted plants (*Capsicum annuum*, *Cyclamen persicum*, *Dendrobium*, *Dendranthema* ×*grandiflora*, *Exacum affine*, *Hibiscus rosa-sinensis*, *Bacchanalia*, *Lagerstroemia indica*, *Lilium*, *Schlumbergera*, *Tulipa gerneriana*), cut flowers (*Eustoma grandiflorum*, *Freesia* ×*hybrida*, *Rosa* ×*hybrida*, *Ornithogalum*), and herbaceous perennials (*Clematis*, *Echinacea*, *Monarda didyma*). Despite the extensive amount of breeding and selection in this wide array of crops, very few species have a classical genetic map although specific genes have been identified in numerous cultivated taxa.

The focus during the mid-twentieth century was on $F_1$ hybrid seed products, since most seed-propagated flowers were open pollination derived (Anderson, 2004; Griesbach, 2006). Of particular importance in the flower breeding scene was Claude Hope, who co-founded PanAmerican Seed in 1945 with Charles Weddle (Martinez, 1997), producing the $F_1$ hybrid *Impatiens walleriana* 'Elfin' series (Uchneat, 2006), while Weddle developed the first $F_1$ *Petunia* ×*hybrida* 'Ballerina' (Griesbach, 2006; Weddle, 1976). Interspecific and intergeneric hybridization was commonly used to derive new flower colors, plant habits, or integrate disease resistance (Anderson and Olsen, 2015).

Early twentieth-century flower breeders also had to either confront or accept societal pressure that flower breeding was 'morally radical' (Taylor, 2014). Undoubtedly the most profound example of this was Luther Burbank, an amazing breeder/selector of horticultural and particularly floricultural crops (Anderson and Olsen, 2015). He became almost god-like and was revered in many circles and was known as the 'Plant Wizard of Santa Rosa' (Anderson and Olsen, 2015), even though his breeding methods were rarely scientific. This intense religious pressure continued until recent times with flower breeders saying they were just 'God's helpers' (J. Oglevee, Oglevee Co., 1982, pers. comm.).

## 4 Early use of biotechnology in marker-assisted breeding

Integration of biochemical and genetic markers that commenced in the early 1980s with the widespread use of starch or polyacrylamide gel electrophoresis (PAGE) to separate compounds, based on molecular weight or electrical charge differences allowed the use of biochemical markers such as isozymes, peroxidases, *Rhizobium*, and seed proteins to chemotype genetic variation in wild species, commercial cultivars, and/or segregating populations (Anderson et al., 2012). Gel banding patterns (Fig. 1; Mejía-Jiménez et al., 1994) could be scored into binary codes (0,1) and analyzed with appropriate computer software to determine genetic relatedness (Fig. 2; Anderson et al., 2012). This became known as 'genetic fingerprinting' and was useful in court cases involving theft of patented materials in the United States (plant patents, utility patents, and plant variety protection) or those protected by plant breeder's rights in the European Union, Canada, and Japan (Aguirre, 2006). It also allowed for the confirmation of new or existing taxonomic classifications, often resulting in either generic or specific epithet name changes, for example, *Coleus* was reclassified into the genus *Solenostemon*, whereas *Belamcanda* became *Iris domestica*, and so on. Isozyme technology is still used in limited arenas, for example, seed testing

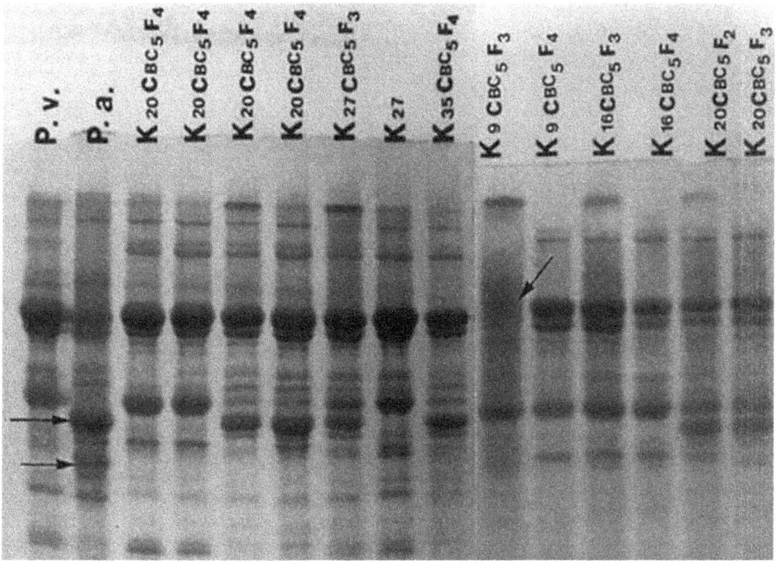


**Figure 1** Seed protein electrophoresis (PAGE) gel of seed proteins in *Phaseolus* populations (Mejía-Jiménez et al., 1994). From left to right: (P.v.) *P. vulgaris* parent (ICA Pijao); (P.a.) *P. acutifolius* (G40001); remaining lanes are congruity backcross-derived F3 and F4 progenies. Arrows show tepary alleles whose introgression is monitored in the congruity hybrids. Note that the phaseolin pattern of CBC 5 shown in the first lane is different from the typical ICA Pijao pattern.

for purity. Regardless of the early molecular techniques employed, breeding programs began amassing arrays of extracted DNA in molecular vaults (−80°C freezers) for subsequent use with advancing technologies. Debener (2002) provided an overview of the molecular markers to assess genetic relatedness. Limitations of these techniques are primarily time and money required to assess the genetic variation and the taxa's genome size (Debener, 2002).

Subsequent technologies have been developed and were/are widely used in numerous floricultural crops for purposes of fingerprinting, gene mapping, and elucidation of genetic variation among/within populations (Debener et al., 2000; Kazuhisa et al., 1998; Wen and Hsiao, 2001). These include a progression of molecular techniques such as RAPDs (randomly amplified polymorphic DNA), RFLPs (random fragment length polymorphisms), AFLPs (amplified fragment length polymorphisms; Roein et al., 2015), ISSRs (inter-simple sequence repeats), QTLs (quantitative trait loci for marker-assisted breeding), SSRs (simple sequence repeats; microsatellites), and SNPs (single nucleotide polymorphisms) (Debener, 2002). Additional techniques, but lesser known and not as widely used, include RAMPO (random amplified microsatellite polymorphisms), where size-specific RAPD products are hybridized with microsatellite repeat probes (Weising et al., 1997), as well as a series of dominant or codominant

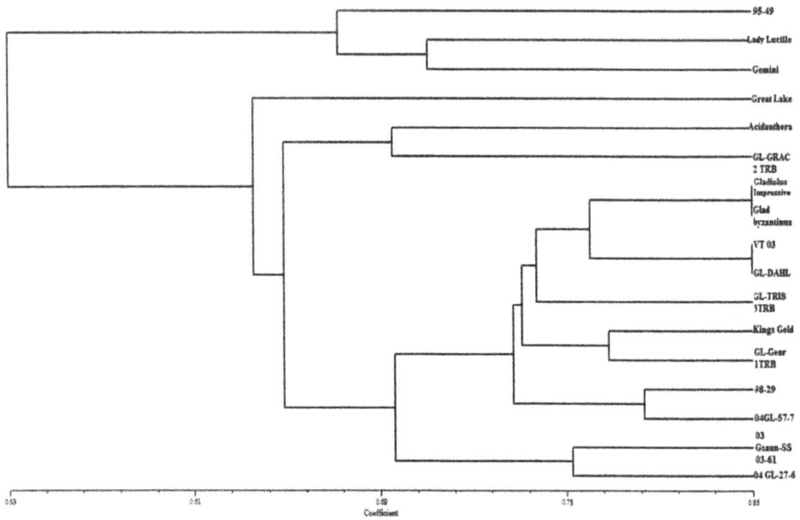


**Figure 2** Dendrograms of relatedness for wild *Gladiolus* spp. (*G. callianthus* (Code: *Acidanthera*), *G. byzantinus, G.* carinatus (1TRB, 2TRB, 3TRB), *G. dalenii* (GL-DAHL), *G. geardii* (GL-Gear), *G. gracilis* (GL-GRAC), *G. saundersii* (03 Gsaun-SS 03-61), *G. tristis* var. *concolor* (GL-TRIS)), five named cultivars (*G. nanus* 'Impressive', *G.* x*grandiflorus* 'Gemini', 'Great Lakes', 'King's Gold', 'Lady Lucille'), and five numbered selections (*G.* x*grandiflorus* 95-49, 98-29, VT03, 04GL-27-6, 04GL-57-7) based on five ISSR primers (UBC 808, 810, 811, 814, 818) (Anderson et al., 2012).

marker types using DNA sequences mapped in the genome, that is, SCAR (sequence characterized amplified regions; Paran and Michelmore, 1993) markers have length differences among amplified fragments, whereas cleaved amplified polymorphic sequences (CAPS) (Konieczyn and Ausubel, 1993) do not have any length differences and single-strand conformation polymorphism (SSCP) (Jordan et al., 1998). All techniques have been used, to varying degrees through time on several, but not all, floricultural crops (Vainstein, 2002). For any molecular marker system to be effective in detecting genetic variation, they must be polymorphic, reproducible, selectively neutral, and relatively rapid for use (Debener, 2002). Since many floricultural crops have inherently high levels of genetic variation, due to polyploidy or interspecific hybridization histories, molecular analyses and/or transformation have not been used as widely as in other crop commodities (Debener, 2002). Cost is often another factor that limits implementation of molecular technology in floricultural crops, since funding for ornamental plants lags significantly behind that of edible grain crops, such as corn (*Zea mays*) and soybeans (*Glycine max*) (Arús, 2000).

In chrysanthemum, for example, a wide range of molecular tools have been used to discover or characterize traits. RAPDs were used for genetic diversity (Wolff and Peters-van Rijn, 1993), whereas RFLPs were used for primer and probe development (Wolff et al., 1993, 1994), genetic fingerprinting (Wolff et al., 1995), and characterization of chimeras or sports (Wolff, 1996). RAPDS were also used to identify clonal integrity of 13 genotypes within a sport family as well as species variation (Wolff and Peters-van Rijn, 1993). Intriguingly, but not too surprising, fragment patterns for genotypes within a sport family were identical. A subsequent study using various molecular techniques found the same results among RAPDs, AFLPs, RFLPs, ISSRs, or hybridization-based DNA fingerprinting (Wolff et al., 1995). They found a high level of genetic variation among cultivars such that as few as two primers were necessary. Until more recent techniques (particularly SNPs) could be implemented, RAPDs were the tool of choice (Anderson, 2006b).

Long generation times in herbaceous perennials, particularly geophytes (plants with underground storage organs, for example, bulbs, corms, tubers, rhizomes, etc.) make the use of marker-assisted selection (MAS) critical in enhancing the rate of cultivar development. Van Tuyl and van Creij (2006) estimated that it takes as long as 25 years after a cross is made before the release of a new *Tulipa* cultivar. MAS has been used in *Lilium* to determine resistance to *Fusarium oxysporum* in the Asiatic lily types (Straathof et al., 1996; Jansen, 1996). The use of RAPD markers resulted in ~3 quantitative trait loci (QTLs) (Straathof et al., 1996), which was later repeated using AFLPs (Van Heusden et al., 2002). Ziang et al. (2003) exploited AFLPs to determine genetic distance threshold for *Dendrobium* clones, derived sports, and siblings. In other crops, AFLPs have been a useful tool to determine the genetic diversity of clones (Douhovnikoff and Dodd, 2003). Zlesak et al. (2007) used AFLPs to distinguish among *Rosa ×hybrida* (Fig. 3) and *Hydrangea macrophylla* cultivars (Fig. 4). However, large genome sizes (~77 pg/2C nucleus) as well as repetitive DNA in *Lilium longiflorum* complicated the use of AFLPs (Fig. 5; Zlesak et al., 2007). Thus, determining genetic variation in species with large genome sizes may not be possible using AFLPs.

As technologies developed, focus switched to PCR-based markers (Van Heusden et al., 2002). The discovery of 12 linkage groups in *Tulipa gesneriana* using AFLPs helped identify markers linked with Tulip Breaking Virus (TBV; Van Heusden et al., 2002). The use of AFLPs hastened the rate and efficiency of cultivar development and has been instrumental in the construction of several genetic maps and an intraspecific linkage map in *Capsicum annuum*, ornamental pepper (Lefebvre et al., 2002; Stummel and Bosland, 2006). AFLPs have also been useful in establishing genetic relatedness among clonal cultivars of various crops, such as rose (Vosman et al., 2004), particularly when differentiating phenotypic markers did not exist. Since most clonal breeding programs actively pursue obtaining mutations of an established clone, thereby creating 'clonal families' or 'mutant groups' whose cultivars vary by flower colors or some other distinguishable traits, it is of interest to protect these

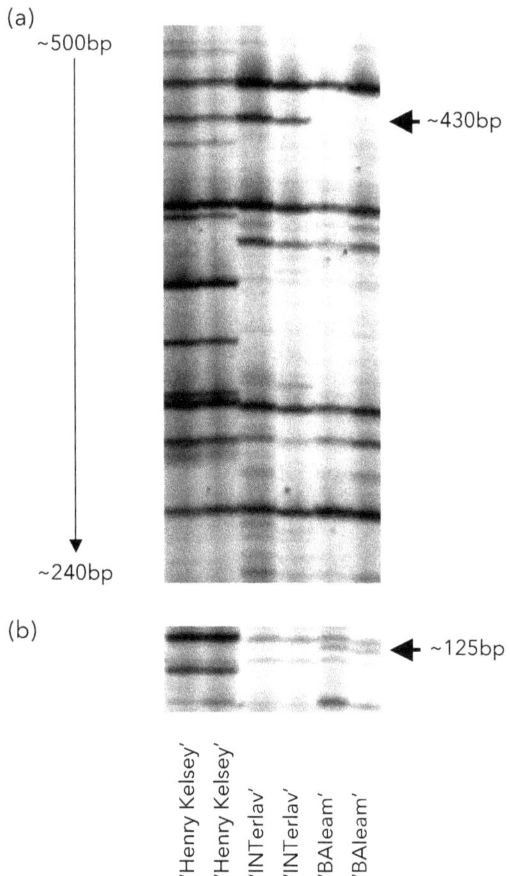


**Figure 3** Sections of AFLP autoradiograms for rose using (A) primer pair E-ACG/M-CAC and (B) E-AGC/M-CAC (Zlesak et al., 2007). Fragments distinguishing 'BAleam' and 'INTerlav' are highlighted with arrows and estimated size.

'essentially derived' materials. In the United States and Europe, protection of each essentially derived cultivar is of substantial economic value when filing plant patents and plant breeder's rights, respectively. For example, in 83 rose cultivars from 13 clonal families, AFLP primer combinations generated hundreds of polymorphic markers (Vosman et al., 2004). This confirmed that AFLPs can discriminate among cultivar pairs within each clonal family, similar to Debener et al.'s (2000) finding with rose seedling populations and sports.

## 5 Flower color

With the exception of foliage plants, herbs, and/or vegetables, which possess other above-ground phenotypes with market value (i.e. leaves), most floricultural

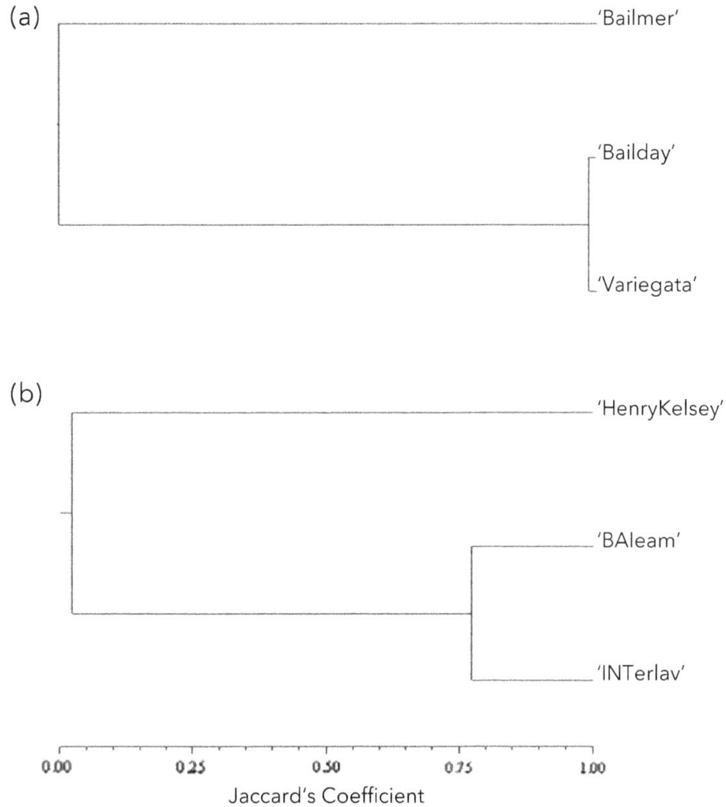


**Figure 4** UPGMA dendrograms constructed from Jaccard's coefficients of similarity from AFLP data for (a) hydrangea (157 AFLP markers) and (b) rose (165 AFLP markers) (Zlesak et al., 2007).

crops have 'flower power' (Anderson, 2006a). Thus, the development and marketing of seed-propagated series of products vary by individual flower colors, with each series' cultivar being the product of a different male onto the same female (seed-producing) plant. Frequently, but not always, each cultivar within a series is a related half-sib family. Thus, changes in flower colors are a critical trait for the market. The most elaborate series of colors, patterns, and petalage occur in well-developed crops that have been bred and marketed for decades or centuries. For example, *Impatiens walleriana* at its hiatus on the market, prior to its demise caused by genome-wide susceptibility to downy mildew (*Plasmopara obducens*), had ~23 specific colors across all series on the market (https://www.syngentaflowers-us.com/). Likewise, *Cyclamen persicum* has >13 named flower colors to describe cultivars of white-pink-red-purple (https://www.syngentaflowers-us.com/).

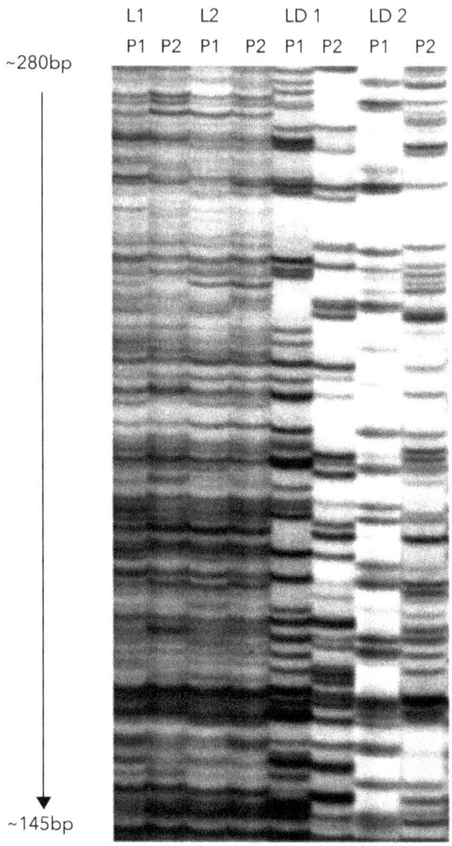


**Figure 5** *Lilium longiflorum* 'Nellie White' AFLP autoradiogram comparing two non-diluted digestions/ligation (L1; L2) and standard dilutions (LD1; LD2) with the standard (E-A/M-C; P1) and a modified preamplification primer pair (E-A/M-CT; P2); amplification primer pair was E-AAC/M-CTAA (Zlesak et al., 2007).

At the molecular level, however, the most comprehensive flower color research in floricultural crops has focused on *Petunia* ×*hybrida* - a model floricultural plant - for molecular research. More recent work has included roses and chrysanthemums (which will be discussed under genome-wide association studies). The petunia flower color work was spawned by classic biochemical work by Drs. Sam Asen and Robert Griesbach's research on flavonoids in the USDA (Asen, 1976) as well as the first transformation of petunia by a Monsanto lab (Fraley et al., 1983). Griesbach (2006) provided an extensive review of *Petunia* floral chemistry including the genetic engineering of flavonoids. Dihydroflavonol reductase (DFR) engineering began in the early 1990s using *Ht1* defective mutants and numerous DFR genes from *Cymbidium*, *Gerbera*, *Rosa*, and *Zea* were introduced therein (*cf.* Greisbach, 2006).

Subsequent study and transformation of petunia flower colors and patterns - along with most other important phenotypic traits - have been continued by several labs across the globe (Gerats and Strommer, 2009), including those at the University of Guelph (Judy Strommer), Radboud University Nijmegen (Tom Gerats), the University of Florida (David Clark), Michigan State University (Ryan Warner), and so on. *Petunia* ×*hybrida* as a model system for understanding floral color, morphology, and other important traits continues unabated at present.

## 6 Flowering genes

Following the identification of the flowering gene pathway and associated genes in *Arabidopsis thaliana* (Simpson and Dean, 2002), the model plant system, numerous flower breeding programs have sought to identify which flowering genes were present in particular crops. In *Chrysanthemum seticuspe*, Hirakawa et al. (2019) showed that *flowering locus T* (*FT*) as well as orthologues, thereof, behave as 'florigens'. A dense linkage map was created for cultivated chrysanthemums. Earlier, *CsAFT*, an antagonist to florigen, had been discovered in a wild *Chrysanthemum* species.

While it follows that many dicotyledonous flower crops would have analogous flowering genes and portions of the *Arabidopsis* pathway, monotoyledonous floricultural crops differ significantly. In studying geophytic *Lilium formosanum* (imbricate bulbs), which flowers in <1 year from seed and lacks dormancy with non-obligate-vernalization-requiring (NV) for flowering, Zlesak and Anderson (2010) reported that NV:OV (OV = obligate vernalization) fit a two-gene model. NV was controlled by two genes *VER1* and *VER2*, with only one dominant allele at either of the loci conferring NV (Zlesak and Anderson, 2010). Both were likely under epistatic genetic control, with *VER2* having the least penetrance. NV, *VER1*, and *VER2* inheritance is similar to flowering in another monocot, spring/winter wheat and its *VRN1* genes. Li et al. (2016) found two vernalization genes (*LoSVP* and *LoVRN1*) in the OV Oriental lily 'Sorbonne' which may be key to flowering. Delayed flowering was postulated to be due to *LoSVP*, while early flowering (NV) may be promoted by *LoVRN1*.

In other geophytes, for example, *Gladiolus* ×*hybridus* (Iridaceae), flowering genes involve a wide range of plant structures, from corm size, ambient temperatures, photoperiod, autonomous pathway, and gibberellins as well as vernalization. Abscisic acid (ABA) upregulates *abscisic acid-intensive 5* (*ABI5*), which is a basic leucine zipper (bZIP) transcription factor involved in seed maturation and upregulation of *FLC* expression. In *Gladiolus* ×*hybridus*, *Abi5 (GhABI5)* induces dormancy in corms, while silencing *GhABI5* promotes sprouting and reduces dormancy (Wu et al., 2015).

## 7 Genetically engineered flowers (GMOs and GEOs)

The invention of the gene gun or biolistic particle delivery system (Sanford et al., 1987), use of microprojectiles (microparticle bombardment), and transformation with *Agrobacterium tumefaciens* using the GUS reporter gene allowed for genetic transformation of crops (Bino et al., 1992; Van der Leede-Plegt et al., 1992; Watad et al., 1998), provided regeneration could be accomplished in vitro. The potential for genetic engineering of floricultural crops has been widely demonstrated, particularly using *Agrobacterium tumefaciens* and *A. rhizogenes* as vectors for genes important to floricultural crops. As a result of this, a US utility patent was issued for transgenic *Impatiens walleriana* in 2000 (Chou, 2000). Other examples include *Agrobacterium rhizogenes*-induced transformations in *Begonia ×hiemalis* integrating the *rolC* gene, which causes increased rooting and/or branching (Kiyokawa et al., 1996). Expression of the GUS reporter gene has been accomplished in *B. ×hiemalis* (Kishimoto et al., 2002). Wilmink et al. (1992, 1995) used a particle delivery system to transform *Tulipa gesneriana* with expression of beta-glucuronidase, although it took seven years from seed before a transgenic plant was found.

Both microprojectiles and *A. tumefaciens* have been used to transform *Lilium* (Bino et al., 1992; Van der Leede-Plegt et al., 1992; Watad et al., 1998) as well as *Rosa ×hybrida* (Dohm et al., 2001), although only a few genes have been introduced into roses (Zlesak, 2006). In orchids, microprojectiles were used to transform *Dendrobium* (Chia et al., 2001). Additionally, *A. tumefaciens* is also effective to transform *Dendrobium* as well as *Phalaenopsis* (Anzai and Tanaka, 2001; Nan et al., 1997; Yu et al., 2001). Regeneration and transformation systems have been developed for a number of floricultural crops, that is, *Chrysanthemum ×grandiflorum* (Bush and Pueppke, 1991; Young et al., 1998). Several labs breeding lisianthus, *Eustoma grandiflora*, have transformed this crop to change flower color, enhance fragrance, and alter time to flowering (Deroles et al., 1993; Handa and Deroles, 2000; Nielsen et al., 2002; Zaccai et al., 2001).

As long as a species can be infected with *A. tumefaciens*, for example, chrysanthemum, as well having stable GUS reporter expression (De Jong et al., 1994), this transformation system is easily adaptable for gene transfer (Miller et al., 1975). *Cyclamen persicum* is one crop example that has been difficult to transform; numerous labs attempted transformation but none were successful until 1999 (Aida et al., 1999). Likewise, ornamental *Capsicum annuum* is also recalcitrant to in vitro regeneration, thus limiting genetic transformations (Stummel and Bosland, 2006).

Genetically modified or engineered (GMOs, GEOs, respectively) crops provide new trait(s) that may benefit numerous crops in floriculture (Anderson and Walker, 2013). With the development of modern-day genetic modification techniques, breeders have been able to incorporate disease and insect resistance (Uzogara, 2000). Most importantly, however, for floricultural crops, flower color changes predominate GMO research (Auer, 2008; Clark et al., 2004; Holton et al. 1993). Secondary traits of interest for GMO cultivar development include fragrance, resistance to abiotic stress, diseases, and pest, as well as extending postharvest vase life (Chandler and Sanchez, 2012). An extensive review of transgenic approaches to produce new mutations in floricultural crops has been already written (Penna et al., 2017). Herein, only specific examples of genetic engineering will be discussed.

A key objective for flower breeders is new flower colors within a series to have a full palette of cultivars available or replace a current cultivar with a better performing one in the same color class (Chandler, 2007). Species' pigmentation pathways and types of pigments (anthocyanins and/or carotenoids) produced are limitations to the number and type of flower colors that can be produced within a species or crop (Tanaka et al., 2008). Frequently, true blue flower color is completely lacking in a species along with other colors, depending on the crop. For example, the three marigold species (*Tagetes erecta*, *T. patula*, *T. tenuifolia*) currently cultivated possess carotenoid pigments, which create the yellows, oranges, and reds typical for the species. An individual participating in W. Atlee Burpee Seed Company's citizen science breeding efforts created 'French Vanilla', the first creamy white type; others soon followed (https://www.burpee.com/flowers/marigolds/marigold-french-vanilla-hybrid-prod000265.html). There are no other colors available in the genus. Thus, GMOs would be the only way to obtain unknown flower colors such as a pink or lavender even though some close derivations exist, for example, 'Strawberry Blonde'.

A limited number of GMO floricultural crops exist in the markets worldwide: nine carnations, one chrysanthemum, and one rose (Anderson and Walker, 2013). Labeling GMO flowers in the United States is not required since they are not consumed as food crops. Thus, most distributors, flower auctions, brokers, wholesalers, floral designers, and consumers are not aware that they exist (Anderson and Walker, 2013). However, GMO restrictions apply throughout much of the rest of the world. Regardless of the country for introduction of a GMO cultivar, stringent, lengthy, and costly regulatory testing and requirements involving trait expression, chances of gene flow into sympatric wild relatives, and risk assessment have to be addressed (Chandler and Brugliera, 2011). This often limits incentives for private and public sector flower breeding programs to develop new GMO cultivars, since new products may require a higher price point to remunerate regulatory approval costs.

Gene transfer between different species, genera, and/or families through genetic engineering is allowed for the creation of transgenic blue roses, chrysanthemums, and carnations as well as annual bedding plants such as petunia (Meyer et al., 1987; Tanaka et al., 2010). For example, in petunia the anthocyanin pigment pathway is well defined both chemically and genetically (Meyer et al., 1987). Pigment derivatives of cyanidin and delphinidin are produced, but none from the pelargonidin pathway, since dihydrokaempferol cannot be reduced by the enzyme dihydroflavonol 4-reductase. At the Max Planck Institute for Plant Breeding Research in Cologne, Germany, Meyer et al. (1987) used a white-flowered mutant, *RL01*, which produced dihydrokaempferol and transferred the *A1* gene from a monocot, *Zea mays*, which resulted in pelargonidin synthesis. Using a vector, p35A1, which had positioned the complementary *A1* DNA in back of the 35S promoter, resulted in red flowers. Subsequent work by Haselmair-Gosch et al. (2018) with the *A1* gene expressed in 'Salmon Ray' created orange petunias 'Viva Orange' and 'Electric Orange' (Fig. 6) that were marketed. Unfortunately, these orange petunias had to be recalled when it was discovered they were transgenic in origin.

Controversy erupted over how plants with the *A1* construct entered petunia breeding programs, despite the US and EU statements that GMO petunias posed no risk to the environment or consumers. After the creation of the first GMO petunia (Meyer et al., 1987), widespread field trials were conducted at

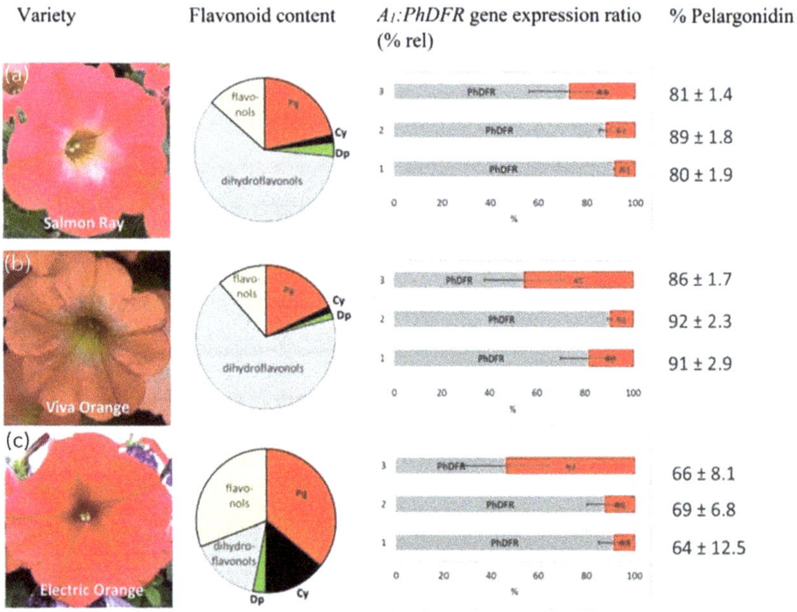


**Figure 6** Flavonoid content, $A_1$:*PhDFR* gene expression ratio and relative pelargonidin content of flowers from *Petunia ×hybrida* (a) 'Salmon Ray', (b) 'Viva Orange', and (c) 'Electric Orange (Haselmair-Gosch et al., 2018). Pie charts denote (bright red: pelargonidin (Pg), black: cyanidin (Cy), green: delphinidin (Dp), gray: dihydroflavonols, yellow: flavonols) of flavonoid contents while the bar charts graphically display the ratio of *A1:PhDFR* gene expression in % pelargonidin for each developmental stage (Stage 1: 0.6-3 cm long buds, 2: 3-5 cm, 3: open flowers).

the Max Planck Institute for Plant Breeding Research and elsewhere in Germany during 1990 and later in the United States. They were used in breeding (Oud et al., 1995; Servick, 2017). Subsequent company buyouts inadvertently ignored the GMO origin in the orange petunias (even though this color does not exist in the wild; David, 2017). Thus, it could have entered the breeding programs deliberately or inadvertently (a less likely scenario of pollen contamination from field trials) and not been noticed since *A1* does not cause orange flowers in other petunia germplasm (Haselmair-Gosch et al., 2018). The long-term impacts of this removal of orange GMO petunias from the market and integration of nonexpressed *A1* into classical petunia breeding programs remains to be determined.

GMO regulation and labeling continue to be debated globally since their inception (Anderson and Walker, 2013). In the United States, the Food and Drug Administration provides guidance on the labeling of GMO and non-GMO food products (Albert, 2009), but little regulation has been implemented for non-edible GMOs in the floral industry. Currently, the largest impediment to adoption of at least some GMO horticultural products is the lack of market acceptance (Clark et al., 2004). Commercialization of GMO flowers is complex and relies on consumer and industry marketplace perceptions and profitability (the GMO has to offer something new, the phenotype of which is readily apparent), which are driven by both the marketplace perception of their value as well as the specific complexities surrounding public debate on the pros and cons of gene technology and its regulation (Chandler, 2007). The GMO carnation introduction in the marketplace in 1996 was not a widely known fact, despite aggressive marketing by Florigene, an Australian subsidiary of Suntory, a Japanese whiskey distillery. The Florigene 'Moon™' carnation series are GMO products, but this aspect is not a central part of marketing (Chandler, 2007) and it is unknown to what extent the floral industry is aware of their GMO nature. This resulted in complete silence by florists and the floral design industry on the matter (Fig. 7), who delighted in having these new colors available without having to dye them (Anderson and Walker, 2013).

## 7.1 Genetically modified carnations

*Dianthus caryophyllus*, carnation, one of the top ten cut flowers on the world market, has had nearly 2 000 years of cultivation due to its lengthy postharvest life (Besemer, 1980; Holley and Baker, 1963; Jawaharlal et al., 2003). The historic production of carnations as cut flowers was traditionally in the United States, Germany, Japan, and the Netherlands, the centers of consumer consumption, but exportation of production technologies in the 1970s moved production to warmer areas near the equator, for example, Colombia, Israel, Kenya, Spain, and the Republic of China (Besemer, 1980; Nau, 2011). In 2015, only 896 000

**Figure 7** An example floral design by an American Institute of Floral Design (AIFD) florist, Jeremy Trentelman, which mentions nothing about the use of Florigene 'MoonVista™' series GMO carnations (https://www.pinterest.com/pin/399272323184495671/).

stems of standard carnations were produced in 15 states accounting for US $273 000 wholesale (w) revenue in the cut flower sector of the horticulture industry (USDA, 2016). Classic carnations are available in a wide color range from classical breeding techniques: white, yellow, and green to reds, pinks, and salmons (Besemer, 1980; Holley and Baker, 1963). Traditional breeding efforts are limited due to the lack of key pigment biosynthesis genes in a particular genetic pathway (Chandler, 2007; Holley and Baker, 1963) prohibiting other colors in many different flower species. For flower colors not commercially available in carnations, wholesale growers (producers), wholesale auction houses, or retail florists use methods such as dip dying, spraying, or vascular uptake to create blue and purple carnations (Besemer, 1980; Hunter, 2000).

The isolation of the gene encoding flavonoid 3′5′-hydroxylase in 1993 (Holton et al., 1993), created the opportunity for creation of GMO-modified *Dianthus caryophyllus* and *Rosa ×hybrida* L. which are now on the market

(Chandler and Sanchez, 2012; Anderson and Walker, 2013). The introduction of the GMO Florigene 'Moon'™ carnations in 1996 (Kikuchi et al., 2008) transformed the commercial flower market by extending the available color line of *Dianthus caryophyllus*. The first GMO carnation was violet-colored, 'Florigene Moondust™', expressing the F3'5'H gene (Kikuchi et al., 2008). Florigene 'Moondust™' was introduced by the Florigene Pty. Ltd. (Florigene Flowers, 2013) in Melbourne, Australia, and was first marketed in Australia in 1996, then in Japan, the United States, and Europe thereafter (Anderson and Walker, 2013). The GMO carnation cultivars sold commercially at this time are Florigene 'Moondust™' as well as 11 newer ones in the Florigene 'MoonVista™' series (Fig. 8; Chandler, 2007). The 12 transgenic carnation cultivars all possess the F3'5'H gene-expressing tints, tones, and shades of violet or purple: 'Moondust™', 'Moonvista™', 'Moonlite™', 'Moonshade™', 'Moonaqua™', 'Moonique™', 'Moonpearl™', 'Moonvelvet™', 'Moonberry™', 'Moontea™', 'Moonburst™', and 'Moonstrike™' (Florigene Flowers, 2019). The series is currently available in Australia, Canada, the European Union, Japan, Russia, the United Arab Emirates, and the United States (Florigene Flowers, 2019).

As GMO carnations began to spread across the US market, the primary difference between GMOs and non-GMO carnations was the pricing differentials each year between classical bred cultivars and transgenic flowers at the wholesale farmgate value (Anderson and Walker, 2013). Table 1 delineates sample prices during select years of their early introduction. By 2009, the difference in wholesale price between more costly GMO and less expensive non-GMO carnations (^$0.12/stem) was half as much as the difference in 2005 (^$0.24/stem) (Table 1; Anderson and Walker, 2013).

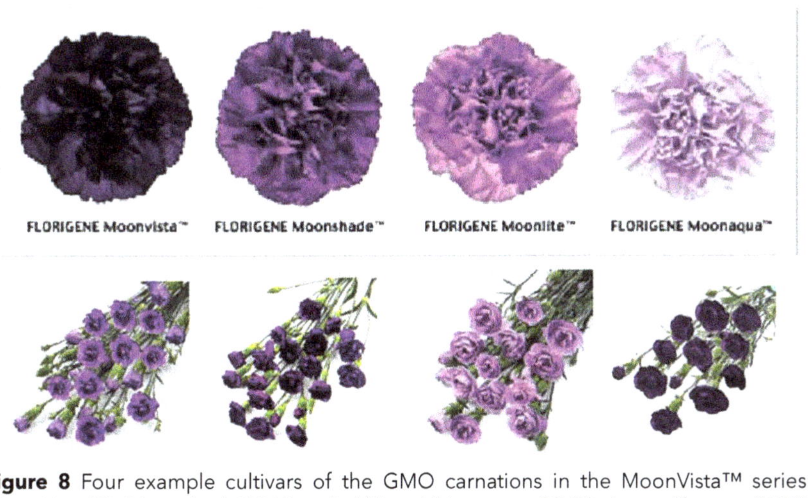


**Figure 8** Four example cultivars of the GMO carnations in the MoonVista™ series: 'MoonVista™', 'Moonshade™', Moonlite™', and 'Moonaqua™' (Florigene Flowers, 2019).

**Table 1** Wholesale price differentials per stem between Non-Genetically Modified Organism (Non-GMO) and GMO standard carnations for 2005, 2007, and 2009, three of the sample years in which this experiment was conducted (Anderson and Walker, 2013)

| Year | Non-GMO standard carnation price/stem | GMO standard carnation price/stem | Price difference between Non-GMO and GMO carnations |
|---|---|---|---|
| 2005 | US $0.33 | US $0.57 | ^US $0.24 |
| 2007 | $0.35 | $0.49 | ^$0.14 |
| 2009 | $0.40 | $0.52 | ^$0.12 |

Data (wholesale) was collected from flowers sold at Koehler & Dramm Wholesale Florists, Minneapolis, MN, USA.

Anderson and Walker (2013) addressed the question: 'how should retailers handle the case of supplying and marketing genetically modified *Dianthus caryophyllus* in the retail setting?' They posited that, while wholesalers and retailers alike were slowly made aware of the GMO carnations, consumers on the other hand were 'relatively uninformed'. Retailer florists faced a 'dilemma' of whether to use the transgenic carnations for these flower colors versus continuing using the classic methods of color modification (dip dyes, sprays, and vascular uptake; Hunter, 2000). Also of interest was to know which was more cost effective and whether there were positive or negative connotations associated with GMO carnations, asking such questions as 'would the use of GMO crops potentially impose a negative opinion of the retail shop? Could the implementation of modified flowers affect the loyalty of certain customers? Are customers willing to pay more for the "Moon™"series or are they satisfied with traditional methods of color alteration? How does the marketing of these flowers differ from classic offerings?' (Anderson and Walker, 2013).

Anderson and Walker (2013) tested the perceptions and opinions of 121 floral design students during 2005–2011 regarding GMO carnations. In each year, students spray painted and/or dip dyed white carnations the colors of the MoonVista™ cultivars and then made a Hogarth (S-curve) design containing standard and miniature MoonVista™ carnations along with a price sheet. Students in 2009 also wrote a marketing paragraph for their design. In each year, students answered an identical question on a subsequent test to garner their position on GMO flowers and what their retail floral shop policy would be regarding customer information (Anderson and Walker, 2013). Student answers were significantly different between groups: not selling any GMO products in their shops (0.8%), giving their customers a GMO/non-GMO option (66.9%), or selling only GMO carnations in that color palette (27.3%). Most (73.6%) would tell their customers upfront about the GMO carnations. Marketing information describing their GMO designs did not highlight the GMO status. These data show a directional opposition to the current florist attitude toward handling GMO cut flower crops and may impact the future industry significantly.

By documenting the 5-year trends in their study, Anderson and Walker (2013) monitored differences in next generation florists on GMO flowers. However, consumers worldwide are not informed regarding GMO cut flowers, so they cannot make decisions regarding purchase options. Future issues that need addressing are consumer reactions (acceptance, boycotts or indifference) as predicated by safety concerns, since GMO food crop safety is a critical issue for regulatory agencies and consumers alike (Anderson and Walker, 2013). In the Australian Agriculture Biotechnology Report, biotech carnations were the first GMO products evaluated by the Gene Technology Regulator to 'pose minimal risks to people or the environment, and are sufficiently safe to be used by anyone without the need for a license' (USDA, 2012). An additional concern is customer loyalty, an important factor in independent retail floral design shops. Loyal customers have long-term relationships with their florists, which would, naturally, engender critical informational transfer on important cut flower factors such as their GMO nature (Anderson and Walker, 2013), although several other technical, economic, and regulatory factors create additional hurdles for GMO use in floricultural crops (Bradford and Alston, 2004).

## 7.2 Genetically modified roses

In 1990, following the successful introduction of highly profitable GMO carnations, Florigene (now Suntory Flowers) continued its transformation to create blue-colored roses (https://www.suntory.com/sic/research/s_bluerose/story/). As with GMO carnations, the idea was to replace traditionally dye-treated white roses to create blue counterparts. Twenty years of transformational research resulted in the Tokyo release of the first-ever 'blue' rose, *Rosa* ×*hybrida* 'Applause', in 2009 (Fig. 9). In actuality, 'Applause' is silver-purple in coloration as opposed to the ultimate goal of a sky blue rose (Fig. 9). Growers were quickly engaged in production, establishing clonal plants to fulfill the anticipated market demand for 'Applause' (Fig. 10). Unquestionably, 'Applause' is the bluest rose to date. Its release caused an international interest in the product and it was subsequently released in the United States and Canada in 2011.

As with other GMO flowers with new colors nonexistent in the crop species, the rose research team was confronted with first isolating the genes to create delphinidin for blue petals along with the transformation development of subsequent regeneration techniques to create plants expressing the gene (Holton et al., 1993). Two 'blue genes' were identified in *Petunia* in 1991, cytochrome P450 type hydroxylase (https://www.suntory.com/sic/research/s_bluerose/story/), which increased production of delphinidin blue pigmentation after insertion via *Agrobacterium*. However, the transformed red

**Figure 9** The first 'blue' GMO rose, 'Applause' released in 2009 to the Japanese cutflower and floral design markets (http://www.suntoryapplause.com).

**Figure 10** The blue GMO rose 'Applause' in production greenhouses for cut roses (https://www.wired.com/2011/09/blue-roses-for-sale/).

roses remained red in color; in the interim, Florigene succeeded in getting delphinidin production in carnations even though it did not work in roses (see above). Researchers downregulated the internal dihydroflavonol 4-reductase (DFR) gene using RNAi which, using a F3′5′H from pansy (*Viola xwittrockiana*), resulted in roses with expressed delphinidin (*cf.* Fig. 11 for the general process).

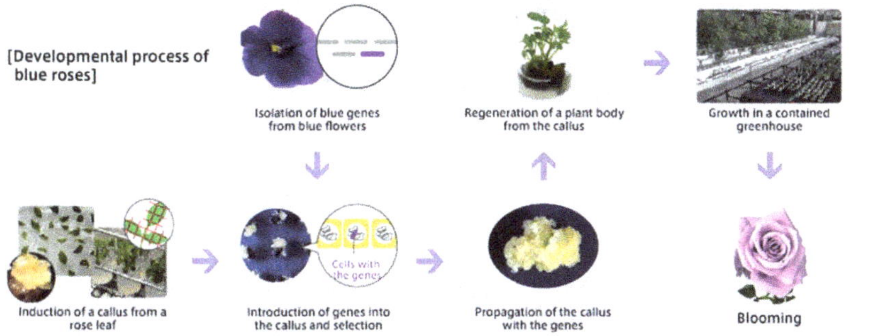


**Figure 11** The developmental process of creating the first blue GMO rose, 'Applause' by Florigene/Suntory (https://www.suntory.com/sic/research/s_bluerose/story/).

Marketing of these blue roses in Japan involved intellectual property (IP) protection costs (https://phys.org/news/2009-10-blue-roses-debut-japan.html). Prices for 'Applause' were priced >10x that of classic rose cutflower stems, ranging from 2000 to 3000 yen per stem (~US$22 and $33, respectively)! Florigene/Suntory 'recommended as a luxurious gift for special occasions such as wedding anniversaries and birthdays' (https://phys.org/news/2009-10-blue-roses-debut-japan.html). The results of significantly higher price points for GMO 'Applause', compared with price point difference in GMO versus non-GMO carnations (Table 1) have yet to be determined, although the higher price of 'Applause' allowed for significant profitability early in the marketing campaign.

### *7.3 Genetically modified chrysanthemums*

Noda et al. (2017) focused on creating the first blue chrysanthemums, *Chrysanthemum ×grandiflorum*. They first used overexpression of flavonoid 3′,5′-hydroxylase (*F3′5′H*) from *Campanula medium*, although the resulting GMO transformants were purple/violet in petal color rather than blue, even though it had worked well in carnations and rose. They postulated that blue flowers would need incorporation of polyacylated anthocyanins with metal complexes to create blue chrysanthemum flowers. Noda et al. (2017) successfully used butterfly pea UDP (uridine diphosphate)-glucose:anthocyanin 3′,5′-*O*-glucosyltransferase gene along with the Canterbury bells *F3′5′H* gene expression (Fig. 12). Copigmentation with flavone glucosides created a blue colored chrysanthemum (Fig. 13). This two-step method has widespread implications for creating other blue ornamental flowers.

## 8 Genome sequencing

Following publication of the *Arabidopsis thaliana* genome sequence in 2000, the advent of whole genome sequencing (WGS), also known as complete-,

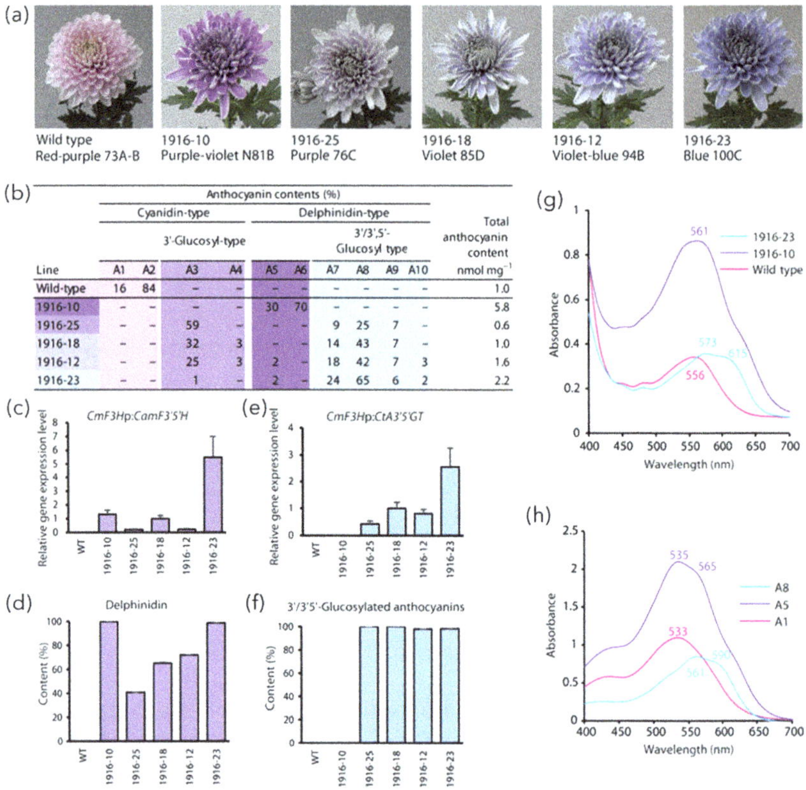


| | Anthocyanin contents (%) | | | | | | | | | | |
|---|---|---|---|---|---|---|---|---|---|---|---|
| | Cyanidin-type | | | | Delphinidin-type | | | | | | Total anthocyanin content |
| | 3'-Glucosyl-type | | | | | | 3'/3',5'-Glucosyl type | | | | |
| Line | A1 | A2 | A3 | A4 | A5 | A6 | A7 | A8 | A9 | A10 | nmol mg$^{-1}$ |
| Wild-type | 16 | 84 | - | - | - | - | - | - | - | - | 1.0 |
| 1916-10 | - | - | - | - | 30 | 70 | - | - | - | - | 5.8 |
| 1916-25 | - | - | 59 | - | - | - | 9 | 25 | 7 | - | 0.6 |
| 1916-18 | - | - | 32 | 3 | - | - | 14 | 43 | 7 | - | 1.0 |
| 1916-12 | - | - | 25 | 3 | 2 | - | 18 | 42 | 7 | 3 | 1.6 |
| 1916-23 | - | - | 1 | - | 2 | - | 24 | 65 | 6 | 2 | 2.2 |

**Figure 12** 'Flower coloration, anthocyanin composition, and transgene expression in GMO chrysanthemum "Sei Arabella": (a) flower colors of wild-type and transgenic lines. Each line presents the color group and number of RHS Color Charts; (b) Anthocyanin composition; (c, d) Relative expression of *CamF3′5′H* (c) and the resultant delphinidin-based anthocyanins (d). Contents of delphinidin-based anthocyanins corresponded well with the expression of *CamF3′5′H*; (e, f) Relative expression of *CtA3′5′GT* (e) and the resultant 3′/3′,5′-glucosylated anthocyanins (f). Contents of 3′/3′,5′-glucosylated anthocyanins corresponded well with the expression of *CtA3′5′GT*; (g) Visible absorption spectra of fresh petals of wild-type (pink), 1916-10 (purple), and 1916-23 (blue) lines; (h) Visible absorption spectra of major anthocyanins derived from pink (A1; wild type), purple-violet (A5), and blue/violet-blue (A8) chrysanthemums in acetate buffer (pH 5.6)' (Noda et al., 2017).

entire-, or full-genome sequencing, became a useful molecular tool to characterize the entire DNA sequence of additional plant species (Hamilton and Buell, 2012). Plant WGS is considerably more difficult than animals due to heterozygosity, polyploidy, highly repetitive sequences, genome duplication, and larger genome sizes (Hamilton and Buell, 2012). Since 2000, the cost of WGS de novo transcriptome assemblages has decreased significantly with the advent of companies providing these services in the market; the competition

**Figure 13** Blue, GMO chrysanthemum flowers were easier to create than researchers had anticipated (https://www.nature.com/news/true-blue-chrysanthemum-flowers-produced-with-genetic-engineering-1.22365).

successfully decreased prices, and small research labs were no longer restricted by needing to purchase sequencing machines. Coordinated improvements in throughput and bioinformatics software capable of running complex algorithms have enabled larger genomes, common in floricultural crops, to be handled efficiently and cost-effectively.

WGS of floricultural crops is in its beginning phase, encompassing a range of crop and commodity types in 15 families, such as 5 species of bedding plants, 3 cutflowers, 3 flowering potted plants, 2 herbaceous perennials, 2 vegetables, as well as 1 water gardening and 1 medicinal crop species (Table 2). Castor bean, *Ricinus communis* (an ornamental foliage bedding plant), was the first floricultural crop with a genome sequence published in 2010 (Chan et al., 2010), followed by numerous others in 2011 – including *Petunia* × *hybrida*, *Solanum lycopersicon*, and *Cannabis sativa* (van Bakel et al., 2011). Others have followed with the most recent WGS of *Chrysanthemum seticuspe* (Hirakawa et al., 2019). These WGS have been useful in a wide variety of fronts, aiding basic scientific research and prompting functional genomic biotechnology tool kits.

While the WGS of *Chrysanthemum nankingense* occurred in 2017 (Table 2), the genome for *C.* × *grandiflorum* has not yet been sequenced. Complications from the hexaploid ($2n = 6x = 54$) nature of *C.* × *grandiflorum*, unstable meiotic pairing (which results in new flower colors in mutants after the gain or loss of a chromosome), self-incompatible highly heterozygous, and heterogeneous nature of this important floricultural crop, necessitated working with simpler species within the *Chrysanthemum* complex. In 2019, a diploid relative, *C. seticuspe*, was proposed to be a model chrysanthemum for hexaploid

**Table 2** List of floricultural crop types and species (families) to date with a whole genome sequence (WGS), the year it was sequenced, the genome size, and the number of predicted genes

| Crop Type | Species (family) | Yr. Sequenced & Genome size | No. of Predicted Genes | Citation |
|---|---|---|---|---|
| Bedding Plants | *Amaranthus hypocondriacus* (Amaranthaceae) | 2016; 403.9Mbp | 23 847 | Clouse et al. (2016) |
| | *Ipomoea nil* (Convolvulaceae) | 2016; 734.6Mbp | 42 783 | Hoshino et al. (2016) |
| | *Mimulus guttatus* (Phrymaceae) | 2013; ~430Mbp | 26 718 | Phytozome v9.1 |
| | *Petunia ×hybrida* (Solanaceae) | 2011 | | The Petunia Platform |
| | *Ricinus communis* (Euphorbiaceae) | 2010; 320Mbp | 31 237 | Chan et al. (2010) |
| Cut flowers | *Chrysanthemum nankingense* (Asteraceae) | 2017; 3.07Gb | 56 870 | Chrysanthemum Genome database |
| | *Linum usitatissimum* (Linaceae) | 2012; ~350Mbp | 43 384 | Wang et al. (2012) |
| | *Rosa chinensis* (Rosaceae) | 2018; 537.4Mbp | | Hilbrand Saint-Oyant et al. (2018) |
| Flowering Potted Plants | | | | |
| | *Chrysanthemum seticuspe* (Asteraceae) | 2019; 3.06Gb | 71 057 | Hirakawa et al. (2019) |
| | *Kalanchoe* (Crassulaceae) | 2013? | | Kalanchoe Genome |
| | *Phalaenopsis equestris* (Orchidaceae) | 2014; 1600Mbp | 29 431 | Cai et al. (2015) |
| Herb. Perennial | *Aquilegia coerulea* (Ranunculaceae) | not published | | Phytozome v9.1 |
| | *Carnegiea gigantean* (Cactaceae) | 2017; 1.40Gb | 28 292 | Copetti et al. (2017) |
| Vegetables | *Capsicum annuum* (Solanaceae) | 2014; ~3.48Gbp | 34 903 | Kim et al. (2014) |
| | *Solanum lycopersicon* (Solanceae) | 2011; ~900 Mbp | 34 727 | Sol Genomics Network |
| Water Gardens | *Nelumbo nucifera* (Nelumbonaceae) | 2013; | | Ming et al. (2013) |
| Medicinal | *Cannabis sativa* (Cannabaceae) | 2011; ~820Mbp | 30 074 | van Bakel et al. (2011) |

*C.* ×*grandiflorum* (Hirakawa et al., 2019). A de novo WGS was created from an $F_6$ inbred generation of a self-compatible strain, AEV2. The genome size from assembled sequences was 2.72 Gb with 354 212 scaffolds which included ~89% of the 3.06 Gb *C. seticuspe* genome (Hirakawa et al., 2019). As many as 71 057 annotated protein-coding genes were postulated. The WGS genome sequences were then used for construction of linkage maps for comparative analysis with *C.* ×*grandiflorum* (van Geest et al., 2017). The power of using WGS to determine linkage groups is useful for QTL mapping in a wide variety of sequenced genomes (Table 2), gene discovery, and use of short-read genome assembly (Hirakawa et al., 2019).

The orchid, *Phalaenopsis equestris*, was the first CAM (crassulacean acid metabolism) plant to have a WGS with a completely sequenced genome (Cai et al., 2015). Since potted orchids are the top flowering potted plant sold in the United States (USDA, 2016) and moth orchids (*Phalaenopsis*) are the most important genus in this category, researchers have used the *P. equestris* WGS to better understand and characterize $C_3$ to CAM in young seedlings (Ping et al., 2018). Prior research had implicated but not proven weak CAM photosynthetic pathways in seedlings. WGS allowed Ping et al. (2018) to determine that the $C_3$ to CAM shift occurred when the first leaf was differentiated in the protocorms. They also discovered that the phosphoenolpyruvate carboxylase kinase (PPCK) gene was responsible for CAM in the seedlings, which will be useful in future studies to aid *Phalaenopsis* micropropagation (Ping et al., 2018).

## 8.1 Genome-wide association studies (GWAS)

Flower breeders will benefit from WGS with the advent of genome-wide-associated sequencing in additional floricultural crops (GWAS; Mitchell-Olds, 2010). Numerous uses of WGS are anticipated to improve cultivated crops as well as conduct basic biological research for important plant functions. For example, cloning/mapping of QTLs identifies the gene(s) association with phenotypes aiding in elucidation of complex traits (van Geest et al., 2017). Study of complex traits for QTL identification have benefited from: (1) GWAS of unrelated genotype populations using SNPs with correlated phenotypes and (2) 'family-based QTL mapping' in complicated pedigrees from differing parents (Mitchell-Olds, 2010). Both techniques have been used for fine-mapping in *Arabidopsis* (Atwell et al., 2010; Kover et al., 2009).

Mapping studies selected genes as most likely candidates controlling traits, but recent developments in GWAS are changing this (Ingvarsson and Street, 2011). Advantages of GWAS include eliminating gene mapping limitations due to gene level higher resolution and associating genetic with phenotypic variation (Brachi et al., 2011). High-density SNPs have been used to find haplotype blocks which are correlated with quantitative traits. Reduced sequencing costs have allowed labs to focus on phenotyping populations in a wide variety of environments, particularly with the use of drones. These developments allow for phenotyping increasingly larger populations to expand our understanding of gene expression's role in phenotypic variability (Ingvarsson and Street, 2011).

Use of GWAS necessitates the development of a genomic map with high-density markers, which is dependent on large population sizes in order to detect recombination events that change phenotypic diversity (Brachi et al., 2011). Genotyping-by-sequencing (GBS) as well as bioinformatics have been useful to find, genotype, and create genomic data in many plants. Both rice (*Oryza sativa*) and corn (*Zea mays*) followed *Arabidopsis* in using GWAS (Brachi et al., 2011). Plant adaptation systems such as in *Aquilegia*, *Helianthus*, and *Mimulus* are predicted to take advantage of GWAS technology (Kramer and Hodges, 2010; Strasburg et al., 2011; Wu et al., 2008). Polyploid crops provide the additional challenge of high heterozygosity. Consequently, few crops have GWAS-related analyses except for cotton (*Gossypium barbadense*; Abdurakhmonov et al., 2009), potato (*Solanum tuberosum*; Lindqvist-Kreuze et al., 2014), and switchgrass (*Panicum virgatum*; Lu et al., 2013).

Most recently, GWAS applications have made inroads for two major polyploid, floricultural crops, that is, roses and chrysanthemums. Construction and development of WGS in diploid relatives of floricultural crops (Table 2) allow for single gene and quantitative traits to be localized (Debener and Linde, 2009), although low genetic variation in populations and crop polyploidy can complicate the process. While the use of AFLPs and SSRs have been used to map the identified QTLs in biparental populations of diploid rose (Crespel et al., 2002; Spiller et al., 2011; Roman et al., 2015), tetraploid cultivated roses necessitate derivation of more markers. GWAS provides advantages by providing higher resolution of marker traits and more alleles per loci, limiting genotyping to candidate genes for specific traits as well as across the entire genome (Schulz et al., 2016). Koning-Boucoiran et al. (2015) used numerous rose-expressed sequence tags (ESTs) to create an Axion WagRhSNP chip (or array) which allowed genotyping with 68 000 SNPs and, thus, GWAS. Schulz et al. (2016) then conducted the first GWAS in roses by examining flower color in the two major classes of pigments (anthocyanins and carotenoids), using scorable and polymorphic SNP markers with the Axion WagRhSNP chip along with AFLPs, SSRs, and candidate gene markers. They found five genomic regions and two clusters associated with anthocyanin and carotenoid contents, respectively, and described several candidate genes (Schulz et al., 2016). This GWAS research represents the very beginning for rose functional genomics and validation experiments with independent populations to identify the causal factor(s) associated with each flower color phenotype. Identified validated markers could then be used to invoke MAS for parental selection to enhance quantitative levels of specific floral pigments. Subsequently, numerous other traits in rose would be valuable to involve in GWAS. Likewise, the development of a new single polymorphism high-density array, generation of transcriptomics data, as well as use of enhanced software packages for ultra-dense diploid/polyploid genetic mapping and dosage scoring can be pursued (Bourke et al., 2017; Hibrand Saint-Oyant et al., 2018; Smulders et al., 2019; Vukosavljev et al., 2016). Storage and availability of big data generated by such endeavors is of necessity to ensure access by basic and applied researchers to continue the momentum in new product development.

Despite the genome complexity of cultivated hexaploid chrysanthemums, linkage mapping has progressed (Van Geest et al., 2017), allowing for QTL detection. Simplex SNPs, present in one of the six homologous chromosomes of a hexaploid, should have simple Mendelian inheritance and were used to generate six homologous linkage maps in hexaploid sweet potato, *Ipomoea batatas* (Shirasawa et al., 2017). Sumitomo et al. (2019) theorized that simplex SNPs would be useful for association analysis, linking phenotypes with simplex loci. This was done in chrysanthemums, where a single gene affects white versus yellow petal coloration and white is dominant to yellow (Langton, 1980; Teynor et al., 1989a,b). A single dominant gene, *CmCCD4a* (carotenoid cleavage dioxygenase), inhibits carotenoids in white petals (Ohmiya et al., 2006). A total of 9219 single-dose SNPs with simplex (1x0) or double simplex (1x1) inheritance, based on the allelic frequencies of SNPs in the $F_1$ hybrids, were used in a GWAS (Sumitomo et al., 2019). The use of simplex and double simplex markers has a stable positioning as they link an allele in one gene to a chromosome; thus, the SNPs were tightly linked to the contributory genes. If, however, duplex, triplex markers had been used, GWAS would have detected numerous false positives and PCR reproducibility would have suffered (Sumitomo et al., 2019). Thus, single-dose SNP marker-based GWAS is a relatively simple means of managing complex heritability in higher ploidy levels of this and other floricultural crops to develop DNA markers.

## 9 Gene editing (CRISPR-Cas9)

The short palindromic repeat (CRISPR)-associated endonuclease-9 (Cas9) is the newest molecular genetic tool, whereby precise edits to a specie's DNA can be done to ultimately alter plant genotypes and phenotypes. Doudna and Charpentier (2014) were able to modify the CRISPR-Cas9 complex to work in higher plants and other organisms to cut DNA in any chosen location. Thereafter, they created CRISPR into a usable version with as many types of editing as possible, stating that 'these results highlight a new era in which genomic manipulation is no longer a bottleneck to experiments, paving the way toward fundamental discoveries in biology, with applications in all branches of biotechnology' (Doudna and Charpentier, 2014). CRIPR-Cas9 has quickly been ranked as a revolutionary tool to quickly effect gene changes with precision. Intriguingly, at least in the United States, the USDA has chosen not to regulate CRISPR (Kwon, 2018) since it, in most instances, does not involve the introduction of foreign DNA into transformed crops. The secretary of agriculture Perdue stated that, 'with this approach, USDA seeks to allow innovation when there is no risk present' (https://www.usda.gov/media/press-releases/2018/03/28/secretary-perdue-issues-usda-statement-plant-breeding-innovation).

In essence, this ruling means that CRISPR is essentially comparable to classical breeding.

Numerous uses for the CRIPR technology include replacing undesirable gene sequences with desired ones, disabling or knocking out gene functions, using the Cas9 enzyme even enhance gene expression (Doudna and Charpentier, 2014). Perhaps even more astounding was the use of CRISPR to add genetic material into the DNA, which led to the new technology of 'gene drive'. This will have an astounding use in floricultural crops since CRISPR will, theoretically, eliminate the need for interspecific hybridization for trait transfer into crop species (which cause hybrid breakdown or incongruity) as well as remove deleterious genes. Due to this recent discovery and with only limited applications in floricultural crops, the extent to which CRISPR will make dramatic and economically viable changes to modern floricultural crops is still unknown.

Japanese researchers from the University of Tsukuba, Yokohama City University, and the National Agriculture and Food Research Organization (NARO) have been the first to use CRISPR in floricultural (ornamental) crops (Watanabe et al., 2017). Since the plant genome of the Japanese morning glory, *Ipomoea nil*, has been sequenced (Table 2) it, along with other crops with sequenced genomes, is one of many potential candidates for CRISPR technology. Watanabe et al. (2017) focused their CRISPR research on an enzyme responsible for stem and flower color, *DFR-B*. Editing out *DFR-B* allowed for instantaneous flower color change from violet to white – a process that would have taken classical breeding multiple years to accomplish.

## 10 Conclusion

The creation of novel phenotypes with economic value and implementing genetic changes to new and old floricultural and ornamental crops will continue to capture the minds of our brilliant researchers throughout the horticultural distribution chain worldwide. It is important that the consuming public continues to recognize the ecological and therapeutic services offered by floriculture products which provide stress relief and therapeutic value for all customers. Coordinated efforts among classic and molecular private and public sector breeding programs will ensure the implementation of cost-effective and efficient crop transformational techniques to continue the evolution of domesticated ornamental crops. These efforts, along with the integration of sustainable production methodology, will ensure that future generations will be able to enjoy ornamental, floricultural crops which can be produced in stressful and unpredictable environmental conditions the planet is experiencing.

## 11 Future trends in research

Undoubtedly, the future of ornamental and flower breeding and genetics is an exciting field with numerous possibilities just over the horizon. The evolution of molecular biotechnology in the past 30–40 years has realized untold challenges and opportunities for floricultural crop improvement. Numerous new and improved series and cultivars are currently on the market as direct testimony to the effectiveness of these technologies. The discovery and implementation of CRISPR-Cas9 as a gene editing tool will undoubtedly reap numerous rewards in numerous floricultural crops. We can only imagine what other possibilities may lie ahead with this tool as well as others that have yet to be invented. The future is bright and will remain one filled with flower and foliage colors, fragrance, and disease/insect resistant crops coupled with other traits that will continue to keep floricultural crops economically viable by providing ecological services in our changing landscapes.

## 12 Where to look for further information

In addition to consulting the citations referenced in this publication, particularly the review articles and monographs focusing on floricultural crop marker-assisted breeding, other options exist for researchers to keep abreast of the constantly changing field of information. Of course, one can use the online search engines freely available, such as Google, to at least obtain general and scientific publications for the keywords of interest. If you are not in academics, become a member of an academic or public library to gain access to the search engines they subscribe to, which will also allow you to gain free access to fee-based online publications. For instance, two serial publications are dedicated to reviews of plant breeding and genetics literature, that is, *Plant Breeding Reviews* (https://www.wiley.com/WileyCDA/Section/id-321445.html) and *Horticultural Reviews* (https://www.wiley.com/WileyCDA/Section/id-321744.html), both of which are published by Wiley. Any researcher can subscribe to receive alerts from scientific journals dedicated to publishing topics of this nature, for example, *HortScience*, *Journal of the American Society for Horticultural Science*, *HortTechnology*, *Acta Horticulturae*, *Euphytica*, *Theoretical and Applied Genetics*, *Sexual Plant Reproduction*, *Plant Biotechnology Journal*, *Nature Genetics*, *DNA Research*, or *Frontiers in Plant Science*.

## 13 References

Abdurakhmonov, I. Y., Saha, S., Jenkins, J. N., Buriev, Z. T., Shermatov, S. E., Scheffler, B. E., Pepper, A. E., Yu, J. Z., Kohel, R. J. and Abdukarimov, A. 2009. Linkage disequilibrium based association mapping of fiber quality traits in *G. hirsutum* L. variety germplasm. *Genetica* 136(3), 401-17. doi:10.1007/s10709-008-9337-8.

Ackerson, C. 1967. Development of the chrysanthemum in China. *Natl. Chrysanth. Soc. Bull.* 23(4), 146-55.

Aguirre, P. 2006. Protection: plant patents, utility patents, plant breeder's rights, trademarks, branding, royalties. In: Anderson, N. O. (Ed.), *Flower Breeding and Genetics: Issues, Challenges and Opportunities for the 21st Century*. Springer, Dordrecht, the Netherlands, pp. 81-112.

Aida, R., Hirose, Y., Kishimoto, S. and Shibata, M. 1999. *Agrobacterium tumefaciens*-mediated transformation of *Cyclamen persicum* Mill. *Plant Sci.* 148(1), 1-7. doi:10.1016/S0168-9452(99)00072-2.

Albert, J. 2009. New technologies and food labeling: the controversy over labeling of foods derived from genetically modified crops. In: Albert, J. (Ed.), *Innovations in Food Labeling. Woodhead Publishing Series in Food Science, Technology and Nutrition* (vol. 184). Woodhead Publishing, Cambridge, UK, pp. 165-84.

Allard, R. W. 1960. *Principles of Plant Breeding*. John Willey and Sons Inc., New York, NY.

Anderson, N. O. 2004. Breeding flower seed crops. In: McDonald, M. and Kwong, F. (Eds), *Flower Seeds*. CABI Publishing, Cambridge, UK, pp. 53-86.

Anderson, N. O. 2006a. *Flower Breeding and Genetics: Issues, Challenges and Opportunities for the 21st Century*. Springer, Dordrecht, the Netherlands, 822p.

Anderson, N. O. 2006b. Chrysanthemum: *Dendranthema* ×*grandiflora* Tzvelv. In: Anderson, N. O. (Ed.), *Flower Breeding and Genetics: Issues, Challenges and Opportunities for the 21st Century*. Springer, Dordrecht, the Netherlands, pp. 389-437.

Anderson, N. O. and Olsen, R. T. 2015. A vast array of beauty: the accomplishments of Luther Burbank, Father of American ornamental breeding. *HortScience* 50(2), 161-88. doi:10.21273/HORTSCI.50.2.161.

Anderson, N. O. and Walker, N. J. 2013. Marketing genetically modified organism carnations by future floral designers: student-designed policy formulation. *HortTechnology* 23(5), 683-8. doi:10.21273/HORTTECH.23.5.683.

Anderson, N. O., Frick, J., Younis, A. and Currey, C. 2012. Heritability of cold tolerance (winter hardiness) in *Gladiolus xgrandiflorus*. In: Abdurakhmonov, I. Y. (Ed.), *Plant Breeding*. IntechOpen. ISBN:978-953-307-932-5.

Anzai, H. and Tanaka, M. 2001. Transgenic *Phalaenopsis* (a moth orchid). In: Bajaj, Y. P. S. (Ed.), *Biotechnology in Agriculture and Forestry, Volume 48: Transgenic Crops III*. Springer-Verlag, New York, NY, pp. 249-64.

Arús, P. 2000. Molecular markers for ornamental breeding. *Acta Hortic.* 508, 91-8. doi:10.17660/ActaHortic.2000.508.11.

Asen, S. 1976. Known factors responsible for the infinite flower color variations. *Acta Hortic.* 63(63), 217-24. doi:10.17660/ActaHortic.1976.63.27.

Atwell, S., Huang, Y. S., Vilhjalmsson, B. J., Willems, G., Horton, M., Li, Y., Meng, D., Platt, A., Tarone, A. M., Hu, T. T., Jiang, R., Muliyati, N. W., Zhang, X., Amer, M. A., Baxter, I., Brachi, B., Chory, J., Dean, C., Debieu, M., de Meaux, J., Ecker, J. R., Faure, N., Kniskern, J. M., Jones, J. D., Michael, T., Nemri, A., Roux, F., Salt, D. E., Tang, C., Todesco, M., Traw, M. B., Weigel, D., Marjoram, P., Borevitz, J. O., Bergelson, J. and Nordborg, M. 2010.

Genome-wide association study of 107 phenotypes in a common set of *Arabidopsis thaliana* inbred lines. *Nature* 465(7298), 627–31. doi:10.1038/nature08800.
Auer, C. 2008. Ecological risk assessment and regulation for genetically-modified ornamental plants. *Crit. Rev. Plant Sci.* 27(4), 255–71. doi:10.1080/07352680802237162.
Bernardo, R. 2014. *Essentials of Plant Breeding*. Stemma Press, Woodbury, MN.
Besemer, S. T. 1980. Carnations. In: Larson, R. A. (Ed.), *Introduction to Floriculture*. Academic Press, New York, NY, pp. 47–79.
Bino, R. J., Van Creijj, M. G. M., Van der Leede-Plegt, L. M., Van Tunen, A. J. and Van Tuyl, J. M. 1992. Application of *in vitro* pollination and fertilization techniques for breeding and genetic manipulations of *Lilium*. In: Creti, M. and Tiezzi, A. (Eds), *Sexual Plant Reproduction*. Springer-Verlag, Berlin, Germany, pp. 127–34.
Bourke, P. M., Arens, P., Voorrips, R. E., Esselink, G. D., Koning-Boucoiran, C. F., Van't Westende, W. P., Santos Leonardo, T., Wissink, P., Zheng, C., van Geest, G., Visser, R. G., Krens, F. A., Smulders, M. J. and Maliepaard, C. 2017. Partial preferential chromosome pairing is genotype dependent in tetraploid rose. *Plant J.* 90(2), 330–43. doi:10.1111/tpj.13496.
Brachi, B., Morris, G. P. and Borevitz, J. O. 2011. Genome-wide association studies in plants: the missing heritability is in the field. *Genome Biol.* 12(10), 232. doi:10.1186/gb-2011-12-10-232.
Bradford, K. J. and Alston, J. M. 2004. Diversity of horticultural biotech crops contributes to market hurdles. *Calif. Agric.* 58(2), 84–5.
Burbank, L. 1914-1915. Luther Burbank, his methods and discoveries and their practical application. Prepared from his original field notes covering more than 100,000 experiments made during forty years devoted to plant improvement. Williams, H. S., Whitson, R. J. and Whitson, J. (Eds). Luther Burbank Press, New York, NY.
Bush, A. L. and Pueppke, S. G. 1991. Cultivar-strain specificity between *Chrysanthemum morifolium* and *Agrobacterium tumefaciens*. *Physiol. Mol. Plant Path.* 39(4), 309–23. doi:10.1016/0885-5765(91)90038-J.
Cai, J., Liu, X., Vanneste, K., Proost, S., Tsai, W. C., Liu, K. W., Chen, L. J., He, Y., Xu, Q., Bian, C., Zheng, Z., Sun, F., Liu, W., Hsiao, Y. Y., Pan, Z. J., Hsu, C. C., Yang, Y. P., Hsu, Y. C., Chuang, Y. C., Dievart, A., Dufayard, J. F., Xu, X., Wang, J. Y., Wang, J., Xiao, X. J., Zhao, X. M., Du, R., Zhang, G. Q., Wang, M., Su, Y. Y., Xie, G. C., Liu, G. H., Li, L. Q., Huang, L. Q., Luo, Y. B., Chen, H. H., Van de Peer, Y. and Liu, Z. J. 2015. The genome sequence of the orchid *Phalaenopsis equestris*. *Nat. Genet.* 47(1), 65–72. doi:10.1038/ng.3149.
Callaway, D. J. and Callaway, M. B. (Eds). 2000. *Breeding Ornamental Plants*. Timber Press, Portland, OR.
Castle, W. E. and Phillips, J. C. 1914. Piebald rats and selection: an experimental test of the effectiveness of selection and of the theory of gametic purity in Mendelian crosses. Carnegie Institution of Washington Publication No. 195.
Chan, A. P., Crabtree, J., Zhao, Q., Lorenzi, H., Orvis, J., Puiu, D., Melake-Berhan, A., Jones, K. M., Redman, J., Chen, G., Cahoon, E. B., Gedil, M., Stanke, M., Haas, B. J., Wortman, J. R., Fraser-Liggett, C. M., Ravel, J. and Rabinowicz, P. D. 2010. Draft genome sequence of the oilseed species *Ricinus communis*. *Nat. Biotechnol.* 28(9), 951–6. doi:10.1038/nbt.1674.
Chandler, S. F. 2007. Genetic modification in floriculture. *Crit. Rev. Plant Sci.* 26(4), 169–97.
Chandler, S. F. and Brugliera, F. 2011. Genetic modification in floriculture. *Biotechnol. Lett.* 33(2), 207–14. doi:10.1007/s10529-010-0424-4.

Chandler, S. F. and Sanchez, C. 2012. Genetic modification the development of transgenic ornamental plant varieties. *Plant Biotechnol. J.* 10(8), 891-903. doi:10.1111/j.1467-7652.2012.00693.x.

Chia, T. F., Lim, A. Y. H., Luan, Y. and Ng, I. 2001. Transgenic *Dendrobium* (Orchid). In: Bajaj, Y. P. S. (Ed.), *Biotechnology in Agriculture and Forestry, Volume 48: Transgenic Crops III*. Springer-Verlag, New York, NY, pp. 95–106.

Chou, T. 2000. Production of transgenic impatiens. US Utility Patent, No. 6,121,511. US Patent Office, Washington DC.

Clark, D., Klee, H. and Dandekar, A. 2004. Despite benefits, commercialization of transgenic horticultural crops lags. *Calif. Agric.* 58(2), 89–98. doi:10.3733/ca.v058n02p89.

Clouse, J. W., Adhikary, D., Page, J. T., Ramaraj, T., Deyholos, M. K., Udall, J. A., Fairbanks, D. J., Jellen, E. N. and Maughan, P. J. 2016. The amaranth genome: genome, transcriptome, and physical map assembly. *Plant Genome* 9(1). doi:10.3835/plantgenome2015.07.0062.

Copetti, D., Búrquez, A., Bustamante, E., Charboneau, J. L. M., Childs, K. L., Eguiarte, L. E., Lee, S., Liu, T. L., McMahon, M. M., Whiteman, N. K., Wing, R. A., Wojciechowski, M. F. and Sanderson, M. J. 2017. Extensive gene tree discordance and hemiplasy shaped the genomes of North American columnar cacti. *Proc. Natl. Acad. Sci. U.S.A.* 114(45), 12003–8. doi:10.1073/pnas.1706367114.

Correns, C. 1908. Die Rolle der männlichen Keimzellen bei der Geschlechtsbestimmung der gynodioecischen Pflanzen. *Ber deut Bot Ges* 26a, 686–701.

Crane, M. B. and Lawrence, W. J. C. 1947. *The Genetics of Garden Plants*. MacMillan & Co., London, UK.

Crespel, L., Chirollet, M., Durel, E., Zhang, D., Meynet, J. and Gudin, S. 2002. Mapping of qualitative and quantitative phenotypic traits in *Rosa* using AFLP markers. *Theor. Appl. Genet.* 105(8), 1207–14. doi:10.1007/s00122-002-1102-2.

Darwin, C. 1865. On the sexual relations of the three forms of *Lythrum salicaria*. *J. Linn. Soc. Bot* 8, 169–96.

Darwin, C. 1868. On the character and hybrid-like nature of the offspring from the illegitimate unions of dimorphic and trimorphic plants. *Proc. Linn. Soc. Bot.* 10(47), 393–437. doi:10.1111/j.1095-8339.1868.tb02030.x.

David, M. 2017. U.S. flower sellers rush to destroy illegal GE petunias. *Sci. Post. Plants Anim. Sci. Policy*. doi:10.1126/science.aal1200.

De Beer, G. 1966. Fundamenta genetica: The revised edition of Mendel's classic paper with a collection of twenty-seven original papers published during the rediscovery era in Brno in 1865. Moravian Museum. Anthropological Publications, Oosterhout, the Netherlands.

Debener, T. 2002. Molecular markers as a tool for analyses of genetic relatedness and selection in ornamentals. In: Vainstein, A. (Ed.), *Breeding for Ornamentals: Classical and Molecular Approaches*. Spring-Science + Business Media, B. V., Dordrecht, the Netherlands, pp. 329–46.

Debener, T. and Linde, M. 2009. Exploring complex ornamental genomes. The rose as a model plant. *Crit. Rev. Plant Sci.* 28(4), 267–80. doi:10.1080/07352680903035481.

Debener, T., Janakiram, T. and Mattiesch, L. 2000. Sports and seedlings of rose varieties analysed with molecular markers. *Plant Breed.* 119(1), 71–4. doi:10.1046/j.1439-0523.2000.00459.x.

De Jong, J., Meterns, M. M. J. and Rademaker, W. 1994. Stable expression of the GUS-reporter in chrysanthemum depends on binary plasmid T-DNA. *Plant Cell Rep.* 14(1), 59–64. doi:10.1007/BF00233300.

Deroles, S. C., Ledger, S. E., Miller, R. M., Davies, K. M. and Given, N. K. 1993. Transformation in *Eustoma grandiflorum* (Lisianthus). In: Bajaj, Y. P. S. (Ed.), *Biotechnology in agriculture and forestry, Volume 22: Plant protoplasts and Genetic Engineering III*. Springer-Verlag, Berlin, Germany, pp. 203–12.

de Vries, H. 1901–1903. *Die mutationstheorie. Versuche und beobachtungen über die entstehung von arten im pflanzenreich*. Veit & Co., Leipzig, Germany.

Dohm, A., Ludwig, C., Schilling, D. and Debener, T. 2001. Transformation of roses with genes for antifungal proteins. *Acta Hortic.* 547(547), 27–33. doi:10.17660/ActaHortic.2001.547.3.

Doudna, J. A. and Charpentier, E. 2014. Genome editing. The new frontier of genome engineering with CRISPR-Cas9. *Science* 346(6213), 1258096. doi:10.1126/science.1258096.

Douhovnikoff, V. and Dodd, R. S. 2003. Intra-clonal variation and a similarity threshold for identification of clones: application to *Salix exigua* using AFLP molecular markers. *Theor. Appl. Genet.* 106(7), 1307–15. doi:10.1007/s00122-003-1200-9.

East, E. M. 1932. Further observations on *Lythrum salicaria*. *Genetics* 17(3), 327–34.

East, E. M. 1934. Norms of pollen-tube growth in incompatible matings of self-sterile plants. *Proc. Natl. Acad. Sci. U.S.A.* 20(4), 225–30. doi:10.1073/pnas.20.4.225.

East, E. M. 1935. Genetic reactions in *Nicotiana*. I. Compatibility. *Genetics* 20(5), 403–13.

Florigene Flowers. 2013. Suntory Holdings Limited, Melbourne. Available at: http://www.florigene.com/.

Florigene Flowers. 2019. Suntory Holdings Limited, Melbourne, Australia. Available at: http://www.florigene.com/.

Fraley, R. T., Rogers, S. G., Horsch, R. B. 1983. Use of a chimeric gene to confer antibiotic resistance to plant cells. In: *Advances in Gene Technology: Molecular Genetics of Plants and Animals*. Miami Winter Symposia 20, pp. 211–21.

Gerats, T. and Strommer, J. (Eds). 2009. *Petunia: Evolutionary, Developmental and Physiological Genetics*. Springer-Verlag, New York, NY.

Gorer, R. 1970. *The Development of Garden Flowers*. Eyre and Spottiswoode, Ltd., Norwich, UK.

Griesbach, R. 2006. Petunia, *Petunia x hybrida*. In: Anderson, N. O. (Ed.), *Flower Breeding and Genetics: Issues, Challenges and Opportunities for the 21st Century*. Springer, Dordrecht, the Netherlands, pp. 301–36.

Hamilton, J. P. and Buell, C. R. 2012. Advances in plant genome sequencing. *Plant J.* 70(1), 177–90. doi:10.1111/j.1365-313X.2012.04894.x.

Handa, T. and Deroles, S. C. 2000. Transgenic *Eustoma grandiflorum* (Lisianthus). In: Fajaj, Y. P. S. (Eds), *Biotechnology in Agriculture and Foresty: Transgenic Crops III*. Springer-Verlag, Berlin, Germany, pp. 107–22.

Hardy, G. H. 1908. Mendelian proportions in a mixed population. *Science* 28(706), 49–50. doi:10.1126/science.28.706.49.

Haselmair-Gosch, C., Miosic, S., Nitarska, D., Roth, B. L., Walliser, B., Paltram, R., Lucaciu, R. C., Eidenberger, L., Rattei, T., Olbricht, K., Stich, K. and Halbwirth, H. 2018. Great sause-small effect: undeclared genetically engineered orange petunias harbor an inefficient dihydroflavonol 4-reductase. *Front. Plant Sci.* 9, 149. doi:10.3389/fpls.2018.00149.

Hibrand Saint-Oyant, L., Ruttink, T., Hamama, L., Kirov, I., Lakhwani, D., Zhou, N. N., Bourke, P. M., Daccord, N., Leus, L., Schula, D., Van de Geest, H., Hesselink, T., Van Laere, K., Debray, K., Balzergue, S., Thouroude, T., Chastellier, A., Jeauffre, J., Voisine, L., Gaillard, S., Borm, T. J. A., Arens, P., Voorrips, R. E., Maliepaard, C., Neu, E., Linde, M., Le Paslier, M. C., Berard, A., Bounon, R., Clotault, J., Choisne, N., Quesneville, H., Kawamura, K., Aubourg, S., Sakr, S., Smulders, M. J. M., Schijlen, E., Bucher, E., Debener, T., De Riek, J. and Foucher, F. 2018. A high-quality genome sequence of *Rosa chinensis* to elucidate ornamental traits. *Nat. Plants* 4(7), 473–84. doi:10.1038/s41477-018-0166-1.

Hirakawa, H., Sumitomo, K., Hisamatsu, T., Nagano, S., Shirasawa, K., Higuchi, Y., Kusaba, M., Koshioka, M., Nakano, Y., Yagi, M., Yamaguchim, H., Taniguchi, K., Nakano, M. and Isobe, S. N. 2019. *De novo* whole genome assembly in *Chrysanthemum seticuspe*, a model species of Chrysanthemums, and its application to genetic and gene discovery analysis. *DNA Res.* 26(3), 195–203.

Holley, W. D. and Baker, R. 1963. *Carnation Production*. W.C. Brown, Dubuque, IA.

Holton, T. A., Brugliera, F., Lester, D. R., Tanaka, Y., Hyland, C. D., Menting, J. G., Lu, C. Y., Farcy, E., Stevenson, T. W. and Cornish, E. C. 1993. Cloning and expression of cytochrome P450 genes controlling flower colour. *Nature* 366(6452), 276–9. doi:10.1038/366276a0.

Hoshino, A., Javakumar, V., Nitasaka, E., Toyoda, A., Noguchi, H., Itoh, T., Shin-I, T., Minakuchi, Y., Koda, Y., Nagano, A. J., Yasugi, M., Honjo, M. N., Kudoh, H., Seki, M., Kamiya, A., Shiraki, T., Carninci, P., Asamizu, E., Nishide, H., Tanaka, S., Park, K. I., Morita, Y., Yokoyama, K., Uchiyama, I., Tanaka, Y., Tabata, S., Shinozaki, K., Hayashizaki, Y., Kohara, Y., Suzuki, Y., Sugano, S., Fujiyama, A., Iida, S. and Sakakibara, Y. 2016. Genome sequence and analysis of the Japanese morning glory *Ipomoea nil*. *Nat. Comm.* 7, Article No. 13295. doi:10.1038/ncomms13295.

Hunter, N. T. 2000. *The Art of Floral Design* (2nd edn.). Delmar Publishing, Albany, NY.

Ingvarsson, P. K. and Street, N. R. 2011. Association genetics of complex traits in plants. *New Phytol.*, Special issue: Featured papers on 'Carbon cycling in tropical ecosystems' 189(4), 909–22. doi:10.1111/j.1469-8137.2010.03593.x.

Jansen, R. C. 1996. A general Monte Carlo method for mapping multiple quantitative trait loci. *Genetics* 142(1), 305–11.

Jawaharlal, M., Ganga, M., Padmadevi, K., Jegadeeswari, V. and Karthikeyan, S. 2003. *A Technical Guide on Carnation*. Tamil Nadu Agricultural University, Tamil Nadu, India.

Jordan, W. C., Foley, K. and Bruford, M. 1998. Single-strand conformation polymorphism (SSCP) analysis. In: Karp, A., Isaac, P. G. and Ingram, D. S. (Eds), *Molecular Tools for Screening Biodiversity*. Chapman and Hall, London, UK.

Kazuhisa, H., Takahasi, H. and Youji, N. 1998. Tracing the parentages of some Oriental hybrid lily cultivars by PCR-RFLP analysis. *J. Jap. Soc. Hort. Sci.* 67(3), 352–9.

Kikuchi, A., Watanabe, K., Tanaka, Y. and Kamada, H. 2008. Recent progress on environmental biosafety assessment of genetically modified tress and floricultural plants in Japan. *Plant Biotechnol.* 25(1), 9–15. doi:10.5511/plantbiotechnology.25.9.

Kim, S., Park, M., Yeom, S. I., Kim, Y. M., Lee, J. M., Lee, H. A., Seo, E., Choi, J., Cheong, K., Kim, K. T., Jung, K., Lee, G. W., Oh, S. K., Bae, C., Kim, S. B., Lee, H. Y., Kim, S. Y., Kim, M. S., Kang, B. C., Jo, Y. D., Yang, H. B., Jeong, H. J., Kang, W. H., Kwon, J. K., Shin, C., Lim, J. Y., Park, J. H., Huh, J. H., Kim, J. S., Kim, B. D., Cohen, O., Paran, I., Suh, M. C., Lee, S. B., Kim, Y. K., Shin, Y., Noh, S. J., Park, J., Seo, Y. S., Kwon, S. Y., Kim, H. A., Park, J. M., Kim, H. J., Choi, S. B., Bosland, P. W., Reeves, G., Jo, S. H., Lee, B. W., Cho, H. T.,

Choi, H. S., Lee, M. S., Yu, Y., Do Choi, Y., Park, B. S., van Deynze, A., Ashrafi, H., Hill, T., Kim, W. T., Pai, H. S., Ahn, H. K., Yeam, I., Giovannoni, J. J., Rose, J. K., Sørensen, I., Lee, S. J., Kim, R. W., Choi, I. Y., Choi, B. S., Lim, J. S., Lee, Y. H. and Choi, D. 2014. Genome sequence of the hot pepper provides insights into the evolution of pungency in *Capsicum* species. *Nat. Genet.* 46(3), 270–8. doi:10.1038/ng.2877.

Kishimoto, S., Aida, R. and Shibata, M. 2002. *Agrobacterium tumefaciens* mediated transformation of *Elatior begonia* (*Begonia x hiemalis* Fotsch). *Plant Sci.* 162(5), 697–703. doi:10.1016/S0168-9452(02)00014-6.

Kiyokawa, S., Kikuchi, Y., Kamada, H. and Harada, H. 1996. Genetic transformation of *Begonia tuberhybrida* by *Ri rol* genes. *Plant Cell. Rep.* 15(8), 606–9. doi:10.1007/BF00232462.

Konieczyn, A. and Ausubel, F. M. 1993. A procedure for mapping *Arabidopsis* mutations using codominant ecotype-specific PCR-based markers. *Plant J.* 4(2), 403–10. doi:10.1046/j.1365-313x.1993.04020403.x.

Koning-Boucoiran, C. F. S., esselink, G. D., Vukosavljev, M., van't Westende, W. P. C., Gitonga, V. W., Krens, F. A., Voorrips, R. E., van de Weg, W. E., Schulz, D., Debener, T., Maliepaard, C., Arens, P. and Smulders, M. J. 2015. Using RNA-Seq to assemble a rose transcriptome with more than 13,000 full-length expressed genes and to develop the WagRhSNP 68K Axiom SNP array for rose (*Rosa* L.). *Front. Plant Sci.* 6, 249. doi:10.3389/fpls.2015.00249.

Kover, P. X., Valdar, W., Trakalo, J., Scarcelli, N., Ehrenreich, I. M., Purugganan, M. D., Durrant, C. and Mott, R. 2009. A multiparent advanced generation inter-cross to fine-map quantitative traits in *Arabidopsis thaliana*. *PLoS Genet.* 5(7), e1000551. doi:10.1371/journal.pgen.1000551.

Kramer, E. M. and Hodges, S. A. 2010. *Aquilegia* as a model system for the evolution and ecology of petals. *Philos. Trans. R. Soc. Lond., Ser. B, Biol. Sci.* 365(1539), 477–90. doi:10.1098/rstb.2009.0230.

Kwon, D. 2018. USDA Will Not Regulate CRISPR-Edited Crops. Restrictions will remain on transgenic plants, which contain artificially inserted genes from other species. *The Scientist*, 02 April 2018. Available at: https://www.the-scientist.com/the-nutshell/usda-will-not-regulate-crispr-edited-crops-29852.

Langton, F. A. 1980. Chimerical structure and carotenoid inheritance in *Chrysanthemum morifolium* (Ramat.). *Euphytica* 29(3), 807–12. doi:10.1007/BF00023228.

Lefebvre, V., Pflieger, S., Thabius, A., Caranta, C., Blattes, A., Chauvet, J. C., Daubeze, A. M. and Palloix, A. 2002. Towards the saturation of the pepper linkage map by alignment of three intraspecific maps including known-function genes. *Genome* 45(5), 839–54. doi:10.1139/g02-053.

Li, W., Liu, X. and Ly, Y. 2016. Transcriptome comparison reveals key candidate genes in response to vernalization of Oriental lily. *BMC Genomics* 17(1), 664. doi:10.1186/s12864-016-2955-0.

Lindqvist-Kreuze, H., Gastelo, M., Perez, W., Forbes, G. A., de Koeyer, D. and Bonierbale, M. 2014. Phenotypic stability and genome-wide association stydy of late blight resistance in potato genotypes adapted to the tropical highlands. *Phytopathology* 104(6), 624–33. doi:10.1094/PHYTO-10-13-0270-R.

Lu, F., Lipka, A. E., Glaubitz, J., Elshire, R., Cherney, J. H., Casler, M. D., Buckler, E. S. and Costich, D. E. 2013. Switchgrass genomic diversity, ploidy, and evolution: novel insights from a network-based SNP discovery protocol. *PLoS Genet.* 9(1), e1003215. doi:10.1371/journal.pgen.1003215.

Mangelsdorf, A. J. and East, E. M. 1927. Studies on the genetics of *Fragaria*. *Genetics* 12(4), 307–39.

Martinez, R. A. 1997. *The Master of Seeds: Life and Work of Claude Hope*. Burpee Books, Warmister, PA.

Mejía-Jiménez, A., Muñoz, C., Jacobsen, H. J., Roca, W. M. and Singh, S. P. 1994. Interspecific hybridization between common and tepary beans: increased hybrid embryo growth, fertility, and efficiency of hybridization through recurrent and congruity backcrossing. *Theor. Appl. Genet.* 88(3–4), 324–31. doi:10.1007/BF00223640.

Mendel, G. 1866. Versuche über Pflanzen-hybriden. *Verh. Naturforschenden Vereines Brünn* 4(1865), 3–47.

Meyer, P., Heidmann, I., Forkmann, G. and Saedler, H. 1987. A new petunia flower colour generated by transformation of a mutant with a maize gene. *Nature* 330(6149), 677–8. doi:10.1038/330677a0.

Miller, H., Miller, J. and Crane, G. 1975. Relative susceptibility of chrysanthemum cultivars to *Agrobacterium tumefaciens*. *Plant Dis. Rep.* 59, 576–81.

Ming, R., VanBuren, R., Liu, Y., Yang, M., Han, Y., Li, L. T., Zhang, Q., Kim, M. J., Schatz, M. C., Campbell, M., Li, J., Bowers, J. E., Tang, H., Lyons, E., Ferguson, A. A., Narzisi, G., Nelson, D. R., Blaby-Haas, C. E., Gschwend, A. R., Jiao, Y., Der, J. P., Zeng, F., Han, J., Min, X. J., Hudson, K. A., Singh, R., Grennan, A. K., Karpowicz, S. J., Watling, J. R., Ito, K., Robinson, S. A., Hudson, M. E., Yu, Q., Mockler, T. C., Carroll, A., Zheng, Y., Sunkar, R., Jia, R., Chen, N., Arro, J., Wai, C. M., Wafula, E., Spence, A., Han, Y., Xu, L., Zhang, J., Peery, R., Haus, M. J., Xiong, W., Walsh, J. A., Wu, J., Wang, M. L., Zhu, Y. J., Paull, R. E., Britt, A. B., Du, C., Downie, S. R., Schuler, M. A., Michael, T. P., Long, S. P., Ort, D. R., Schopf, J. W., Gang, D. R., Jiang, N., Yandell, M., dePamphilis, C. W., Merchant, S. S., Paterson, A. H., Buchanan, B. B., Li, S. and Shen-Miller, J. 2013. Genome of the long-living sacred lotus (*Nelumbo nucifera* Gaertn.). *Genome Biol.* 14(5), R41. doi:10.1186/gb-2013-14-5-r41.

Mitchell-Olds, T. 2010. Complex-trait analysis in plants. *Genome Biol.* 11(4), 113. doi:10.1186/gb-2010-11-4-113.

Nan, G. -L., Tang, C. S., Kuehnle, A. R. and Kado, C. I. 1997. *Dendrobium* orchid contains an inducer of *Agrobacterium* virulence genes. *Phsiol. Mol. Plant Path.* 51(6), 391–9. doi:10.1006/pmpp.1997.0128.

Nau, J. 2011. *Ball Red Book* (18th edn.). Ball Publishing, Chicago, IL.

Nielsen, K. M., Deroles, S. C., Markham, K. R., Bradley, M. J., Podivinsky, E. P. and Manson, D. 2002. Antisense flavonol synthase alters copigmentation and flower color in Lisianthus. *Mol. Breed.* 9(4), 217–29. doi:10.1023/A:1020320809654.

Noda, N., Yoshioka, S., Kishimoto, S., Nakayama, M., Douzono, M., Tanaka, Y. and Aida, R. 2017. Generation of blue chrysanthemums by anthocyanin B-ring hydroxylation and glucosylation and its coloration mechanism. *Sci. Adv.* 3(7), e1602785. doi:10.1126/sciadv.1602785.

Ohmiya, A., Kishimoto, S., Aida, R., Yoshioka, S. and Sumitomo, K. 2006. Carotenoid cleavage dioxygenase (*CmCCD4a*) contributes to white color formation in chrysanthemum petals. *Plant Physiol.* 142(3), 1193–201. doi:10.1104/pp.106.087130.

Oud, J. S. N., Schneiders, H., Kool, A. J. and Van Grinsven, M. Q. J. M. 1995. Breeding of transgenic orange *Petunia hybrida* varieties. *Euphytica* 84(3), 175–81. doi:10.1007/BF01681809.

Paran, I. and Michelmore, R. W. 1993. Development of reliable PCR-based markers linked to downy mildew resistance genes in lettuce. *Theor. Appl. Genet.* 85(8), 985–93. doi:10.1007/BF00215038.

Penna, S., Ganapathi, T. R., Ghag, S. and Jain, S. M. 2017. Genetic modifications of horticultural plants by induced mutations and transgenic approach. *Acta Hortic.* 1187, 22. doi:10.17660/ActaHortic.2017.1187.22.

Ping, C. Y., Chen, F. C., Cheng, T. C., Lin, H. L., Lin, T. S., Yang, W. J. and Lee, Y. I. 2018. Expression profiles of phosphoenolpyruvate carboxylase and phosphoenolpyruvate carboxylase kinase genes in *Phalaenopsis*, implications for regulating the performance of crassulacean acid metabolism. *Front. Plant Sci.* 9, 1587. doi:10.3389/fpls.2018.01587.

Purugganan, M. D. and Fuller, D. Q. 2009. The nature of selection during plant domestication. *Nature* 457(7231), 843–8. doi:10.1038/nature07895.

Reynolds, J. and Tampion, J. 1983. *Double Flowers: a Scientific Study. Scientific and Academic Editions*. Van Nostrand Reinhold Co., New York, NY.

Roein, Z., Asil, M. H. and Sabouri, A. 2015. Identification of AFLP markers associated with flowering time and ornamental traits in *Chrysanthemum*. *Iran. J. Genet. Plant Breed.* 4(2), 37–49.

Roman, H., Rapicault, M., Miclot, A. S., Larenaudie, M., Kawamura, K., Thouroude, T., Chastellier, A., Lemarquand, A., Dupuis, F., Foucher, F., Loustau, S. and Hibrand-Saint Oyant, L. 2015. Genetic analysis of the flowering date and number of petals in rose. *Tree Genet. Genomes* 11(4), 85. doi:10.1007/s11295-015-0906-6.

Sanford, J. C., Klein, T. M., Wolf, E. D. and Allen, N. 1987. Delivery of substances into cells and tissues using a particle bombardment process. *Part. Sci. Technol.* 5(1), 27–37. doi:10.1080/02726358708904533.

Schlegel, R. H. J. 2018. *History of Plant Breeding*. CRC Press, Boca Raton, FL.

Schulz, D. F., Schott, R. T., Voorrips, R. E., Smulders, M. J. M., Linde, M. and Debener, T. 2016. Genome-wide association analysis of the anthocyanin and carotenoid contents of rose petals. *Front. Plant Sci.* 7, 1798. doi:10.3389/fpls.2016.01798.

Schweinfurth, G. 1919. Pflanzenbilder im Tempel von Karnak. In: Engler, A. (Ed.), *Botanische Jahrbücher LV*, pp. 464–80.

Servick, K. 2017. The strange case of the orange petunias. *Science* 356(6340), 792. doi:10.1126/science.356.6340.792.

Shirasawa, K., Tanaka, M., Takahata, Y., Ma, D., Cao, Q., Liu, Q., Zhai, H., Kwak, S. S., Cheol Jeong, J., Yoon, U. H., Lee, H. U., Hirakawa, H. and Isobe, S. 2017. A high-density SNP genetic map consisting of a complete set of homologous groups in autohexaploid sweetpotato (*Ipomoea batatas*). *Sci. Rep.* 7, 44207. doi:10.1038/srep44207.

Simpson, G. C. and Dean, C. 2002. Arabidopsis, the Rosetta stone of flowering time? *Science* 296(5566), 285–9. doi:10.1126/science.296.5566.285.

Smulders, M. J. M., Arens, P., Bourke, P. M., Debener, T., Linde, M., De Riek, J., Leus, L., Ruttink, T., Baudino, S., Hibrant Saint-Oyant, L., Clotault, J. and Foucher, F. 2019. In the name of the rose: a roadmap for rose research in the genome era. *Hortic. Res.* 6, 65. doi:10.1038/s41438-019-0156-0.

Solecki, R. S. 1975. Shanidar IV, a Neanderthal flower burial in northern Iraq. *Science* 190(4217), 880–1. doi:10.1126/science.190.4217.880.

Spiller, M., Linde, M., Hibrand-Saint, L., Tsai, C. J., Byrne, D. H., Smulders, M. J., Foucher, F. and Debener, T. 2011. Towards a unified genetic map for diploid roses. *Theor. Appl. Genet.* 122(3), 489–500. doi:10.1007/s00122-010-1463-x.

Straathof, Th. P., Van Tuyl, J. M., Dekker, B., Can Windern, M. J. M. and Sandbrink, J. M. 1996. Genetic analysis of inheritance of partial resistance to *Fusarium oxysporum* in Asiatic hybrids of lily using RAPD markers. *Acta Hort.* 414, 209–18.

Strasburg, J. L., Kane, N. C., Raduski, A. R., Bonin, A., Michelmore, R. and Rieseberg, L. H. 2011. Effective population size is positively correlated with levels of adaptive divergence among annual sunflowers. *Mol. Biol. Evol.* 28(5), 1569–80. doi:10.1093/molbev/msq270.

Stummel, J. R. and Bosland, P. W. 2006. Ornamental pepper: *Capsicum annuum*. In: Anderson, N. O. (Ed.), *Flower Breeding and Genetics: Issues, Challenges and Opportunities for the 21st Century*. Springer, Dordrecht, the Netherlands, pp. 561–99.

Sumitomo, K., Shirasawa, K., Isobe, S., Hirakawa, H., Hisamatsu, T., Nakano, Y., Yagi, M. and Ohmiya, A. 2019. Genome-wide association study overcomes the genomic complexity in autohexaploid chrysanthemum and tags SNP markers onto the flower color genes. *Sci. Rep.* 9(1), 13947. doi:10.1038/s41598-019-50028-z.

Tanaka, Y., Sasaki, N. and Ohmiya, A. 2008. Biosynthesisof plantpigments: anthocyanins, betalainsandcarotenoids. Plant 54, 733–49. doi:10.1111/j.1365-313X.2008.03447.x.

Taylor, J. M. 2014. *Visions of Loveliness: Great Flower Breeders from the Past*. The Ohio State University Press, Columbus, OH.

Teynor, T. T., Widmer, R. E., Ascher, P. D. and Luby, J. J. 1989a. Inheritance of flower color in *Dendranthema grandiflora* Tzvelev. (*Chrysanthemum morifolium* Ramat.) using cultivars and inbreds. I. Plastid pigmentation. *Euphytica* 42(3), 199–207. doi:10.1007/BF00034455.

Teynor, T. T., Widmer, R. E., Ascher, P. D. and Luby, J. J. 1989b. Inheritance of flower color in *Dendranthema grandiflora* Tzvelev. (*Chrysanthemum morifolium* Ramat.) using cultivars and inbreds. II. Vacuole pigmentation. *Euphytica* 42(3), 297–305. doi:10.1007/BF00034468.

Turelli, M. 2016. Edward East on the Mendelian basis of quantitative trait variation. *Genetics* 204(4), 1321–3. doi:10.1534/genetics.116.196915.

Uchneat, M. 2006. Impatiens, *Impatiens wallerana*. In: Anderson, N. O. (Ed.), *Flower Breeding and Genetics: Issues, Challenges and Opportunities for the 21st Century*. Springer, Dordrecht, the Netherlands, pp. 277–99.

United States Department of Agriculture. 2012. *Australia Agricultural Biotechnology Annual Report 2012*. National Agricultural Statistics Service (NASS), Washington DC.

United States Department of Agriculture. 2016. *Floriculture Crops 2015 Summary*. National Agricultural Statistics Service (NASS), Washington DC.

Uzogara, S. G. 2000. The impact of genetic modification of human foods in the 21st century: a review. *Biotechnol. Adv.* 18(3), 179–206. doi:10.1016/s0734-9750(00)00033-1.

Vainstein, A. 2002. *Breeding for Ornamentals: Classical and Molecular Approaches*. Springer, Dordrecht, the Netherlands. doi:10.1007/978-94-017-0956-9.

van Bakel, H., Stout, J. M., Cote, A. G., Tallon, C. M., Sharpe, A. G., Hughes, T. R. and Page, J. E. 2011. The draft genome and transcriptome of *Cannabis sativa*. *Genome Biol.* 12(10), R102. doi:10.1186/gb-2011-12-10-r102.

Van der Leede-Plegt, L. M., Van der Ven, B. C. E., Bino, R. J., Van der Salm, T. P. M. and Van Tunen, A. J. 1992. Introduction and differential sue of various promoters in pollen grains of *Nictoiana glutinosa* and *Lilium longiflorum*. *Plant Cell Rep.* 11(1), 20–4. doi:10.1007/BF00231833.

van Geest, G., Bourke, P. M., Voorrips, R. E., Marasek-Ciolakowska, A., Liao, Y., Post, A., van Meeteren, U., Visser, R. G. F., Maliepaard, C. and Arens, P. 2017. An ultra-dense integrated linkage map for hexaploid chrysanthemum enables multi-allelic QTL analysis. *Theor. Appl. Genet.* 130(12), 2527–41. doi:10.1007/s00122-017-2974-5.

Van Heusden, A. W., Jongerius, M. C., Van Tuyl, J. M., Straathof, T. P. and Mes, J. J. 2002. Molecular assisted breeding for disease resistance in lily. *Acta Hortic.* 572(572), 131-8. doi:10.17660/ActaHortic.2002.572.15.

Van Tuyl, J. M. and van Creij, M. G. M. 2006. Tulip: *Tulipa gesneriana* and *T.* hybrids. In: Anderson, N. O. (Ed.), *Flower Breeding and Genetics: Issues, Challenges and Opportunities for the 21st Century*. Springer, Dordrecht, the Netherlands, pp. 623–41.

Van Tuyl, J. M., Arens, P., Miller, W. B. and Anderson, N. O. 2014. The role of ornamentals in human life. In: Dixon, G. R. and Aldous, D. E. (Eds), *Horticulture. Plants for People and Places* (vol. 1). Springer Science, Dordrecht, the Netherlands. Chapter 11. doi:10.1007/978-94-017-8578-5_11.

Vosman, B., Visser, D., Rouppe van der Voort, J. R., Smulders, M. J. M. and van Eeuwijk, F. 2004. The establishment of 'essential derivation' among rose varieties, using AFLP. *Theor. Appl. Genet.* 109(8), 1718–25. doi:10.1007/s00122-004-1809-3.

Vukosavljev, M., Arens, P., Voorips, R. E., van't Westende, W. P., Esselink, G. D., Bourke, P. M., Cox, P., van de Weg, W. E., Visser, R. G., Maliepaard, C. and Smulders, M. J. 2016. High-density SNP-based genetic maps for the parents of an outcrosses and a selfed tetraploid garden rose cross, inferred from admixed progeny using the 68k rose SNP array. *Hortic. Res.* 3, 16–52.

Wallace, A. R. 1858. On the tendency of varieties to depart indefinitely from the original type. *Linnaean Soc. Lond. Proc.* 3, 53–62.

Wang, Z., Hobson, N., Galindo, L., Zhu, S., Shi, D., McDill, J., Yang, L., Hawkins, S., Neutelings, G., Datla, R., Lambert, G., Galbraith, D. W., Grassa, C. J., Geraldes, A., Cronk, Q. C., Cullis, C., Dash, P. K., Kumar, P. A., Cloutier, S., Sharpe, A. G., Wong, G. K., Wang, J. and Deyholos, M. K. 2012. The genome of flax (*Linum usitatissimum*) assembled *de novo* from short shotgun sequence reads. *Plant J.* 72(3), 461–73. doi:10.1111/j.1365-313X.2012.05093.x.

Watad, A. A., Yun, D. J., Matusmoto, T., Niu, X., Wu, Y., Kononwicz, A. K., Bressan, R. A. and Hasegawa, P. M. 1998. Microprojectile bombardment-mediated transformation of *Lilium longiflorum*. *Plant Cell Rep.* 17(4), 262–7. doi:10.1007/s002990050389.

Watanabe, K., Kobayashi, A., Endo, M., Sage-Ono, K., Toki, S. and Ono, M. 2017. CRISPR/Cas9-mediated mutagenesis of the *dihydroflavonol-4-reductase-B (DFR-B)* locus in the Japanese morning glory *Ipomoea* (*Pharbitis*) *nil*. *Sci. Rep.* 7(1), Article Number: 10028. doi:10.1038/s41598-017-10715-1.

Weddle, C. 1976. Petunias. In: Mastaler, L. (Ed.), *Bedding Plants*. Penn. State Manual, State College, PA, pp. 252–65.

Weinberg, W. 1908. Über den Nachweis der Vererbung beim Menschen Jahr. *Ver.f. Vaterland Nat. Würz.* 64, 369–82.

Weising, K., Wkinter, P., Hüttel, B. and Kahl, G. 1997. Microsatellite markers for molecular breeding. *J. Crop Prod.* 1(1), 113–43. doi:10.1300/J144v01n01_06.

Wen, C. S. and Hsiao, J. Y. 2001. Altitudinal genetic differentiation and diversity of Taiwan lily (*Lilium longiflorum* var. *formosanum*; Liliaceae) using rapid markers and morphological characters. *Int. J. Plant Sci.* 162(2), 287–95. doi:10.1086/319588.

Wilmink, A., Van de Ven, B. C. E. and Dons, J. J. M. 1992. Expression of the GUS-gene in the monocot tulip after introduction by particle bombardment and *Agrobacterium*. *Plant Cell Rep.* 11(2), 76-80. doi:10.1007/BF00235257.

Wilmink, A., Van de Ven, B. C. E., Custers, J. B. M., Van Tuyl, J. M., Eikelboom, W. and Dons, J. J. M. 1995. Genetic transformation in *Tulipa* species (tulips). In: Bajaj, Y. P. S. (Ed.), *Biotechnology in Agriculture and Forestry, Volume 34: Protoplasts and Genetic Engineering VI*. Springer, Berlin, Heidelberg, pp. 289-98.

Wolff, K. 1996. RAPD analysis of sporting and chimerism in chrysanthemum. *Euphytica* 89(2), 159-64. doi:10.1007/BF00034601.

Wolff, K. and Peters-van Rijn, J. 1993. Rapid detection of genetic variability in chrysanthemum (*Dendranthema grandiflora* Tzvelv.) using random primers. *Heredity* 71(4), 335-41. doi:10.1038/hdy.1993.147.

Wolff, K., Schoen, E. D. and Peters-van Rijn, J. P. 1993. Optimizing the generation of random amplified polymorphic DNAs in chrysanthemum. *Theor. Appl. Genet.* 86(8), 1033-7. doi:10.1007/BF00211058.

Wolff, K., Peters-van Rijn, J. and Hofstra, H. 1994. RFLP analysis in chrysanthemum. I. Probe and primer development. *Theor. Appl. Genet.* 88(3-4), 472-8. doi:10.1007/BF00223663.

Wolff, K., Zietkiewicz, E. and Hofstra, H. 1995. Identification of chrysanthemum cultivars and stability of DNA fingerprint patterns. *Theor. Appl. Genet.* 91(3), 439-47. doi:10.1007/BF00222971.

Wu, C. A., Lowry, D. B., Cooley, A. M., Wright, K. M., Lee, Y. W. and Willis, J. H. 2008. *Mimulus* is an emerging model system for the integration of ecological and genomic studies. *Heredity* 100(2), 220-30. doi:10.1038/sj.hdy.6801018.

Wu, J., Seng, S., Sui, J., Vonapartis, E., Luo, X., Gong, B., Liu, C., Wu, C., Liu, C., Zhang, F., He, J. and Yi, M. 2015. *Gladiolus hybridus* ABSCISIC ACID INSENSITIVE 5 (GhABI5) is an important transcription factor in ABA signaling that can enhance Gladiolus corm dormancy and Arabidopsis seed dormancy. *Front. Plant Sci.* 6, 960. doi:10.3389/fpls.2015.00960.

Young, K. J., Park, S. J., Um, B. Y., Pak, C. H., Chung, Y. S. and Shing, J. S. 1998. Transformation of chrysanthemum by *Agrobacterium tumefaciens* with three different types of vectors. *J. Korean Soc. Hort. Sci.* 39(3), 360-6.

Yu, H., Yang, S. H. and Goh, C. J. 2001. *Agrobacerium*-mediated transformation of a *Dendrobium* orhid with the class I *Knox* gene *DOH1*. *Plant Cell Rep.* 20(4), 301-5. doi:10.1007/s002990100334.

Zaccai, M., Lewinsohn, E. and Pinchersky, E. 2001. Modifying lisianthus traits by genetic engineering. *Acta Hortic.* 552(552), 137-42. doi:10.17660/ActaHortic.2001.552.14.

Ziang, N., Hon, Y. and Lam-Chan, L. T. 2003. Genetic analysis of tropical orchid hybrids (Dendrobium) with fluorescence amplified fragment-length polymorphism (AFLP). *J. Am. Soc. Hortic. Sci. Am. Soc. Hortic. Sci.* 128(5), 731-5.

Zlesak, D. C. 2006. Rose: *Rosa ×hybrida*. In: Anderson, N. O. (Ed.), *Flower Breeding and Genetics: Issues, Challenges and Opportunities for the 21st Century*. Springer, Dordrecht, the Netherlands, pp. 695-738.

Zlesak, D. and Anderson, N. O. 2010. Inheritance of non-obligate vernalization requirement for flowering in *Lilium formosanum* Wallace. *Isr. J. Plant Sci.* 57(4), 315-27.

Zlesak, D., Bradeen, J. and Anderson, N. O. 2007. The use of AFLP markers to resolve clonal origin and integrity in rose, hydrangea, and lily. *Floric. Ornament. Biotechnol.* 1(1), 51-60.

# Chapter 6

# The use of gene-editing techniques in breeding improved ornamentals

*Bruno Trevenzoli Favero, Josefine Nymark Hegelund and Henrik Lütken, University of Copenhagen, Denmark*



## 1 Introduction

The ornamental plant market is of global economic significance with Europe being the major player where the turnover from the Netherlands reached €4.7 billion in 2017 (FloraHolland, 2018). There is a continued quest for cultivars with new and improved characteristics, for example, petal/foliage colors, enhanced scent, improved plant architecture, biotic/abiotic stresses and extended postharvest life, for example, tolerance to the phytohormone ethylene (Azadi et al., 2016). Although, the transfer of new traits is achievable with conventional and mutational breeding it encounters restrictions. For example, hybridization barriers prevent natural introgression of traits from distantly related species (Kuligowska et al., 2016; Shibata, 2008; Teixeira da Silva et al., 2011). A way to circumvent this issue has been obtained in the form of genetic engineering through guided mutations, which is synergistically driven by the development of genome-sequencing initiatives of ornamental plants (as reviewed by Azadi et al., 2016).

Novel-targeted delivery of new traits had an impressive development after the clustered regularly interspaced short palindromic repeats (CRISPR) technique advent, revolutionizing the molecular breeding possibilities. Since the first publication exploiting the CRISPR system for genome editing in 2013 (as reviewed by Eş et al., 2019), the number of published articles has steadily grown

http://dx.doi.org/10.19103/AS.2020.0066.07

reaching over 17 000 in less than 7 years (Pubmed query: CRISPR, Updated on 17 February 2020). This chapter first describes the background for the genome-editing technology, with specific emphasis on the CRISPR technology followed by its applicability in plants. Subsequently, the use in horticulture is depicted through a case study targeting the potted plant *Campanula*. In this example, an experiment was carried out in the attempt to produce plants less sensitive to ethylene. A step-by-step genetic-editing method will be outlined. Finally, the ornamental perspective uses and regulations are described.

## 1.1 Genome-editing background

The aim to more precisely control DNA modifications has led to the development of genome-editing technologies. Precise genome editing refers to the use of engineered sequence specific nucleases (SSN) to modify targeted-genome sequences (Gaj et al., 2013; Kemp et al., 2017). The central idea behind these technologies is creating double stranded breaks (DSB) in genomic DNA and induce cellular repair by either of two different mechanisms: homology directed repair (HDR) or non-homologous end-joining (NHEJ) (Fig. 1) (Xiong et al., 2015).

The advantage of NHEJ is its error-prone feature, thus the DNA lesions will be repaired imperfectly, generating insertions or deletions (indels) within the target sequence (Wyman and Kanaar, 2006). Indels can generate a frameshift mutation, disrupting important functional domains, causing, for example, a gene knockout (Zhou et al., 2014).

On the other hand, the HDR pathway mediates insertion of specific sequences in the targeted locus by using a homologous donor template, of either endogenous or exogenous origin (Gratz et al., 2014; Liang et al., 2017).

The idea of using engineered nuclease systems, such as CRISPR, was preceded by other systems like zinc-finger nucleases (ZFNs) and later, transcription activator-like effector nucleases (TALENs). These systems overcame some of the limitations of approaches like homologous recombination and RNA interference (Boettcher and McManus, 2015; Gaj et al., 2013; Kim and Kim, 2014). Through the use of tailored DNA-binding domains fused to nucleases, the system is capable of recognizing specific DNA sequences and introduces DSBs.

However, the valuable target precision facilitated by the ZFN and TALEN methodologies is hampered by their elaborate and time-consuming procedures. The size of the delivered cassette can negatively affect the cloning and delivery steps, where the typical TALEN cDNA is approximately 3 kb whereas the ZFN cDNA is approximately 1 kb. Furthermore, both systems require dimerization, which increases the final effective size while limited by the transfer capacity of effective adenovirus delivery systems, that is, $< 5$ kb (Ellis et al., 2013; Gupta

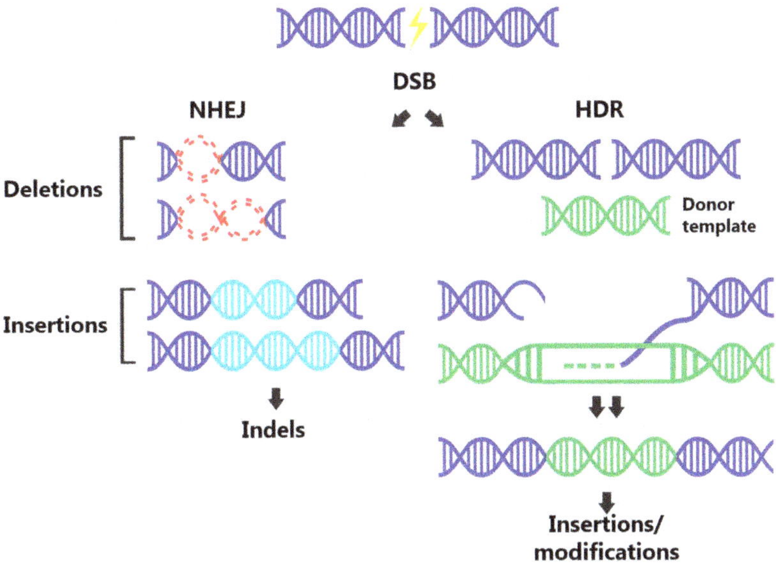


**Figure 1** The double-stranded break (DSB) repair mechanisms, that is, non-homologous end joining (NHEJ) or homology-directed repair (HDR). The result changes depending on the donor template presence; when absent, the NHEJ pathway will prevail generating small insertions or deletions (indels). While in its presence, the donor template with extended homology arms, leads to introduction of single or multiple inserts to correct or replace existing gene sections (HDR).

and Musunuru, 2014). Moreover, a high risk of sequence rearrangement and truncation in TALEN systems are associated with the repetitive DNA-binding domains (Holkers et al., 2013) which require extensive transcriptional and translational optimizations considering, for example, codon usage or promoter design (Liu et al., 2015).

The precise genome-editing field was revolutionized by the appearance of CRISPR-associated protein 9 (Cas9) technology (Cong et al., 2013). This technology was modified based on archaea and bacteria defense mechanism first hypothesized by Mojica et al. (2005) and proposed by Makarova et al. (2006). The identification of gene regions with small direct repeats in prokaryotes, that is, CRISPRs, has been known since 1987 (Ishino et al., 1987) and it is shared among multiple species (Mojica et al., 2000). These repetitive regions are interspaced by non-repetitive regions where in the upstream region nuclease homology is always found (i.e. *CRISPR-associated* (*CAS*) genes) (Jansen et al., 2002). In addition, the non-repetitive regions are derived from bacteriophages and are dissimilar even between closely related species (Makarova et al., 2006). Thus, leading to the adaptive immunity hypothesis involving CRISPR regions and nucleases, where foreign bacteriophage DNA, that is, fragment/spacer, is integrated into the prokaryote genome, preventing a future attack by the same phage strain (Barrangou et al., 2007).

The spacer sequences are complementary to previously encountered viral DNA (Marraffini and Sontheimer, 2008) and are located upstream of a conserved protospacer adjacent motif (PAM) (Sternberg et al., 2014). In a next viral infection event, bacterial transcription of a CRISPR RNA (crRNA) library matches the invading DNA and a variety of defense systems can result in DNA inactivation. The studies of defense system type II in prokaryotes paved the way to the precise genome editing possibilities. In defense type II, the crRNA and the trans-activating RNA (tracrRNA) form the duplex at the target spacer plus PAM leading to dsDNA cleavage by the Cas9 protein (Fig. 2) (Sorek et al., 2013). It was early noticed that by exchanging the crRNA complementarity, the whole complex could precisely edit different sequences. Moreover, the type II is suited for use in genome editing because it has a single protein performing the targeting and cutting (Garneau et al., 2010).

A system optimization toward a simpler programmable RNA-guided DNA endonuclease later arose with the crRNA and tracrRNA fusion into a single-guide RNA (sgRNA), rather than two (Gasiunas et al., 2012).

## 1.2 Precise genome editing in plants

So far, most research utilizing the CRISPR/Cas9 system has been conducted in prokaryote and mammalian cells, however many researchers within the

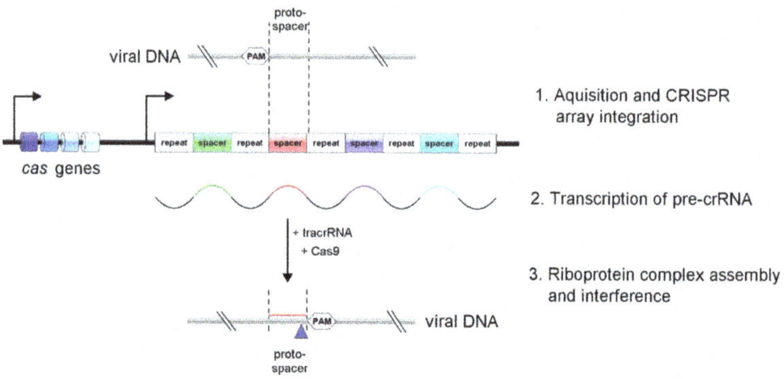


**Figure 2** Steps of the CRISPR/Cas9 adaptive immune system in prokaryotes. (1) During bacteriophage attack, the foreign DNA is injected into the cell and small pieces of direct repeats are integrated in the Clustered Regularly Interspaced Short Palindromic Repeats (CRISPR) loci. (2) Upon subsequent invasion, the bacterial cells transcribe and process crRNAs that form riboprotein complexes with the transactivating RNA and the Cas-proteins. (3) These complexes enable the direct targeting by Cas nucleases toward viral DNA inducing its cleavage (blue triangle) 3 nucleotide upstream of the protospacer adjacent motif (PAM) site.

field of plant research have recognized the potential for the technique. Some approaches focus on the application of the system, that is, the 'dead' Cas9 (dCas9) where the nuclease activity is not present, but the targeted aspects of the system continues to be present (Eid et al., 2018). This can lead to applications like reversible gene silencing in *Staphylococcus aureus* (Zhao et al., 2017), replication fork arrest for DNA replication studied (Whinn et al., 2019) or base editing as it does not require DSB. In addition, transcriptional activator or repressor can be fused to dCas9 (Hille and Charpentier, 2016) whereas other studies are focusing on optimizing or expanding the functionalities of the method. The CRISPR/Cas9 system comprises multiple important elements, many of which can be constantly optimized to increase the efficiency of the system.

Due to the fact that both the protein (Cas9) and the RNA (sgRNA) are part of the expression cassette, two separate promoters are needed to ensure accurate timing of expression of either *Cas9* or sgRNA. The expression of *Cas9* is often controlled by the constitutive promoter *Cauliflower Mosaic Virus 35S* (*35S*). However, some studies have experimented with alternative promoters to obtain tissue specific expression of both *Cas9* and sgRNA (Table 1). Examples such as the *ICU2p* and *EC1.2* promoters with activity in frequently dividing cells, that is, callus and meristems, have been shown to increase efficiency and result in fewer chimeric T1 plants (Hyun et al., 2015; Wang et al., 2015). Previous studies have successfully utilized the RNA polymerase III promoters, *AtU6* and *AtU6-26* to transcribe sgRNAs in various plants, for example, *Nicotiana benthamiana* and *Solanum lycopersicum*, showing that these promoters can facilitate transcription in other dicots. The CRISPR/Cas9 system has also been modified to employ multiple sgRNAs in a single vector - a technique that often results in higher efficiency and enables the testing of multiple sgRNAs at once (Li et al., 2013; Ma et al., 2015). This could save time as the efficiencies of sgRNAs are not entirely predictable.

Precise DNA insertion and replacement provide new possibilities in plant breeding. Higher plants are often referred to as being recalcitrant to homologous recombination (HR) (Eckardt, 2006; Hanin and Paszkowski, 2003). The nickase Cas9 proteins in which one of the two catalytic domains, HNH or RuvC, have been inactivated by deletions allow single-stranded cleavage as opposed to the DSB of the wild type Cas9 (Trevino and Zhang, 2014). It has been shown that nickases may be used to enhance DNA insertions through HDR and decrease off-target effects (Fauser et al., 2014). Furthermore, recent studies have reported the successful use of gemini viruses' vectors, a technique that has shown to increase the efficiency of HR-mediated insertions in plants. The rationale behind usage of gemini viruses as vectors is their capability to infect many host species across numerous families while requiring a single protein (RepA for master viruses) for replication initiation in the host cell

**Table 1** Examples of studies using the novel precise genome-editing tool CRISPR/Cas9 in ornamental plants

| Plant species | Cas9 promoter | sgRNA promoter | Delivery method | Traits/targets | References |
|---|---|---|---|---|---|
| *Chrysanthemum morifolium* | *PcUbi* | *AtU6-26* | *A. tumefaciens* | Decreased fluorescence | Kishi-Kaboshi et al. (2017) |
| *Dendrobium officinale* | C3H, C4H, 4CL, CCR, IRX | *OsU3* | *A. tumefaciens* | Lignocellulose biosynthesis pathway | Kui et al. (2017) |
| *Ipomoea nil* | *pDE* | *AtU6* | *A. tumefaciens* | Decreased petal senescence | Shibuya et al. (2018) |
| *Ipomoea nil* 'Violet' | *PcUbi* | *AtU6* | *A. tumefaciens* | Flower color change | Watanabe et al. (2017) |
| *Ipomoea nil* 'AK77' | *PcUbi* | *AtU6* | *A. tumefaciens* | Flower color change | Watanabe et al. (2018) |
| *Petunia hybrida* Mitchell diploid | *35S* | *AtU6* | *A. tumefaciens* | Albino | Zhang et al. (2016) |
| *Petunia hybrida* 'Madness' series | - | - | Transfection | - | Subburaj et al. (2016) |
| *Torenia fournieri* | 35S | *AtU6-26* | *A. tumefaciens* | Flower color modification | Nishihara et al. (2018) |
| *Trapogon* spp. | *AtUBQ10* | *AtU6-1, AtU6-26, AtU6-29* | *A. tumefaciens* | Polyploidy studies | Shan et al. (2018) |

Key elements have been highlighted, such as promoters and method of delivery.

(Baltes et al., 2014). In addition, its replication mechanism involves HR and interferes with the host cell cycle, reverting it to the S-phase, which is ideal for HR (Richter et al., 2016). Moreover, geminiviruses display high-replication efficiency producing replicons at elevated rates, consequently increasing SSNs while improving targeting efficiency (Čermák et al., 2015; Hanley-Bowdoin et al., 2013). Moreover, the delivery methods vary from particle bombardment to *Agrobacterium*-mediated methods (Table 1). Optimal strategies for delivery and transformation are highly dependent on plant species and may present an obstacle when working with non-model plant species.

## 2 Applications in horticulture: a case study on Campanula

The first reports for the CRISPR genome-editing technique mainly covered model plants, for example, *Arabidopsis thaliana* and *N. benthamiana*, but recent articles include major agricultural crops, for example, *Brassica oleracea* var. *capitata* (Ma et al., 2019), *Glycine max* (Bao et al., 2019), *Oryza sativa* (Pathak et al., 2019) among others and several ornamental species (Table 1). The development and potential of precise genome editing is progressing at a rapid pace and the applicability of the system to new plant species follows.

For the potted plant and cut flower industry, flower longevity is pivotal and of great importance for the customer appeal and satisfaction (Ferrante et al., 2015). Exposure to the phytohormone ethylene leads to senescence symptoms like leaf, bud and flower drop in many plants which significantly impair the plant product. Hence, ethylene insensitivity is a desirable trait to target in breeding of ornamental plants. Because of this, ethylene biosynthesis and the signal transduction pathway are of great interest for researchers working with ornamentals. Previous studies have shed light on the correlation of mutations in genes associated with the ethylene signal transduction pathway and ethylene insensitivity. This has inspired various studies in which genetic transformation has been utilized to introduce ethylene insensitivity in plants. In Europe, the use of transgenic plants may reduce the commercial market potential for such cultivars. Precise genome editing may provide a new and in some cases, less controversial way of modifying plant traits in existing cultivars. Ethylene signaling is involved in a wide array of biological functions and completely interrupting ethylene signaling may therefore not be fortuitous for plant breeders, as it will have consequences for other plant functions (Little et al., 2009). Consequently, it is important to consider the implications of targeting various ethylene-associated genes. This section will elaborate on scientific literature within this topic and propose strategic targets for precise gene knockout in *Campanula* mediated by CRISPR/Cas9.

To inhibit senescence and increase longevity of ornamental plants, producers often apply chemical inhibitors such as 1-methylcyclopropene

(1-MCP), which acts as a competitive ligand and prevents ethylene signal transduction. However, increased awareness of conscientious use of chemicals has incited research into alternative solutions (Lütken et al., 2012). Ethylene signal transduction is perceived by ER-membrane embedded receptors that act as negative regulators of ethylene responses in the absence of ethylene (Fig. 3) (Hua and Meyerowitz, 1998). When ethylene is perceived, the receptors disassociate from the Constitutive Triple Response factor 1 (CTR1) (Fig. 3). CTR1 is a ser/thr kinase repressor that constitutively represses ethylene insensitive 2 (EIN2) by targeting it for proteolytic degradation (Huang et al., 2003). When

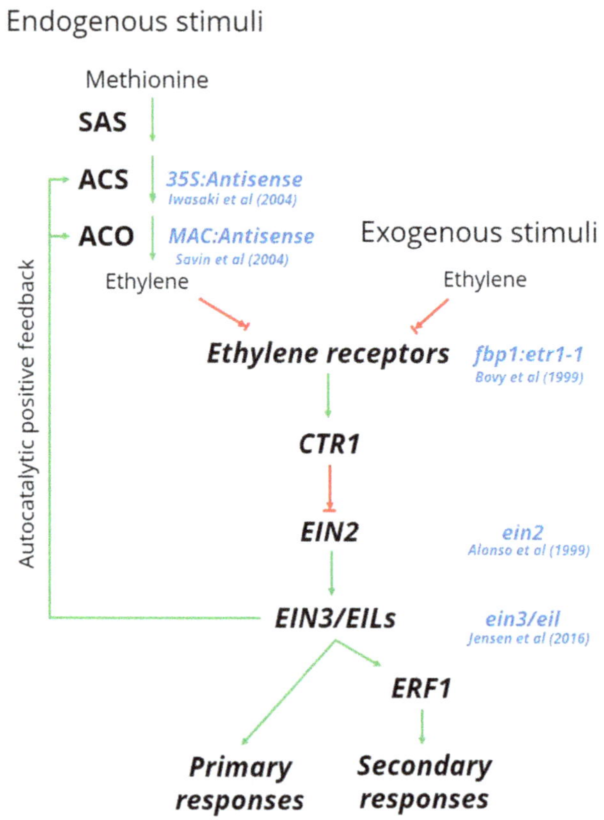


**Figure 3** Ethylene synthesis and signaling targets for molecular modification. Key components of the ethylene biosynthesis and simplified signaling pathways are shown in black. Positive and negative regulations are shown in green and red, respectively. Molecular modification strategies are shown in blue. SAS: S-adenosylmethionine synthase; ACS: aminocyclopropane carboxylate synthase; *35S*: constitutive promoter; ACO: aminocyclopropane carboxylic acid oxidase; *MAC*: constitutive promoter; *fbp1*: flower specific promoter; CTR1: constitutive triple response 1; *EIN2*: *ethylene insensitive 2*; *EIN3*: *ethylene insensitive 3* / EILs: EIN3-like proteins; ERF1: ethylene response factor 1.

ethylene is perceived, the C-terminal end of EIN2 is cleaved and translocated to the nucleus where it serves as a positive regulator of ethylene responses. This is likely achieved by stabilization of transcription factors such as EIN3/EILs, by inhibiting EIN3 binding F-box 1 and 2 (EBF1/2) mediated degradation (An et al., 2010). The EIN3 and EILs transcription factors initiate downstream ethylene responses (Chao et al., 1997; Yamasaki et al., 2005) (Fig. 3). Previous studies have utilized transgenic approaches to alter ethylene perception, leading to decreased ethylene sensitivity in various ornamental plants, for example, *Campanula* and *Kalanchoë* (Fig. 3) (Sanikhani et al., 2008; Sriskandarajah et al., 2007). However, these studies rely on transgenic expression of missense-mutated ethylene response 1 (*etr1*) ethylene receptors from *A. thaliana* making them fall within the scope of GMO legislation (Lütken et al., 2012).

## 2.1 Targets for precise genome editing

Based on the scientific literature discussed in the previous section there are multiple candidate targets for precise genome editing that may result in decreased ethylene sensitivity. However, because ethylene signaling is involved in many important developmental functions it is important to consider the consequence of altering ethylene signaling. The optimal target for genetic modification should result in reduced ethylene-induced flower senescence to prevent bud and flower drop, while having minimal effect on other ethylene-induced responses.

It can be hypothesized that interference in later steps is preferable to early steps as this may result in fewer undesired pleiotropic effects. RNA interference of ACS and ACO genes have previously shown success in reducing endogenous ethylene production and extending vase life of *D. caryophyllus* cut flowers from regenerated lines harboring the ACS sense and ACS/ACO antisense constructs (Iwazaki et al., 2004; Savin et al., 1995). One of the problems with that approach was that it interferes early in the pathway and may result in lack of proper ethylene signaling in other biological functions, for example, low yield in regeneration after transformation. Another issue is that it does not inhibit exogenous stimuli.

Ethylene insensitivity may also be achieved by overexpression of receptors or by knockout of *EIN2*, these approaches have some drawbacks. Constitutive overexpression may have a greater chance of having pleiotropic effects as ethylene signaling is interfered equally throughout all developmental stages and downstream responses. Previous studies have therefore used a flower specific promoter (*FBP1*) from petunia (Bovy et al., 1999) (Fig. 3). By using a *fbp1::etr1-1* mutant, Winkelmann et al. (2016), showed that the use of a tissue specific promoter was effective in elevating gene expression in the perianth of *Burrageara* orchids. In comparison to untransformed plants, the vase life was extended 7 days and the bud drop initiation was delayed up to 19 days in the transgenic plants. This is however not a suitable target for targeted knockout by precise genome editing since single mutations in ethylene receptors often results in little to no phenotype (Hua and Meyerowitz, 1998). Mutagenic knockout of EIN2 may be even more troublesome as it resulted in complete ethylene insensitivity in *A. thaliana* (Alonso et al., 1999) (Fig. 3). Downstream transcription factors may also have multiple response genes related to abiotic stresses and be involved in various developmental stages (reviewed by Debbarma et al., 2019). However, results from the recent study by Jensen et al. (2016) correlated a natural mutation in *EIL2* in the commercially available *C. medium* with reduced ethylene-induced senescence. This is a very interesting target for precise genome editing as it suggests that a mutation in orthologous genes in related species may result in similar phenotypes.

The study by Jensen et al. (2016) correlated an ethylene-insensitive phenotype in *C. medium* with a 7 bp frameshift mutation in the *Cmeil2* gene, sharing homology to EIN3/EIL transcription factors (Figs. 4 and 5). In that study,

```
CfEIL1_KX058427.1     CTCGCTTTTATCAGCTTTGATGCAGCATTGTGACCCTCCTCAGAGAAGGTTCCCTCTTGA   474
CmEIL1_KX058429.1     CTCGCTTTTATCAGCTTTGATGCAGCATTGTGACCCTCCTCAGAGAAGGTTCCCTCTTGA   474
CpEIL1b_KX058426.1    CTCACTTTTATCAGCTCTGATGCAGCATTGTGACCCTCCGCAAAGGCGGTTCCCTCTAGA   474
CpEIL1a_KX058425.1    CTCACTTTTATCAGCTCTGATGCAGCATTGTGACCCTCCTCAGAGACGGTTCCCTCTAGA   477
CfEIL2_KX058428.1     TTCGTTATTATCTGCTCTCATGCAGCATTGTGATCCTCCCCAGAGGCGTTTCCCTCTGGA   480
CmEIL2_KX058430.1     TTTGTTATTATCTGCTCTC-------ATTGTGATCCTCTACAGAGGCGTTTCCCTCTCGA   473
                       *   * ***** *** *       ******* ****  ** **  * ******** **
```

**Figure 4** Nucleotide sequence alignment of *Campanula EIL* sequences. Sequence alignment shows the targeted view showing the alignment of *ethylene insensitive 3-like (EIL)* sequences surrounding the 7 bp deletion in *Cmeil2*. Abbreviations: *C. medium* (Cm), *C. formanekiana* (Cf), *C. portenschlagiana* (Cp) and asterisk (*) indicates alignment consensus. A 7 bp deletion in *Cmeil2* results in loss of gene function. The alignment was produced in Clustal Ω (McWilliam et al., 2013) and the NCBI accession numbers are listed on the figure.

```
                                                  Proline rich region           BDIII
AtEIN3/183-255    DTTLGSLLSALMQHCDPPQRRFPLEKGVPPPWWPNGKEDWWPQLGLPKDQ--GPAPYKKPHDLKKAWKVGVLTAV
CmEIL1/134-196    DTTLGSLLSALMQHCDPPQRRFPLEKGVAPPWWPKGDEEWWAQLGLPKDQ--GPPPYKKPHDLKK----------
Cmeil2/177-236    DTTLGSLLSSLIVILPRGVSLWRKVFRPRGCPRGRRNGGLNHVCRKNKGRAHRLTRSLM----------------
CfEIL1/134-196    DTTLGSLLSALMQHCDPPQRRFPLEKGVAPPWWPKGDEEWWAQLGLPKDQ--GPPPYKKPHDLKK----------
CfEIL2/177-251    DTTLGSLLSALMQHCDPPQRRFPLEKGVPPPWWPIGTEEWWPQLGLQKDQGAGPPPYKKPHDLKKAWKVGVLTAV
CpEIL1a/135-197   DTTLGSLLSALMQHCDPPQRRFPLEKGVAPPWWPNGDEEWWAQLGLPKDQ--GPPPYKKPHDLKK----------
CpEIL1b/134-196   DTTLGSLLSALMQHCDPPQRRFPLEKGVAPPWWPNGDEEWWPQLGLPKDQ--GPPPYKKPHDLKK----------
                                                                              *
```

**Figure 5** Translated amino acid sequence alignment of AtEIN3 and EIL sequences from *Campanula* species. Partial EIL nucleotide sequences were translated into amino acid sequences and aligned to the annotated EIN3 sequence from *Arabidopsis thaliana*. The red-shaded region represents the proline-rich region, and the black box shows the basic domain III both of which are thought to be part of the DNA-binding domain of the transcription factors. The asterisk marks an important Lys-245 to Asn mutation found in ein3-3 which impairs EIN3 function. The red-highlighted sequence shows the nonsense sequence following a 7 bp deletion found in the nucleotide sequence of *C. medium eil2*. Alignment was made using Clustal Omega and Jalview. Sequences were derived from (Chao et al., 1997; Jensen et al., 2016).

senescence was assessed in response to exposure to ethylene in *C. formanekiana* and *C. medium*. RT-PCR and RT-qPCR revealed that *Cmeil2* was transcribed 100-fold less than *CfEIL2*. The mutation in *Cmeil2* introduced a premature stop codon resulting in a truncated mRNA sequence. *CfEIL* sequences were constitutively expressed in the four-flower developmental stages investigated, whereas *Cmeil2* was only found in very low amounts relative to *CfEIL2* transcripts. In the wild type scenario, an ethylene response would result in stabilization of the transcription factors, thereby eliciting ethylene responses.

However, in the case of *Cmeil2* this will not happen as the protein function has likely been lost (Fig. 4). This would explain the ethylene-insensitive phenotype of *C. medium*. Jensen et al. (2016) also isolated two ~600 bp CpEIL sequences from *C. portenschlagiana* which were named *CpEIL1a* and *b* as they share highest similarity to *CfEIL1* and *CmEIL1* (Fig. 4). These sequences differ by the presence of a *DdeI* restriction site in the *CpEIL1a*.

## 2.2 Approaches for defining targets in Campanula portenschlagiana

The *Cmeil2* mutation presents a beacon for developing *Campanula* plants with reduced ethylene sensitivity. Unfortunately, *C. medium* is not an important horticultural ornamental plant product, however, *C. portenschlagiana* is a key ornamental plant, with more than 20 million potted plants produced annually in Denmark (Floradania, 2015) This section outlines a strategy describing the overall steps (Table 2) that we have pursued so far for targeting silencing of *EIL1A* and *B* in *C. portenschlagiana*.

### 2.2.1 Designing gRNAs

Designing several gRNAs (the sequence specificity is conferred by a ~20 bp) for each target gene is needed to direct the nuclease activity for the respective

**Table 2** General steps for genome editing of plants

| Step | Task | Outcome |
|---|---|---|
| 1 | Designing gRNAs | Target-specific sequences |
| 2 | Creating an expression cassette | Gene construct |
| 3 | *Agrobacterium tumefaciens* transformation with destination vector | Delivery system |
| 4 | Plant transformation, selection using antibiotic marker and regeneration | Putatively edited plants |
| 5 | Molecular characterization | Edited plant |
| 6 | Outcrossing | Marker and *Cas9* free plants |

genomic target to be modified. In the current case study, six gRNAs were designed to introduce a frameshift mutation in the putative *EIL1A* and *B* sequences from *C. portenschlagiana* (Fig. 6a). They were designed using the BroadInstitute online sgRNA designer tool (Doench et al., 2016), where a list based on the algorithm scoring system was generated. The gRNA sequences were selected by both high algorithm score and when disrupting a restriction enzyme (RE) site. The gRNA spacer sequences were appended with an initial G if not present before, as this is required for RNA polymerase III transcription. The gRNAs are shown in Fig. 6b in terms of the activating key component in the CRISPR/Cas9 genome-editing system. Once designed, the sense and antisense (reverse complemented) gRNAs were ordered as small nucleotides (approximately 20 bp plus restriction site sequences). The oligonucleotide pairs were annealed in a thermocycler by ramping the equimolar mixture to 25°C from 95°C at 5°C $min^{-1}$ to create the dsDNA oligonucleotides.

### 2.2.2 Creating an expression cassette

In the current study, the cloning procedure was divided into two steps, first the dsDNA oligonucleotides were cloned into the entry vector pENTR-CRISPR (Li et al., 2013), creating the expression cassette. Subsequently, *Escherichia*

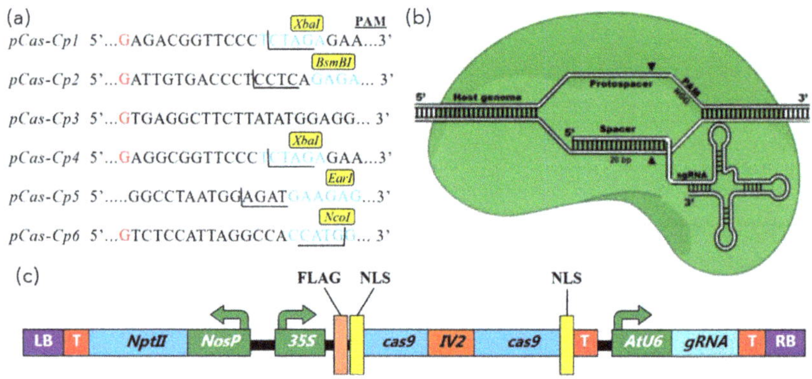


**Figure 6** Schematic overview of the CRISPR/Cas9 system design. (a) The six different guide RNA (gRNA) designs used in this case study, illustrated by their DNA targets. Red G's represent appended guanines, blue sequences represents restriction enzyme (RE) recognition sites, black lines show RE cut sites. Yellow boxes show RE enzymes for each sequence. (b) Overview of the CRISPR/Cas9 genome-editing system. The system is activated and directed by the sgRNA. The system introduces blunt DSBs in the targeted DNA sequence (black triangles, Kemp et al., 2017). (c) Schematic overview of the expression cassette design: Kanamycin selection in plants by *NosP::NptII*; *35S::Cas9* with a potato *IV2* intron and flanked by FLAG-tag and nuclear localization signals (NLS); *AtU6::*gRNA; the expression cassette is flanked by left and right border sequences (LB and RB).

*coli* was transformed and positive clones were selected by ampicillin. The transcription of *SpCas9* from *Streptococcus pyogenes* is plant codon optimized and controlled by the constitutive *35S* promoter whereas the gRNA is controlled by the polymerase III promoter, *AtU6*, from *A. thaliana* (Fig. 6c). *SpCas9* has been appended with nuclear localization signals at both terminal ends and a FLAG-tag at the 5′ end of the *SpCas9*. A potato intron, *IV2*, has been inserted in the *SpCas9* sequence to inhibit adverse effects of the *SpCas9* in bacterial hosts during cloning (Li et al., 2013).

Secondly, the expression cassette was transferred into a binary destination vector (in this case *pK2GW7*) (Karimi et al., 2002). The reaction is driven by recombination and followed the Invitrogen™ Gateway™ LR-Clonase™ II (Thermo Fischer Scientific Inc.) instructions. Subsequently, *E. coli* was transformed and positive clones were selected by spectinomycin. The destination vector confers kanamycin resistance in plants via the *neomycin phosphotransferase II (NptII)* gene as selective marker and contains the left and right boarders (LB/ RB) necessary for *Agrobacterium*-mediated transformation (Fig. 6c). To confirm the destination vector *pENTR-CRISPR:pK2GW7* construction, a double RE assay with *EcoRI* and *HindIII* was made. Successful expression vector constructs should exhibit two bands, that is, 900 and 500 bp.

### 2.2.3 *Agrobacterium tumefaciens* transformation with destination vector

*A. tumefaciens* strain GV2850 was transformed according to standard protocols with three sequenced-confirmed plasmids *pENTR-CRISPR:pK2GW7* of each gDNA target from the previous step and positive colonies selected with 50 mg $L^{-1}$ rifampicin / 75 mg $L^{-1}$ spectinomycin.

### 2.2.4 Plant transformation and regeneration

The *A. tumefaciens* harboring the respective *pENTR-CRISPR:pK2GW7* of each gDNA target was cultured in liquid media using the same abovementioned antibiotics. Petioles of *C. portenschlagiana* derived from axenic culture were inoculated and transformed according to Hegelund et al. (2017) with minor modifications.

### 2.2.5 Molecular analysis

The first molecular analyses were initiated in young shoots. Firstly, PCR was conducted with primers targeting *Cas9* in order to identify transformed tissue (Fig. 7a). Similarly, PCR with primers targeting the endogenous *Actin* gene were used to verify presence of plant DNA (Fig. 7a). Subsequently, PCR was

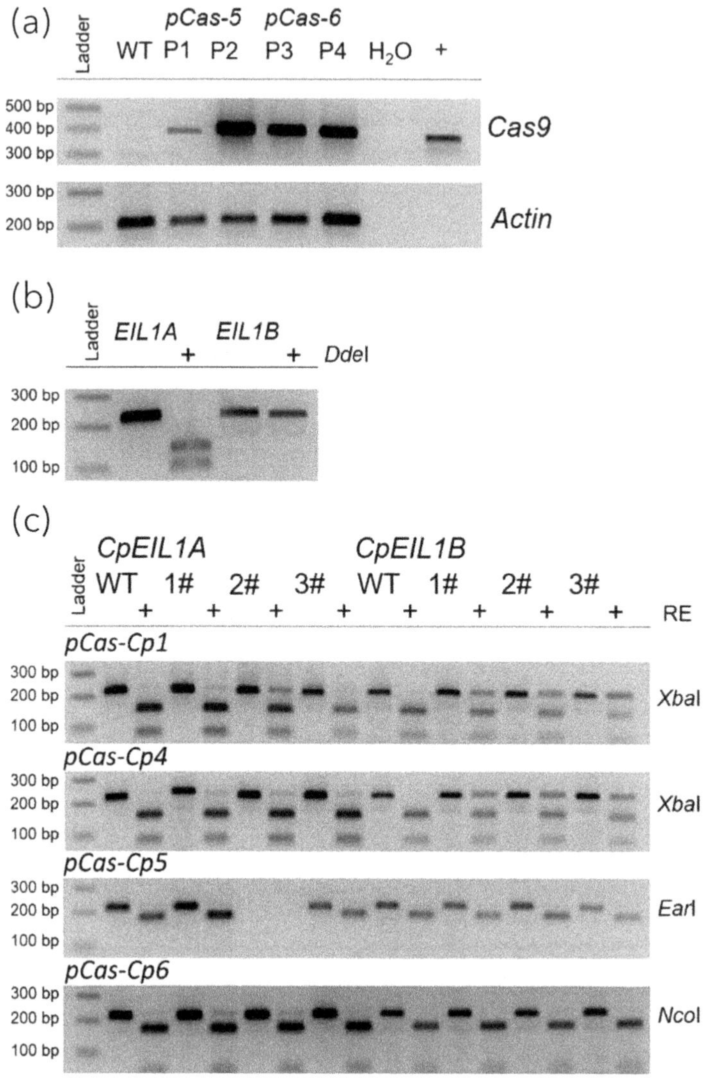


**Figure 7** Molecular analysis of transformed plants and tissue. (a) PCR amplification of *Cas9* and *Actin* to test for the presence of expression cassette in four green regenerated shoots (P1-4). WT and water were used as negative controls, *pENTR-CRISPR* was used as positive control. (b) Specific primers for the CRISPR/Cas9-targeted *EIL1A* and *B* sequences were designed. To verify specificity amplicons from WT DNA template was amplified with both primer sets and digested with *Dde*I, which only has a recognition site in *EIL1A*. Digested samples were run next to undigested samples in equal concentrations. (c) *EIL1A* and *B* amplicons were amplified from three representative calli transformations, with four different gRNA designs (*pCas-Cp1-4*). Amplicons were digested with enzymes predicted to be disrupted from CRISPR/Cas9-mediated mutations. Resistant bands indicate CRISPR/Cas9 activity. Restriction digests and undigested (+) amplicons were analyzed in equal concentration. WT amplicons were used to check for complete digestion.

conducted using primers flanking the gRNA. These amplicons were digested with RE with cutting sites predicted to be disrupted from the CRISPR/Cas9 mediated mutations. Hence, resistant bands indicate CRISPR/Cas9 activity (Fig. 7b). Restriction digests (in this case with *Dde*I) were analyzed next to undigested amplicons in equal concentration. Simultaneously, WT amplified amplicons were used to check for complete digestion. In the case study, *EIL1A* and *B* PCR amplicons from four regenerated plants and from 12 calli samples representing in total four separate gRNA designs, *pCas*-[*Cp1, Cp4, Cp5 and Cp6*] were analyzed by RE digest, to indicate RE motif-disrupting indels (Vouillot et al., 2015). *EIL1* amplicons from regenerated plants showed no RE-resistant bands, suggesting that no CRISPR/Cas9 mediated indels have been produced (data not shown). Also, no indels were observed in calli tissue from transformants of *pCas-Cp5* and *pCas-Cp6*. However, calli tissue transformed with *pCas-Cp1* and *pCas-Cp4* showed RE-resistant bands in *EIL1A* and *B* amplicons (Fig. 7c). These results indicate CRISPR/Cas9-mediated indel mutations in targeted *CpEIL1* sequences in calli transformed with *pCas-Cp1+4*. Following regeneration of plantlets, indels should be reconfirmed by RE digests and verified by sequencing.

### 2.2.6 Outcrossing

Once a genome-edited plant has been obtained, the next step to consider would be the generation of a *Cas9* and marker-free plant. This is achievable by outcrossing of both the antibiotic resistance genes as well as the transgenic *Cas9* gene (Lawrenson et al., 2015; Zhou et al., 2015).

## 3 Prospective uses and regulation of ornamental plants

The potential benefits of novel transgenic ornamental plants may provide faster implemented diversity in floral and foliage color, architecture and tolerance to stresses (Wolt et al., 2016). However, its applicability is highly dependent on how it is perceived legally, but there is generally no consensus whether the genome-edited plants should be assessed according to the current GMO legislation (Sprink et al., 2016). In the United States, the USDA-APHIS provides a regulatory omission to genetically engineered plants under the regulations at 7 CFR part 340 (USA, 2018). Hilscher et al. (2017) showed an updated list of genome-edited crops approved, including maize with high amylopectin content, white button mushroom with no browning among others.

In the European Union, similar decisions were firstly made and released by Germany, confirming that gRNAs are not recombinant DNA, and the Swedish Board of Agriculture, stated in 2015 the equivalency of CRISPR/Cas9 to mutagenesis as no foreign DNA would remain in plants. However, the Court of Justice of the European Union ruled against the initial understandings of the breakthrough technique and decided all products CRISPR-derived should fall under the 2001 Directive (European Union, 2001, 2018).

In Brazil, the National Biosafety Technical Commission (CNTBio) approved Normative 16 (Brazil, 2018) indicating a case-by-case assessment and if no transgene is inserted, the crop will not be subjected to GMO regulation. Similar regulations were previously approved by Argentina in 2015 (Orroño and Vesprini, 2018) and Chile in 2017 (Metje-Sprink et al., 2019) where the central factor is the lack of recombinant DNA in the final product for the non-GMO classification. Further policy alignment in the Latin America region has been signaled by Paraguay and Uruguay, these countries have intention to coordinate efforts in expanding Argentina, Brazil and Chile's case-by-case and genome-editing friendly legislation (CAS, 2018).

## 4 Conclusion

The recent and fast developing CRISPR/Cas9 technique is capable of precise genome editing in plants. The system key factor is specificity and it is defined by a 20 bp long sequence, that is, sgRNA, upstream a PAM sequence, which will guide the Cas9 protein to the DNA target creating a DSB. The DNA damage will be repaired by either NHEJ or HDR. The first can be used to generate knockout mutants and develop further knowledge on specific and combined gene functions, but the later has also the great potential of inserting genes through the HDR repair mechanism. The method has been applied in several agricultural plants and its potential in ornamental species is emerging. Improved tolerance toward the phytohormone ethylene is a key target in breeding as hormones greatly impact the quality of many ornamental plants. A mutation present in *Campanula*, leading to decreased ethylene sensitivity can serve as a platform for CRISPR/Cas9-mediated longevity of ornamental plants.

## 5 Acknowledgements

We thank Oliver Kemp, Bent L. Petersen and Renate Müller for the valuable input and fruitful discussion; Svenning Rune Möller for performing the cloning work that provided the *pENTR-CRISPR* vector; Gartneriet PKM A/S for providing *C. portenschlagiana* stock plants. Vinicius P. Trevenzoli is acknowledged for Fig. 1 design. This work was supported by The Danish Council for Independent Research (Grant n. DFF-7017-00197) and the Brazilian National Council for Scientific and Technological Development (Grant n. CNPq-201196/2015-8).

## 6 Where to look for further information

In terms of CRISPR future research it is clear that enormous advancements are made by the day, therefore it is pivotal to maintain yourself updated and a simple way of doing it is to create citation or topic alerts in well-established research databases, for example Scopus, Science Direct, Web of Knowledge and Google Scholar. The experimental workflow will vary depending on the ultimate goal, for instance, if you detected a gene mutation that resulted in a phenotype that is beneficial, the experiment will pursue a CRISPR construct targeting the reproduction of a similar mutation (Table 2). Another strategy is to induce indels in a group of genes to understand the regulatory loop by generating knockout mutants. Additionally, the off-targets prediction tools vastly improved since the advent of gene editing by CRISPR, but always check for recent updates or novel prediction algorithms. Please consult the list of references, there are further readings to complement this book chapter.

## 7 References

Alonso, J. M., Hirayama, T., Roman, G., Nourizadeh, S. and Ecker, J. R. 1999. EIN2, a bifunctional transducer of ethylene and stress responses in *Arabidopsis*. *Science (New York, N.Y.)* 284(5423), 2148–52. doi:10.1126/science.284.5423.2148.

An, F., Zhao, Q., Ji, Y., Li, W., Jiang, Z., Yu, X., Zhang, C., Han, Y., He, W., Liu, Y., Zhang, S., Ecker, J. R. and Guo, H. 2010. Ethylene-induced stabilization of ETHYLENE INSENSITIVE3 and EIN3-LIKE1 is mediated by proteasomal degradation of EIN3 binding F-box 1 and 2 that requires EIN2 in *Arabidopsis*. *The Plant Cell* 22(7), 2384–401. doi:10.1105/tpc.110.076588.

Azadi, P., Bagheri, H., Nalousi, A. M., Nazari, F. and Chandler, S. F. 2016. Current status and biotechnological advances in genetic engineering of ornamental plants. *Biotechnology Advances* 34(6), 1073–90. doi:10.1016/j.biotechadv.2016.06.006.

Baltes, N. J., Gil-Humanes, J., Cermak, T., Atkins, P. A. and Voytas, D. F. 2014. DNA replicons for plant genome engineering. *The Plant Cell* 26(1), 151–63. doi:10.1105/tpc.113.119792.

Bao, A., Chen, H., Chen, L., Chen, S., Hao, Q., Guo, W., Qiu, D., Shan, Z., Yang, Z., Yuan, S., Zhang, C., Zhang, X., Liu, B., Kong, F., Li, X., Zhou, X., Tran, L. P. and Cao, D. 2019. CRISPR/Cas9-mediated targeted mutagenesis of GmSPL9 genes alters plant architecture in soybean. *BMC Plant Biology* 19(1), 131. doi:10.1186/s12870-019-1746-6.

Barrangou, R., Fremaux, C., Deveau, H., Richards, M., Boyaval, P., Moineau, S., Romero, D. A. and Horvath, P. 2007. CRISPR provides acquired resistance against viruses in prokaryotes. *Science (New York, N.Y.)* 315(5819), 1709–12. doi:10.1126/science.1138140.

Boettcher, M. and McManus, M. T. 2015. Choosing the right tool for the job: RNAi, TALEN, or CRISPR. *Molecular Cell* 58(4), 575–85. doi:10.1016/j.molcel.2015.04.028.

Bovy, A. G., Angenent, G. C., Dons, H. J. M. and van Altvorst, A. 1999. Heterologous expression of the *Arabidopsis etr1-1* allele inhibits the senescence of carnation flowers. *Molecular Breeding* 5(4), 301–8. doi:10.1023/A:1009617804359.

Brazil. 2018. Resolução Normativa No 16, de 15 de janeiro de 2018. CTNBio, Brasilia. Available at: http://ctnbio.mcti.gov.br/resolucoes-normativas/-/asset_publisher/OgW431Rs9dQ6/content/resolucao-normativa-no-16-de-15-de-janeiro-de-2018.

CAS. 2018. Consejo Agropecuario do Sur. Available at: http://consejocas.org/wp-content/uploads/2018/09/XXXVI-RO-CAS-Declaración-II.-Técnicas-de-Edición-Génica.pdf.

Čermák, T., Baltes, N. J., Čegan, R., Zhang, Y. and Voytas, D. F. 2015. High-frequency, precise modification of the tomato genome. *Genome Biology* 16(1), 232. doi:10.1186/s13059-015-0796-9.

Chao, Q., Rothenberg, M., Solano, R., Roman, G., Terzaghi, W. and Ecker, J. R. 1997. Activation of the ethylene gas response pathway in Arabidopsis by the nuclear protein ETHYLENE-INSENSITIVE3 and related proteins. *Cell* 89(7), 1133–44. doi:10.1016/s0092-8674(00)80300-1.

Cong, L., Ran, F. A., Cox, D., Lin, S., Barretto, R., Habib, N., Hsu, P. D., Wu, X., Jiang, W., Marraffini, L. A. and Zhang, F. 2013. Multiplex genome engineering using CRISPR/Cas systems. *Science* 339(6121), 819–23. doi:10.1126/science.1231143.

Debbarma, J., Sarki, Y. N., Saikia, B., Boruah, H. P. D., Singha, D. L. and Chikkaputtaiah, C. 2019. Ethylene response factor (ERF) family proteins in abiotic stresses and CRISPR-Cas9 genome editing of ERFs for multiple abiotic stress tolerance in crop plants: a review. *Molecular Biotechnology* 61(2), 153–72. doi:10.1007/s12033-018-0144-x.

Doench, J. G., Fusi, N., Sullender, M., Hegde, M., Vaimberg, E. W., Donovan, K. F., Smith, I., Tothova, Z., Wilen, C., Orchard, R., Virgin, H. W., Listgarten, J. and Root, D. E. 2016. Optimized sgRNA design to maximize activity and minimize off-target effects of CRISPR-Cas9. *Nature Biotechnology* 34(2), 184–91. doi:10.1038/nbt.3437.

Eckardt, N. A. 2006. Homologous recombination in higher plants: clues from fasciata1-4, a new chromatin formation mutant of *Arabidopsis*. *The Plant Cell* 18(10), 2417–8. doi:10.1105/tpc.106.048231.

Eid, A., Alshareef, S. and Mahfouz, M. M. 2018. CRISPR base editors: genome editing without double-stranded breaks. *Biochemical Journal* 475(11), 1955–64. doi:10.1042/BCJ20170793.

Ellis, B. L., Hirsch, M. L., Porter, S. N., Samulski, R. J. and Porteus, M. H. 2013. Zinc-finger nuclease-mediated gene correction using single AAV vector transduction and enhancement by Food and Drug Administration-approved drugs. *Gene Therapy* 20(1), 35–42. doi:10.1038/gt.2011.211.

Eş, I., Gavahian, M., Marti-Quijal, F. J., Lorenzo, J. M., Mousavi Khaneghah, A., Tsatsanis, C., Kampranis, S. C. and Barba, F. J. 2019. The application of the CRISPR-Cas9 genome editing machinery in food and agricultural science: current status, future perspectives, and associated challenges. *Biotechnology Advances* 37(3), 410–21. doi:10.1016/j.biotechadv.2019.02.006.

European Union. 2001. DIRECTIVE 2001/18/EC OF THE EUROPEAN PARLIAMENT AND OF THE COUNCIL on the deliberate release into the environment of genetically modified organisms. DIRECTIVE 2001/18/EC, Brussels. Available at: https://eur-lex.europa.eu/legal-content/EN/TXT/?uri=celex:32001L0018.

European Union. 2018. Judgment of the EU court C528/16. Available at: http://curia.europa.eu/juris/document/document_print.jsf?docid=204387&text=&dir=&doclang=EN&part=1&occ=first&mode=lst&pageIndex=0&cid=5226816.

Fauser, F., Schiml, S. and Puchta, H. 2014. Both CRISPR/Cas-based nucleases and nickases can be used efficiently for genome engineering in Arabidopsis thaliana. *The Plant Journal: for Cell and Molecular Biology* 79(2), 348–59. doi:10.1111/tpj.12554.

Ferrante, A., Trivellini, A., Scuderi, D., Romano, D. and Vernieri, P. 2015. Post-production physiology and handling of ornamental potted plants. *Postharvest Biology and Technology* 100, 99–108. doi:10.1016/j.postharvbio.2014.09.005.

Floradania. 2015. Top 10 over danske kulturer 2014. Available at: http://floradania.dk/branchenyt/top-10/.

FloraHolland. 2018. FloraHolland. Available at: http://jaarverslag.royalfloraholland.com/?_ga=2.20167931.291158191.1550304215-258211969.1550304215#/feiten-en-cijfers/omzet?_k=ix6uc2 (accessed on 11 March 2019).

Gaj, T., Gersbach, C. A. and Barbas, C. F. 2013. ZFN, TALEN, and CRISPR/Cas-based methods for genome engineering. *Trends in Biotechnology* 31(7), 397–405. doi:10.1016/j.tibtech.2013.04.004.

Garneau, J. E., Dupuis, M. È, Villion, M., Romero, D. A., Barrangou, R., Boyaval, P., Fremaux, C., Horvath, P., Magadán, A. H. and Moineau, S. 2010. The CRISPR/Cas bacterial immune system cleaves bacteriophage and plasmid DNA. *Nature* 468(7320), 67–71. doi:10.1038/nature09523.

Gasiunas, G., Barrangou, R., Horvath, P. and Siksnys, V. 2012. Cas9-crRNA ribonucleoprotein complex mediates specific DNA cleavage for adaptive immunity in bacteria. *Proceedings of the National Academy of Sciences of the United States of America* 109(39), E2579–86. doi:10.1073/pnas.1208507109.

Gratz, S. J., Ukken, F. P., Rubinstein, C. D., Thiede, G., Donohue, L. K., Cummings, A. M. and O'Connor-Giles, K. M. 2014. Highly specific and efficient CRISPR/Cas9-catalyzed homology-directed repair in *Drosophila*. *Genetics* 196(4), 961–71. doi:10.1534/genetics.113.160713.

Gupta, R. M. and Musunuru, K. 2014. Expanding the genetic editing tool kit: ZFNs, TALENs, and CRISPR-Cas9. *Journal of Clinical Investigation* 124(10), 4154–61. doi:10.1172/JCI72992.

Hanin, M. and Paszkowski, J. 2003. Plant genome modification by homologous recombination. *Current Opinion in Plant Biology* 6(2), 157–62. doi:10.1016/s1369-5266(03)00016-5.

Hanley-Bowdoin, L., Bejarano, E. R., Robertson, D. and Mansoor, S. 2013. Geminiviruses: masters at redirecting and reprogramming plant processes. *Nature Reviews. Microbiology* 11(11), 777–88. doi:10.1038/nrmicro3117.

Hegelund, J. N., Lauridsen, U. B., Wallström, S. V., Müller, R. and Lütken, H. 2017. Transformation of *Campanula* by wild type *Agrobacterium rhizogenes*. *Euphytica* 213(2), 1–9. doi:10.1007/s10681-017-1845-0.

Hille, F. and Charpentier, E. 2016. CRISPR-Cas: biology, mechanisms and relevance. *Philosophical Transactions of the Royal Society Series B: Biological Sciences* 371(1707), 20150496. doi:10.1098/rstb.2015.0496.

Hilscher, J., Bürstmayr, H. and Stoger, E. 2017. Targeted modification of plant genomes for precision crop breeding. *Biotechnology Journal* 12(1). doi:10.1002/biot.201600173.

Holkers, M., Maggio, I., Liu, J., Janssen, J. M., Miselli, F., Mussolino, C., Recchia, A., Cathomen, T. and Gonçalves, M. A. 2013. Differential integrity of TALE nuclease genes following adenoviral and lentiviral vector gene transfer into human cells. *Nucleic Acids Research* 41(5), e63. doi:10.1093/nar/gks1446.

Hua, J. and Meyerowitz, E. M. 1998. Ethylene responses are negatively regulated by a receptor gene family in *Arabidopsis thaliana*. *Cell* 94(2), 261–71. doi:10.1016/s0092-8674(00)81425-7.

Huang, Y., Li, H., Hutchison, C. E., Laskey, J. and Kieber, J. J. 2003. Biochemical and functional analysis of CTR1, a protein kinase that negatively regulates ethylene signaling in *Arabidopsis*. *The Plant Journal: for Cell and Molecular Biology* 33(2), 221–33. doi:10.1046/j.1365-313x.2003.01620.x.

Hyun, Y., Kim, J., Cho, S. W., Choi, Y., Kim, J. S. and Coupland, G. 2015. Site-directed mutagenesis in *Arabidopsis thaliana* using dividing tissue-targeted RGEN of the CRISPR/Cas system to generate heritable null alleles. *Planta* 241(1), 271–84. doi:10.1007/s00425-014-2180-5.

Ishino, Y., Shinagawa, H., Makino, K., Amemura, M. and Nakata, A. 1987. Nucleotide sequence of the iap gene, responsible for alkaline phosphatase isozyme conversion in *Escherichia coli*, and identification of the gene product. *Journal of Bacteriology* 169(12), 5429–33. doi:10.1128/jb.169.12.5429-5433.1987.

Iwazaki, Y., Kosugi, Y., Waki, K., Yoshioka, T. and Satoh, S. 2004. Generation and ethylene production of transgenic carnations harboring ACC synthase cDNA in sense or antisense orientation. *System* 6(2), 67–71.

Jansen, R., Embden, J. D. A. van, Gaastra, W. and Schouls, L. M. 2002. Identification of genes that are associated with DNA repeats in prokaryotes. *Molecular Microbiology* 43(6), 1565–75. doi:10.1046/j.1365-2958.2002.02839.x.

Jensen, L., Hegelund, J. N., Olsen, A., Lütken, H. and Müller, R. 2016. A natural frameshift mutation in *Campanula* EIL2 correlates with ethylene insensitivity in flowers. *BMC Plant Biology* 16(1), 117. doi:10.1186/s12870-016-0786-4.

Karimi, M., Inzé, D. and Depicker, A. 2002. GATEWAYTM vectors for *Agrobacterium*-mediated plant transformation. Trends in Plant Sciences 7(5), 193–5. doi:10.1016/s1360-1385(02)02251-3.

Kemp, O., Favero, B. T., Hegelund, J. N., Møller, S. R., Müller, R., Petersen, B. L. and Lütken, H. 2017. Modification of ethylene sensitivity in ornamental plants using CRISPR/Cas9. *Acta Horticulturae* 1167(1167), 271–80. doi:10.17660/ActaHortic.2017.1167.40.

Kim, H. and Kim, J. S. 2014. A guide to genome engineering with programmable nucleases. *Nature Reviews. Genetics* 15(5), 321–34. doi:10.1038/nrg3686.

Kishi-Kaboshi, M., Aida, R. and Sasaki, K. 2017. Generation of gene-edited chrysanthemum morifolium using multicopy transgenes as targets and markers. *Plant and Cell Physiology* 58(2), 216–26. doi:10.1093/pcp/pcw222.

Kui, L., Chen, H., Zhang, W., He, S., Xiong, Z., Zhang, Y., Yan, L., Zhong, C., He, F., Chen, J., Zeng, P., Zhang, G., Yang, S., Dong, Y., Wang, W. and Cai, J. 2017. Building a genetic manipulation tool box for orchid biology: identification of constitutive promoters and application of CRISPR/Cas9 in the orchid, *Dendrobium officinale*. *Frontiers in Plant Science* 7(January), 1–13. doi:10.3389/fpls.2016.02036.

Kuligowska, K., Lütken, H. and Müller, R. 2016. Towards development of new ornamental plants: status and progress in wide hybridization. *Planta* 244(1), 1–17. doi:10.1007/s00425-016-2493-7.

Lawrenson, T., Shorinola, O., Stacey, N., Li, C., Østergaard, L., Patron, N., Uauy, C. and Harwood, W. 2015. Induction of targeted, heritable mutations in barley and *Brassica oleracea* using RNA-guided Cas9 nuclease. *Genome Biology* 16, 258. doi:10.1186/s13059-015-0826-7.

Li, J. F., Norville, J. E., Aach, J., McCormack, M., Zhang, D., Bush, J., Church, G. M. and Sheen, J. 2013. Multiplex and homologous recombination-mediated genome editing in *Arabidopsis* and *Nicotiana benthamiana* using guide RNA and Cas9. *Nature Biotechnology* 31(8), 688–91. doi:10.1038/nbt.2654.

Liang, X., Potter, J., Kumar, S., Ravinder, N. and Chesnut, J. D. 2017. Enhanced CRISPR/Cas9-mediated precise genome editing by improved design and delivery of gRNA, Cas9 nuclease, and donor DNA. *Journal of Biotechnology* 241, 136–46. doi:10.1016/j.jbiotec.2016.11.011.

Little, H. A., Grumet, R. and Hancock, J. F. 2009. Modified ethylene signaling as an example of engineering for complex traits: secondary effects and implications for environmental risk assessment. *HortScience* 44(1), 94–101. doi:10.21273/HORTSCI.44.1.94.

Liu, J., Gaj, T., Yang, Y., Wang, N., Shui, S., Kim, S., Kanchiswamy, C. N., Kim, J. S. and Barbas, C. F. 2015. Efficient delivery of nuclease proteins for genome editing in human stem cells and primary cells. *Nature Protocols* 10(11), 1842–59. doi:10.1038/nprot.2015.117.

Lütken, H., Clarke, J. L. and Müller, R. 2012. Genetic engineering and sustainable production of ornamentals: current status and future directions. *Plant Cell Reports* 31(7), 1141–57. doi:10.1007/s00299-012-1265-5.

Ma, X., Zhang, Q., Zhu, Q., Liu, W., Chen, Y., Qiu, R., Wang, B., Yang, Z., Li, H., Lin, Y., Xie, Y., Shen, R., Chen, S., Wang, Z., Chen, Y., Guo, J., Chen, L., Zhao, X., Dong, Z. and Liu, Y. G. 2015. A robust CRISPR/Cas9 system for convenient, high-efficiency multiplex genome editing in monocot and dicot plants. *Molecular Plant* 8(8), 1274–84. doi:10.1016/j.molp.2015.04.007.

Ma, C., Zhu, C., Zheng, M., Liu, M., Zhang, D., Liu, B., Li, Q., Si, J., Ren, X. and Song, H. 2019. CRISPR/Cas9-mediated multiple gene editing in Brassica oleracea var. capitata using the endogenous tRNA-processing system. *Horticulture Research* 6(1), 20. doi:10.1038/s41438-018-0107-1.

Makarova, K. S., Grishin, N. V., Shabalina, S. A., Wolf, Y. I. and Koonin, E. V. 2006. A putative RNA-interference-based immune system in prokaryotes: computational analysis of the predicted enzymatic machinery, functional analogies with eukaryotic RNAi, and hypothetical mechanisms of action. *Biology Direct* 1, 7. doi:10.1186/1745-6150-1-7.

Marraffini, L. A. and Sontheimer, E. J. 2008. CRISPR interference limits horizontal gene transfer in staphylococci by targeting DNA. *Science* 322(5909), 1843–5. doi:10.1126/science.1165771.

McWilliam, H., Li, W., Uludag, M., Squizzato, S., Park, Y. M., Buso, N., Cowley, A. P. and Lopez, R. 2013. Analysis tool web services from the EMBL-EBI. *Nucleic Acids Research* 41(W1), W597–600. doi:10.1093/nar/gkt376.

Metje-Sprink, J., Menz, J., Modrzejewski, D. and Sprink, T. 2019. DNA-free genome editing: past, present and future. *Frontiers in Plant Science* 9, 1957. doi:10.3389/fpls.2018.01957.

Mojica, F. J. M., Díez-Villaseñor, C., Soria, E. and Juez, G. 2000. Biological significance of a family of regularly spaced repeats in the genomes of Archaea, Bacteria and mitochondria. *Molecular Microbiology* 36(1), 244–6. doi:10.1046/j.1365-2958.2000.01838.x.

Mojica, F. J. M., Díez-Villaseñor, C., García-Martínez, J. and Soria, E. 2005. Intervening sequences of regularly spaced prokaryotic repeats derive from foreign genetic elements. *Journal of Molecular Evolution* 60(2), 174–82. doi:10.1007/s00239-004-0046-3.

Nishihara, M., Higuchi, A., Watanabe, A. and Tasaki, K. 2018. Application of the CRISPR/Cas9 system for modification of flower color in *Torenia fournieri*. *BMC Plant Biology* 18(1), 331. doi:10.1186/s12870-018-1539-3.

Orroño, D. I. and Vesprini, F. 2018. Directives and requirements for genetically modified (GM) crop regulation in Argentina. *Crop Breeding and Applied Biotechnology* 18(3), 301–8. doi:10.1590/1984-70332018v18n3r44.

Pathak, B., Zhao, S., Manoharan, M. and Srivastava, V. 2019. Dual-targeting by CRISPR/Cas9 leads to efficient point mutagenesis but only rare targeted deletions in the rice genome. *3 Biotech* 9(4), 158. doi:10.1007/s13205-019-1690-z.

Richter, K. S., Serra, H., White, C. I. and Jeske, H. 2016. The recombination mediator RAD51D promotes geminiviral infection. *Virology* 493, 113–27. doi:10.1016/j.virol.2016.03.014.

Sanikhani, M., Mibus, H., Stummann, B. M. and Serek, M. 2008. *Kalanchoe blossfeldiana* plants expressing the *Arabidopsis etr1-1* allele show reduced ethylene sensitivity. *Plant Cell Reports* 27(4), 729–37. doi:10.1007/s00299-007-0493-6.

Savin, K. W., Baudinette, S. C., Graham, M. W., Michael, M. Z., Nugent, G. D., Lu, C., Chandler, S. F. and Cornish, E. C. 1995. Antisense ACC oxidase RNA delays carnation petal senescence. *HortScience* 30(5), 970–2. doi:10.21273/HORTSCI.30.5.970.

Shan, S., Mavrodiev, E. V., Li, R., Zhang, Z., Hauser, B. A., Soltis, P. S., Soltis, D. E. and Yang, B. 2018. Application of CRISPR/Cas9 to Tragopogon (Asteraceae), an evolutionary model for the study of polyploidy. *Molecular Ecology Resources* 18(6), 1427–43. doi:10.1111/1755-0998.12935.

Shibata, M. 2008. Importance of genetic transformation in ornamental plant breeding. *Plant Biotechnology* 25(1), 3–8. doi:10.5511/plantbiotechnology.25.3.

Shibuya, K., Watanabe, K. and Ono, M. 2018. CRISPR/Cas9-mediated mutagenesis of the EPHEMERAL1 locus that regulates petal senescence in Japanese morning glory. *Plant Physiology and Biochemistry: PPB* 131(April), 53–7. doi:10.1016/j.plaphy.2018.04.036.

Sorek, R., Lawrence, C. M. and Wiedenheft, B. 2013. CRISPR-mediated adaptive immune systems in bacteria and archaea. *Annual Review of Biochemistry* 82, 237–66. doi:10.1146/annurev-biochem-072911-172315.

Sprink, T., Eriksson, D., Schiemann, J. and Hartung, F. 2016. Regulatory hurdles for genome editing: process- vs. product-based approaches in different regulatory contexts. *Plant Cell Reports* 35(7), 1493–506. doi:10.1007/s00299-016-1990-2.

Sriskandarajah, S., Mibus, H. and Serek, M. 2007. Transgenic *Campanula carpatica* plants with reduced ethylene sensitivity. *Plant Cell Reports* 26(6), 805–13. doi:10.1007/s00299-006-0291-6.

Sternberg, S. H., Redding, S., Jinek, M., Greene, E. C. and Doudna, J. A. 2014. DNA interrogation by the CRISPR RNA-guided endonuclease Cas9. *Nature* 507(7490), 62–7. doi:10.1038/nature13011.

Subburaj, S., Chung, S. J., Lee, C., Ryu, S. M., Kim, D. H., Kim, J. S., Bae, S. and Lee, G. J. 2016. Site-directed mutagenesis in Petunia × hybrida protoplast system using direct delivery of purified recombinant Cas9 ribonucleoproteins. *Plant Cell Reports* 35(7), 1535–44. doi:10.1007/s00299-016-1937-7.

Teixeira da Silva, J. A., Chin, D. P., Van, P. T. and Mii, M. 2011. Transgenic orchids. *Scientia Horticulturae* 130(4), 673–80. doi:10.1016/j.scienta.2011.08.025.

Trevino, A. E. and Zhang, F. 2014. Genome editing using Cas9 nickases. *Methods in Enzymology* 546, 161–74. doi:10.1016/B978-0-12-801185-0.00008-8.

USA. 2018. USDA APHIS 7 CFR Part 340 – introduction of organisms and products altered or produced through genetic engineering which are plant pests or which there is reason to believe are plant pests. APHIS 7 CFR Part 340. Available at: https://www.ecfr.gov/cgi-bin/text-idx?SID=ac22214a42bbdaf700cddb95e4e140a4&mc=true&node=pt7.5.340&rgn=div5#se7.5.340_11.

Vouillot, L., Thelie, A. and Pollet, N. 2015. Comparison of T7E1 and surveyor mismatch cleavage assays to detect mutations triggered by engineered nucleases. *G3* 5(3), 407–15. doi:10.1534/g3.114.015834.

Wang, Z. P., Xing, H. L., Dong, L., Zhang, H. Y., Han, C. Y., Wang, X. C. and Chen, Q. J. 2015. Egg cell-specific promoter-controlled CRISPR/Cas9 efficiently generates homozygous mutants for multiple target genes in *Arabidopsis* in a single generation. *Genome Biology* 16(1), 144. doi:10.1186/s13059-015-0715-0.

Watanabe, K., Kobayashi, A., Endo, M., Sage-Ono, K., Toki, S. and Ono, M. 2017. CRISPR/Cas9-mediated mutagenesis of the *dihydroflavonol-4-reductase-B* (DFR-B) locus in the Japanese morning glory *Ipomoea (Pharbitis) nil*. *Scientific Reports* 7(1), 10028. doi:10.1038/s41598-017-10715-1.

Watanabe, K., Oda-Yamamizo, C., Sage-Ono, K., Ohmiya, A. and Ono, M. 2018. Alteration of flower colour in *Ipomoea nil* through CRISPR/Cas9-mediated mutagenesis of *carotenoid cleavage dioxygenase 4*. *Transgenic Research* 27(1), 25–38. doi:10.1007/s11248-017-0051-0.

Whinn, K. S., Kaur, G., Lewis, J. S., Schauer, G. D., Mueller, S. H., Jergic, S., Maynard, H., Gan, Z. Y., Naganbabu, M., Bruchez, M. P., O'Donnell, M. E., Dixon, N. E., van Oijen, A. M. and Ghodke, H. 2019. Nuclease dead Cas9 is a programmable roadblock for DNA replication. *Scientific Reports* 9(1), 13292. doi:10.1038/s41598-019-49837-z.

Winkelmann, T., Warwas, M., Raffeiner, B., Serek, M. and Mibus, H. 2016. Improved postharvest quality of inflorescences of *fbp1::etr1-1* transgenic *Burrageara* 'Stefan Isler lava flow'. *Journal of Plant Growth Regulation* 35(2), 390–400. doi:10.1007/s00344-015-9545-2.

Wolt, J. D., Wang, K. and Yang, B. 2016. The regulatory status of genome-edited crops. *Plant Biotechnology Journal* 14(2), 510–8. doi:10.1111/pbi.12444.

Wyman, C. and Kanaar, R. 2006. DNA double-strand break repair: All's Well That Ends Well. *Annual Review of Genetics* 40(1), 363–83. doi:10.1146/annurev.genet.40.110405.090451.

Xiong, J. S., Ding, J. and Li, Y. 2015. Genome-editing technologies and their potential application in horticultural crop breeding. *Horticulture Research* 2(February), 15019. doi:10.1038/hortres.2015.19.

Yamasaki, K., Kigawa, T., Inoue, M., Yamasaki, T., Yabuki, T., Aoki, M., Seki, E., Matsuda, T., Tomo, Y., Terada, T., Shirouzu, M., Tanaka, A., Seki, M., Shinozaki, K. and Yokoyama, S. 2005. Solution structure of the major DNA-binding domain of *Arabidopsis thaliana* ethylene-insensitive3-like3. *Journal of Molecular Biology* 348(2), 253–64. doi:10.1016/j.jmb.2005.02.065.

Zhang, B., Yang, X., Yang, C., Li, M. and Guo, Y. 2016. Exploiting the CRISPR/Cas9 system for targeted genome mutagenesis in petunia. *Scientific Reports* 6(1), 20315. doi:10.1038/srep20315.

Zhao, C., Shu, X. and Sun, B. 2017. Construction of a gene knockdown system based on catalytically inactive ("dead") Cas9 (dCas9) in *Staphylococcus aureus*. *Applied and Environmental Microbiology* 83(12). doi:10.1128/AEM.00291-17.

Zhou, H., Liu, B., Weeks, D. P., Spalding, M. H. and Yang, B. 2014. Large chromosomal deletions and heritable small genetic changes induced by CRISPR/Cas9 in rice. *Nucleic Acids Research* 42(17), 10903–14. doi:10.1093/nar/gku806.

Zhou, X., Jacobs, T. B., Xue, L. J., Harding, S. A. and Tsai, C. J. 2015. Exploiting SNPs for biallelic CRISPR mutations in the outcrossing woody perennial Populus reveals 4-coumarate:CoA ligase specificity and redundancy. *The New Phytologist* 208(2), 298–301. doi:10.1111/nph.13470.

# Chapter 7

# Advances in abiotic stress-resistant varieties of ornamentals

*Qiansheng Li and Mengmeng Gu, Texas A&M AgriLife Extension Service, USA*



## 1 Introduction

Plants face various abiotic stresses, both above and below ground, including drought, salinity, cold, heat, waterlogging, high light intensity, UV-radiation, nutrient imbalance and soil compaction (Roy et al., 2011). Strategies for improving abiotic stress tolerance of ornamentals are very different from those used on agricultural crops. Agronomic crops tend to include a very limited number of high-yield species, and thus have very limited genetic diversity. To ensure food security, breeding of agronomic crops has also focused more on developing agricultural crop varieties that can both tolerate abiotic stresses and maintain optimum yield (Roy et al., 2011).

In contrast, the number of ornamental species and their wild relatives is estimated to be 85000–99000 (Long et al., 2018), providing much greater genetic diversity than in agronomic crops. Given this enormous genetic diversity, there is a general rule of 'right plant, right place' for selecting ornamental plants for a particular landscape (Beck, 2013), matching ornamental plants to local environmental conditions. In addition, under the intensive management conditions of commercial ornamental plant production, particularly in greenhouses equipped with irrigation, fertilization and climate control systems,

http://dx.doi.org/10.19103/AS.2020.0066.08

abiotic stresses can be reduced or minimized. As a result, the development of abiotic stress-resistant varieties in ornamental plants has not received the same attention as it has for agricultural crops (Mou, 2011).

The primary focus in ornamental plant breeding is the novel aesthetic characteristics (e.g. novel flower patterns or colors, flower size, foliar characteristics, etc.). Some of these novel aesthetic characteristics, such as variegation in foliage, may reduce plant vigor and thus reduce the tolerance to abiotic stresses. Many ornamental plants are prized for their foliar variegation (Li et al., 2007), but the partial loss of photosynthetically active surfaces in variegated leaves affects absorption and utilization of light and therefore net photosynthesis, potentially reducing the growth and abiotic stress tolerance. Some ornamental varieties with aesthetic properties (e.g. variegated foliage) induced by plant viruses may show less vigor. While causing no adverse symptoms in these plants, such viruses may cause severe disease in other ornamentals or agricultural crops, including reduced stress tolerance (Valverde et al., 2012).

Recently there has been more focus in ornamental breeding on the importance of tolerance to abiotic stresses. Climate change has caused more frequent weather events such as drought (e.g. the 2011–2014 drought in California and the 2011 drought in Texas), flooding (e.g. Hurricane Harvey in 2018 in Houston, Texas) and record high temperatures throughout the world. The impact of global warming can be seen in the way that many horticultural species have been found to shift their geographic ranges northward. In one study, of 357 native European plant species investigated, 73% were being sold hundreds or even thousands of kilometers north of their natural geographic range (Van der Veken et al., 2008). Climate change and urban heat island effects will significantly affect phenological patterns (Luo et al., 2007), flowering and floral development. It has also been shown that vulnerability to one type of abiotic stress may reduce the ability of a plant to withstand another kind of stress, as shown in the case of the effect of drought stress on the freezing tolerance of *Viola* (Oraee et al., 2018). As a result, developing varieties tolerant to abiotic stresses is becoming another focus for ornamental plant breeders (Kuligowska et al., 2016).

Improving abiotic stress tolerance in ornamental crops is challenging using conventional breeding (Dhakar, 2017). Plant responses to abiotic stresses are complex, making it harder to screen for stress-tolerant varieties. A diverse germplasm collection is the backbone of any program to improve stress tolerance (Mou, 2011). Higher ploidy levels are found in plants in colder habitats with a wider range of average annual high and low temperatures (Paule et al., 2017). In ornamental plant breeding, increased abiotic stress tolerance has been associated with techniques such as in vitro induced polyploidy, for example, against drought, cold, and salinity in *Chrysanthemum nankingense*

(Liu et al., 2018) and against drought in *Spathiphyllum* (Van Laere et al., 2011). Plant tissue culture techniques exploit natural somaclonal variations, which can be further enhanced by applying various physical and chemical mutagen treatments.

With the recent rapid development of gene mapping and gene-editing technology, molecular breeding using specific trait markers has great potential for providing cultivars with abiotic stress tolerance (Dhakar, 2017; Roy et al., 2011). For example, the gene CdICE1 from *Chrysanthemum dichrum* has been linked to tolerance of *C. grandiflorum* to low temperature/freezing, drought and salinity (Chen et al., 2012). The genes MtDREB1C and RcXET in rose (*Rosa chinensis*) are associated with both freezing and drought tolerance (Chen et al., 2016a). This chapter focuses on the recent advances in abiotic stress-tolerant varieties through both conventional and molecular breeding.

## 2 Improving resistance to low-temperature stress

Low temperature is a common environmental stress factor, which seriously affects plant growth and is an important factor limiting the plant's northernmost distribution. Low-temperature stresses include chilling (non-freezing; >0°C) and freezing (< 0°C). Chilling usually causes foliage injury, especially in tropical plants (Li et al., 2008). Symptoms of foliage injury include yellowing (chlorosis), water-soaked patches and finally wilting (Fig. 1). Freezing stress is one of the

**Figure 1** The leaves of tropical foliage plant *Philodendron bipennifolium* 'Aurea' showed water-soaked patches and will finally be wilted after chilled at 4°C for a few hours.

most important limiting factors determining the ecological distribution and production of species (Körner et al., 2016). Under freezing stress, the apical buds can be damaged (Fig. 2), and the loss of apical dominance results in changes in growth patterns, subsequently changing a plant's architecture (Charrier et al., 2015). Freezing may cause complete loss of above-ground part of herbaceous and woody ornamental plants, and sometimes even the below-ground part. Winter survival, especially among woody evergreens, also requires tolerance of light stress, which can result in photo-oxidative damage at cold temperatures when the biochemistry of photosynthesis is compromised (Arora and Rowland, 2011). The root temperature can also affect light stress (Franco et al., 2011).

**Figure 2** The apical buds and young shoots of tropical evergreen oak *Quercus austrocochinchinensis* planted in the northern region of subtropics were damaged during −5°C freezing event.

Breeding new varieties with low-temperature tolerance, based on assessment of chilling and freezing risks, is thus critical for ornamental plant development. Chilling tolerance can vary significantly from cultivar to cultivar. As an example, the chilling sensitivity of eight most popular *Dieffenbachia* cultivars was ranked: 'Tropic Honey' > 'Sterling' > 'Carina' > 'Octopus' > 'Camille' > 'Camouflage' > 'Star Bright' > 'Panther'. The origin of the most chilling tolerant cultivar 'Panther' is unknown, but is possibly a chilling-tolerant wild variety, while 'Star Bright' is an interspecific hybrid. Both could be used for breeding of new chilling-tolerant cultivars (Li et al., 2008). In another example, the cultivars 'Red Missile' and 'Salsa Yellow' are considered as cold-tolerant ornamental pepper cultivars. They are the potential candidates in breeding programs to develop new ornamental and vegetable pepper genotypes for high- and low-temperature tolerance (Gajanayake et al., 2011).

Wild species usually have high tolerance to cold, and hybridization with wild species is often used for cold tolerance breeding. In roses, *R. rugosa* is well known for its high resistance to environmental stresses and especially to sub-zero temperatures. Over a three-year trial, rose cultivars 'Dagmar Hastrup' and 'John Cabot' were found to be the most cold-hardy genotypes (Ouyang et al., 2019). Field trials are the most direct, but time-consuming way to test cold hardiness. Physiological parameters are suitable for screening survival temperature tolerance. When an injury index is combined with genotypes with known cold hardiness in screening programs, it can lead to more rapid screening of new genotypes (Ouyang et al., 2019).

Specific gene expression related to low-temperature stress tolerance has been investigated to accelerate molecular breeding. Cold tolerance of chrysanthemum at the vegetative and flowering stages was controlled by two pairs of major genes that exhibited additivity-dominance-epistasis effects and the heritability of the major genes were 88.56% and 65.86%, respectively (Ao et al., 2019). A genome-wide transcript profile of *C.grandiflorum* var. Jinba has helped to identify the molecular mechanisms of low-temperature stress tolerance in *C. grandiflorum*. These studies highlight candidate genes for chilling-tolerance and freezing-tolerance molecular breeding in chrysanthemum (Wang et al., 2018). Genes with cold-responsive transcription factors or involved in cold sensor or signal transduction and in the stabilization of the plasma membrane and osmosensing response were discovered in *Camellia japonica* var. *decumbens*, which is a highly resistant variety (Wu et al., 2019).

Modern molecular techniques have helped to accelerate breeding of water lily for cold-stress tolerance, which has been a target for breeding for more than 150 years. To improve cold-stress tolerance of tropical water lilies (including *Brachyceras*, *Anecphya*, *Lotos* and *Hydrocallis*), a gene encoding choline oxidase (CodA) initiated by a cold stress-inducible promoter was inserted into a tropical water lily through pollen-tube transformation. Six independent transgenic lines were tested, which expanded the cultivating area of tropical water lilies, from 24.3°N latitude to 30.3°N latitude (Yu et al., 2018). Similarly, the PlGPAT gene has been identified as playing an important role in regulating cold resistance in *Paeonia lactiflora*, and can be used to improve cold resistance using transgenic technologies (Li et al., 2018). The MtDREB1C gene, isolated from *Medicago truncatula*, and initiated by the arabidopsis rd29A promoter, has a significantly enhanced freezing tolerance in transgenic Chinese rose without any visible morphological or developmental abnormality (Chen et al., 2010). Petunia has been transformed via overexpression of the p5cs gene to produce commercial varieties that are more tolerant to cold stress than the wild types (Jamshidnia et al., 2018). The mei flower (*Prunus mume*) is an important landscape plant in China, but low-cold tolerance severely limits its use in North China. Three PmCBF genes from a cold-tolerant cultivar *P. mume* 'Xue Mei' have been isolated. Their role in enhancing freezing tolerance has been linked to enhanced reactive oxygen species (ROS) scavenging (Peng et al., 2016). Dehydrin genes were cloned from cold-tolerant *P. mume* 'Beijingyudie', and overexpressed in tobacco. The transgenic tobacco plants showed higher cold tolerance through lower relative content of malondialdehyde, relative electrolyte leakage and higher relative content of water than control plants (Bao et al., 2017). The serine/threonine protein kinase (CIPK) gene family provides the key candidate genes and a theoretical basis for cold resistance molecular-assisted breeding in *P. mume* (Li et al., 2019).

## 3 Improving resistance to high-temperature stress

Heat stress negatively affects plant growth, development, physiological processes and yield, particularly, as it is often accompanied by stress from drought (Fahad et al., 2017). Heat stress often causes poor seedling growth, leaf senescence or chlorosis, reduced flower yield and undesirable flower characteristics (such as smaller flowers) (Fig. 3). As in the case of chilling and freezing tolerance, tolerance to heat stress varies amongst varieties. In the case of *Gerbera jamesonii*, for example, 'Meihongheixin' has relatively high heat resistance, 'Shijihong' and 'Linglong' are moderately heat resistant, 'Dadifen' and 'Taiyangfengbao' are moderately heat sensitive, while 'Beijixing' is highly heat sensitive (Chen et al., 2016b). Similarly, studies under controlled environment conditions have found that the heat tolerance of nine marigold (*Tagetes erecta* and *T. patula*) cultivars decreased in the order of 'Giant', 'Gate Gold', 'Chokdee', 'Discovery', 'Janie', 'Bonanza', 'Little Hero', 'Great Hero' and 'Durango', with the first three recommended for cultivation in areas with high temperatures during the growth season (Tian et al., 2015).

The physiological and biochemical reactions to heat stress are active research areas, and molecular approaches are being adopted for developing

**Figure 3** The heat stress accompanying with high light caused foliar yellowing, foliar burning and smaller flower of *Anthurium*.

high-temperature tolerance in plants (Bunty et al., 2018). Heat shock genes (HSGs), which encode heat shock proteins (HSPs), are a key part of a plant's response to heat stress (Hasanuzzaman et al., 2013). The HSPs and heat shock transcription factors were both identified in *P. mume* (Wan et al., 2019). It has been suggested that the CmCPL1 gene-encoding enzyme RNAPII CTD phosphatase plays an important role in the tolerance of chrysanthemum to heat stress (Qi et al., 2018a). Identifying such genes and transcription factors can then be used in breeding. As an example the CmDREB6 gene has been cloned from the chrysanthemum cultivar 'Jinba' to create two transgenic lines with an overexpression of CmDREB6 to provide the enhanced heat tolerance (Du et al., 2018).

## 4 Improving resistance to drought

Increased evapotranspiration thus occurs under high temperatures, a direct result of global warming, causing more frequent and more intense periods of drought. Water shortages in many areas, therefore, are becoming more frequent with local governments restricting irrigation, thus exposing plants to drought stress (Domenghini et al., 2013). Selection of drought-tolerant species is one solution to adapt the water shortages in landscapes and home gardens.

Plants have evolved a series of mechanisms at the morphological, physiological, biochemical, cellular and molecular levels to overcome drought stress (Fang and Xiong, 2015). Plants have four basic mechanisms to survive from drought stress: drought avoidance, drought tolerance, drought escape and drought recovery (Fang and Xiong, 2015).

Many succulent plants (Fig. 4) are very drought tolerant and water-use efficient as they have a crassulacean acid metabolism photosynthetic pathway. For example, sedum species have a wide geographic distribution from temperate to subtropical regions. They primarily grow in arid environments on shallow soils on stony or gravel stands. The sedum species are slow-growing plants and can survive for a long period without water (Kozminska et al., 2019). They are frequently used as ground cover or in rock gardens, and on 'green roofs' (Fig. 5), where drought is a norm (Domenghini et al., 2013). The sedum has the ability to switch from the $C_3$ photosynthetic pathway to the crassulacean acid metabolism photosynthetic pathway when growing under water deficit conditions (Domenghini et al., 2013). A relatively high degree of genetic variability exists within the genus with *S. spurium* exhibiting the highest tolerance to water deficit (Kozminska et al., 2019). It is therefore important to compare not just different species but also different genotypes/ecotypes within a single species such as *Sedum*.

As with other types of abiotic stress, species and varieties vary in their degree of drought tolerance. The research on ornamental shrubs has shown, for example, that *Laurus nobilis* and *Thunbergia erecta* are less sensitive to drought stress than *Callistemon citrinus, Pittosporum tobira* and *Viburnum tinus* (Toscano et al., 2014). A study of the drought tolerance of twelve cultivars of three ornamental *Tagetes* species (*T. patula, T. tenuifolia* and *T. erecta*) found

**Figure 4** Many succulent plants are very drought tolerant and water-use efficient.

**Figure 5** Drought is very normal for plants on 'green roofs'.

that *T. erecta* was more drought tolerant than *T. patula* and *T. tenuifolia* (Cicevan et al., 2016). An investigation of twenty birch genotypes found that *Betula nigra* 'Cully', *B.* 'Royal Frost', *B. pendula*, *B. nigra* 'BNMTF', *B. davurica* and *B. papyrifera* could survive drought conditions in the Southern United States, but that water deficit reduced the final tree height and trunk diameter (Gu et al., 2007a). The research has investigated the role of ethylene in causing water-deficit, stress-induced leaf abscission and highlighted the need to investigate other hormonal factors such as abscisic acid (ABA) or cytokinin (Gu et al., 2007b).

ABA acts as a mediator in plant responses to a range of stresses including drought stress (Ali et al., 2017). Drought stress induces the activation of genes associated with ABA perception and biosynthesis as well as the ethylene pathway (Toscano et al., 2019). Rose RhNAC3 protein could mediate ABA signaling both in rose and in *A. thaliana*, and thus increase drought tolerance (Jiang et al., 2014). The high expression of DREB1A, DREB2A, CuZnSOD, NCED3A and NCED3B transcripts in CmWRKY10-overexpressed transgenic chrysanthemum plants provided strong evidence that drought tolerance mechanism was associated with ABA pathway (Jaffar et al., 2016). Expression of CmMYB2 in chrysanthemum leaves was up-regulated in response to drought, salinity and cold stress, and it increased plant sensitivity to ABA and reduced stomatal aperture (Shan et al., 2012). The pepper ethylene-responsive transcription factor CaAIEF1 (*Capsicum annuum* ABA-induced ERF 1) positively regulates the drought stress response and the ABA signaling in pepper (Hong et al., 2017). PeCHYR1, a ubiquitin E3 ligase isolated from *Populus euphratica*, plays a crucial role in enhancing drought tolerance via ABA-induced stomatal closure caused by hydrogen peroxide production in transgenic poplar plants. Besides ABA, salicylic acid (SA) and jasmonic acid (JA) are also involved in inducing tolerance to drought stress through modulation of physiological and biochemical processes in plants (Ali et al., 2017; Khan et al., 2015). The significant increase of ABA, JA and brassinosteroids was observed in Kentucky bluegrass with an extended drought period (Chen et al., 2019). Osmolytes also perform an important function in the adjustment of plants against drought stresses through osmotic adjustment (Singh et al., 2015).

Natural genetic diversity exists among genotypes with different behavior in terms of stress tolerance, with effects upon the expression of adaptive morphological and/or physiological target traits (Guerra et al., 2015). Breeding methods to improve drought resistance could include interspecific hybridization (between cultivars and their wild species), genetic modification and polyploidy. As an example of the first approach, six interspecific hybrid lines between *Chrysanthemum grandiflorum* and *C. indicum* have produced plants with improved drought tolerance (Sun et al., 2010). Polyploids can also improve abiotic stress tolerance. The tetraploid *Spathiphyllum* plants were found to be more resistant to drought stress compared with diploid plants (Van Laere et al., 2011). The transcriptome sequences can provide novel insights into chrysanthemum responses to dehydration stress and offer candidate genes or markers that can be used to breed drought-tolerant chrysanthemum cultivars (Xu et al., 2013). The overexpression of Sorghum bicolor Sloan-Kettering retrovirus interacting protein (SbSKIP) in *Petunia hybrida* could promote the growth of lateral roots to enhance drought tolerance (Qi et al., 2018b). Some microRNAs (miRNAs) are functionally conserved across plant species and are regulated by drought stress. Reprogramming of gene expression by miRNAs could be the potential strategies and directions for future in the improvement of drought tolerance (Ferdous et al., 2015).

## 5 Improving resistance to waterlogging

Waterlogging, causing decreased availability of oxygen to the submerged organs of plants, may negatively affect plant development. Waterlogging stress often negatively affects root growth, and finally causes root rot after a long-time stress (Fig. 6). Waterlogging causes anaerobic (lack of oxygen) respiration with a by-product, ethanol, and the accumulation of ethanol over time can cause harm to plant organs. Deficient aeration and low oxygen diffusion affect rhizogenesis and other aspects of root function (Franco et al., 2011). Adaptive responses to waterlogging may include the formation of aerenchyma tissues in the root cortex, the development of adventitious roots near the soil surface and an increase in the root diameter (Franco et al., 2011).

Waterlogging stress is among the most common abiotic stressors in *Chrysanthemum*. Forty-five molecular markers associated with waterlogging tolerance in chrysanthemum have been detected, and four favorable parental

**Figure 6** Long-time waterlogging caused the rot of roots and rhizome of *Primulina* 'Kazu'.

lines are identified as potential donors for improving waterlogging tolerance (Su et al., 2016). The comparison of the transcriptomes of the waterlogging-tolerant cultivar 'Nannongxuefeng' and the waterlogging-sensitive cultivar 'Qinglu' indicated that the waterlogging-tolerant cultivar had a higher ethylene production under waterlogging and reoxygenation conditions. This provides a mechanism for understanding the response to waterlogging and a guide for breeding more tolerant varieties (Zhao et al., 2018). It has been suggested that higher cDNA of 1-aminocyclopropane-1-carboxylate oxidase expression contributes to higher accumulation of ethylene in the waterlogging-tolerant species *C. zawadskii* (Zhang et al., 2018).

## 6 Improving tolerance to saline conditions

Salinity in soils is one of the major abiotic constraints on plant productivity (Acosta-Motos et al., 2017). It has been estimated that salinity affects 6% of the world's total land area. In addition, the scarcity of water has led to an increasing use of wastewater for landscape irrigation, which often has high salt levels (Cassaniti et al., 2012; Niu and Cabrera, 2010). Compared with agricultural crops, a few studies have investigated the developing salt-tolerant ornamental crops.

Research on the mechanisms of salt stress suggests that it decreases the leaf growth and chlorophyll content, and inhibits plant photosynthesis (Cassaniti et al., 2012; Gao et al., 2012; Niu and Cabrera, 2010), causing the accumulation of ROS in the cytoplasm, which triggers oxidative stress. Salinity imposes two types of stress on roots: osmotic stress resulting from the lowered water potential in the root medium, and ionic stress induced by changes in the concentrations of specific ions in the root medium and inside the root tissue. Root growth is usually less affected by salinity than shoot growth (Franco et al., 2011; Acosta-Motos et al., 2017). Plants have evolved mainly two types of tolerance mechanisms based on either limiting the entry of salt by the roots, or controlling its concentration and distribution (Hanin et al., 2016). The ABA-dependent signaling pathway has been suggested as the main pathway mediating salt stress response in plants such as *Rosa chinensis* (Tian et al., 2018). Two pathways (plant hormone signal transduction and glutathione metabolism) were also found to be active in salt stress responses (Tian et al., 2018).

An overview of the challenges of growing ornamental plants in brackish water is provided by Cassaniti et al. (2013), who identified the need to identify halophytes and salt-tolerant varieties. Halophytes are plants that naturally inhabit saline environments and have evolved with many strategies to cope with seawater and high concentrations of salts. It has been suggested that mechanisms to cope with salt stress include processes such as limiting ion uptake by roots and transport into leaves, selective exclusion of ions, compartmentalization of ions at the cellular and whole plant level, synthesis of compatible solutes, changes in the photosynthetic pathway, alteration in membrane structure, induction of antioxidant enzymes to scavenge free oxygen radicals and altered synthesis of plant hormones (Al Hassan et al., 2017).

Many halophytic species are already in use as ornamentals but more could be selected. Most suitable species came from the Chenopodiaceae, particularly *Limonium*. Plant genotypes with thicker cuticles, like 'blue' conifer cultivars, or plants with very hirsute leaves evaporate less water (thus intake less water) and may aid salt tolerance (Cassaniti et al., 2013). *Sesuvium portulacastrum* is a pioneer plant species, and is used for sand-dune fixation, desalination and phytoremediation along coastal regions. It can grow at a high salinity of 1 000 mM NaCl without any toxic symptoms on leaves (Lokhande et al., 2013). *Pistacia lentiscus* is a suitable ornamental species for landscapes in arid and saline areas. Salt tolerance in *P. lentiscus* has been associated with restricted uptake of $Cl^-$ and its storage in roots (Alvarez et al., 2018).

*Iris halophila* is a perennial ornamental halophyte distributed mainly on grasslands, slopes and saline-alkaline soils in China and Russia. Its ability to adapt to saline conditions has been attributed to genes related to ion transport and the ROS scavenging system (Liu et al., 2018). The study of transcriptomes helps to clarify the regulatory network involved in the salt toxicity response of *I. halophila* and helps in the development of salt-tolerant varieties (Liu et al., 2018). *Iris lactea* var. *chinensis* is also a halophyte with high salt tolerance. Research by Tang et al. (2018) has explored the role of the transcription factor gene IlWRKY2 in determining salt tolerance, providing a target for future breeding.

In identifying salt-tolerant varieties, the first step is to group tolerance into four categories (based on measuring plant growth and visual symptoms of stress such as foliar damage): highly tolerant, tolerant, moderately tolerant and sensitive (Cassaniti et al., 2012). Research has suggested the use of an index based on growth data and visual inspection to better evaluate levels of salt tolerance compared to aesthetic quality (Niu and Cabrera, 2010; de Oliveira et al., 2018). This approach has been used to evaluate the relative salt tolerance of bedding plants such as angelonia (*Angelonia angustifolia*), ornamental pepper (*Capsicum annuum*), helenium (*Helenium amarum*), licorice plant (*Helichrysum petiolatum*), and plumbago (*Plumbago auriculata*), *Achillea millefolium*, *Gaillardia aristata*, *Lantana × hybrida*, *Lonicera japonica*, *Rosmarinus officinalis* and lupines such as Texas bluebonnet (*Lupinus texensis*) and Big Bend bluebonnet (*L. havardii*) (Niu et al., 2010).

Other studies have ranked the relative salt tolerance of jessamine (*Cestrum* 'Orange Peel'), Mexican hummingbird bush (*Dicliptera suberecta*), flame acanthus (*Anisacanthus quadrifidus* var. wrightii), rock rose (*Pavonia lasiopetala*) and bluebeard (*Caryopteris × clandonensis* 'Dark Knight') (Wu et al., 2016a). Sun et al. (2015) have assessed the salt tolerance of Texas Superstar® perennials such as *Malvaviscus arboreus var. drummondii*, *Ruellia* 'Katie Blue', Salvia 'Henry Duelberg', and *Verbena* 'Blue Princess', *Phlox* 'John Fanick', *P.* 'Texas Pink' and *Salvia leucantha*. Texas betony (*Stachys coccinea*) has been found to be the most salt tolerant among *Lamiaceae* ornamental species (Wu et al., 2016b). Wu et al. (2016b) have also ranked the salt tolerance of ten aster perennials: *Conoclinium greggii*, *Viguiera stenoloba*, *Chrysactinia Mexicana*, *Melampodium leucanthum*, *Tagetes lemmonii*, *Tetraneuris scaposa* and *Leucanthemum × superbum* 'Becky').

Among roses, 'Belinda's Dream', 'Climbing Pinkie', Mrs. Dudley Cross', 'Reve d'Or' and 'Sea Foam' have been found to be the most salt-tolerant cultivars, whereas 'Cecile Brunner', 'Else Poulsen', 'Madame Antoine Mari', 'Perle d'Or', 'Spice' and 'Souvenir de St. Anne's' were the least salt-tolerant among the eighteen rose cultivars investigated (Cai et al., 2014). In rose rootstocks, *Rosa × fortuniana* was more salt-tolerant than *R. odorata* which was more salt-tolerant than *R. multiflora* (Niu et al., 2008). *Rosa × fortuniana* accumulated the least salt (Niu and Rodriguez, 2008).

Salinity is also a problem in coastal gardens and landscapes, where plants are damaged by salt-containing aerosols originating from the sea (Ferrante et al., 2011). Most of the research conducted on salinity tolerance concerns saline soil or saline irrigation, and there is little information available on foliar salt tolerance. Farieri et al. (2016) have evaluated the effects of seawater aerosol on ornamental shrubs in the Mediterranean environment.

Tissue culture techniques have been used to develop new salt-tolerant varieties. One salt-tolerant chrysanthemum (*C. morifolium*) variant (E2) has been developed through in-vitro mutagenesis using ethylmethane sulfonate (EMS) as the chemical mutagen (Hossain et al., 2006). Two salt-tolerant mutation lines of *Petunia* × atkinsiana have also been produced using EMS treatment (Krupa-Malkiewicz et al., 2017).

## 7 Improving tolerance to poor soil conditions

Soil pH is a key factor affecting plant health since it affects the solubility and availability of minerals. High concentrations of calcium carbonate ($CaCO_3$) and hydrogen carbonate ($HCO_3^{\ -}$) in alkaline soils limit the availability of iron (Fe) and the availability of other micronutrients such as manganese (Mn), zinc (Zn) and copper (Cu), showing the symptom of foliar chlorosis (Fig. 7).

Plants in Ericaceae family generally do not grow well in lime soils with high pH. Wang et al. (2018b) have analyzed specimens of *Rhododendron* to identify a group of taxa with potential lime tolerance, providing valuable resources for more detailed evaluation of lime tolerance and in breeding. Similar work on azalea has been undertaken, identifying a diploid Japanese azalea *R. indicum* 'Osakazuki' as a resource for breeding lime-tolerant varieties (Demasi et al., 2017). Seedling phenotyping protocols based on leaf-area changes were developed for quick screening of new germplasm to assess pH adaptability in deciduous azalea (Susko et al., 2018). Low pH could be a problem for some ornamentals. Enhanced low-pH stress tolerance was found in anthocyanin-enriched transgenic petunia lines (PM2, PM6, and PM8) by overexpressing the RsMYB1 gene (Naing et al., 2018).

**Figure 7** Chlorosis of *Quercus stewardiana*'s young leaves caused by high soil pH.

Ornamentals may also be exposed to heavy metal-contaminated soils. One coping response is the activation of antioxidative enzymes counteracting ROS produced by heavy metal stress (Ashrafzadeh and Leung, 2015). Plants, which cannot develop an efficient defense mechanism, can hardly survive in heavy metal-polluted soils. The tolerance to heavy metals may vary greatly among species and cultivars. It has been found, for example, *Lagerstroemia indica* is more tolerant to Cadmium (Cd) than *L. fauriei* and is an excellent species for remediating Cd-contaminated soils (Wang et al., 2016). *Chlorophytum comosum* also has high tolerance to Cd stress with similar potential in the treatment of Cd-contaminated soils (Wang et al., 2012). In *Petunia*, RsMYB1 has been identified as a potential dual-function gene that can have an impact on the improvement of anthocyanin production and heavy metal stress tolerance (Ai et al., 2018).

## 8 Future trends and conclusion

Due to the large number of ornamental plants (and their enormous genetic diversity), planting the right plants (through screening) in the right place is still the best way to cope with abiotic stresses. Breeding for stress tolerance will likely continue to be of low priority, compared to ornamental characteristics. Conventional crossing techniques typically take 8–10 years to produce a new stress-resistant variety. As a result, the use of modern breeding techniques to confer resistance to abiotic and biotic stresses has started to attract more attention (Noman et al., 2017). However, stress-tolerant variety breeding continues to be limited by the complexity of stress responses and their relationship with other physiological and developmental processes. Breeding for stress resistance in ornamental plants is particularly difficult due to the small number of resistance genes identified so far (Azadi et al., 2016). Although there have been several examples of successful genetic transformation to confer abiotic stress resistance, problems still exist with the efficiency of these methods and regeneration into whole plants (Palmer et al., 2017).The molecular analysis of gene insertion sites and sequencing flanking regions remain technically difficult (Chandler and Brugliera, 2011). The financial barriers to market entry for genetically modified varieties of ornamental crops, such as costs of regulatory approval, are also a problem. As a result no new cultivars have yet been made commercially available using this method (da Silva et al., 2015).

Garden petunias are very popular bedding plants. Petunia is also a model species in plant molecular biology and genetics (Vandenbussche et al., 2016). Petunia is particularly easy to transform and is suited to biochemical analysis (Vandenbussche et al., 2016). A complete *P. hybrida* chloroplast genome sequence has recently been reported, providing the foundation for future molecular breeding (Wong et al., 2019). Autophagy, regulated by autophagy-related genes (ATG), has also been identified as a factor in abiotic stress response in petunia, providing a potential target for breeding (Pillajo et al., 2018). Recently, CRISPR/Cas9 technology has emerged as a powerful new technique for targeted genome modification in petunia, suggesting a possible way forward for the future (Zhang et al., 2016).

## 9 Where to look for further information

Phytozome, the Plant Comparative Genomics portal of the Department of Energy's Joint Genome Institute, provides JGI users and the broader plant science community a hub for accessing, visualizing and analyzing JGI-sequenced plant genomes, as well as selected genomes and datasets that have been sequenced elsewhere. There are some ornamental plants species in the database, such as *Kalanchoe fedtschenkoi, Kalanchoe laxiflora, Populus trichocarpa, Trifolium pratense.* The specific abiotic stress genes can be searched in the database:

https://phytozome.jgi.doe.gov/.

The Genome Database for Rosaceae is a curated and integrated web-based relational database providing genomics, genetics and breeding data for Rosaceae, which is developed by Mainlab Bioinformatics at Washington State University:

https://www.rosaceae.org.

The Solanaceae Genomics Network, a database of Solanaceae genomic and phenotypic data and tools, was developed by Mueller's team from the Boyce Thompson Institute for Plant Research and Cornell University. This database includes the genome sequences of ornamental plants *Petunia axillaris* and *P. inflata*:

https://solgenomics.net/.

The Carnation DB provides the sequence analysis of the Genome of Carnation (*Dianthus caryophyllus L.*):

http://carnation.kazusa.or.jp/index.html.

Breeders can search specific abiotic resistance genes in these databases. With the development of molecular biology, more and more genome databases will be established.

## 10 References

Acosta-Motos, J. R., Ortuno, M. F., Bernal-Vicente, A., Diaz-Vivancos, P., Sanchez-Blanco, M. J. and Hernandez, J. A. 2017. Plant responses to salt stress: adaptive mechanisms. *Agronomy* 7(1). doi:10.3390/agronomy7010018.

Ai, T. N., Naing, A. H., Yun, B. W., Lim, S. H. and Kim, C. K. 2018. Overexpression of RsMYB1 enhances anthocyanin accumulation and heavy metal stress tolerance in transgenic petunia. *Frontiers in Plant Science* 9, 1388. doi:10.3389/fpls.2018.01388.

Al Hassan, M., Estrelles, E., Soriano, P., López-Gresa, M. P., Bellés, J. M., Boscaiu, M. and Vicente, O. 2017. Unraveling salt tolerance mechanisms in halophytes: a comparative study on four Mediterranean *Limonium* species with different geographic distribution patterns. Frontiers in Plant Science, 8, 1438. doi:10.3389/fpls.2017.01438.

Ali, F., Bano, A. and Fazal, A. 2017. Recent methods of drought stress tolerance in plants. *Plant Growth Regulation* 82(3), 363–75. doi:10.1007/s10725-017-0267-2.

Alvarez, S., Rodriguez, P., Broetto, F. and Sanchez-Blanco, M. J. 2018. Long term responses and adaptive strategies of *Pistacia lentiscus* under moderate and severe deficit irrigation and salinity: osmotic and elastic adjustment, growth, ion uptake and photosynthetic activity. *Agricultural Water Management* 202, 253-62. doi:10.1016/j.agwat.2018.01.006.

Ao, N., Ma, J., Xu, T., Su, J., Yang, X., Guan, Z., Fang, W., Chen, F. and Zhang, F. 2019. Genetic variation and QTL mapping for cold tolerance in a chrysanthemum F1 population at different growth stages. *Euphytica* 215(5), 88. doi:10.1007/s10681-019-2412-7.

Arora, R. and Rowland, L. J. 2011. Physiological research on winter-hardiness: deacclimation resistance, reacclimation ability, photoprotection strategies, and a cold acclimation protocol design. *HortScience*, 46, 1070–1078. doi:10.21273/HORTSCI.46.8.1070

Ashrafzadeh, S. and Leung, D. M. W. 2015. *In vitro* breeding of heavy metal-resistant plants: a review. *Horticulture, Environment, and Biotechnology* 56(2), 131–6. doi:10.1007/s13580-015-0128-8.

Azadi, P., Bagheri, H., Nalousi, A. M., Nazari, F. and Chandler, S. F. 2016. Current status and biotechnological advances in genetic engineering of ornamental plants. *Biotechnology Advances* 34(6), 1073–90. doi:10.1016/j.biotechadv.2016.06.006.

Bao, F., Du, D., An, Y., Yang, W., Wang, J., Cheng, T. and Zhang, Q. 2017. Overexpression of Prunus mume dehydrin genes in tobacco enhances tolerance to cold and drought. *Frontiers in Plant Science* 8, 151. doi:10.3389/fpls.2017.00151.

Beck, T. 2013. Right plant, right place: biogeography and plant selection. In: Beck, T. (Ed.) *Principles of Ecological Landscape Design*. Island Press, Washington DC, pp. 1–5. doi:10.5822/978-1-61091-199-3_0

Bunty, S., Singh, V., Sheoran, R., Thakral, N. and Anu, N. 2018. Heat stress in plants. *Annals of Biology* 34, 70–8.

Cai, X. Y., Sun, Y. P., Starman, T., Hall, C. and Niu, G. H. 2014. Response of 18 earth-kind (R) rose cultivars to salt stress. *HortScience* 49, 544–9.

Cassaniti, C., Romano, D. and Flowers, T. J. 2012. The response of ornamental plants to saline irrigation water. In: Garcia-Garizabal, I and Abrahao, R. (Ed.) *Irrigation-Water Management, Pollution and Alternative Strategies*. InTech. doi:10.5772/31787

Cassaniti, C., Romano, D., Hop, M. E. C. M. and Flowers, T. J. 2013. Growing floricultural crops with brackish water. *Environmental and Experimental Botany* 92, 165–75. doi:10.1016/j.envexpbot.2012.08.006.

Chandler, S. F. and Brugliera, F. 2011. Genetic modification in floriculture. *Biotechnology Letters* 33(2), 207–14. doi:10.1007/s10529-010-0424-4.

Charrier, G., Ngao, J., Saudreau, M. and Ameglio, T. 2015. Effects of environmental factors and management practices on microclimate, winter physiology, and frost resistance in trees. *Frontiers in Plant Science* 6, 259. doi:10.3389/fpls.2015.00259.

Chen, J.-R., Lü, J.-J., Liu, R., Xiong, X.-Y., Wang, T.-X., Chen, S.-Y., Guo, L.-B. and Wang, H.-F. 2010. Dreb1C from Medicago truncatula enhances freezing tolerance in transgenic *M. truncatula* and China Rose (*Rosa chinensis* Jacq.). *Plant Growth Regulation* 60(3), 199–211. doi:10.1007/s10725-009-9434-4.

Chen, L., Chen, Y., Jiang, J., Chen, S., Chen, F., Guan, Z. and Fang, W. 2012. The constitutive expression of *Chrysanthemum dichrum* ICE1 in *Chrysanthemum grandiflorum* improves the level of low temperature, salinity and drought tolerance. *Plant Cell Reports* 31(9), 1747–58. doi:10.1007/s00299-012-1288-y.

Chen, J. R., Chen, Y. B., Ziemianska, M., Liu, R., Deng, Z. N., Niedzwiecka-Filipiak, I., Li, Y. L., Jiao, J. X. and Xiong, X. Y. 2016a. Co-expression of MtDREB1C and RcXET enhances stress tolerance of transgenic China Rose (*Rosa chinensis* Jacq.). *Journal of Plant Growth Regulation* 35(2), 586–99. doi:10.1007/s00344-015-9564-z.

Chen, W., Zhu, X., Han, W., Wu, Z. and Lai, Q. 2016b. Morphological, physiological and biochemical responses of Gerbera cultivars to heat stress. *Korean Journal of Horticultural Science and Technology* 34, 1–14. doi:10.12972/kjhst.20160018.

Chen, Y., Chen, Y., Shi, Z., Jin, Y., Sun, H., Xie, F. and Zhang, L. 2019. Biosynthesis and signal transduction of ABA, JA, and BRs in response to drought stress of Kentucky bluegrass. *International Journal of Molecular Sciences* 20(6), 1289. doi:10.3390/ijms20061289.

Cicevan, R., Al Hassan, M., Sestras, A. F., Prohens, J., Vicente, O., Sestras, R. E. and Boscaiu, M. 2016. Screening for drought tolerance in cultivars of the ornamental genus *Tagetes* (Asteraceae). *PeerJ* 4, e2133. doi:10.7717/peerj.2133.

Da Silva, J. A. T., Dobránszki, J., Zeng, S., Winarto, B., Lennon, A. M., Jaufeerally-Fakim, Y. and Christopher, D. A. 2015. Genetic transformation and molecular research in Anthurium: progress and prospects. *Plant Cell, Tissue and Organ Culture (PCTOC)* 123(2), 205–19. doi:10.1007/s11240-015-0832-1.

Demasi, S., Caser, M., Handa, T., Kobayashi, N., De Pascale, S. and Scariot, V. 2017. Adaptation to iron deficiency and high pH in evergreen azaleas (*Rhododendron* spp.): potential resources for breeding. *Euphytica* 213(7). doi:10.1007/s10681-017-1931-3.

De Oliveira, E. V., De Lacerda, C. F., Neves, A. L. R., Gheyi, H. R., Oliveira, D. R., De Oliveira, F. Í. F. and Viana, T. V. D. 2018. A new method to evaluate salt tolerance of ornamental plants. *Theoretical and Experimental Plant Physiology* 30(3), 173–80. doi:10.1007/s40626-018-0112-7.

Dhakar, S. 2017. Breeding for abiotic stress tolerance in ornamental crops: a review. *Chemical Science Review and Letters* 6, 1549–54.

Domenghini, J. C., Bremer, D. J., Fry, J. D. and Davis, G. L. 2013. Prolonged drought and recovery responses of Kentucky bluegrass and ornamental groundcovers. *HortScience* 48(9), 1209–15. doi:10.21273/HORTSCI.48.9.1209.

Du, X., Li, W., Sheng, L., Deng, Y., Wang, Y., Zhang, W., Yu, K., Jiang, J., Fang, W., Guan, Z., Chen, F. and Chen, S. 2018. Over-expression of chrysanthemum CmDREB6 enhanced tolerance of chrysanthemum to heat stress. *BMC Plant Biology* 18(1), 178. doi:10.1186/s12870-018-1400-8.

Fahad, S., Bajwa, A. A., Nazir, U., Anjum, S. A., Farooq, A., Zohaib, A., Sadia, S., Nasim, W., Adkins, S., Saud, S., Ihsan, M. Z., Alharby, H., Wu, C., Wang, D. and Huang, J. 2017. Crop production under drought and heat stress: plant responses and management options. *Frontiers in Plant Science* 8, 1147. doi:10.3389/fpls.2017.01147.

Fang, Y. and Xiong, L. 2015. General mechanisms of drought response and their application in drought resistance improvement in plants. *Cellular and Molecular Life Sciences* 72(4), 673–89. doi:10.1007/s00018-014-1767-0.

Farieri, E., Toscano, S., Ferrante, A. and Romano, D. 2016. Identification of ornamental shrubs tolerant to saline aerosol for coastal urban and peri-urban greening. *Urban Forestry and Urban Greenery* 18, 9–18. doi:10.1016/j.ufug.2016.02.014.

Ferdous, J., Hussain, S. S. and Shi, B. J. 2015. Role of micro RNA s in plant drought tolerance. *Plant Biotechnology Journal* 13(3), 293–305. doi:10.1111/pbi.12318.

Ferrante, A., Trivellini, A., Malorgio, F., Carmassi, G., Vernieri, P. and Serra, G. 2011. Effect of seawater aerosol on leaves of six plant species potentially useful for ornamental purposes in coastal areas. *Scientia Horticulturae* 128(3), 332–41. doi:10.1016/j.scienta.2011.01.008.

Franco, J. A., Bañón, S., Vicente, M. J., Miralles, J. and Martínez-Sánchez, J. J. 2011. Review Article: Root development in horticultural plants grown under abiotic stress conditions – a review. *The Journal of Horticultural Science and Biotechnology* 86(6), 543–56. doi:10.1080/14620316.2011.11512802.

Gajanayake, B., Trader, B. W., Reddy, K. R. and Harkess, R. L. 2011. Screening ornamental pepper cultivars for temperature tolerance using pollen and physiological parameters. *HortScience* 46(6), 878–84. doi:10.21273/HORTSCI.46.6.878.

Gao, J., Gu, M., Niu, G. and Chen, Y. 2012. Effects of Salinity on Three *Pennisetum* Cultivars. *Journal of Food, Agriculture and Environment*. 10(3–4), 1005–1007.

Gu, M., Robbins, J. A. and Rom, C. R. 2007a. Early landscape performance of 20 field-grown birch genotypes at two locations in Arkansas, US and response to irrigation. *Arboriculture and Urban Forestry* 33, 275–82.

Gu, M., Robbins, J. A. and Rom, C. R. 2007b. The role of ethylene in water-deficit stress responses in *Betula papyrifera* marsh. *HortScience* 42(6), 1392-5. doi:10.21273/HORTSCI.42.6.1392.

Guerra, D., Crosatti, C., Khoshro, H. H., Mastrangelo, A. M., Mica, E. and Mazzucotelli, E. 2015. Post-transcriptional and post-translational regulations of drought and heat response in plants: a spider's web of mechanisms. *Frontiers in Plant Science* 6, 57. doi:10.3389/fpls.2015.00057.

Hanin, M., Ebel, C., Ngom, M., Laplaze, L. and Masmoudi, K. 2016. New insights on plant salt tolerance mechanisms and their potential use for breeding. *Frontiers in Plant Science* 7, 1787. doi:10.3389/fpls.2016.01787.

Hasanuzzaman, M., Nahar, K., Alam, M. M., Roychowdhury, R. and Fujita, M. 2013. Physiological, biochemical, and molecular mechanisms of heat stress tolerance in plants. *International Journal of Molecular Sciences* 14(5), 9643–84. doi:10.3390/ijms14059643.

He, F., Wang, H. L., Li, H. G., Su, Y., Li, S., Yang, Y., Feng, C. H., Yin, W. and Xia, X. 2018. PeCHYR1, a ubiquitin E3 ligase from *Populus euphratica*, enhances drought tolerance via ABA-induced stomatal closure by ROS production in *Populus*. *Plant Biotechnology Journal*, 16, 1514–1528. doi:10.1111/pbi.12893.

Hong, E., Lim, C. W., Han, S. W. and Lee, S. C. 2017. Functional analysis of the pepper ethylene-responsive transcription factor, CaAIEF1, in enhanced ABA sensitivity and drought tolerance. *Frontiers in Plant Science* 8, 1407. doi:10.3389/fpls.2017.01407.

Hossain, Z., Mandal, A. K., Datta, S. K. and Biswas, A. K. 2006. Development of NaCl-tolerant strain in *Chrysanthemum morifolium* Ramat. through *in vitro* Mutagenesis. *Plant Biology* 8(4), 450–61. doi:10.1055/s-2006-923951.

Jaffar, M. A., Song, A., Faheem, M., Chen, S., Jiang, J., Liu, C., Fan, Q. and Chen, F. 2016. Involvement of CmWRKY10 in drought tolerance of chrysanthemum through the ABA-signaling pathway. *Int. J. Mol. Sci.* 17(5), 193–207. doi:10.3390/ijms17050693.

Jamshidnia, M., Ghobadi, C., Tabatabaei, B. E. S. and Yamchi, A. 2018. Over expression of p5cs gene to increase cold stress tolerance in Iranian species of *Petunia*. *Plant Tissue Culture and Biotechnology* 28(1), 35–44. doi:10.3329/ptcb.v28i1.37196.

Jiang, G., Jiang, X., Lü, P., Liu, J., Gao, J. and Zhang, C. 2014. The rose (*Rosa hybrida*) NAC transcription factor 3 gene, RhNAC3, involved in ABA signaling pathway both in rose and *Arabidopsis*. *PLoS ONE*, 9, e109415. doi:10.1371/journal.pone.0109415.g003.

Khan, M. I. R., Fatma, M., Per, T. S., Anjum, N. A. and Khan, N. A. 2015. Salicylic acid-induced abiotic stress tolerance and underlying mechanisms in plants. *Frontiers in Plant Science* 6, 462. doi:10.3389/fpls.2015.00462.

Körner, C., Basler, D., Hoch, G., Kollas, C., Lenz, A., Randin, C. F., Vitasse, Y. and Zimmermann, N. E. 2016. Where, why and how? Explaining the low-temperature range limits of temperate tree species. *Journal of Ecology* 104(4), 1076–88. doi:10.1111/1365-2745.12574.

Kozminska, A., Al Hassan, M., Wiszniewska, A., Hanus-Fajerska, E., Boscaiu, M. and Vicente, O. 2019. Responses of succulents to drought: comparative analysis of four *Sedum* (*Crassulaceae*) species. *Scientia Horticulturae* 243, 235–42. doi:10.1016/j.scienta.2018.08.028.

Krupa-Malkiewicz, M., Kosatka, A., Smolik, B. and Sedzik, M. 2017. Induced Mutations through EMS treatment and *in vitro* screening for salt tolerance plant of *Petunia x atkinsiana* D. Don. *Notulae Botanicae Horti Agrobotanici Cluj-Napoca* 45(1), 190–6. doi:10.15835/nbha45110578.

Kuligowska, K., Lütken, H. and Müller, R. 2016. Towards development of new ornamental plants: status and progress in wide hybridization. *Planta* 244(1), 1–17. doi:10.1007/s00425-016-2493-7.

Li, Q., Chen, J., McConnell, D. B. and Henny, R. J. 2007. A simple and effective method for quantifying leaf variegation. *HortTechnology* 17(3), 285–8. doi:10.21273/HORTTECH.17.3.285.

Li, Q. S., Chen, J. J., Stamps, R. H. and Parsons, L. R. 2008. Variation in chilling sensitivity among eight Dieffenbachia cultivars. *HortScience* 43(6), 1742-5. doi:10.21273/HORTSCI.43.6.1742.

Li, X. T., Liu, P., Yang, P. P., Fan, C. Z. and Sun, X. M. 2018. Characterization of the glycerol-3-phosphate acyltransferase gene and its real-time expression under cold stress in *Paeonia lactiflora* Pall. *PLoS ONE* 13(8), e0202168. doi:10.1371/journal.pone.0202168.

Li, P., Zheng, T., Li, L., Zhuo, X., Jiang, L., Wang, J., Cheng, T. and Zhang, Q. 2019. Identification and comparative analysis of the CIPK gene family and characterization of the cold stress response in the woody plant *Prunus mume*. *PeerJ* 7, e6847. doi:10.7717/peerj.6847.

Liu, Q. Q., Tang, J., Wang, W. L., Zhang, Y. X., Yuan, H. Y. and Huang, S. Z. 2018. Transcriptome analysis reveals complex response of the medicinal/ornamental halophyte *Iris halophila* Pall. to high environmental salinity. *Ecotoxicology and Environmental Safety* 165, 250–60. doi:10.1016/j.ecoenv.2018.09.003.

Liu, S., Chen, S., Chen, Y., Guan, Z., Yin, D. and Chen, F. 2011. In vitro induced tetraploid of Dendranthema nankingense (Nakai) Tzvel. shows an improved level of abiotic stress tolerance. *Scientia Horticulturae*, 127, 411–419. doi:10.1016/j.scienta.2010.10.012.

Lokhande, V. H., Gor, B. K., Desai, N. S., Nikam, T. D. and Suprasanna, P. 2013. *Sesuvium portulacastrum*, a plant for drought, salt stress, sand fixation, food and phytoremediation. A review. *Agronomy for Sustainable Development* 33(2), 329–48. doi:10.1007/s13593-012-0113-x.

Long, C., Chen, Z., Zhou, Y. and Long, B. 2018. The role of biodiversity and plant conservation for ornamental breeding. In: Van Huylenbroeck, J. (Ed.), *Ornamental Crops*. Handbook of Plant Breeding (vol. 11). Springer, Cham, Switzerland.

Luo, Z., Sun, O. J., Ge, Q., Xu, W. and Zheng, J. 2007. Phenological responses of plants to climate change in an urban environment. *Ecological Research* 22(3), 507–14. doi:10.1007/s11284-006-0044-6.

Mou, B. 2011. Improvement of horticultural crops for abiotic stress tolerance: an introduction. *HortScience* 46(8), 1068–9. doi:10.21273/HORTSCI.46.8.1068.

Naing, A. H., Lee, D. B., Ai, T. N., Lim, K. B. and Kim, C. K. 2018. Enhancement of low pH stress tolerance in anthocyanin-enriched transgenic petunia overexpressing RsMYB1 gene. *Frontiers in Plant Science* 9. doi:10.3389/fpls.2018.01124.

Niu, G. and Cabrera, R. I. 2010. Growth and physiological responses of landscape plants to saline water irrigation: a review. *HortScience* 45(11), 1605–9. doi:10.21273/HORTSCI.45.11.1605.

Niu, G. and Rodriguez, D. S. 2008. Responses of growth and ion uptake of four rose rootstocks to chloride- or sulfate-dominated salinity. *Journal of the American Society for Horticultural Science* 133(5), 663–9. doi:10.21273/JASHS.133.5.663.

Niu, G., Rodriguez, D. S. and Aguiniga, L. 2008. Effect of saline water irrigation on growth and physiological responses of three rose rootstocks. *HortScience* 43(5), 1479. doi:10.21273/HORTSCI.43.5.1479.

Niu, G., Rodriguez, D. S. and Gu, M. 2010. Salinity tolerance of *Sophora secundiflora* and *Cercis canadensis* var. *mexicana*. *HortScience* 45(3), 424–7. doi:10.21273/HORTSCI.45.3.424.

Noman, A., Aqeel, M., Deng, J., Khalid, N., Sanaullah, T. and Shuilin, H. 2017. Biotechnological advancements for improving floral attributes in ornamental plants. *Frontiers in Plant Science* 8, 530. doi:10.3389/fpls.2017.00530.

Oraee, A., Tehranifar, A., Nezami, A. and Shoor, M. 2018. Effects of drought stress on cold hardiness of non-acclimated viola (*Viola x wittrockiana* 'Iona Gold with Blotch') in controlled conditions. *Scientia Horticulturae* 238, 98–106. doi:10.1016/j.scienta.2018.04.027.

Ouyang, L., Leus, L. and Van Labeke, M. C. 2019. Three-year screening for cold hardiness of garden roses. *Scientia Horticulturae* 245, 12–8. doi:10.1016/j.scienta.2018.10.003.

Parmar, N., Singh, K. H., Sharma, D., Singh, L., Kumar, P., Nanjundan, J., Khan, Y. J., Chauhan, D. K. and Thakur, A. K. 2017. Genetic engineering strategies for biotic and abiotic stress tolerance and quality enhancement in horticultural crops: a comprehensive review. *3 Biotech*, 7, 239. doi:10.1007/s13205-017-0870-y.

Paule, J., Wagner, N. D., Weising, K. and Zizka, G. 2017. Ecological range shift in the polyploid members of *Fosterella* (Bromeliaceae). *Annals of Botany* 120(2), 233–43. doi:10.1093/aob/mcw245.

Peng, T., Guo, C., Yang, J., Xu, M., Zuo, J., Bao, M. and Zhang, J. 2016. Overexpression of a Mei (*Prunus mume*) CBF gene confers tolerance to freezing and oxidative stress in *Arabidopsis*. *Plant Cell, Tissue and Organ Culture (PCTOC)* 126(3), 373–85. doi:10.1007/s11240-016-1004-7.

Pillajo, J. O. Q., Chapin, L. J. and Jones, M. L. 2018. Senescence and abiotic stress induce expression of autophagy-related genes in petunia. *Journal of the American Society for Horticultural Science* 143(2), 154–63. doi:10.21273/JASHS04349-18.

Qi, Y., Liu, Y., Zhang, Z., Gao, J., Guan, Z., Fang, W., Chen, S., Chen, F. and Jiang, J. 2018a. The over-expression of a chrysanthemum gene encoding an RNA polymerase II CTD phosphatase-like 1 enzyme enhances tolerance to heat stress. *Horticulture Research* 5, 37. doi:10.1038/s41438-018-0037-y.

Qi, Y., Yao, X. Z., Zhao, D. G. and Lu, L. T. 2018b. Overexpression of SbSKIP, A Pre-mRNA splicing factor from *Sorghum bicolor*, enhances root growth and drought tolerance in *Petunia hybrida*. *Scientia Horticulturae* 240, 281–7. doi:10.1016/j.scienta.2018.06.056.

Roy, B., Noren, S. K., Mandal, A. B. and Basu, A. K. 2011. Genetic engineering for abiotic stress tolerance in agricultural crops. *Biotechnology (Faisalabad)* 10(1), 1–22. doi:10.3923/biotech.2011.1.22.

Shan, H., Chen, S., Jiang, J., Chen, F., Chen, Y., Gu, C., Li, P., Song, A., Zhu, X., Gao, H., Zhou, G., Li, T. and Yang, X. 2012. Heterologous expression of the chrysanthemum R2R3-MYB transcription factor CmMYB2 enhances drought and salinity tolerance, increases hypersensitivity to ABA and delays flowering in *Arabidopsis thaliana*. *Molecular Biotechnology* 51(2), 160–73. doi:10.1007/s12033-011-9451-1.

Singh, M., Kumar, J., Singh, S., Singh, V. P. and Prasad, S. M. 2015. Roles of osmoprotectants in improving salinity and drought tolerance in plants: a review. *Reviews in Environmental Science and Bio/Technology* 14(3), 407–26. doi:10.1007/s11157-015-9372-8.

Soto, A., Hernandez, L. and Quiles, M. J. 2014. High root temperature affects the tolerance to high light intensity in *Spathiphyllum* plants. Plant Sci, 227, 84–9. doi:10.1016/j.plantsci.2014.07.004.

Su, J., Zhang, F., Li, P., Guan, Z., Fang, W. and Chen, F. 2016. Genetic variation and association mapping of waterlogging tolerance in chrysanthemum. *Planta* 244(6), 1241–52. doi:10.1007/s00425-016-2583-6.

Sun, C.-Q., Chen, F.-D., Teng, N.-J., Liu, Z.-L., Fang, W.-M. and Hou, X.-L. 2010. Interspecific hybrids between *Chrysanthemum grandiflorum* (Ramat.) Kitamura and *C. indicum* (L.) Des Moul. and their drought tolerance evaluation. *Euphytica* 174(1), 51–60. doi:10.1007/s10681-009-0117-z.

Sun, Y., Niu, G. and Perez, C. 2015. Relative salt tolerance of seven Texas Superstar® perennials. *HortScience* 50(10), 1562–6. doi:10.21273/HORTSCI.50.10.1562.

Susko, A. Q., Rinehart, T. A., Bradeen, J. M. and Hokanson, S. C. 2018. An evaluation of two seedling phenotyping protocols to assess pH adaptability in deciduous azalea (*Rhododendron* sect. *Pentanthera G.* Don). *HortScience* 53(3), 268–74. doi:10.21273/HORTSCI12520-17.

Tang, J., Liu, Q. Q., Yuan, H. Y., Zhang, Y. X., Wang, W. L. and Huang, S. Z. 2018. Molecular cloning and characterization of a novel salt-specific responsive WRKY transcription factor gene IlWRKY2 from the halophyte *Iris lactea var. chinensis*. Genes & Genomics, 40, 893-903. doi:10.1007/s13258-018-0698-9.

Tian, Z. G., Yang, Y. and Wang, F. 2015. A comprehensive evaluation of heat tolerance in nine cultivars of marigold. *Horticulture, Environment, and Biotechnology* 56(6), 749–55. doi:10.1007/s13580-015-0077-2.

Tian, X. M., Wang, Z. Y., Zhang, Q., Ci, H. C., Wang, P. S., Yu, L. and Jia, G. X. 2018. Genome-wide transcriptome analysis of the salt stress tolerance mechanism in *Rosa chinensis*. *PLoS ONE* 13(7), e0200938. doi:10.1371/journal.pone.0200938.

Toscano, S., Scuderi, D., Giuffrida, F. and Romano, D. 2014. Responses of Mediterranean ornamental shrubs to drought stress and recovery. *Scientia Horticulturae* 178, 145-53. doi:10.1016/j.scienta.2014.08.014.

Toscano, S., Ferrante, A. and Romano, D. 2019. Response of Mediterranean ornamental plants to drought stress. *Horticulturae* 5(1), 6. doi:10.3390/horticulturae5010006.

Valverde, R. A., Sabanadzovic, S. and Hammond, J. 2012. Viruses that enhance the aesthetics of some ornamental plants: beauty or beast? *Plant Disease* 96(5), 600-11. doi:10.1094/PDIS-11-11-0928-FE.

Vandenbussche, M., Chambrier, P., Rodrigues Bento, S. and Morel, P. 2016. Petunia, your next supermodel? *Frontiers in Plant Science* 7, 72. doi:10.3389/fpls.2016.00072.

Van Der Veken, S., Hermy, M., Vellend, M., Knapen, A. and Verheyen, K. 2008. Garden plants get a head start on climate change. *Frontiers in Ecology and the Environment* 6(4), 212-6. doi:10.1890/070063.

Van Laere, K., FrançA, S. C., Vansteenkiste, H., Van Huylenbroeck, J., Steppe, K. and Van Labeke, M.-C. 2011. Influence of ploidy level on morphology, growth and drought susceptibility in *Spathiphyllum wallisii*. *Acta Physiologiae Plantarum* 33, 1149-56.

Wan, X., Yang, J., Guo, C., Bao, M. and Zhang, J. 2019. Genome-wide identification and classification of the Hsf and sHsp gene families in *Prunus mume*, and transcriptional analysis under heat stress. *PeerJ* 7, e7312. doi:10.7717/peerj.7312.

Wang, Y., Yan, A., Dai, J., Wang, N. and Wu, D. 2012. Accumulation and tolerance characteristics of cadmium in *Chlorophytum comosum*: a popular ornamental plant and potential Cd hyperaccumulator. *Environmental Monitoring and Assessment* 184(2), 929-37. doi:10.1007/s10661-011-2010-1.

Wang, Y. X., Gu, C. H., Bai, S. B., Sun, Z. B., Zhu, T. T., Zhu, X. D., Grit, D. H. and Tembrock, L. R. 2016. Cadmium accumulation and tolerance of *Lagerstroemia indica* and *Lagerstroemia fauriei* (Lythraceae) seedlings for phytoremediation applications. *International Journal of Phytoremediation* 18(11), 1104-12. doi:10.1080/15226514.2016.1183581.

Wang, K., Bai, Z. Y., Liang, Q. Y., Liu, Q. L., Zhang, L., Pan, Y. Z., Liu, G. L., Jiang, B. B., Zhang, F. and Jia, Y. 2018. Transcriptome analysis of chrysanthemum (*Dendranthema grandiflorum*) in response to low temperature stress. *BMC Genomics* 19(1), 319. doi:10.1186/s12864-018-4706-x.

Wong, J., Mudd, E. A., Hayes, A. and Day, A. 2019. The chloroplast genome sequence of the ornamental plant *Petunia hybrida*. *Mitochondrial DNA Part B* 4(1), 249-50. doi:10.1080/23802359.2018.1547136.

Wu, S., Sun, Y. and Niu, G. 2016a. Morphological and physiological responses of nine ornamental species to saline irrigation water. *HortScience* 51(3), 285-90. doi:10.21273/HORTSCI.51.3.285.

Wu, S., Sun, Y., Niu, G., Pantoja, G. L. G. and Rocha, A. C. 2016b. Responses of six Lamiaceae landscape species to saline water irrigation. *Journal of Environmental Horticulture* 34, 30-5.

Wu, Y., Müller, M., Bai, T., Yao, S., Gailing, O. and Liu, Z. 2019. Transcriptome profiling in *Camellia japonica* var. *decumbens* for the discovery of genes involved in chilling tolerance under cold stress. *Annals of Forest Research* 62(1), 51-68. doi:10.15287/afr.2018.1311.

Xu, Y., Gao, S., Yang, Y., Huang, M., Cheng, L., Wei, Q., Fei, Z., Gao, J. and Hong, B. 2013. Transcriptome sequencing and whole genome expression profiling of chrysanthemum under dehydration stress. *BMC Genomics* 14, 662. doi:10.1186/1471-2164-14-662.

Yu, C. W., Qiao, G. R., Qiu, W. M., Yu, D. B., Zhou, S. R., Shen, Y., Yu, G. C., Jiang, J., Han, X. J., Liu, M. Y., Zhang, L. S., Chen, F., Chen, Y. C. and Zhuo, R. Y. 2018. Molecular breeding of water lily: engineering cold stress tolerance into tropical water lily. *Horticulture Research* 5, 73. doi:10.1038/s41438-018-0086-2.

Zhang, B., Yang, X., Yang, C., Li, M. and Guo, Y. 2016. Exploiting the CRISPR/Cas9 system for targeted genome mutagenesis in petunia. *Scientific Reports* 6, 20315. doi:10.1038/srep20315.

Zhang, Z., Luan, D., Chen, G., Xiao, F., Yin, D., Ni, D., Li, X. and Qi, L. 2018. Isolation and characterization of a waterlogging-responsive gene involved in ethylene biosynthesis in chrysanthemum. *Acta Physiologiae Plantarum* 40(6), 100. doi:10.1007/s11738-018-2667-y.

Zhao, N., Li, C. W., Yan, Y. J., Cao, W., Song, A. P., Wang, H. B., Chen, S. M., Jiang, J. F. and Chen, F. D. 2018. Comparative transcriptome analysis of waterlogging-sensitive and waterlogging-tolerant *Chrysanthemum morifolium* Cultivars under waterlogging stress and reoxygenation conditions. *International Journal of Molecular Sciences* 19(5), 21. doi:10.3390/ijms19051455.

# Chapter 8

## Improving nutrient management in the cultivation of ornamental plants in greenhouse, container and field production

*John Majsztrik, Clemson University, USA; and James S. Owen Jr., USDA-ARS, USA*



### 1 Introduction

Historically, plants have provided food, fibre and medicine since the beginning of our civilization (Zhou 1994). As food production has largely been outsourced to specialist producers, particularly in more industrialized nations, many people interact with plants in their environment only through ornamental plants in their urban or residential landscapes. The production of ornamental plants, or plants used for their aesthetic purposes, is a big business. Ornamental production is a leading agricultural commodity in the United States, accounting for $16.5 billion of yearly sales (8.5% of total sales), and is ranked fifth for plant commodities (US Department of Agriculture 2019). The top five US states for ornamental production in 2017 were California, Florida, Michigan, Texas and North Carolina and accounted for over $3.0 billion in sales (US Department

http://dx.doi.org/10.19103/AS.2020.0066.11

of Agriculture 2019). Globally, ornamental plant purchases are estimated to reach $57.4 billion by 2024 (https://www.dccchina.org/2019/02/ornamental-plants-chinese-importers/). These figures do not include add-on services such as plant installation, maintenance, and residential and commercial irrigation systems which add to the overall value of this industry. Taken together, the ornamental industry has a large impact on the global economy.

Proper nutrition in containerized and field grown plants is essential for maximized growth and profitability. Macronutrients and micronutrients along with substrate or soil amendments ensure that plants remain healthy and continue to grow optimally in both mineral soils and soil-less substrates. Nutrient management coupled with precise irrigation are key factors in growing healthy high-quality plants for the ornamental plant industry. The following topics will be discussed in this chapter: irrigation and water quality concerns for growers; the different types of fertilizer and when they should be used; the impact of soils and substrates on fertilizers; and the impact of production systems on fertilizer longevity.

Several factors can influence the production of ornamental plants including, but not limited to, laws and political pressure, climate variability, access to markets, availability of raw and processed materials (i.e. peat and bark used to make substrate mixes, petroleum for containers, and minerals and petroleum to manufacture fertilizers), labour availability and cost, and water availability to name just a few. Many of these factors are for the most part out of the control of the nurseries producing the plants, but there are few key factors that growers can and should control, including optimizing irrigation and fertilizer efficiency in their operation, which are the focus of this chapter.

Environmental horticulture, including nursery and greenhouse production, occurs at varying levels of infrastructure, based on the production method and schedule. This mainly has to do with the intensity of production (person hours per unit area per week) which impacts how many people are needed for a given size production facility. Greenhouses have the highest production intensity, followed by container, bare root and field production. For example, a 4-acre greenhouse typically employs a higher number of workers than a 4-acre outdoor container operation, with a 4-acre field operation requiring the fewest number of workers. This does not take such factors as the type of plants and containers, automation, worker efficiency and so on into account.

Plants can be grouped based on their production needs, such as irrigation requirements, plant or container size, pest and disease problems, and nutrient use. Every operation groups plants differently, with no one right or wrong way to do so. In some instances, grouping by plant nutrient requirement (low, medium or high) makes sense, particularly for liquid feed (fertigation) scenarios, and a variety of factors should be taken into account when grouping plants whenever possible.

## 2 Irrigation and water quality

Since ornamental plants are typically multi-week to multi-year crops, having consistent access to high-quality water is critical. Irrigation water quality and quantity are two very important factors not only for current production, but also for future growth. Water quality refers to water pH, buffering capacity, soluble salts [electrical conductivity (EC)] and the presence of pathogens (plant diseases) or pesticide residues, all of which can negatively impact plant growth. Water quantity refers to the amount of acceptable quality water that is available for production of plants including irrigation, pesticide and soluble-fertilizer mixing, worker safety and cooling. An important aspect of water quantity is having adequate water availability during droughts or other long-term events such as a malfunctioning pump or a damaged water reservoir. Being water limited, either by municipal pumping restrictions, low-water quality, limited well production or limited storage capacity for droughts can have a negative impact on plant production. An extreme example of water scarcity is having to bring in water by truck to save the plants under production during a drought or other low-water situations.

An important aspect of water quality is water pH which is critical for plant growth, disinfectant effectiveness (i.e. chlorine) and for efficacy of some agrochemicals. Potential (aka Power) hydrogen (pH) indicates the concentration of $H^+$ ions that are measured on a negative logarithmic scale from 0 to 14. A pH of 7 is neutral with values greater than 7 being basic (i.e. alkaline) and below 7 being acidic. Each 1-unit change results in a 10-fold change in the number of $H^+$ ions. There is an equal and opposite effect on hydroxide ($OH^-$) ions (pOH).

Increasing the temperature results in lower pH in pure water (6.14 at 100°C), whereas higher pH will be observed at lower temperatures (7.47 at 0°C) when compared to ambient pH 7.0 at 25°C. Root zone pH has an impact on plant-nutrient uptake as can be seen in Fig. 1. Soil, soil-less substrate and irrigation water should be adjusted to a pH between 5.5 and 7.0 for optimal growth for most plants (Peterson 1982; Marschner et al. 1987). If the pH is outside of those ranges, symptoms of nutrient deficiency can be seen even if nutrients are present in the soil (Argo 2003). There are some plants that have a preferred pH range outside of 5.5–7.0 such as rhododendrons, azaleas and many bog plants. Additionally, water temperature should be neither too hot nor cold with additional consideration for temperature when producing crops susceptible to developing leaf spot such as foliage plants (Robbins 2018).

If bicarbonate ($HCO_3^-$) or carbonate ($CO_3^{-2}$) ions are present in water (collectively called alkalinity) they increase the ability of water to stay at a near neutral or alkaline pH, which is called buffering capacity. At higher alkalinity levels, calcium (Ca) and magnesium (Mg) can become insoluble and precipitate, which may lead to emitter clogging in drip and micro-irrigation

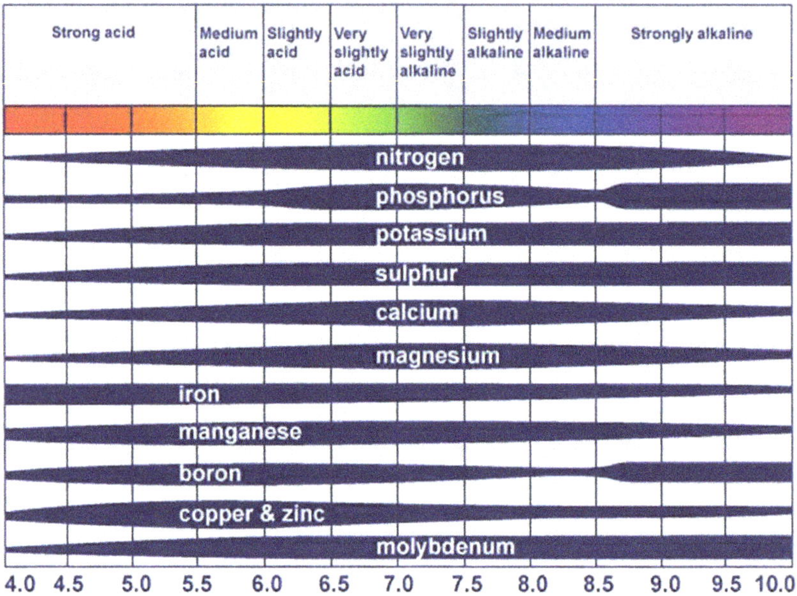


**Figure 1** Nutrient availability by pH for roots in mineral soils. Source: adapted from Roques et al. (2013) used with permission.

systems. The pH of unbuffered or low-alkalinity irrigation water will have little effect on soil or substrate pH, unless it is very acidic as a result of acid injection before application. Alkaline water however will act similar to lime, potentially increasing the pH of the substrate over time. Raising soil/substrate pH could have a negative effect on nutrient uptake if not adequately controlled.

The source of water is a major factor in water quality. Some sources such as deep wells typically have consistent high-quality water that needs no additional treatment before use. However, problems can arise in deep wells if iron (Fe), boron (B) or sodium (Na) concentrations are high. A pH less than 5 and low oxygen levels increase the Fe salts dissolved in groundwater (Anon. 2013). The presence of iron-fixing bacteria allows Fe to remain soluble and mobile in irrigation water, which does not harm plants, but may stain production areas or leaves a blue tint once it leaves the emitter and is oxidized.

The total soluble salts should be monitored regularly through routine testing. Table 1 provides general recommendations for how often different water sources should be tested. Sodium occurring naturally or as a result of salt water intrusion can affect plant water availability or Ca and Mg uptake. High hardness can potentially lead to clogged emitters, as previously stated, or indicate high alkalinity. The sodium adsorption ratio (SAR) is a diagnostic tool that takes into account the inter-dependent relationship of Na, Ca and Mg, helping identify potential issues with alkalinity or Na levels. Well water may have other minerals

**Table 1** Minimum recommended frequency of source water testing for various types of water sources. It is recommended that you test for pH, electrical conductivity (EC) and alkalinity. Testing for pathogens for any type of recycled or surface water is also recommended.

| Water source | Testing frequency (minimum frequency) |
|---|---|
| Deep well | Annually |
| Shallow well | 4 times a year |
| Surface water | 4 times a year |
| Recycling pond | Monthly |
| Closed loop (ebb and flood) | Weekly |

including, but not limited to, chloride, B or fluoride that can cause toxicity and other issues, but these problems are not widespread. Continued water quality analysis from a reputable lab can help diagnose problems as they arise.

Soluble salts can impact the resale value of plants due to staining, residue deposits or foliar damage, without necessarily impacting plant quality. For instance, the common abiotic condition of iron staining in plants occurs when the reduced iron form (ferrous, $Fe^{+2}$) of acidic groundwater is oxidized in the presence of air to an insoluble (ferric $Fe^{+3}$) form on the surface of overhead-irrigated plants (Zinati and Shuai 2005). Other salts, like calcium bicarbonate, can leave unsightly white films. The majority of times these do not affect plant growth and development. However, some minerals can cause foliar damage, especially in arid climates where rapidly evaporated water facilitates foliar accumulation.

Surface water sources include streams, rivers or tail-water collection reservoirs. Surface waters are influenced by seasonal variations, and water quality is dependent on upstream users and non-point source contaminants. Thus, pH, total soluble salts or (EC), agrichemicals, pathogens, and temperature flux should be monitored. Streams or rivers should be monitored both when entering and exiting the property or at the point of water withdrawal. The degree of irrigation source variability may result in a need for additional monitoring and equipment to sanitize water, control pH or blend water sources to increase water quality.

Recycled tail-water (i.e. runoff), water from reservoirs, may be laden with mineral nutrients. These recycled nutrients should be considered as part of the fertility programmes during production, although they may be removed by chlorine (Cl) or other disinfectants (Copes et al. 2004; Meador and Fisher 2013). Salt concentrations should be continuously monitored via inline EC measurements and with routine water quality analysis identifying specific mineral nutrients and their concentration.

Higher concentrations of nitrogen (N) and phosphorus (P) in reservoirs make them more prone to algae and subsequent diurnal pH swings due to carbon dioxide concentrations. During the day, algae remove carbon dioxide from the water and produce oxygen, increasing the pH due to the removal of carbon dioxide/carbonic acid from the water. During the night, algae continue to undergo cellular respiration and produce carbon dioxide, which is turned into carbonic acid, and reduces water pH (Picot et al. 1993). This is further affected by the volume:surface area ratio of the reservoir, water buffering capacity, concentration and type of algae, and temperature. Algae can clog filtration equipment, reduce nutrient concentrations, and cause downstream effects such as fish kills and anoxic conditions in all surface waters. Additional concerns with recycling excess fertilizers are their removal during filtration (mainly sediment-bound P) and their reaction and degradation with chlorine and other sanitizing agents, which both removes the nutrients and reduces the effectiveness of the sanitizer.

Recycled water from reservoirs can become a means of pathogen dispersal, which is why sanitation is always recommended for application of surface waters to plants. The concentration of plant pathogens can be reduced by maximizing the distance water travels before the irrigation intake; for instance, ensuring the tail-water travels through vegetative buffers or settling ponds before entering the reservoir and placing the irrigation intake furthest from the returning water. All of these factors can impact not only the type of plants that recycled water can be used on (pH or salt-sensitive, propagation vs. rooted cutting vs. established plant), but also what treatment (chlorination, acid injection, dilution, filtering) is necessary before application. It is recommended that you test irrigation water sources at a minimum frequency as stated in Table 1 for pathogens and other parameters.

## 3 Types of production systems

### 3.1 Open air nursery

We define open air nurseries as operations that are growing containerized plants which do not have any climate modification. The only exception for open air production and climate modification are: (1) hoophouses for overwintering containerized plants and (2) shade structures for plants that cannot be grown in full sun. Plants that are grown in open air nurseries are typically irrigated by both rainfall and irrigation.

Rainfall is beneficial in that it saves the grower from having to irrigate when sufficient rainfall is received, and it can refill water-storage structures onsite (cisterns, reservoirs etc.). This is particularly helpful in the western United States, where municipal water tends to have high levels of salt (measured as electrical conductivity or EC) which can cause plant damage if it is not diluted with low EC water. The major disadvantage of rainfall is that it can cause nutrient leaching and pesticide runoff. Nutrients can be quickly leached out of the container since there is limited ion-exchange capacity (Argo 1998; Broschat 2005; Silber 2019) and water storage in the substrate. These fertilizers, particularly N and P, can then move in runoff water or sediment into surface waters or leach into groundwater. This is less of an issue with controlled-release fertilizer (CRF) since only a small amount of fertilizer is available from the prills at any time (Hoskins et al. 2014).

Open air nurseries typically irrigate via overhead (impact or spray), spray stake, or drip systems. Plant materials grown in container sizes under 19 L (5 gallon) are typically overhead irrigated (see https://www.cleanwater3.org/growertools.asp for an interception efficiency calculator). Larger container sizes are more efficiently irrigated with spray stakes or drip emitters, due to the large amount of space between containers. Water is wasted when it lands between containers (Fig. 2). However, overhead irrigation allows for a wide variety of material (container sizes, different plant heights etc.) to be irrigated with the same infrastructure, with the major disadvantage being varying interception efficiency or percentage of water that is received by the plant depending on container size and spacing. Drip irrigation and spray stake irrigation are highly efficient, but much less flexible. There is the added problem of required labor to ensure that each spray stake is working correctly to avoid plant death and minimize leaching.

One of the major differences between open air nurseries and greenhouse and field production is that plants in open air nurseries are typically overwintered under protective cover (Mathers 2003) in temperate regions that have cold winters. Overwintering protective structures are typically unheated, or in colder areas heated to maintain a minimum temperature. The structure or covering provides protection for the plant's root and shoots. As

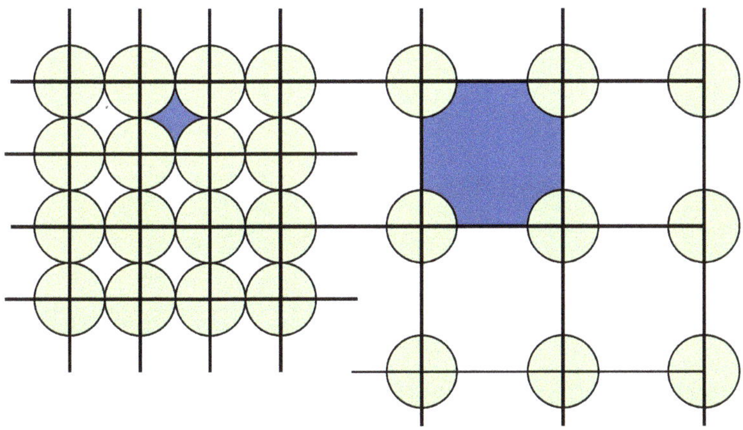

**Figure 2** Interception efficiency of overhead irrigation using two container spacings. Container tight spacing (left) results in high interception efficiency, while the wider container spacing (right) results in low interception efficiency. It is recommended that wider container spacings (typically above 19 L (5 gallon) containers) be irrigated with spray or drip irrigation to maximize water-use efficiency.

the plants go into dormancy in the fall, they are placed, typically can-tight, into a protective structure. Plants will typically be overhead irrigated as needed within the structure using sprinklers that run along the centre of the structure. Hoop houses covered in white or clear plastic can be used to 'force' or aid in breaking plant dormancy early while remaining protected from frost or damaging temperatures. The ability to force plants to break dormancy more quickly compared with passively covered or field-grown material of the same species, can allow growers to reach markets earlier, particularly if they are to be shipped to warmer climates or to promote flowering.

### 3.2 Pot-in-pot production

A special type of container production is pot-in-pot production where containerized plants are grown in the ground. The system consists of a socket pot which is a plastic container that is permanently installed in the ground and used to nest a second container or pot containing the substrate and plant (Fig. 3). Pot-in-pot production requires capital costs for extensive field renovation to ensure adequate drainage and water conveyance. This is accomplished by precision grading of the site and using gravel and corrugated pipes installed below the socket pots to minimize standing water in and around the containers. Therefore, this production type is typically limited to woody plants produced in large containers.

Pot-in-pot production provides a unique environment and multiple opportunities when compared to traditional above ground, open-air container production of woody ornamentals. Plants' roots in the ground remain cooler in the summer and warmer in the winter (London et al. 1998). Drip or spray stake irrigation is often installed so plants are irrigated efficiently. This includes being able to supplement with liquid fertilizer during times of rapid growth or when CRF has been exhausted, which is typical later in the growing season. The

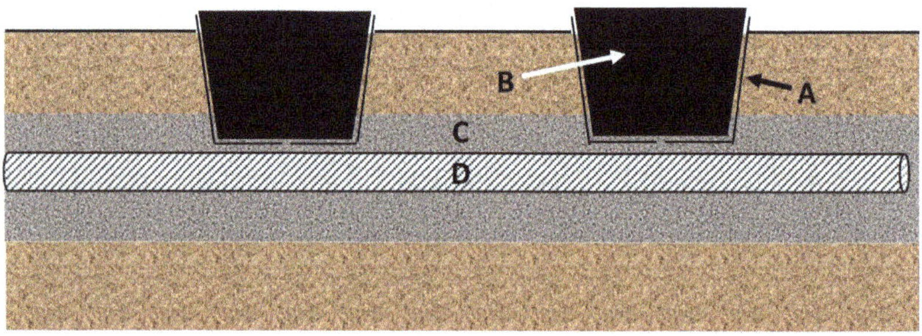


**Figure 3** Schematic design of a pot-in-pot container operation, which is a type of container production where plants are grown in the ground instead of above ground. This system consists of a socket pot (A), plant container (B), along with gravel (C) and a corrugated pipe (D) for drainage.

potential for tip over in tall plant materials is significantly reduced or eliminated in pot-in-pot systems. Additionally, it has been shown that pot-in-pot production can produce saleable plants more quickly than container production or field production (Adrian et al. 1998).

The disadvantages to pot-in-pot are that the production area is taken up long term and only one size containers [i.e. 56 L (15)] can be used. These limitations restrict changes to the product mix in the future without extensive renovation. Similarly, since distances between containers are fixed, there is the potential for wasted space when either small material is kept in a pot-in-pot area for several years until it reaches a saleable size or more compact material is grown. Both of these scenarios would decrease the benefits of pot-in-pot production.

### 3.3 Greenhouse production

Greenhouse production in many ways is similar to open air container production. Plants are grown in containers with a soil-less substrate and irrigated and fertilized to maintain maximum growth. Greenhouse operations utilize a controlled environment that can be actively heated and to some degree cooled. Rainfall is commonly excluded from plants, however it may be harvested and used to extend water supply. The more controlled greenhouse environment is beneficial for allowing more refined production techniques that limit leaching of water and recirculated nutrients.

Most greenhouse operations deliver water-soluble mineral nutrients coupled with irrigation (i.e. fertigation) using micro-, drip- or sub-irrigation. Liquid fertilizers have a number of advantages compared to CRF for greenhouse production. Producing herbaceous and other short-rotation crops with injected fertilizers allows growers the flexibility to quickly adjust fertilizer rates (e.g. 100 vs. 200 mg/L) or type (e.g. ammoniacal vs. nitrate N), or eliminate (i.e. clear water) fertilizer as needed (Kläring 2001; Bergstrand 2017). This is not possible with CRF which continuously releases nutrients over a period of weeks or months once placed in the container. This gives greenhouse operations a higher degree of control to speed up or slow down growth for these short-term crops. Different types of fertilizer are often used depending on the time of year, pH of the water, species being grown and growth stage. Multiple solutions with different rates or types can be used either on the same production area or in different areas depending on crop needs.

A common concern with greenhouse fertigation due to the absence of flushing rains, low leaching fractions, and higher fertilizer levels are a build-up of salt (i.e. EC) over time in the container. Salt stress can lead to slowed plant growth, tissue damage, or death. Damage from salts is particularly problematic when plants are produced 'dry' where substrates are maintained at a relatively low moisture level. As water is removed from the container through evapotranspiration, the salts become more concentrated which make it more difficult for the plants to remove water via osmosis, and can lead to wilting. Salt or EC levels in the substrate should be monitored using the pour-thru method or similar non-destructive sampling technique (Cavins et al. 2000; Bilderback 2001). Electrical conductivity in containers can be controlled by leaching out excess salts with clear water, optimizing the fertility regime or irrigating more frequently. On the other hand, more frequent irrigation can lead to disease or insect problems, be wasteful, and push or leach nutrients out of the container. Some work has been done using irrigation sensor networks that have showed good promise for limiting leaching, while producing high-quality plants (Lea-Cox et al. 2013).

Plant growth regulators (PGRs), which are synthetic plant hormones, have been used for decades to control plant growth in select plants such as poinsettia. These PGRs are very effective even at very low concentrations. The major concern with their use is that if water is recycled at an operation, the PGRs may decrease growth of non-target plants. It has been found that activated carbon (activated charcoal) is effective in removing PGRs from runoff water so it can be more effectively recycled (Grant et al. 2018). A newer technique, precise control of deficit irrigation using sensor networks, has been shown to be just as effective as PGRs in regulating plant height over time of certain taxa (Alem et al. 2015).

### 3.4 Field production

Field production of ornamental plants is the oldest type of production system. Field-grown ornamental materials are based in mineral soils, and plants are typically grown in good-quality loam-containing soil. Unlike container production which does not interact with soils, field producers cannot easily change their soils, so soil characteristics are a major consideration for suitability of production. Irrigation in field production is typically applied in the first few years to supplement rainfall until plants are established and may be applied after establishment during drought conditions. Fertilizers, with the exception of Nitrogen (N) and sulphur (S), are applied before planting based on soil test(s) as a broadcast application and incorporated as needed. Nitrogen and S are applied during production as either a side-banded dry fertilizer or as water-soluble fertilizer applied through drip irrigation (i.e. fertigation). The land associated with field production is less expensive due to less capital inlay. Seedlings and liners can be produced at relatively high-density plantings, because they are short-term crops grown for two years or less. On the other hand, field production is also used to produce large diameter trees or larger shrubs for upscale homeowner or commercial markets compared with container production. This requires more space and time for trees to reach saleable size.

Major benefits of field production are the ability to plant high-density liner beds and the nutrient and water buffering capacity of soil requiring less water and mineral nutrient inputs. Once the plant material reaches saleable size, they can be held there for an extended period of time by reducing irrigation and fertilization inputs to lengthen their market window. There is also no risk of blow-over once plants have sufficiently rooted, which is an issue in the production of large containerized trees and shrubs. There are several drawbacks of field production including the large amount of land required to produce trees and shrubs compared with greenhouse and container production. Larger trees and shrubs are typically planted at final spacing (unless alternating plants are harvested at a smaller size), harvest is limited to specific times of the year, and specialized equipment is needed for harvest and moving trees and shrubs during harvest and often through the final installation for larger specimens. Additionally, open spaces exist for several years where weeds must be controlled, and fertilizer and water are wasted until the plant roots have colonized those spaces and the ground is shaded by the canopy, and again when trees are harvested in a row. During plant establishment sediment runoff needs to be controlled to minimize soil loss.

## 4 Mineral nutrients

The elements that plants need for survival are acquired in two different ways - primarily through the roots (fertilizers) and also via stomates (carbon, oxygen and hydrogen). Nitrogen is found at the highest levels in plants after C, H and O. Nitrogen is an essential element for making chlorophyll, as well as amino acids, which are the building blocks of all proteins. It is essential for making ATP (adenosine triphosphate) which is a way of storing energy for cells and DNA for cellular reproduction.

Nitrogen compounds include ammonium ($NH_4^+$) ions, nitrate ($NO_3^-$) ions and urea ($CH_4N_2O$). Nitrogen fertilizers are typically applied as urea ($NH_2 \rightarrow NH_3 \rightarrow NH_4$), ammonium, or nitrate salts. In soil-based systems, 95–99% of the N in the soil is in organic forms such as plant material, organic matter, or in living soil microorganisms such as bacteria. This organic N is not directly available to plants, but some can be converted to available forms by specific microorganisms. A very small amount of organic N may exist in soluble organic compounds, such as urea, that may be slightly available to plants. The majority of plant-available nitrogen (also called mineral nitrogen) is in the dissociated inorganic forms of ammonium and nitrate. These ions can bind to cation (+) and to a much lesser degree anion (−) exchange sites in soils. Nitrate ions are dissolved in soil water or precipitated as soluble salts under dry conditions.

Nitrate is very mobile in both soil-less substrates and mineral soils because of low anion exchange capacity. Nitrate moves through the soil or substrates and with surface water much more quickly than either phosphorus or potassium. Because it is so mobile, N typically needs to be applied more frequently than P or K. Nitrogen also undergoes denitrification by microorganisms under anoxic (low oxygen) conditions in both soils and soil-less substrates where N is used as an electron acceptor instead of oxygen (Martin et al. 1999; Amha and Bohne 2011). Although denitrification reduces N runoff, it can also deplete N in the root zone, particularly in soils or substrates that remain water saturated.

Phosphorus (P) is required in plants for a number of essential functions including production of ATP (energy storage), respiration, photosynthesis, RNA and DNA synthesis, and cell division. Phosphorus is available to plants as two forms of orthophosphate, which are pH dependent: either as primary orthophosphate ($H_2PO_4^-$) at pH < 7.0 or as secondary orthophosphate ion ($HPO_4^{-2}$) at pH > 8.0. Between pH 7 and 8 both forms exist. In addition to synthetic P fertilizers, P is mined and applied as calcium phosphate or refined to make other mineral nutrient salts such as mono-ammonium phosphate (MAP; $NH_4H_2PO_4$).

Phosphorus (P) is not typically found dissolved in water in soil, but instead is bound to clay and other minerals with high anion exchange capacities in soil systems. However, P, dominantly present as an oxyanion ($PO_4^{-3}$), remains mobile or partially mobile in soil-less culture as dictated by physical and chemical amendments that facilitate precipitation or formation of P-complexes (Shreckhise et al. 2019). Once leached from the container, P typically moves bound to sediment. As plants take up P from the soil solution, a relatively small amount of labile phosphorus replaces depleted solution P. The environmental concern for P is that it is typically a limiting nutrient in most freshwater systems (Hansen et al. 2002; Majsztrik and Lea-Cox 2013). The addition of P from sediment runoff causes algal blooms, which can lead to anoxic conditions when the algae die en masse. Bound P is difficult to remove from biological systems, and work is underway to reduce P loading in receiving waters through a number of different means including plant uptake and adsorption of P before it leaves the site (Majsztrik et al. 2017).

Potassium (K) is the last of the three macronutrients that plants need in relatively high amounts. Potassium is used for enzyme activation, movement of water, nutrients and carbohydrates in the plant, ATP production and maintaining turgor pressure. Like P, most K found in soils is not plant available. As plants and other organisms in the soil use labile K, more becomes available to keep the soil solution system in equilibrium. Potassium is less of a problem for algal blooms and negative impacts on surface waters since typically it is not a limiting nutrient in freshwater systems.

Secondary nutrients which include Mg, Ca and S are essential to crop growth but are commonly added with soil or substrate amendments or as the balancing salt in macronutrient or micronutrient fertilizers. Micronutrients, including B, Cl, copper (Cu), Fe, manganese (Mn), molybdenum (Mo), nickel (Ni) and zinc (Zn), are necessary for plant growth but are needed in smaller amounts (1–10% of macronutrients). Positively charged cations (Ca, Cu, Fe, Mn, Mg, S, and Zn) will readily bind to soil particles that have cation exchange capacity sites. Cations tend to be deficient in calcareous soils and soils with high organic matter where chelation may occur. Anions (S, B, Cl, Mo) will quickly leach out of the plant root zone, similar to N, and need to be consistently supplied for sustained growth either as fertilizer or through decomposing organic matter. It is important to reiterate that soil-less substrates have limited anion and cation exchange capacity and most nutrients will readily leach from these substrates. Inherent nutrients provided by soil-less substrate components such as peat, bark, wood or coir is short-lasting and provides little to none of the mineral nutrients needed over the duration of the crop production period.

Micronutrients can either be supplied as chelated or non-chelated forms. Chelation is the process of binding ions or molecules to metals. This weak bond provides an organic coating to the micronutrients which help them to be better absorbed by the plant (Baligar et al. 2001). This is particularly important in soils or substrates that are basic since Fe, Mn, Zn, and Cu can react to form insoluble substances which are then not available to plants. Chelation stops these reactions from occurring at higher pH, and ensures that these micronutrients are plant available. Plants are able to easily take up these chelated micronutrients, and once they are in the plant, the chelate is released (Wallace et al. 1957). Chelated nutrients can be absorbed through the waxy cuticle since the organic molecule can penetrate this layer, while the inorganic forms of these nutrients cannot (Modaihsh 1997).

## 5 Fertilizers

Continuous fertilization throughout the production in soil-less substrates is necessary to supply adequate mineral nutrients to the crop. Fertilizers are often supplemented in field production systems based on soil tests and how much biomass the grower is trying to produce on the plants. Fertilizers may be withheld or reduced in field stock that is at or nearing saleable size, while containerized plants need constant supplies of fertilizer for optimal appearance. Fertilizers are available in several different forms including a single nutrient fertilizer (e.g. N or $P_2O_5$ only), incomplete nutrient formulations [e.g. N and potassium ($K_2O$)] and complete fertilizer formulations (i.e. N, $P_2O_5$, $K_2O$). Complete formulations, with or without micronutrients, are typically applied in open air container production systems, while greenhouse and field operations may apply any of the above depending on plant need, soil or substrate extract tests results, and grower preference.

Water-soluble fertilizers are mineral nutrients that readily solubilize in water and move into the root zone with water as the carrier. The length of time that mineral nutrients reside in the root zone is dependent on several factors, but most importantly rainfall or irrigation. The problem with water-soluble fertilizers is that they can be easily leached from the root zone before uptake, particularly in outdoor container production open to rainfall. Mineral nutrient laden runoff or sediment can move into surface or ground-waters and degrade water quality. Water-soluble, and controlled-release, fertilizers either can include or exclude S-balanced or chelated micronutrients. Historically, soluble fertilizers were used in overhead irrigation of open-air containerized plants but that practice has mostly stopped with the widespread adoption of controlled-release fertilizers. Water-soluble and SRFs are typically applied directly to the soil surface in field production.

Mineral nutrients can be encapsulated and applied in the form of a slow release fertilizer (SRF) or CRF (Table 2) (Cabrera 1996). Mineral nutrient release from SRF is based on a combination of microbial activity, soil moisture, temperature, or pH (Guertal 2009). Slow release fertilizers typically have a less predictable release characteristic than CRFs; however, the mechanism of release varies, potentially providing unique opportunities for growers depending on crop, specific usage and production cycle.

Slow-release fertilizer longevity is measured in weeks or up to a few months at most. Slow-release fertilizers are used in field production or blended with CRF as an upfront, short-term nutrient source. Technologies include isobutylidene diurea (IBDU), methylene ureas or urea formaldahydes, and sulphur-wax-coated granules (Table 2). Furthermore, SRF technology continues to develop, including an increase in the effectiveness of urease inhibitors for field applications (Upadhyay 2012). Urease inhibitors help to reduce the volatilization of nitrogen when urea is broken down into ammonia. Urea is a cost-effective N source that can be quickly lost due to ammonia volatilization under the right conditions (warm or hot moist soils) when applied to the soil or substrate surface.

Controlled release fertilizers are water-soluble fertilizers which are coated in a polymer or resin. Each coated particle is referred to as a 'prill'. A CRF can be homogenous in which the entire product analysis is included in a given prill, or heterogeneous in which individual mineral nutrient content of the prills varies, but when individual prills are blended they represent the labelled analysis. Heterogenous mixes may include water-soluble or slow-release macronutrients, and may include or exclude micronutrients. The thickness and pore size of the

**Table 2** Major types of slow (SRF) or controlled (CRF) release fertilizers currently available along with their main characteristics

| Type of coating | Polymer-based (CRF) coating | Sulphur-based coating | Urea-based isobutylidene diurea (IBDU) coating | Urea-based (ureaform and Methylene urea) coating |
|---|---|---|---|---|
| Examples | Polyon, ICL | Poly SCU, SCU | Promate, progrow | Florikan |
| Release control | Temperature | Temperature, soil moisture, microbial breakdown | Temperature, soil moisture | Temperature, soil moisture, microbial breakdown, pH, organic matter |
| Longevity | 2-14 months | 2 months | Prill size dependent | Ureaform: 1 year Methylene urea: 10-12 weeks |

prill coating controls the amount of water entering the prill. Once inside the prill the water incrementally dissolves the fertilizer, which in turn diffuses out of the prill via a concentration gradient. By varying the size and number of holes in the prill, as a function of thickness, the manufacturer can regulate the longevity of mineral nutrient release. Mineral nutrient release is dependent on temperature and the presence of water. The labelled longevity is typically based on 21-26°C (70-80°F) with higher temperatures releasing mineral nutrients faster and lower temperatures extending release. Different longevities of CRF are sold with products typically lasting for 2-14 months. Lastly, the fertilizer's chemical and physical formulation, and the delivery method of the fertilizer impact the rate and total change in soil or substrate pH, with the effect typically being greater in unbuffered or poorly buffered soil-less culture systems.

## 6 Fertilizer delivery

Fertilizers can be effectively delivered in a number of ways in ornamental production. In the case of liquid feeding (fertigation), fertilizers are dissolved in stock tanks as a concentrate and then injected directly into the irrigation system. Injection or delivery to irrigation lines is accomplished through the use of three main types of injectors: venturi, positive displacement, or pump heads. It is important to mention that if sanitizing agents are used (chlorine, ozone etc.) they need to have sufficient contact time of at least 2 min before fertilizers are injected to maximize their efficacy.

Venturi-type injectors use a difference in pressure to draw the concentrated fertilizer solution into the irrigation line. These injectors are simple, inexpensive and require little maintenance. On the negative side, they are susceptible to pressure fluctuations, which vary in terms of the injection ratios. In addition, venturi injectors require a low injection ratio (around 1:16) and that can require a larger stock tank needing a more frequent mixing of stock solution (compared to a 1:100 or 1:500 ratio). These types of injectors are not readily used in commercial ornamental production, but would be suitable for some applications.

Positive displacement injectors are the most widely used injector types in ornamental production (Pennisi and Kessler 2017). They use a piston which is driven by the water passing through the injector to deliver a known volume of concentrated fertilizer solution into the irrigation lines per piston pulse. Pump heads work similarly but are not dependent on flow, and work well for irrigation systems with flow rates exceeding 40–100 gallons per minute (GPM; 151–379 litres per minute, LPM). Fertilizer ratios can typically be adjusted from 1:50 to 1:1000 depending on the unit. These types of units are less sensitive to variations in water pressure when compared with venturi-type injectors, but are limited by the minimum and maximum flow rates, although the units typically offer a wide range of flow.

Application of fertilizers in the field is typically accomplished via broadcast (applied uniformly to the surface), side-dress (applied to the side of the plant row near the base of the stems) methods or through surface or sub-surface fertigation. Broadcast application of fertilizer, with lime or elemental S, is recommended for pre-plant application and incorporation in the Ap horizon to meet any deficiencies identified by soil test(s), using the appropriate extraction method for a soil type at given pH. For example, a Mehlich or Bray extract would be appropriate for acidic soil Utisols, whereas an Olsen extract would be appropriate for calcareous soils. Side-dress applications or drip fertigation are used to apply soil-mobile nutrients (like N and S) directly to the crop roots during active growth. Other mobile or partially mobile mineral nutrients can be applied by these methods, such as K, or Mn if crop deficiency is recognized or the pH is unsuitable for readily available metal species (e.g. Mn in red maple production). These more-direct methods of application reduce annual use of fertilizers, maximize nutrient crop-use efficiency, and reduce weed pressure.

There are several different options in regard to the placement of CRF in containers. Typically, growers will uniformly incorporate CRF and lime to adjust the pH into the soil-less substrate at the time of potting for ease of application. The theory is that since roots are spread throughout the container, the fertilizer should be too. The problem is that fertilizers located towards the bottom of the container will not encounter roots early in production before they leach out with irrigation or rainfall.

The other typical practice is top-dressing fertilizers, where CRF is added to the top of the substrate. This is typical for plants that are overwintered in the same container, or the CRF that was incorporated has been exhausted (i.e. run out). This top-dressed CRF typically lasts longer than incorporated CRF. This is because there is less moisture moving through the prill, and the fertilizer is only moved out of the prill during irrigation and rainfall events, compared with the incorporated fertilizer that is surrounded by a moist substrate prolonging the release after irrigation or rainfall. The release of CRFs, regardless of application method, in greenhouse operations is slower or less quick to release than open air production systems because they have more controlled substrate temperatures, lack rainfall, and will typically leach less water. A drawback of the use of top-dressed CRF is the risk of being blown off plants by wind, which not only removes the nutrients from the container, but also releases those nutrients directly into the environment.

## 7 Soils and substrates

The combination of soil or substrate physical and chemical characteristics regulates mineral nutrient (ion or metal), retention, availability and toxicity. The cation and anion exchange capacity of soils and media is determined by organic matter, minerology, elemental composition, ionic radii, pH and in some cases oxidation-reduction potential (Eh). Container crops are conventionally produced in well-drained systems with a pH between 4.5 and 6.5 for soil-less substrate and pH < 7 for most soil series where possible (Argo 2003).

Soils are comprised dominantly of inorganic minerals with organic matter ranging from 0 to 4%; whereas soil-less substrates are primarily organic materials comprised mainly of lignin or partially decomposed organic matter. Permeable, well-drained soils are preferred for field grown ornamental crops. The desired soil texture will vary based on the crop being produced. For example, a heavier soil will be desired for in-ground production versus more friable soil for perennial plants to be bare-root shipped upon harvest.

The gravitational potential existing in the container (equal to the height of the container) is not enough to overcome the suction imposed on water by the small diameter pores when using soil (Mastalerz 1977; Bilderback and Cartwright 1980). Thus, the total porosity (air and water) of soil-less substrates ranges from 60 to 90% by volume (Fonteno 1996) to alleviate any container effect or drainage issues. Minimum air space is typically maintained at greater than 20% (by vol.) when the substrate is at container capacity (i.e. maximum water-holding capacity; 30–70% by vol.). Although not recommended, soils are still used in some container production operations, especially in arid regions where water or substrate components are limited or in areas where porous soils are abundant (e.g. Oregon or Hawaii).

Soil-less substrates continue to be improved to balance air-filled porosity, water-holding capacity, weight, sustainability/availability of supply, and nutrient retention. Substrate manufacturers have to balance these factors and have different substrate mixes that prioritize one or a few of these parameters to the detriment of the others. Typical substrate components include peat, vermiculite, perlite, sand, and pine bark. Newer substrates include coir (coconut), rice hulls, peanut shells and milled whole trees (Chavez et al. 2008; Wright et al. 2008; Jackson et al. 2009; Vaughn et al. 2011). These newer substrates are typically made from waste products from other industries and as such are able to expand the useful life of other traditional substrates (e.g. peat). Major concerns with these newer substrates are their impact on current growing practices, continued availability, reproducibility and cost, including distance and weight when shipping.

## 8 Environmental impacts of runoff

Although there is no question of the benefits of fertilizers for optimizing plant growth, there is a downside to their use. As mentioned previously, agricultural irrigation and fertilization practices contribute to algal blooms, sedimentation and environmental degradation. A less discussed problem is groundwater and other subsurface water contamination. There are reports of groundwater levels of N in excess of the 10 mg/L (10 ppm) limit allowed by drinking water regulations, likely due to a number of factors including agriculture (Ward et al. 2018). Elevated levels of P and K are not typical in groundwater, but they are typical in surface waters since they are sediment bound and can lead to algal blooms.

Limiting water and sediment movement on- and off-site could be beneficial to an operation and its surrounding environment. Remediating agricultural runoff water for reuse or release off-site would have benefits for both the grower and the environment (White 2013; Majsztrik et al. 2019). Recaptured water and nutrients could be used on-site as long as it is treated for pathogens, and potentially pesticides before reuse. Some work is being done on using nutrients in irrigation ponds to grow plants that can be sold. This would use non-productive area to increase sales and may be a viable option for some growers with substantial environmental benefits (White 2013; Lamm et al. 2017; Majsztrik et al. 2017).

## 9 Future trends and conclusion

It is likely that future coating technologies will more effectively release nutrients when plants are actively growing. This will allow for more efficient fertilizer use, better plant growth and fewer environmental impacts. Better irrigation control through mesh networks that employ real-time monitoring of substrate or soil moisture will increase the time that fertilizers are in the root zone along with a number of other benefits including reduced water use, reduced disease presence, increased plant growth and increased plant quality (Belayneh et al.

2013; Lea-Cox et al. 2013; Belayneh and Lea-Cox 2015; Saavoss et al. 2016). There are also likely to be better matching of fertilizers to species, growth pattern and weather for increased fertilizer performance.

Growing ornamental plants is far from simple given the diversity in the species and cultivars in production, container sizes, substrate types, irrigation systems and fertilizer formulations available. We started this chapter with irrigation water quality and quantity since this is the basis for successful ornamental production. Next we discussed the types of production systems for ornamental production before delving into mineral nutrients and their role in plants. This provided the basis for different fertilizers and how they can be delivered most effectively to plants. These concepts were tied back to the system by discussing soil-less substrates and soils. We hope that this chapter has provided you with useful information regardless of your area of study or level of expertise.

## 10 Where to look for further information

Further information can be found in the references, but also in the following resources:

- Bilderback, T., Boyer, C., Chappell, M., Fain, G., Fare, D., Gilliam, C., Jackson, B., Lea-Cox, J., LeBude, A., Niemiera, A., Owen, J., Ruter, J., Tilt, K., Warren, S., White, S., Whitwell, T., Wright, R. and Yeager, T. 2013. *Best Management Practices: Guide for Producing Nursery Crops* (3rd edn.). Southern Nursery Association, 02 November 2015. Available at: http://contents.sna.org/bmpv30.html.
- Fulcher, A., LeBude, A. V., Owen, J. S., White, S. A. and Beeson, R. C. 2016. The next ten years: strategic vision of water resources for nursery producers. *HortTechnology* 26, 121–32.
- Majsztrik, J. C., Ristvey, A. G. and Lea-Cox, J. D. 2011. Water and nutrient management in the production of container-grown ornamentals. *Horticultural Reviews* 38, 253–97.

## 11 References

Adrian, J. L., Montgomery, C. C., Behe, B. K., Duffy, P. A. and Tilt, K. M. 1998. Cost comparisons for infield, above ground container and pot-in-pot production systems. *Journal of Environmental Horticulture* 16, 65-8.

Agner, H. and Schenk, M. K. 2006. Nitrogen emissions ($N_2O$+$N_2$) and redox potential of a peat medium during pot plant cultivation. *European Journal of Horticultural Science* 71, 237-9.

Alem, P., Thomas, P. A. and van Iersel, M. W. 2015. Controlled water deficit as an alternative to plant growth retardants for regulation of poinsettia stem elongation. *HortScience* 50(4), 565–9. doi:10.21273/HORTSCI.50.4.565.

Amha, Y. and Bohne, H. 2011. Denitrification from the horticultural peats: effects of pH, nitrogen, carbon, and moisture contents. *Biology and Fertility of Soils* 47(3), 293-302. doi:10.1007/s00374-010-0536-y.

Anon. 2013. Iron staining caused by irrigation systems. Water Quality Protection Note 21. Department of Water, Government of Western Australia.

Argo, W. R. 1998. Root medium chemical properties. *HortTechnology* 8(4), 486–94. doi:10.21273/HORTTECH.8.4.486.

Argo, B. 2003. Understanding pH managment and plant nutrition. *Journal of the International Phalaenopsis Alliance* 13, 1–5.

Baligar, V. C., Fageria, N. K. and He, Z. L. 2001. Nutrient use efficiency in plants. *Communications in Soil Science and Plant Analysis* 32(7–8), 921–50. doi:10.1081/CSS-100104098.

Belayneh, B. E. and Lea-Cox, J. D. 2015. Using sensor networks to maximize irrigation water use efficiency in strawberry production. In: *VIII International Symposium on Irrigation of Horticultural Crops* 1150, pp. 399–406.

Belayneh, B. E., Lea-Cox, J. D. and Lichtenberg, E. 2013. Costs and benefits of implementing sensor-controlled irrigation in a commercial pot-in-pot container nursery. *HortTechnology* 23(6), 760–9. doi:10.21273/HORTTECH.23.6.760.

Bergstrand, K.-J. I. 2017. Methods for growth regulation of greenhouse produced ornamental pot-and bedding plants–a current review. *Folia Horticulturae* 29(1), 63-74. doi:10.1515/fhort-2017-0007.

Bilderback, T. 2001. Using the pourthru procedure for checking EC and pH for nursery crops. NC State University Department of Horticultural Science Horticulture Information Leaflet 401, p. 6.

Bilderback, T. and Cartwright, E. 1980. Influence of four commercial fertilizers applied to pine bark and hardwood bark media on growth, foliar N, soil N, pH and soluble salts of four nursery species. *Buxus microphylla*, *Ilex* sp., *Cupressocyparis leylandii*. *Annual Report-Southern Nurserymen's Association Research Conference*, USA.

Broschat, T. K. 2005. Rates of ammonium nitrogen, nitrate nitrogen, phosphorus, and potassium from two controlled release fertilizers under different substrate environments. *HortTechnology* 15(2), 332–5. doi:10.21273/HORTTECH.15.2.0332.

Cabrera, R. I. 1996. Using slow- and controlled-release fertilizers in container nursery crops. *Rutgers Cooperative Research and Extension*, pp. 1–4.

Cavins, T. J., Whipker, B. E., Fonteno, W. C., Harden, B., McCall, I. and Gibson, J. L. 2000. Monitoring and managing pH and EC using the pour thru extraction method. NC State University Department of Horticultural Science Horticulture Information Leaflet 590, 17p.

Chavez, W., Di Benedetto, A., Civeira, G. and Lavado, R. 2008. Alternative soilless media for growing *Petunia×hybrida* and *Impatiens wallerana*: physical behavior, effect of fertilization and nitrate losses. *Bioresource Technology* 99(17), 8082–7. doi:10.1016/j.biortech.2008.03.063.

Copes, W. E., Chastaganer, G. A. and Hummel, R. L. 2004. Activity of chlorine dioxide in a solution of ions and pH against *Thielaviopsis basicola* and *Fusarium oxysporum*. *Plant Disease* 88(2), 188–94. doi:10.1094/PDIS.2004.88.2.188.

Fonteno, W. 1996. *Water, Media and Nutrition for Greenhouse Crops*.

Grant, G. A., Fisher, P. R., Barrett, J. E. and Wilson, P. C. 2018. Removal of paclobutrazol from irrigation water using granular-activated carbon. *Irrigation Science* 36(3), 159-66. doi:10.1007/s00271-018-0572-1.

Guertal, E. A. 2009. Slow-release nitrogen fertilizers in vegetable production: a review. *HortTechnology* 19(1), 16-9. doi:10.21273/HORTSCI.19.1.16.

Hansen, N. C., Daniel, T. C., Sharpley, A. N. and Lemunyon, J. L. 2002. The fate and transport of phosphorus in agricultural systems. *Journal of Soil and Water Conservation* 57, 408-25.

Hoskins, T., Owens, J. and Niemiera, A. 2014. Controlled-release fertilizer placement affects the leaching pattern of nutrients from nursery containers during irrigation. HortScience 49(10), 1341-5.

Jackson, B. E., Wright, R. D. and Alley, M. M. 2009. Comparison of fertilizer nitrogen availability, nitrogen immobilization, substrate carbon dioxide efflux, and nutrient leaching in peat-lite, pine bark, and pine tree substrates. *HortScience* 44(3), 781-90. doi:10.21273/HORTSCI.44.3.781.

Kläring, H.-P. 2001. Strategies to control water and nutrient supplies to greenhouse crops. A review. *Agronomie* 21(4), 311-21. doi:10.1051/agro:2001126.

Lamm, A. J., Warner, L. A., Martin, E. T., White, S. A. and Fisher, P. 2017. Enhancing extension programs by discussing water conservation technology adoption with growers. *Journal of Agricultural Education* 58(1), 251-66. doi:10.5032/jae.2017.01251.

Lea-Cox, J. D., Bauerle, W. L., van Iersel, M. W., Kantor, G. F., Bauerle, T. L., Lichtenberg, E., King, D. M. and Crawford, L. 2013. Advancing wireless sensor networks for irrigation management of ornamental crops: an overview. *HortTechnology* 23(6), 717-24. doi:10.21273/HORTTECH.23.6.717.

London, J. B., Fernandez, R. T., Young, R. E. and Christenbury, J. D. 1998. Media temperature in above-ground and in-ground pot-in-pot container systems. *HortScience* 33(3), 512a-512. doi:10.21273/HORTSCI.33.3.512a.

Majsztrik, J. C. and Lea-Cox, J. D. 2013. Water quality regulations in the Chesapeake Bay: working to more precisely estimate nutrient loading rates and incentivize best management practices in the nursery and greenhouse industry. *HortScience* 48(9), 1097-102. doi:10.21273/HORTSCI.48.9.1097.

Majsztrik, J. C., Fernandez, R. T., Fisher, P. R., Hitchcock, D. R., Lea-Cox, J., Owen, J. S., Oki, L. R. and White, S. A. 2017. Water use and treatment in container-grown specialty crop production: a review. *Water, Air, and Soil Pollution* 228(4), 151. doi:10.1007/s11270-017-3272-1.

Majsztrik, J. C., Behe, B., Hall, C. R., Ingram, D. L., Lamm, A. J., Warner, L. A. and White, S. A. 2019. Social and economic aspects of water use in specialty crop production in the USA: a review. *Water* 11(11), 2337. doi:10.3390/w11112337.

Marschner, H., Römheld, V. and Cakmak, I. 1987. Root-induced changes of nutrient availability in the rhizosphere. *Journal of Plant Nutrition* 10(9), 1175-84. doi:10.1080/01904168709363645.

Martin, T. L., Kaushik, N. K., Trevors, J. T. and Whiteley, H. R. 1999. Review: denitrification in temperate climate riparian zones. *Water, Air, and Soil Pollution* 111(1/4), 171-86. doi :10.1023/A:1005015400607.

Mastalerz, J. W. 1977. *The Greenhouse Environment*. John Wiley and Sons, New York, NY.

Mathers, H. M. 2003. Summary of temperature stress issues in nursery containers and current methods of protection. *HortTechnology* 13(4), 617-24. doi:10.21273/HORTTECH.13.4.0617.

Meador, D. P. and Fisher, P. R. 2013. Ammonium in nutrient solutions decreases free chlorine concentration from sodium hypochlorite. *HortScience* 48(10), 1304–8. doi:10.21273/HORTSCI.48.10.1304.

Modaihsh, A. S. 1997. Foliar application of chelated and non-chelated metals for supplying micronutrients to wheat grown on calcareous soil. *Experimental Agriculture* 33(2), 237–45. doi:10.1017/S001447979700001X.

Pennisi, B. and Kessler, R. 2017. Fertilizer injectors: selection, maintenance and calibration. University of Georgia Cooperative Extension Bulletin 1237, 16p.

Peterson, J. C. 1982. Effects of pH upon nutrient availability in a commercial soilless root medium utilized for floral crop production. *Research Circular* 268, 15–19.

Picot, B., Moersidik, S., Casellas, C. and Bontoux, J. 1993. Using diurnal variations in a high rate algal pond for management pattern. *Water Science and Technology* 28(10), 169–75. doi:10.2166/wst.1993.0227.

Robbins, J. A. 2018. Irrigation water for greenhouses and nurseries. University of Arkansas Extension FSA6061.

Roques, S., Kendall, S., Smith, K., et al. 2013. A review of the non-NPKS nutrient requirements of UK cereals. Research Review 78. HGCA-AHDB, Kenilworth, UK.

Saavoss, M., Majsztrik, J., Belayneh, B., Lea-Cox, J. and Lichtenberg, E. 2016. Yield, quality and profitability of sensor-controlled irrigation: a case study of snapdragon (*Antirrhinum majus* L.) production. *Irrigation Science* 34(5), 409–20. doi:10.1007/s00271-016-0511-y.

Shreckhise, J. H., Owen, J. S., Eick, M. J., Niemiera, A. X., Altland, J. E. and White, S. A. 2019. Dolomite and micronutrient fertilizer affect phosphorus fate in pine bark substrate used for containerized nursery crop production. *Soil Science Society of America Journal* 83(5), 1410–20. doi:10.2136/sssaj2018.12.0493.

Silber, A. 2019. Chemical characteristics of soilless media. In: Raviv, M., Lieth, J. H. and Bar-Tal, A. (Eds), *Soilless Culture* (2nd edn.). Elsevier, Boston, MA, pp. 113–48. Chapter 4.

Upadhyay, L. S. B. 2012. Urease inhibitors: a review. *Indian Journal of Biotechnology* 11(4), 381–8.

US Department of Agriculture. 2019. 2017 census of agriculture: summary and state data. United States Department of Agriculture and National Agricultural Statistics Service.

Vaughn, S. F., Deppe, N. A., Palmquist, D. E. and Berhow, M. A. 2011. Extracted sweet corn tassels as a renewable alternative to peat in greenhouse substrates. *Industrial Crops and Products* 33(2), 514–7. doi:10.1016/j.indcrop.2010.10.034.

Wallace, A., Shannon, L. M., Lunt, O. R. and Impey, R. L. 1957. Some aspects of the use of metal chelates as micronutrient fertilizer sources. *Soil Science* 84(1), 27–42. doi:10.1097/00010694-195707000-00004.

Ward, M. H., Jones, R. R., Brender, J. D., de Kok, T. M., Weyer, P. J., Nolan, B. T., Villanueva, C. M. and van Breda, S. G. 2018. Drinking water nitrate and human health: an updated review. *International Journal of Environmental Research and Public Health* 15(7), 1557. doi:10.3390/ijerph15071557.

White, S. A. 2013. Wetland technologies for nursery and greenhouse compliance with nutrient regulations. *HortScience* 48(9), 1103–8. doi:10.21273/HORTSCI.48.9.1103.

Wright, R. D., Jackson, B. E., Browder, J. F. and Latimer, J. G. 2008. Growth of chrysanthemum in a pine tree substrate requires additional fertilizer. *HortTechnology* 18(1), 111–5. doi:10.21273/HORTTECH.18.1.111.

Zhou, W. Z. 1994. The role of horticulture in human history and culture. *Horticulture in Human life, Culture and Environment* 391(391), 41–52. doi:10.17660/ActaHortic.1995.391.3.

Zinati, G. and Shuai, X. 2005. Management of iron in irrigation water. NJ Experiment Agricultural Experiment Station. Rutgers Cooperate Research and Extension.

# Part 2

## Cultivation techniques

# Chapter 9

# Advances in irrigation practices and technology in ornamental cultivation

*John D. Lea-Cox, University of Maryland, USA*



## 1 Introduction

Two issues will likely dominate irrigation and water management for ornamental cultivation in the coming years, namely the decreasing availability of fresh water for agricultural use and environmental concerns that will increasingly limit the runoff of water with any contaminants from commercial operations. Both issues are the subject of increasing scrutiny and regulation in the United States and Europe, and many operations are addressing these issues through an array of better management practices. Both objectives could be better achieved by increasing the precision of irrigation applications and understanding how we can better time the frequency and duration of irrigation events, not only to optimize plant growth but also to limit leaching and runoff.

This chapter will address these systemic issues, to increase the efficiency of irrigation applications, to reduce runoff volume and limit contaminant load, and to make the most effective use of any capture or remediation capacity. After assessing current irrigation systems used in ornamental production, the chapter assesses key challenges for improving irrigation efficiency, starting with systems design and maintenance. It then reviews sensor-based technologies

http://dx.doi.org/10.19103/AS.2020.0066.12

for irrigation scheduling before discussing irrigation data management and decision support systems.

## 2 Ornamental production systems and water use

Ornamental production systems are typically very intensive, with high plant densities and fast growth rates, requiring high inputs of resources, time, and labor (Lea-Cox and Ross, 2001; Majsztrik et al., 2018). Access to reliable sources of good-quality water is not only a principal requirement for most ornamental operations but essential for container production in nurseries and greenhouses, where daily irrigation applications are required except when supplemented by rainfall. Commercial operation locations are often initially chosen because of access to good water resources, but with periodic droughts, urbanization, and increasing restrictions on water use, producers are faced with increasing water security issues, even in 'water-rich' areas (Ristvey et al., 2019).

The US nursery industry consolidated slightly between 2008 and 2013, with 2% fewer operations on 13% fewer acres, most likely due to the economic downturn during those years. And, yet, overall water use by the nursery industry remained the same, but with 12% more of that water coming from groundwater resources (Fulcher et al., 2016). Water regulations vary greatly between states in the United States. Ornamental producers in some eastern states (under riparian water rights law) are only required to report usages if volumes applied exceed a daily threshold, and yet in other states, 'reasonable use' is not quantified, nor are there limitations on water use for agricultural producers. In western states, the volume of available water depends on the seniority of the users within the watershed (under prior appropriation rights), which typically favors agriculture over urban consumers, since agriculture often has a senior claim. However, laws and regulations currently limit nursery water consumption in California, Florida, North Carolina, Texas, and Oregon and are expected to become more stringent in other states, in the face of changing water availability (Fulcher et al., 2016).

Despite relatively lax federal oversight, growers in many states are typically more attuned to local water scarcity issues in terms of limited well- or surface-water volumes, combined with increasing pumping costs and water quality issues. Growers see sensor-based systems and other new technologies as a way to monitor the water use of ornamental crops and increase the use of existing water resources, but as with any technology, many are concerned about cost and reliability. These concerns also include the technical expertise to manage these technologies, in terms of hardware and software, and the scalability and ease-of-use of any technological system (Majsztrik et al., 2013a). Growers also expect a reasonable return on investment for any technology; site-specific returns are almost always a function of the plants being grown and the cost, availability, and quality of local water resources (Belayneh et al., 2013; Saavoss et al., 2016).

In 1998, over half of the irrigation water used by both open and protected horticulture was applied with overhead sprinkler systems (Lea-Cox et al., 2001). With better data from the 2013 Farm and Ranch Irrigation survey (US Department of Agriculture, 2014), we calculated that 40.0% of total US irrigated horticulture (530 779 acres = 214 799 Ha) was in protected and open production of nursery crops, while an additional 8.9% of this acreage was devoted to floriculture and bedding crop production. Of this total acreage, 68.1% of irrigated horticultural crops were overhead irrigated, while 25.1% were irrigated with low-volume or sub-irrigation systems and 1.4% were hand-watered. The remaining 5% of acreage was flood irrigated, presumably for nut and other specialty crops such as cranberry. It is therefore likely that a large proportion of ornamental nursery crops are still irrigated using overhead irrigation systems.

Most commercial greenhouse operations typically use precision irrigation methods, with some level of environmental monitoring and control. They are usually more labor-intensive and apply less water per unit area than open-air container or pot-in-pot nurseries (Bailey et al., 1999). A higher level of management and control can lead to higher efficiencies in water use, but this is highly dependent on a number of site-specific variables, including the type of irrigation system used (e.g. boom, drip emitters, micro-emitters, or ebb and flood), system design, and irrigation scheduling decisions (good management). All these factors contribute to high efficiency. Drip and micro-emitters typically require higher-quality water (obtained through filtration and sanitation) and regular maintenance to avoid emitter clogging and subsequent plant loss or damage (Burnett and van Iersel, 2018).

Container-nursery operations grow plants on various surfaces (bare ground, gravel, landscape fabric, etc.) that are often graded to remove any standing water from production areas (Fig. 1). Overhead irrigation is typically applied using impact or wobbler sprinkler heads. Larger containers (typically >7 Gal = 26.5 L) are often irrigated more precisely, using drip emitters or spray-stakes. Above this container size, overhead irrigation is typically not employed, due to very low interception and high runoff rates. Another distinct advantage of micro-irrigation is the potential to implement fertigation (direct injection of water-soluble fertilizer), which is significantly more efficient (allowing for split applications and precise timing to support growth flushes and avoiding leaching) and much less costly than the use of slow-release fertilizers with these large plants (Majsztrik et al., 2017).

Overhead irrigation in nursery production, while economically advantageous from a labor and mechanization standpoint, is inherently inefficient because of the space needed between individual plants to allow for uniform light and plant growth (Fulcher et al., 2016). Based on irrigation system

**Figure 1** Well-designed nursery production area, with gravel beds allowing for rapid drainage through underground pipes (foreground). This then drains to the side sediment channel/containment pond shown in Fig. 2.

design recommendations (Aldrich and Bartok, 1994), water applications using overhead sprinkler irrigation can exceed 20 000 Gal/Acre (180 000 L/Ha) per day, with 10–50% of that running directly from less pervious production areas (Berghage et al., 1999). Theoretical interception efficiencies (Ross et al., 2001) for a 5-Gal (19 L) spaced container configuration are typically less than 28%, but interception efficiencies for these containers were actually between 50% and 64% for *Azalea* and *Ilex crenata*, respectively, when canopies were large (Ristvey, 2004), due to canopy interception and stem flow. Beeson and Yeager (2003) reported that irrigation application efficiencies decrease quadratically with increase in spacing between containers, and optimal efficiencies are achieved when plants are grown at the minimum spacing required for ensuring commercial quality.

Most commercial nursery and greenhouse producers face the conundrum of applying water precisely but economically to plants, within the context of addressing runoff and environmental issues. Fulcher et al. (2016) noted that there are few economic alternatives to using overhead irrigation for container sizes less than 5-Gal (19 L) capacity. And yet, the authors noted that open floriculture and herbaceous perennial growers are increasingly using drip irrigation systems for 1- and 2-Gal (3.9–7.5 L) containers, driven by perceived reductions in disease pressure, lack of consumer tolerance for foliar residues, together with decreasing production inputs for low-margin crops. Pot-in-pot woody perennial nurseries are also realizing additional economic benefits of precisely irrigating smaller (3-5 Gal; 11-19 L) container sizes, including the ability to precisely apply soluble fertilizers in low concentrations to match specific plant growth rates. Labor availability and timeliness of irrigation applications are often the over-riding factors for optimizing plant growth rates while reducing environmental impacts. Given the widespread use of overhead irrigation in nurseries, the capture and recycling of irrigation runoff (i.e. the 'first flush' of runoff) using containment ponds remains the most viable economic alternative to using fresh potable water, especially for production of small container crops. This requires that the growing operation has the available space for these structures, which should be specifically designed and engineered for this purpose (Ross, 2008d; Fig. 2).

The capture of stormwater to supplement water supply (in addition to production runoff) is limited by the climate, topographic limitations, and the regulation of waterways. In California, state regulations require growers to capture and retain water on site if it is not of high enough quality to discharge. However, discharge can occur if runoff treatment and remediation are performed to prevent water body impairment (Board, 2015). If growers are going to invest in the infrastructure required for remediation and treatment, it is logical to reuse this recycled water for irrigation, to reduce cost and reduce the need for other water sources, particularly in a dry climate (Pitton et al., 2018). The authors noted from their comprehensive cost analysis that capital costs associated with construction of recycled water systems were

**Figure 2** Side sediment channel that settles sediment and detritus in runoff water from production areas prior to it entering the main irrigation pond. This settling area should also be designed for routine cleaning by farm equipment.

high, but were offset by rebates from the water provider and a grant from the National Resource Conservation Service (NRCS) in this specific case. However, subsequent operational costs (e.g. electricity for pumping) are low, which reduces the marginal cost of recycled water in future years (Pitton et al., 2018). Moreover, this strategy insulates the operation from reduced water availability and/or future price increases, increasing overall water security. The need to increase water recycling and effectively capture greater proportions of rain events will be essential for nurseries in regions that will experience increasing droughts as a result of increasing climatic variability.

Compounding the irrigation runoff situation is the use of agrochemicals in ornamental production (White et al., 2019). Since nursery and greenhouse production systems are input intensive and water is an effective transport mechanism, they have the potential to contribute agrochemical loading to the surrounding environment, if appropriate management and water-control structures are not in place. Quantifying the range of fertilizer and irrigation application rates used by the ornamental nursery and greenhouse industry is challenging because of the variety of species, production systems, and cultural management techniques employed (Majsztrik et al., 2011). More recently, Majsztrik et al. (2018) provided a break-down of self-reported annual fertilization rates for greenhouse, container-nursery, and field-nursery operations in Maryland, by major categories of plant and container sizes. Field-nursery operations were found to have very low annual nutrient ($N - P_2O_5 - K_2O$) application rates, with container-nursery operations having the greatest inputs. Application rates for greenhouse operations were less than container nurseries, but notably, greenhouses typically produce three to five crop cycles on the same area per year, whereas container-nursery operations typically only produce one. The authors noted that it was highly likely that growers fertilizing at rates within the highest quartile could significantly decrease those rates without negatively impacting plant growth, as other growers were producing similar plants of similar quality with substantially less fertilizer in the same region. Reducing fertilization rates in ornamental operations provides cost savings from reduced fertilizer use and an environmental benefit from reduced nutrient leaching. This is particularly true for greenhouse and container-nursery production (Bayer et al., 2015) and to a lesser degree for field-nursery production.

Maryland probably has the most restrictive nutrient management regulations in the United States at present (Lea-Cox and Ross, 2001; Majsztrik et al., 2011), although many states are implementing various types of regulations, as previously noted. Lea-Cox et al. (2001) developed the first comprehensive water and nutrient planning process for nursery and greenhouse systems in the United States. This process was developed in response to the Maryland Water Quality Improvement Act of 1998, a set of regulations that were intended to reduce nonpoint nutrient loading from agricultural operations into the Chesapeake Bay (Majsztrik and Lea-Cox, 2013). To date, over 350 commercial operations have written and implemented these water and nutrient management plans, using a comprehensive suite of best management practices (Yeager et al., 2007; Bilderback et al., 2013). Lea-Cox et al. (2002) also provide an overview of how this integrated water and nutrient management planning process can estimate total maximum daily loading (TMDL) nutrient loads for these horticultural operations by quantifying site, fertilization, and water application practices in this risk assessment process (Lea-Cox et al., 2001). Many growers have found that this risk assessment process has revealed many inefficiencies in their irrigation and nutrient management practices; by formally assessing their practices, growers have found many ways to economize and save on production costs, by implementing site-specific best management practice recommendations (Lea-Cox and Ross, 2018).

## 3 Key challenges for improving irrigation efficiency: systems design and maintenance

As noted, managing irrigation water application efficiency provides us with a prime opportunity for improving plant growth while reducing agrochemical leaching and runoff issues from intensive production systems. In this section, the various opportunities and challenges for implementing different practices and technologies will be reviewed, starting with systems design and maintenance.

Most irrigation systems have distinct advantages and disadvantages, depending on the specific production environment and the type of ornamental plants being grown. Irrigation systems can be divided into two basic categories:

1. Systems that apply water *precisely* to individual plants or containers, for example, drip, micro-sprinkler, but which can also include hand-watering and overhead boom applications to densely-packed plants.
2. Large-volume applications, for example, overhead and boom irrigation systems, ebb, and flood-floors.

Furthermore, all types of irrigation technology can be used within 'open' production environments (i.e. where runoff is not contained or recycled) or in 'closed' systems, where some form of capture, remediation and/or recycling of irrigation and/or rainfall is involved. Whether an operation employs 'closed' and/or 'open' production systems is typically a function of local water availability and regulatory issues, although oftentimes, growers implement practices for other reasons, such as labor availability.

Majsztrik et al. (2017) provide a comprehensive recent view of water use and treatment in nursery and greenhouse production. Grower reluctance to capture and reuse runoff is typically associated with concerns about the opportunity cost of lost production area, the installation and management costs of containment ponds, and water treatment technology (Majsztrik et al., 2017). Primary concerns are the reintroduction and management of disease-causing organisms (Lea-Cox and Ross, 2014), the phytotoxicity of pesticide, herbicide, or growth regulator runoff (Abdi and Fernandez, 2019), or other land-use restrictions (high water table, steep slopes, etc.). Nevertheless, many growers recognize the dangers of inaction and the impact of water security on their long-term profitability (Ristvey et al., 2019).

When one considers the efficiency of any irrigation system, two factors are paramount - system design and maintenance. Both factors are synergistic in that neglecting either greatly amplifies their impact on overall system efficiency.

### 3.1 Irrigation system design

Good irrigation system design is the prime opportunity for achieving high water application efficiency in any operation. Without a fundamental knowledge of irrigation system hydraulics (i.e. the interactive relationship between water flow, friction, and pressure; Ross, 2008a), the poor design of suitable infrastructure (i.e., adequate sizing of pipes, emitters and pumps, etc.) is often the greatest contributor to overall system inefficiency. Poor irrigation design inevitably leads to poor application uniformity of water; poor uniformity then leads to either over-application of irrigation water to large areas, to ensure that dry areas are adequately watered, or it requires additional labor and resources where supplemental hand-watering is employed (Ross, 2008b). Irrigation systems can also deteriorate over time due to poor water quality or by simple expansion of the operation, when main lines are not upgraded and replaced to maintain adequate volume and pressure to new production areas.

When building, modifying, or operating any part of an irrigation system, one should always consider the effect of that modification *on the entire system*, as the balance of flows and pressures must be maintained. Three overall design criteria should be kept in mind:

1 *Calculating the adequate availability of water:* An operation's water supply may be limited to a minimum flow rate for year-round delivery. Groundwater from wells may be reasonably reliable all year, but surface-water volumes typically fluctuate seasonally or even daily. The limits on the minimum water supply must be calculated and primary/backup pumps must be sized for the available flow. Most notably, seasonal minimum flow rates will most likely occur at the time of maximum irrigation demand.
2 *Irrigation system capacity:* An irrigation system is usually designed to irrigate a maximum number of 'zones' that can be irrigated simultaneously, without exceeding the total available water. Each irrigation zone must

be able to supply the maximum volume of water needed by a mature crop on a hot, summer day; this volume can also be divided into several cycles, which can also improve efficiency (Tyler et al., 1996). An irrigation system should not be expected to operate 24 h a day to meet crop needs; some maintenance downtime should be included in the daily schedule. Additional capacity should also be designed into the system for extreme weather conditions (Ross, 2008b).

3 *Distribution uniformity:* Uniformity refers to the extent to which all plants in an irrigation zone receive the same amount of water over the same period of time. Although perfect uniformity cannot be achieved in practice, a high level of water application uniformity is desired so that crop growth is itself uniform.

Pressure losses in the pipe system, variability in emitters or sprinklers, variability in container size (rooting volume), substrate water-holding capacity, and plant water demand all contribute to variability of water use in the production area. Uniformity can be improved by investing in more expensive irrigation equipment, such as larger diameter pipes and more sprinklers to achieve better precision of supply to individual containers or by improving the overlap of watering patterns. This investment may not be justified by small increases in uniformity, but the more uniform the irrigation system, the less over-irrigation is required. And increasing uniformity always leads to reduced leaching of nutrients, runoff from daily irrigation events, and less stress on any remediation, containment or recycling infrastructure.

### 3.2 Irrigation system maintenance

Many growers also tend to neglect or ignore irrigation system maintenance. Irrigation systems audits (Ross, 2008c) are cost-effective and should be performed on a routine (bi-annual) basis to understand where irrigation system inefficiencies exist. Ensuring that emitters are clean and positioned correctly should be a part of *daily* system maintenance, but the measurement of emitter discharge on a block-by-block basis is typically more revealing of systematic losses in efficiency and forms one of the many better irrigation management practices (Yeager et al., 2007; Bilderback et al., 2013).

Before an audit is conducted, a maintenance check should be done on the system to ensure it is working as well as possible. Records of any irrigation component purchase and installation may be useful in determining the initial specifications of a component and are always helpful in understanding where/when changes have occurred. Wear on a pump may be evidence of a reduced water flow or reduced pressure. Of course, wear on nozzles will show an increase in flow rate and a higher demand on the pump to maintain the required pressure. A visual inspection of the irrigation system will show if all the required components are in place and in good apparent condition. Crop appearance is a good indicator of non-uniform application of water, but regular audits in early spring allow growers to quantify system performance and to pre-empt any losses due to poor system function or a lack of uniformity before crop losses become evident (Ross, 2008c).

Pressure gauges and flow meters at critical points in the nursery or greenhouse also allow for pre-emptive diagnosis of systemic problems, particularly where real-time monitoring is employed. As sensors and technology become more robust and easier to deploy (retrofit) in operations, many growers/managers are recognizing the benefits of having a sensing capability as part of their ongoing maintenance strategy, especially where labor availability and time management issues are critical. Automation has long been integral to greenhouse management, but there is an ever-increasing demand for systems that can provide this capability for the irrigation of large-scale ornamental operations (Lea-Cox et al., 2018). With increasing water scarcity and cost, often coupled with declining water quality, water conservation in states like California (Newman, 2018) is becoming a requirement for staying economically viable and thriving within a more regulated environment. However, it should be emphasized that investing in sensing technology will not compensate for poor irrigation system design, a lack of system maintenance, or poor management.

## 4 Sensor-based technologies for irrigation scheduling

### 4.1 Soil-moisture approaches

Automated irrigation systems are a core technology for most ornamental operations, designed to provide scalable and economic (easy to manage) methodologies to apply water to plants. As water availability and quality have become more of an issue, the addition of technologies that are capable of providing real-time feedback on soil/substrate water content and availability have become more widely used, for both monitoring and, more recently, for automated (non-human) control systems (Lea-Cox et al., 2013). The development of precision (low-volume) irrigation systems has played a major role in reducing the total water applied to high-value crops. However, the need to optimize crop yield has increased the necessity for new methods to achieve accurate irrigation scheduling and control (Jones, 2008). For high-value horticultural crops, there is significant interest in using decision irrigation (i.e. irrigating based on information such as soil-moisture content) to precisely apply water while improving crop yield, reducing nutrient loss and disease pressure. Therefore, achievement of optimal irrigation will depend not only on the use of precision irrigation systems, but also on the development of easy-to-use decision tools which can help the farmer monitor and automate irrigation scheduling (Lea-Cox, 2012).

Various soil irrigation scheduling technologies and methods have been used for decades and have been the subject of much research that is summarized by reviews on measuring soil moisture (e.g. Dane and Topp, 2002; Bitelli, 2011), plant physiological measurements (e.g. Jones, 2004; Cifre et al., 2005), or water-balance calculations (e.g. Allen et al., 1998). The conventional sensor-based approach has typically scheduled irrigation events on the basis of soil-moisture status, whether using direct soil-moisture measurements with capacitance or time-domain reflectometry sensors (Topp and Davis, 1985; Smith and Mullins, 2001), tensiometers (Smajstrla and Harrison, 1998), or soil-moisture water-balance methods using daily ET estimates (Allen et al., 1998). Some automated greenhouse systems have used load cells for the estimation of daily plant water use (Raviv et al., 2000). However, these load cell systems must be programmed to accurately correct for increasing total plant mass over the crop cycle or to adjust to changes in wind and temperature changes, if deployed in outdoor environments (Burnett and van Iersel, 2018). Jones (2004) provides a still-relevant summary of the advantages and disadvantages of many sensor techniques that are used for irrigation scheduling in horticultural applications. The variety of soil-moisture sensors (tensiometric, neutron, resistance, heat dissipation, psychrometric, or dielectric) has evolved rapidly in the past 25 years. But unraveling this variety of sensing methods can be daunting for growers, since each sensor typically has specific strengths and weaknesses in a specific situation, especially considering cost and ease-of-use.

Unfortunately, for ornamental production, many of these historical reviews do not address the use of sensors in soilless substrates, which have much greater porosity than most soils, which affects the precision of many soil capacitance and matric potential sensors. A specific example of this is the use of tensiometers (similarly gypsum blocks), which have long been used to measure matric potential in soils (Smajstrla and Harrison, 1998) and in soilless substrates (Burger and Paul, 1987). Although tensiometers have been valuable research tools, they have not been adopted widely in greenhouse and nursery production, mainly because of the maintenance problems with using them in highly porous substrates. Tensiometers rely on direct contact between the porous ceramic tip and substrate moisture. If the substrate dries or the tensiometer is disturbed, this contact may be disrupted. Air then enters and breaks the water column in the tensiometer, resulting in unreliable readings (Zazueta et al., 1994). The recent introductions by various manufacturers of mini-tensiometers with automatic refill devices have greatly improved the reliability of these measurements, but the high cost of these sensors remains an issue for widespread deployment in most commercial operations. Nevertheless, the measurement and integration of soil matric potential into irrigation scheduling is vitally important, since soil volumetric water content (VWC) measures total water moisture, not available soil/substrate moisture. This is especially important when we are considering alleviating plant water stress, often at soil-moisture tensions that are much lower than typically appreciated by researchers and growers alike (see Section 6).

### 4.2 Crop evapotranspiration (ET) approaches

Jones (2008) noted that an ideal automated irrigation system should be (1) sensitive to small changes, whether in terms of soil-moisture content, evaporative demand, or plant response; (2) able to respond to plant water demand in real time, allowing for continual monitoring and maintenance of optimal water status with changing weather conditions; (3) readily adaptable to different crops, growth stages, and environments without the need for extensive modification; (4) robust and reliable; (5) user-friendly, requiring little user training; (6) capable of automation, thus reducing labor requirements; and (7) low cost, both in terms of purchase price and maintenance costs.

To make real progress in applying irrigation at the frequency and duration required to optimize application efficiencies and growth, we need to couple our ability to precisely apply water, with a knowledge of soil/substrate water availability, informed by microclimatic conditions (e.g. rainfall, vapor pressure deficit, and daily light integral) in near-real time. We can take one of two approaches: First, we could use ET-based scheduling approaches, which have real value for large-scale applications such as landscape irrigation, or a hybrid approach where a diverse array of crops are being grown.

Water budget calculations are relatively easy to implement for scheduling irrigation applications to agronomic crops (or landscape applications), since there are relatively simple algorithms available to calculate crop evapotranspiration ($E_T$), using Penman-Monteith (Allen et al., 1998) or other modeling methods. These models use local meteorological station or pan evaporation data (Fereres et al., 2003) and calculate the reference daily $E_T$ ($ET_0$), which is multiplied by an empirically determined crop coefficient (Kc, crop water use/ETo) for each crop to adjust irrigation applications. At present there are good estimates of Kc values for most horticultural crops, even though most research has been conducted on the major field crops (Allen et al., 1998).

Previous studies with a variety of crop, ornamental, and turf species have reported that the use of appropriate scheduling methods and precision irrigation technologies can save significant amount of water, while maintaining or increasing yield and product quality (Bacci et al., 2008; Beeson and Brooks, 2008; Blonquist et al., 2006; Fereres et al., 2003). Many of these empirical approaches have successfully incorporated environmental variables into various models to further increase the precision of irrigation scheduling (e.g.

Treder et al., 1997). Since crop coefficients depend on rapidly changing growing conditions (Kim and van Iersel, 2011; Schuch and Burger, 1997), continuous functions have been developed for ornamental production to estimate daily crop water use (ETc) rates from ETo (Beeson, 2005, 2010; Grant et al., 2012; Irmak, 2005; Pardossi et al., 2009; Warsaw et al., 2009a,b). For example, Beeson (2010) developed ET functions that use projected plant area and a growth index to estimate ETc from ETo. Million et al. (2011), using a partial cover function developed by Ritchie (1972) for field crops, estimated ETc as a function of ETo and leaf area index. But inaccuracies remain in measuring actual water use in container nurseries. Unlike in field crop situations, partial cover in container nurseries can result in significant canopy temperature increases, when solar radiation heats non-radiating black container sidewalls and ground-cloth surfaces (Million et al., 2011).

The ETc-based approach tends to be more precise when it is combined with regular soil/substrate moisture to integrate daily Ks values instead of seasonal Kc values (Lea-Cox et al., 2019) or combined with leaching fraction monitoring techniques (Million and Yeager, 2019), which can help reset the model (e.g. after irrigation or rainfall). Fernandez et al. provided data to devise a method to effectively scale this approach to a wide range of individual crop species within a nursery, using 'Irrigation functional groups' (IFGs), based on daily water use (Fig. 3). To schedule irrigation for a wide range of crops in the same IFG, but which are at different phenological growth stages, the measured water use of indicator plants could be adjusted to calculate irrigation requirements of crops in production areas using a limited number of sensors in a 'monitoring block', similar to the approach adopted by Belayneh et al. (2013).

Daily or weekly water use would be compared to calculated ETo, allowing the development of a database with Kc values for each IFG as a function of growing degree days and/or daily light integral. From this type of data, a software program such as Sensorweb™ (Kohanbash et al., 2013) could use a site-specific Kc database to then determine the scheduling of functional groups of plants within a nursery, providing a simple, adaptive approach to scheduling precision irrigation for a large number of species in large-scale operations. A particular strength of combining water-balance and volumetric soil-based approaches is that they address not only scheduling issues about 'when to irrigate' but also about 'how much to apply'. But although this approach may be appropriate for soil-based irrigation scheduling (at time-scales of 1–3 days), it is important to ensure that these calculations can be performed more rapidly for greenhouse and container-nursery operations, where irrigating four to six times per day may be required to maintain available water in small root volumes on hot, sunny, or windy days (Tyler et al., 1996). We therefore need to ensure that our sensing and computational frequency can operate systems at least on an hourly basis, for them to be useful for these growers (Kohanbash et al., 2013).

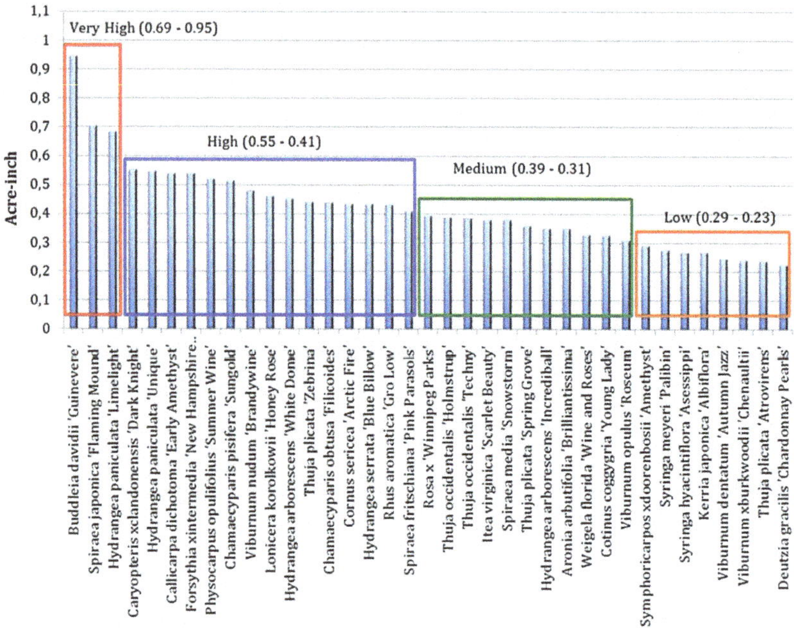


**Figure 3** Irrigation functional groups based on daily water use averaged over a season. Source: adapted from Fernandez (2018).

Most importantly, we need to simplify our approaches and technologies for any approach to be practical and likely to be adopted by growers (Majsztrik et al., 2013b). For this reason, van Iersel, Lea-Cox, Fulcher, and other research groups have focused their efforts over the last decade or so on understanding the use and precision of soil capacitance sensors in soilless substrates, to provide simpler, more cost-effective, and grower-friendly approaches. For example, van Iersel and his group showed with various studies (Nemali et al., 2007; Burnett and van Iersel, 2008; Kim and van Iersel, 2009; van Iersel et al., 2010, 2011) that automated irrigation using soil-moisture sensors allows for the very precise irrigation of greenhouse crops in soilless substrates, with significant decreases in water applications. The Managing Irrigation and Nutrition through Distributed Sensing (MINDS) project illustrated multiple economic benefits accruing from the use of sensors and automated control in ornamental production (Lea-Cox et al., 2013). The ability to maintain substrate moisture contents at very precise levels advances our capability to use precision irrigation scheduling for regulated deficit irrigation (RDI) techniques (Jones, 2004), to increase fruit crop quality (Fereres et al., 2003) and aid in precision nutrient (Ristvey et al., 2004; Bayer et al., 2015) and disease management (Lea-Cox and Ross, 2014). Nevertheless, challenges remain with the integration and scaling of sensor-based systems for use in commercial operations.

## 5 Irrigation data management and decision support systems

If we consider the bigger picture, irrigation management in a commercial operation is a complex scheduling task that requires an integration of zonal water requirements, in terms of the species being grown and the stage of crop development, both of which are dependent on site-specific environmental and management factors. Add to this the total number of species grown in a 'typical' nursery or greenhouse operation (oftentimes >250 species; Majsztrik et al., 2011), the variety of container sizes and substrate formulations used (affecting rooting volume and water-holding capacity), and it becomes obvious why irrigation scheduling in these types of operations oftentimes devolves to a convenient time-based routine (Lea-Cox, 2012), without much consideration for the actual daily water use by the plant. It is surprising how many 'advanced' irrigation scheduling systems automate irrigation cycles *only* on the basis of time, without any direct feedback to help increase the precision and timeliness of those cycles. Thus, even with programmable systems, the scheduling decision is largely dependent on human intelligence and experience and the time taken to evaluate crop water use and integrate other sources of information. It is therefore unlikely that any system that relies on human intelligence will reliably provide any further increases in irrigation efficiency. The reasons for this are quite simple: growers are time-limited, typically risk-averse, and the variable cost of water and issues associated with a lack of irrigation efficiency are typically very low for most ornamental operations.

To provide any real benefit to growers, sensor-based information systems should be designed to efficiently gather and graphically illustrate data to add value to human-decisions or fully automate the decision-making (action) process for irrigation.

The information path that any sensor-based system should provide a grower is:

Data ⇒ Information ⇒ Knowledge ⇒ Action

Collecting data with sensors is relatively easy, but data are only reliable if the limitations of those sensors are understood by human operators. Processing data into usable information is also relatively easy with good software, but poor-quality data leads to poor information. van Iersel et al. (2013) provide an excellent review of the most common environmental and soil-based monitoring sensors that can be used for irrigation and other monitoring purposes in ornamental operations. Optimizing the value of any information requires not only selecting the best sensor(s) for a particular purpose (not only optimizing cost, but also reliability), but also determining the optimal number of sensors to be deployed (to capture variability) and assuring that any data collected is as accurate and precise as possible.

When faced with a purchasing decision, it is easy to forget that the *quality of data* is paramount to the precision of the information gathered. Poor-quality data is often a function of unreliable (poor-quality) sensors, and/or poor installation or inadequate maintenance. However, if these factors are optimized, the use of various sensing approaches can result in large increases in irrigation efficiency and profitability for ornamental operations (Belayneh et al., 2013; Saavoss et al., 2016; Fernandez, 2018).

## 5.1 Precision, accuracy, and repeatability

All sensors have measurement errors and it is important to understand the different limitations that affect measurements. There are three important sources of measurement error (Fig. 4):

1. *Accuracy:* how close is the measured value to the true value being measured,
2. *Precision* (also called repeatability): how similar are the measurements if they are measured multiple times, and
3. *Resolution:* what is the smallest change in the physical characteristic that can be detected by the sensor.

## 5.2 Sensor reliability and cost

The ideal sensor combines good accuracy with good precision and high resolution, but in many cases, this increase in sensor performance is not justified by the large increase in price. However, it is important to consider what minimum specifications a sensor needs to have to be useful for a particular purpose. For example, a soil-moisture sensor used for research

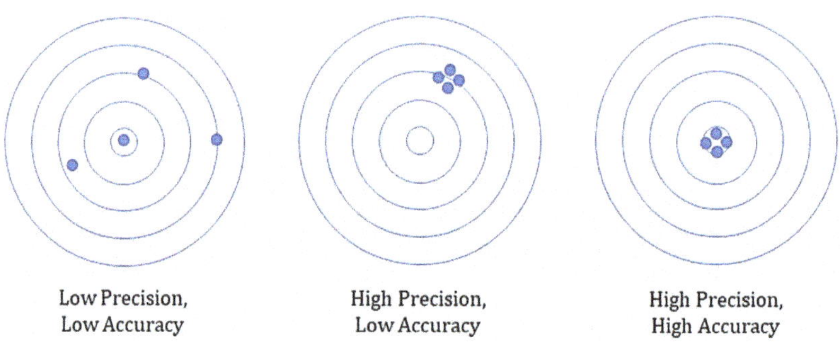


**Figure 4** Graphical presentation of differences between precision and accuracy.

purposes needs to be accurate, have good precision, and needs to have adequate resolution. However, a soil-moisture sensor that is used for irrigation control purposes does not necessarily have to have the same accuracy, as long as the readings are repeatable and have adequate resolution. If the sensor can reliably detect changes in soil/substrate water content, the sensor can be used for irrigation control, even if the readings are not accurate (van Iersel et al., 2013).

### 5.3 Sources of variability

There are many sources of variability associated with measuring soil moisture. Main factors are the age and stage of growth of the plant/crop being grown, the substrate or soil type, root density, and the unknown factor of sensor placement within the root zone/substrate. Plant water use is in turn affected by daily environmental conditions, including light, temperature, vapor pressure deficit, and how plant stoma adjust dynamically to these fluxes. A great deal of positional variability in the substrate can be removed by placing sensors at standardized depths in the root zone (i.e. 15 cm and 30 cm (6 and 12 in below an irrigation emitter)). The average VWC of a number of 15-cm sensors can be used to both initiate and stop an irrigation event (as the majority of roots are from 0 cm to 15 cm deep), whereas the 30-cm sensor provides information about whether the root zone is adequately irrigated, as well as information about EC and nutrient leaching.

### 5.4 Harnessing variability to improve reliability

It is also important to consider the frequency of data collection, which depends on the precision of the sensor, as well as a knowledge of the environment being sensed (van Iersel et al., 2013). Repeatability can be improved by (1) increasing the number of sensors and (2) collecting data more frequently, using both to calculate a running average for irrigation control decisions. Although more frequent measurements do not improve the performance of the sensor itself, averaging multiple sensors can remove the inevitable natural variation when biological environments are sensed. The optimal sampling frequency also depends on how dynamic the changes to be measured are, for example, substrate water content of large plants in relatively small containers should be measured more often than soil water content of small plants grown under field conditions.

### 5.5 Transforming information into knowledge

Most irrigation software at present are limited by a poor understanding of factors that influence water availability, particularly with respect to soils and soilless substrates. To illustrate, the total water-holding capacity of a specific soil or substrate type is influenced by a combination of soil physical properties ('texture') and drainage (influenced by topography, hard pans, rock particles, etc.). In general, this is typically much lower for sandy soils than for silt loams or clay soils (Fig. 5). Plant available water (PAW) is typically defined as the difference in the percentage of VWC between field capacity (FC) (or container-capacity; FC ≅ CC) and the proportion of soil moisture that corresponds to crop water stress.

From this, adjustments in soil-moisture thresholds can be made for scheduling irrigation events, based on capacitance sensor measurements. Handreck and Black (2010) coined the terms easily available water (EAW), readily available water (RAW), and water buffering capacity, such that WBC = EAW + RAW (Fig. 5). WBC is the amount of water that can be easily extracted by a plant, which for most soilless substrates is typically 25–35% of PAW (Fernandez, 2018). When we now consider the relationship between the water content (VWC, %) and the availability of that moisture to a plant (matric potential, kPa), it can been seen from Fig. 6a and b that this relationship is typically non-linear, and oftentimes plant water availability can be 10–20 times less than the point typically regarded as permanent wilting point in soils (Fig. 5).

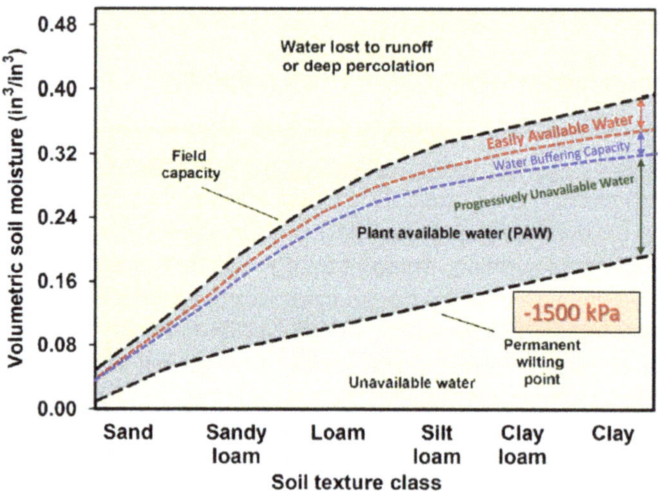


**Figure 5** General relationship between plant-available moisture (PAW) as a function of FC, permanent wilting point, and soil texture class. Source: modified from Zotarelli et al. (2010). The proportion of EAW, water-buffering capacity (WBC), and progressively unavailable water (PUW) is illustrated as a proportion of PAW (refer to Fig. 6a and b).

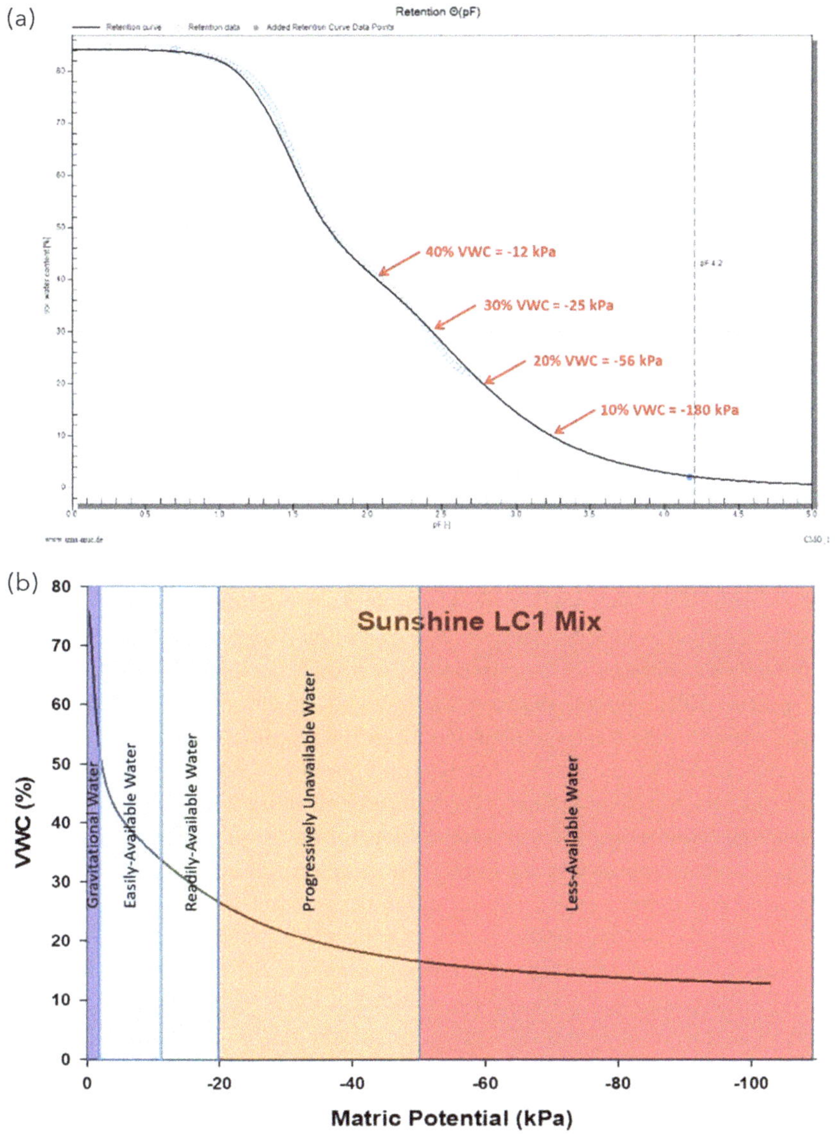


**Figure 6** The relationship between substrate water content (%) and matric potential for a commonly used peat-based substrate (LC-1, Sunshine Mix). (a) A van-Genuchten curve fitted to a substrate water retention curve ($n$ = 3), generated by a Hyprop Apparatus (Meter-Group, Pullman, WA) annotated with VWC (%); (b) the same (smoothed) van-Genuchten curve, illustrating the proportion of easily-available (−2 to −10 kPa), readily-available (−10 to −20 kPa), progressively unavailable (−20 to −50 kPa), and mostly unavailable water (>−50 kPa) in an LC-1 peat-based substrate.

## 6 Developing a universal irrigation protocol: a case study

Developing a universal irrigation protocol offers us a simple, practical way to integrate measuring soil water content (% VWC) with matric potential, by equating the 'maximum allowable depletion' (MAD) to the proportion of soil moisture defined by EAW + RAW. MAD is defined as the ideal operating range of soil water content for irrigation management (Zotarelli et al., 2010). In theory, irrigations are initiated at the lower RAW threshold *only* for the length of time that it takes to replace this soil water in the root zone, without greatly exceeding FC. This avoids over-application of water that leads to water and nutrients leaching out of the root zone. In practice, we have found that RAW is also influenced by rooting depth and the sensitivity of the crop to water stress at different ontogenic stages (e.g. vegetative canopy development versus flowering and fruit set in strawberry; B. Belayneh, unpublished data).

Lea-Cox et al. (2018) investigated the optimization of a sensor-based irrigation protocol for a large-scale cut-flower operation in southern California, where a network of wireless cellular dataloggers (EM50G; Meter-Group, Pullman, WA) and associated sensors were deployed to provide managers with soil moisture (% VWC) and electrical conductivity (EC) data, to aid in day-to-day irrigation and salinity management decisions. The primary objective was to achieve a mandated 25% cut in water use (due to an intense 4-year regional drought in CA), without affecting the yield or quality of various cut-flower crops, many of which are grown in long-lived (4–7 years) perennial blocks. The company had been using drip irrigation and precision irrigation techniques for over 20 years, but had not been able to further optimize their water use, due to difficulties in gathering and interpreting data for the large number of flowering crops grown, together with assessing topographical effects on soil hydrology and variations in daily environmental conditions (specifically light and VPD).

Since soil-moisture contents can be significantly influenced by site-specific conditions, we found that the in situ measurement of FC in each field was vital for accurately estimating the range of RAW in each block. We followed the procedure of Zotarelli et al. (2010) to practically determine the FC for each sensor datalogger installation (Fig. 7).

Soil-moisture threshold alerts for each block were created in Sensorweb™ (Lea-Cox et al., 2018). When the average set-point VWC was reached, the system texted the cell phone of the irrigation manager, who then checked the data to schedule the next irrigation. In this manner, growers were able to easily fine-tune irrigation durations to not exceed FC and reduce the frequency of irrigation events, according to daily plant water use and crop development. Figure 8 provides another example of this protocol for Myrtle, a deep-rooted perennial crop. Note that although the threshold irrigation value of 28% VWC

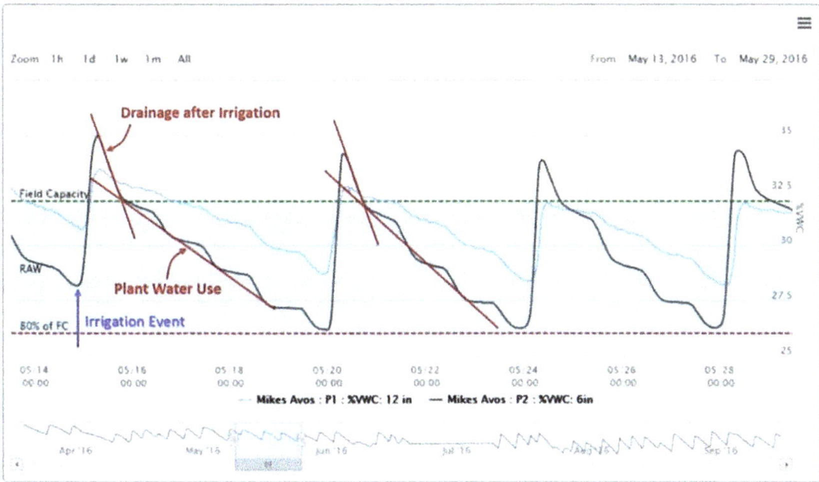


**Figure 7** Estimation of field capacity (FC) from in situ soil sensor placement (after Zotarelli et al., 2010). FC was estimated from the slope intercept of the drainage and plant water use (red lines), transposed onto the 15-cm (6 in) soil-moisture data after an irrigation (or rain event). In this specific case, 15-cm sensor dynamics were used, the depth at which the majority of the plant roots are found. Daily plant water uptake (reduction in soil moisture) can be clearly seen, particularly at the 15-cm depth (black line). Deeper sensors (30 cm; blue line) illustrate slower crop water uptake.

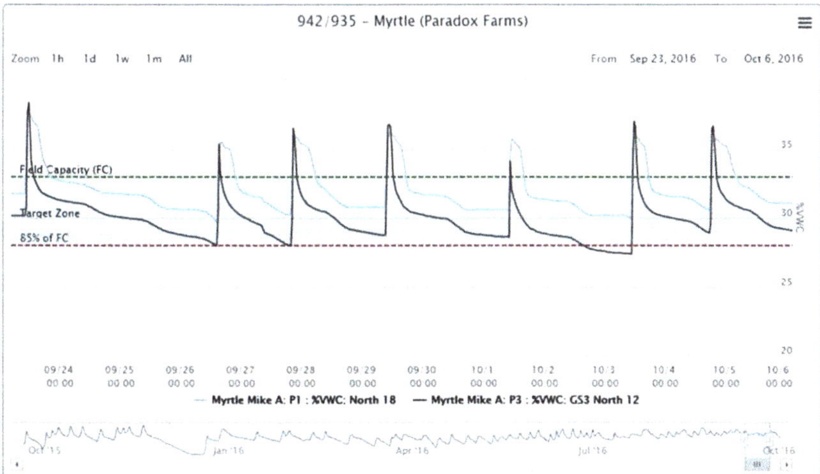


**Figure 8** Irrigation scheduling of a Myrtle (*Myrtus communis*) crop, using a threshold of 85% of FC (= 28% soil VWC) as the irrigation set-point.

(85% of FC) was fixed, the time between irrigations varies, according to daily environmental conditions. Daily plant water dynamics can also be clearly seen, particularly at 30 cm (12-in depth; black line).

Prior to implementing this sensor-based irrigation strategy, irrigation managers would typically apply irrigation 'sets' of 2-h duration, two to three times a week, based on season and the stage of crop development. After applying the previously mentioned sensor-based protocol, irrigation durations were reduced to 1-h durations, except when a leaching irrigation was required (EC data not shown). The frequency of irrigations was dictated primarily by daily environmental conditions, but also by the state of crop development. Water savings of about 40% were achieved with *Solidago* in 2014 and 50–60% for Myrtle through the fall/winter of 2015, compared to the water applied to these crops in previous years. In this way, the farm was able to achieve their targeted irrigation reductions, without compromising crop yield or quality (Lea-Cox et al., 2018).

The tangible benefits of increased irrigation efficiency include decreased reliance on existing water resources (Pitton et al., 2018), reduced pumping costs (Belayneh et al., 2013) improved crop uniformity (Chappell et al., 2013), and reduced nutrient leaching (Ristvey et al., 2004). The use of sensing techniques in ornamental production has resulted in better timing and frequency of irrigation applications, reducing total water use by 6–75%, with similar reductions in runoff water generated and 38–74% lower nutrient loading in the runoff (Lea-Cox et al., 2013; Pershey et al., 2015; Tyler et al., 1996; Warsaw et al., 2009a,b).

With better information, growers also have the ability to further refine their growing practices for increases in yield and quality. Bayer et al. (2014) found that sensor-based irrigation strategies can greatly reduce leaching of nutrients, cutting the required fertilizer applications by 50%. Majsztrik et al. (2013a) and Saavoss et al. (2016) demonstrated that more timely irrigation decisions in greenhouse production increased the yield and quality of snapdragon (cut-flowers) by 30%, depending on season and cultivar. Plant growth regulators (PGRs) are widely used in ornamental horticulture to control plant size. Alem et al. (2015) showed that the use of a controlled water deficit is an effective, non-chemical alternative to the use of PGRs in poinsettia production. The use of non-chemical growth regulation could also be used for marketing purposes, since consumer concern over the use of agro-chemicals is steadily increasing.

In addition to the direct benefits, it is the intangible benefits of improved irrigation practices that are likely to have the greatest economic impacts for growers. Chappell et al. (2013) showed that with sensor-based irrigation, disease-related losses in *Gardenia* were reduced from 30% to virtually zero and the production cycle was shortened from 14 months to 8 months, with consequent reductions in inputs (labor, fertilizer, fungicides, etc.). Overall, this resulted in a 256% increase in annualized profit (Lichtenberg et al., 2013), with a payback period of less than 1 month on the sensor network (approximately $6000). Although perhaps unusual, this study illustrates the synergistic economic benefits of increases in efficiency, yield, and disease reduction as well as increased turnover of production space.

Additionally, many producers are constrained by the capacity of their well or, more often, by the time it takes to irrigate any given area of land. Belayneh et al. (2013) showed that sensor-based automated irrigation reduced the total maximum time taken by the producer to irrigate the nursery. Although the direct savings in water costs were relatively small for this grower, the increased irrigation efficiency allowed him to increase production area by an additional 30 acres with the same amount of available water, simply by giving the grower the ability to irrigate only those plants that required water each day.

For most ornamental growers, the cost of water is very low compared to other variable costs, such as labor and environmental risk-reduction practices. The automation and control of irrigation is likely to reduce operational fixed costs by reducing the need for one or two full-time irrigation manager positions (Majsztrik et al., 2013b). For many ornamental operations, this could amount to between $50 000 and $75 000 per year. However, it is unlikely that these jobs would be lost, since lower-skilled positions would be replaced by higher-skilled (and higher paid) irrigation management positions. With better information, irrigation managers are likely to make much better and more timely irrigation decisions and translate that knowledge into better economic outcomes.

## 7 Future trends

We have a great number of tools and strategies that are available to increase the efficiency of our irrigation practices, as exemplified by many research studies and best management practices developed over the past 20–30 years. However, changes in practice at the local level are likely to be driven by a mixture of regulations (most likely due to concerns with environmental runoff of agrochemicals), the occurrence of drought and other climatic changes, and the availability and cost of fresh water resources. The cost of water used to be considered a primary driver of change, but it is apparent that there are many intangible benefits of not over-irrigating (reductions in disease and reduced crop production times) apart from the direct benefits of reducing runoff and reducing the size of capture and remediation systems. The greatest return on investment for most growers will likely be gained from good irrigation system design and regular maintenance. But for growers who do need to increase their overall water availability to ensure sustainability, we have a number of excellent resources on all of these subjects that are freely available.

Despite these advances in our knowledge, challenges remain. As scientists and extension specialists, we need to create more elegant (simpler, more effective) tools for growers to implement better practices at minimal cost. Software tools that integrate multiple data sources (as exemplified in the case study) will help increase irrigation efficiency beyond simple scheduling with a time clock, which is still the predominant practice. Integrating methods to cost-effectively measure matric potential and reduce plant water stress in the variety of soilless substrates the industry uses - and for the range of species grown - is a particular challenge. Nevertheless, using common-sense approaches by integrating both sensing and modeling approaches (perhaps by using IFGs) will help growers achieve even greater water use efficiencies for minimal cost.

## 8 Where to look for further information

- Clean WateR$^3$ project is focused on research and outreach to help growers 'reduce', 'remediate', and 'recycle' irrigation water. (https://www.cleanwater3.org/about.asp).
- Green Industry Knowledge Center for Water, Nutrient and Plant Health Management: 26 online learning modules at: http://waternut.org/moodle.
- Southern Nursery Association Best Management Practices: A Guide for Producing Nursery Crops. http://contents.sna.org/bmpv30.html.
- *Water, Root Media and Nutrient Management for Greenhouse Crops*. K. M. Williams, S. Mangiafico and D. Merhaut. (Eds.). University California Press, Publication 3551. (https://anrcatalog.ucanr.edu/Details.aspx?itemNo=3551).

## 9 References

Abdi, D. E. and Fernandez, R. T. 2019. Workshop: reducing water and pesticide movement in nursery production. *HortTechnology* 29(6), 730–5. doi:10.21273/HORTTECH04298-19.

Aldrich, R. A. and Bartok Jr., J. W. 1994. Greenhouse Engineering. NRAES-33. Natural Resource, Agriculture and Engineering Service, Cooperative Extension, Ithaca, NY.

Alem, P., Thomas, P. A. and van Iersel, M. W. 2015. Controlled water deficit as an alternative to plant growth retardents for regulation of poinsettia stem elongation. *HortScience* 50(4), 565–9. doi:10.21273/HORTSCI.50.4.565.

Allen, R. G., Pereira, L. S., Raes, D. and Smith, M. 1998. Crop evapotranspiration (guidelines for computing crop water requirements). FAO Irrigation and Drainage Paper No. 56. United Nations Food and Agriculture Organization, Rome, Italy. Available at: http://www.fao.org/3/X0490E/x0490e00.htm (accessed on 16 March 2020).

Bacci, L., Battista, P. and Rapi, B. 2008. An integrated method for irrigation scheduling of potted plants. *Sci. Hortic.* 116(1), 89–97. doi:10.1016/j.scienta.2007.11.005.

Bailey, D., Bilderback, T. D. and Bir, D. 1999. Water considerations for container production of plants. NC State University Department of Horticultural Science Horticulture Information Leaflet 557, 11p.

Bayer, A., Whitaker, K., Chappell, M., Ruter, J. and van Iersel, M. W. 2014. Effect of irrigation duration and fertilizer rate on plant growth, substrate EC, and leaching volume. *Acta Hortic.* 1034, 477–84.

Bayer, A. J., Ruter, J. and van Iersel, M. W. 2015. Optimizing irrigation and fertilization of *Gardenia jasminoides* for good growth and minimal leaching. *HortScience* 50(7), 994–1001. doi:10.21273/HORTSCI.50.7.994.

Beeson Jr., R. C. 2005. Modeling irrigation requirements for landscape ornamentals. *HortTechnology* 15, 6–23.

Beeson Jr., R. C. 2010. Modeling actual evapotranspiration of *Viburnum odoratissimum* during production from rooted cuttings to market size plants in 11.4-L containers. *HortScience* 45(8), 1260–4. doi:10.21273/HORTSCI.45.8.1260.

Beeson Jr., R. C. and Brooks, J. 2008. Evaluation of a model based on ETo for precision irrigation using overhead sprinklers during nursery production of *Ligustrum japonica* grown in 11-L containers. *Acta Hortic.* 792, 85–90.

Beeson Jr., R. C. and Yeager, T. H. 2003. Plant canopy affects sprinkler application efficiency of container-grown ornamentals. *HortScience* 38, 1373–7.

Belayneh, B. E., Lea-Cox, J. D. and Lichtenberg, E. 2013. Costs and benefits of implementing sensor-controlled irrigation in a commercial pot-in-pot container nursery. *HortTechnology* 23(6), 760–9. doi:10.21273/HORTTECH.23.6.760.

Berghage, R. D., MacNeal, E. P., Wheeler, E. F. and Zachritz, W. H. 1999. 'Green' water treatment for the green industries: opportunities for biofiltration of greenhouse and nursery irrigation water and runoff with constructed wetlands. *HortScience* 34(1), 50–4. doi:10.21273/HORTSCI.34.1.50.

Bilderback, T., Boyer, C., Chappell, M., Fain, G., Fare, D., Gilliam, C., Jackson, B. E., Lea-Cox, J., LeBude, A. V., Niemiera, A., Owen, J., Ruter, J., Tilt, K., Warren, S., White, S., Whitewell, T., Wright, R. and Yeager, T. 2013. *Best Management Practices: Guide for Producing Nursery Crops* (3rd edn.). Southern Nursery Association, Acworth, GA.

Bitelli, M. 2011. Measuring soil water content: a review. *HortTechnology* 21(3), 293–300. doi:10.21273/HORTTECH.21.3.293.

Blonquist, J. M., Jones, S. B. and Robinson, D. A. 2006. Precise irrigation scheduling for turfgrass using a subsurface electromagnetic soil moisture sensor. *Agric. Water Manag.* 84(1–2), 153–65. doi:10.1016/j.agwat.2006.01.014.

Board, C. S. W. R. C. 2015. Irrigated lands regulatory program. Available at: http://www.swrcb.ca.gov/waterissues/programs/agriculture/docs/aboutagwaivers.pdf (accessed on 10 April 2019).

Burger, D. W. and Paul, J. L. 1987. Soil moisture measurements in containers with solid state electronic tensiometers. *HortScience* 22, 309–10.

Burnett, S. E. and van Iersel, M. W. 2008. Morphology and irrigation efficiency of *Gaura lindheimeri* grown with capacitance sensor-controlled irrigation. *HortScience* 43(5), 1555–60. doi:10.21273/HORTSCI.43.5.1555.

Burnett, S. E. and van Iersel, M. W. 2018. Water, irrigation systems, and their control. In: Williams, K. M., Mangiafico, S. and Merhaut, D. (Eds), *Water and Nutrient Management for Greenhouse Crops*. University of California Press, CA, pp. 11–45. Chapter 2.

Chappell, M., Dove, S. K., van Iersel, M. W., Thomas, P. A. and Ruter, J. 2013. Implementation of wireless sensor networks for irrigation control in three container nurseries. *HortTechnology* 23(6), 747–53. doi:10.21273/HORTTECH.23.6.747.

Cifre, J., Bota, J., Escalona, J. M., Medrano, H. and Flexas, J. 2005. Physiological tools for irrigation scheduling in grapevine (*Vitis vinifera* L.). *Agric. Ecosyst. Environ.* 106(2–3), 159–70. doi:10.1016/j.agee.2004.10.005.

Dane, J. H. and Topp, G. C. 2002. *Methods of Soil Analysis. Part 4: Physical Methods*. Soil Science Society of America, Madison, WI.

Fereres, E., Goldhamer, D. A. and Parsons, L. R. 2003. Irrigation water management of horticultural crops. *HortScience* 38(5), 1036–42. doi:10.21273/HORTSCI.38.5.1036.

Fernandez, R. T. 2018. Keeping nutrients in their place: irrigation management to enhance nutrient retention in container production. *Acta Hortic.* 1212(1212), 89–96. doi:10.17660/ActaHortic.2018.1212.18.

Fulcher, A., LeBude, A. V., Owen, J. S., White, S. A. and Beeson Jr., R. C. 2016. The next ten years: strategic vision of water resources for nursery producers. *HortTechnology* 26(2), 121–32. doi:10.21273/HORTTECH.26.2.121.

Grant, O. M., Davies, M. J., Longbottom, H. and Harrison-Murray, R. 2012. Evapotranspiration of container ornamental shrubs: modelling crop-specific factors for a diverse range of crops. *Irrig. Sci.* 30(1), 1–12. doi:10.1007/s00271-010-0258-9.

Handreck, K. A. and Black, N. 2010. *Growing Media for Ornamental Plants and Turf* (4th edn.). University of New South Wales Press, Sydney, Australia.

Irmak, S. 2005. Crop evapotranspiration and crop coefficients of *Viburnum odoratissimum* (Ker.-Gawl). *Appl. Eng. Agric.* 21(3), 371–81. doi:10.13031/2013.18456.

Jones, H. G. 2004. Irrigation scheduling: advantages and pitfalls of plant-based methods. *J. Exp. Bot.* 55(407), 2427–36. doi:10.1093/jxb/erh213.

Jones, H. G. 2008. Irrigation scheduling – comparison of soil, plant and atmosphere monitoring approaches. *Acta Hort.* 792, 391–403.

Kim, J. and van Iersel, M. W. 2009. Daily water use of abutilon and lantana at various substrate water contents. *Proc. S. Nurs. Assn. Res. Conf.* 54, 12–16.

Kim, J. and van Iersel, M. W. 2011. Slowly-developing drought stress increases photosynthetic acclimation of *Catharanthus roseus*. *Physiol. Plant.* 143(2), 166–77. doi:10.1111/j.1399-3054.2011.01493.x.

Kohanbash, D., Kantor, G., Martin, T. and Crawford, L. 2013. Wireless sensor network design for monitoring and irrigation control: user-centric hardware and software development. *HortTechnology* 23(6), 725–34. doi:10.21273/HORTTECH.23.6.725.

Lea-Cox, J. D. 2012. Using wireless sensor networks for precision irrigation scheduling. In: Kumar, M. (Ed.), *Problems, Perspectives and Challenges of Agricultural Water Management*. InTech Press, Rijeka, Croatia, pp. 233–58. Chapter 12.

Lea-Cox, J. D. and Ross, D. S. 2001. A review of the Federal Clean Water Act and the Maryland Water Quality Improvement Act: the rationale for developing a water and nutrient management planning process for container nursery and greenhouse operations. *J. Environ. Hortic.* 19, 226–9.

Lea-Cox, J. D. and Ross, D. S. 2014. Water management to minimize pathogen movement in containerized production systems. In: Hong, C. H., Moorman, G. W. and Wohanka, W. (Eds), *Biology, Detection and Management of Plant Pathogens in Irrigation Water*. American Phytopathology Society, St. Paul, MN, pp. 377–87. Chapter 32.

Lea-Cox, J. D. and Ross, D. S. 2018. Managing water and nutrients to reduce environmental impact. In: Williams, K. M., Mangiafico, S. and Merhaut, D. (Eds), *Water and Nutrient Management for Greenhouse Crops*. University of California Press, CA, pp. 273–88. Chapter 16.

Lea-Cox, J. D., Ross, D. S. and Teffeau, K. M. 2001. A water and nutrient management planning process for container nursery and greenhouse production systems in Maryland. *J. Environ. Hortic.* 19, 230–6.

Lea-Cox, J. D., Ross, D. S., Ristvey, A. G. and Murray, J. D. 2002. Estimating nitrogen and phosphorus total maximum daily loads for container nursery and greenhouse production systems. *Proceedings of the Total Maximum Daily Load Environmental Regulations Conference*, Forth Worth, Texas, pp. 466–71.

Lea-Cox, J. D., Bauerle, W. L., van Iersel, M. W., Kantor, G. F., Bauerle, T. L., Lichtenberg, E., King, D. M. and Crawford, L. 2013. Advancing wireless sensor networks for irrigation management of ornamental crops: an overview. *HortTechnology* 23(6), 717–24. doi:10.21273/HORTTECH.23.6.717.

Lea-Cox, J. D., Williams, J. and Mellano, M. A. 2018. Optimizing a sensor-based irrigation protocol for a large-scale cut-flower operation in southern California. *Acta Hortic.* 1197, 219–25.

Lea-Cox, J. D., Belayneh, B. E., Starry, O. and DeStefano, D. 2019. Monitoring urban landscapes to measure ecosystem services. *Proc. S. Nurs. Assoc. Res. Conf.* 64 (in Press).

Lichtenberg, E., Majsztrik, J. and Saavoss, M. 2013. Profitability of sensor-based irrigation in greenhouse and nursery crops. *HortTechnology* 23(6), 770–4. doi:10.21273/HORTTECH.23.6.770.

Majsztrik, J. C. and Lea-Cox, J. D. 2013. Water quality regulations in the Chesapeake Bay: working to more precisely estimate nutrient loading rates and incentivized best management practices in the nursery and greenhouse industry. *HortScience* 48(9), 1097–102. doi:10.21273/HORTSCI.48.9.1097.

Majsztrik, J. C., Ristvey, A. G. and Lea-Cox, J. D. 2011. Water and nutrient management in the production of container-grown ornamentals. *Hort. Rev.* 38, 253–97.

Majsztrik, J. C., Lichtenberg, E. and Saavoss, M. 2013a. Ornamental grower perceptions of wireless irrigation sensor networks: results from a national survey. *HortTechnology* 23(6), 775–82. doi:10.21273/HORTTECH.23.6.775.

Majsztrik, J. C., Price, E. W. and King, D. M. 2013b. Environmental benefits of wireless sensor-based irrigation networks: case-study projections and potential adoption rates. *HortTechnology* 23(6), 783–93. doi:10.21273/HORTTECH.23.6.783.

Majsztrik, J. C., Fernandez, R. T., Fisher, P. R., Hitchcock, D. R., Lea-Cox, J. D., Owen Jr., J. S., Oki, L. R. and White, S. A. 2017. Water use and treatment in containerized specialty crop production: a review. *Water Air Soil Pollut.* 228(4), 151. doi:10.1007/s11270-017-3272-1.

Majsztrik, J. C., Ristvey, A. G., Ross, D. S. and Lea-Cox, J. D. 2018. Comparative water and nutrient applications among ornamental operations in Maryland. *HortScience* 53(9), 1364–71. doi:10.21273/HORTSCI12685-18.

Million, J. B. and Yeager, T. H. 2019. Testing an automated irrigation system based on leaching fraction testing and weather in a container nursery. *HortTechnology* 29(2), 114–21. doi:10.21273/HORTTECH04213-18.

Million, J. B., Ritchie, J. T., Yeager, T. H., Larsen, C. A., Warner, C. D. and Albano, J. P. 2011. CCROP – simulation model for container-grown nursery plant production. *Sci. Hortic.* 130(4), 874–86. doi:10.1016/j.scienta.2011.08.030.

Nemali, K. S., Montesano, F., Dove, S. K. and van Iersel, M. W. 2007. Calibration and performance of moisture sensors in soilless substrates: $ECH_2O$ and Theta probes. *Sci. Hortic.* 112(2), 227–34. doi:10.1016/j.scienta.2006.12.013.

Newman, J. P. 2018. Water issues and irrigation management–a California perspective. Sidebar 2.1. In: Williams, K. M., Mangiafico, S. and Merhaut, D. (Eds), *Water and Nutrient Management for Greenhouse Crops*. University of California Press, CA, p. 18.

Pardossi, A., Incrocci, I., Incrocci, G., Tognoni, F. and Marzialetti, P. 2009. What limits and how to improve water use efficiency in outdoor container cultivation of ornamental nursery stocks. *Acta Hortic.* 843, 73–80.

Pershey, N. A., Cregg, B. M., Andresen, J. A. and Fernandez, R. T. 2015. Irrigating based on daily water use reduces nursery runoff volume and nutrient load without reducing growth of four conifers. *HortScience* 50(10), 1553–61. doi:10.21273/HORTSCI.50.10.1553.

Pitton, B. J. L., Hall, C. R., Haver, D. L., White, S. A. and Oki, L. R. 2018. A cost analysis for using recycled irrigation runoff water in container nursery production: a Southern California nursery case study. *Irrig. Sci.* 36(4–5), 217–26. doi:10.1007/s00271-018-0578-8.

Raviv, M., Lieth, J. H. and Wallach, R. 2000. Effect of root-zone physical properties of coir and UC mix on performance of cut rose (cv. Kardinal). *Acta Hortic.* 554, 231–8.

Ritchie, J. T. 1972. A model for predicting evaporation from a low crop with incomplete cover. *Water Resour. Res.* 8(5), 1204–13. doi:10.1029/WR008i005p01204.

Ristvey, A. G. 2004. Water and nutrient dynamics in container-nursery production systems, PhD Diss. University of Maryland, College Park, MD.

Ristvey, A. G., Belayneh, B. E. and Lea-Cox, J. D. 2019. A comparison of irrigation-water containment methods and management strategies between two ornamental production systems, to minimize water security threats. *Water.* 11, 2588. https://doi.org/10.3390/w11122558.

Ristvey, A. G., Lea-Cox, J. D. and Ross, D. S. 2004. Nutrient uptake, partitioning and leaching losses from container-nursery production systems. *Acta Hortic.* 630(630), 321–8. doi:10.17660/ActaHortic.2004.630.40.

Ross, D. S. 2008a. Basic hydraulics: irrigation management. In: Lea-Cox, J. D., Ross, D. S. and Zhao, C. (Eds), *Green Industry Knowledge Center for Water and Nutrient Management Learning Modules*. University of Maryland, College Park, MD. Available at: http://www.waternut.org/moodle/course/view.php?id=16.

Ross, D. S. 2008b. Irrigation system design and components. In: Lea-Cox, J. D., Ross, D. S. and Zhao, C. (Eds), *Green Industry Knowledge Center for Water and Nutrient Management Learning Modules*. University of Maryland, College Park, MD. Available at: http://www.waternut.org/moodle/course/view.php?id=19.

Ross, D. S. 2008c. Irrigation system audits. In: Lea-Cox, J. D., Ross, D. S. and Zhao, C. (Eds), *Green Industry Knowledge Center for Water and Nutrient Management Learning Modules*. University of Maryland, College Park, MD. Available at: http://www.waternut.org/moodle/course/view.php?id=26.

Ross, D. S. 2008d. Containment basin design. In: Lea-Cox, J. D., Ross, D. S. and Zhao, C. (Eds), *Green Industry Knowledge Center for Water and Nutrient Management Learning Modules*. University of Maryland, College Park, MD. Available at: http://www.waternut.org/moodle/course/view.php?id=29.

Ross, D. S., Lea-Cox, J. D. and Teffeau, K. M. 2001. The importance of water in the nutrient management process. *Proc. S. Nurs. Assoc. Res. Conf.* 46, 588–91.

Saavoss, M., Majsztrik, J. C., Belayneh, B. E., Lea-Cox, J. D. and Lichtenberg, E. 2016. Yield, quality, and profitability of sensor-controlled irrigation: a case study of snapdragon (*Antirrhinum majus* L.) production. *Irrig. Sci.* 34(5), 409–20. doi:10.1007/s00271-016-0511-y.

Schuch, U. K. and Burger, D. W. 1997. Water use and crop coefficients of woody ornamentals in containers. *J. Amer. Soc. Hortic. Sci.* 122(5), 727–34. doi:10.21273/JASHS.122.5.727.

Smajstrla, A. G. and Harrison, D. S. 1998. Tensiometers for soil moisture measurement and irrigation scheduling. Univ. Florida Inst. Food Agr. Serv. Ext. Circ. 487.

Smith, K. A. and Mullins, C. 2001. *Soil and Environmental Analysis: Physical Methods* (2nd edn.). Marcel Decker, New York, NY.

Topp, G. C. and Davis, J. L. 1985. Time-domain reflectometry (TDR) and its application to irrigation scheduling. *Adv. Irrig.* 3, 107–27. doi:10.1016/B978-0-12-024303-7.50008-X.

Treder, J., Matysiak, B., Nowak, J. and Treder, W. 1997. Evapotranspiration and potted plants water requirements as affected by environmental factors. *Acta Hortic.* 449(449), 235–40. doi:10.17660/ActaHortic.1997.449.33.

Tyler, H. S., Warren, S. L. and Bilderback, T. E. 1996. Cyclic irrigation increases irrigation application efficiency and decreases ammonium losses. *J. Environ. Hortic.* 14, 194–8.

US Department of Agriculture. 2014. *2013 Farm and Ranch Irrigation Survey*. National Agricultural Statistics Service, Washington DC. Available at: https://www.nass.usda.gov/Publications/AgCensus/2012/Online_Resources/Farm_and_Ranch_Irrigation_Survey/(accessed on 28 March 2019).

van Iersel, M. W., Dove, S., Kang, J. G. and Burnett, S. E. 2010. Growth and water use of petunia as affected by substrate water content and daily light integral. *HortScience* 45(2), 277–82. doi:10.21273/HORTSCI.45.2.277.

van Iersel, M. W., Dove, S. and Burnett, S. E. 2011. The use of soil moisture probes for improved uniformity and irrigation control in greenhouses. *Acta Hortic.* 893(893), 1049–56. doi:10.17660/ActaHortic.2011.893.119.

van Iersel, M. W., Chappell, M. and Lea-Cox, J. D. 2013. Sensors for improved efficiency of irrigation in greenhouse and nursery production. *HortTechnology* 23(6), 735–46. doi:10.21273/HORTTECH.23.6.735.

Warsaw, A. L., Fernandez, R. T., Cregg, B. M. and Andresen, J. A. 2009a. Water conservation, growth, and water use efficiency of container-grown woody ornamentals irrigated based on daily water use. *HortScience* 44(5), 1308–18. doi:10.21273/HORTSCI.44.5.1308.

Warsaw, A. L., Fernandez, R. T., Cregg, B. M. and Andresen, J. A. 2009b. Container-grown ornamental plant growth and water runoff nutrient content and volume under four irrigation treatments. *HortScience* 44(6), 1573–80. doi:10.21273/HORTSCI.44.6.1573.

White, S. A., Owen, J. S., Majsztrik, J. C., Oki, L. R., Fisher, P. R., Hall, C. R., Lea-Cox, J. D. and Fernandez, R. T. 2019. Greenhouse and Nursery Water Management Characterization and Research Priorities. *Water* 11, 2338. https://doi.org/10.3390/w11112338.

Zazueta, F. S., Yeager, T. H., Valiente, J. I., Brealey, J. A. 1994. A modified tensiometer for irrigation control in potted ornamental production. *Proc. Soil Crop Sci. Soc. Fla.* 53, 36–9.

Zotarelli, L., Dukes, M. D. and Morgan, K. T. 2010. Interpretation of soil moisture content to determine soil field capacity and avoid over-irrigating sandy soils using soil moisture sensors. IFAS Publication AE460. University of Florida Cooperative Extension Service, Gainesville, FL, 4p. Available at: http://edis.ifas.ufl.edu/ae460.

# Chapter 10

# Advances in protected cultivation of ornamentals

*James E. Faust, Clemson University, USA*



## 1 Introduction

The twenty-first century has witnessed an increasingly mature ornamentals industry in North America and Europe. After decades of continual growth, demand and prices for ornamental plants have flattened while operating costs have continued to increase. The resulting smaller profit margins have dictated a changing approach to business. The industry has entered an era of consolidation, where fewer, but much larger, greenhouse operations exist. Mid-sized businesses have been bought out or have become contract growers for the large growers that are the sole suppliers to the national retailers. The ownership of businesses in the ornamentals sector, particularly among the breeding companies, is evolving from family-owned operations to corporate entities, including private equity investors.

The need for increased production efficiency through reduced labor inputs and the influx of investment dollars have driven the development and purchase of new equipment, facilities, and technologies that provide automated and robotic systems to perform the tasks essential to plant production. These systems' heightened requirements for uniform plant materials challenge the grower's ability to precisely manipulate growth of millions of plants at a time.

http://dx.doi.org/10.19103/AS.2020.0066.13

Tightening markets have also sharpened retailer focus. Today's retailers expect better quality, better performance, and more exact timing, and they are presenting growers with ever-more-precise specifications for crops and delivery dates. Consequently, growers' skill at manipulating growth and flowering is under continual assessment. Plants must meet the specified height, width, flower number, and flower size on an exact, predetermined shipping date to be deemed a successful crop. There is little room for error, and profit margins are slim.

High-precision growing requires a reliable supply of clean (disease- and pest-free), young plants. Young-plant suppliers of tissue-culture plantlets and unrooted cuttings have largely relocated to countries that offer a stable economic situation, a climate that requires minimal energy costs for greenhouse production and an ample supply of agricultural workers. The quality of the stock plant production greenhouses built over the past decade in countries like Ethiopia, Nicaragua, and El Salvador rivals those found anywhere in the world, and the tissue-culture labs in China, Vietnam, and India are state of the art. From start to finish, the cultivation of ornamentals across the world and throughout the supply chain has become a highly technical undertaking.

The end consumer's expectations of agricultural businesses in general have grown. While the focus on ornamental plants is less keen than on food crops, consumers expect that the plants they purchase do not have residues that are dangerous to themselves, to their pets, or to the pollinators in their gardens. Many growers have responded by voluntarily participating in certification programs that seek to promote sustainable practices, such as reducing chemical inputs. Some certification programs also address social issues surrounding workers' well-being. Consumers expect growers to be good stewards of natural resources by seeking out and using renewable resources and materials while striving to reduce their carbon footprint.

Maintaining a profitable ornamentals business in this era requires unprecedented depth of knowledge, skill, and adaptability. An understanding of the production challenges facing today's ornamentals industry helps us appreciate the key issues horticultural scientists find themselves tasked with solving. These issues fall into three categories which are addressed in this chapter:

- Utilization of light-emitting diodes (LEDs) to create new, energy-efficient lighting systems to grow plants in novel ways;
- Manipulation of plant growth and development through the management of the greenhouse light environment, improved cutting production and propagation systems, and new applications for plant growth regulators; and
- Development of sustainable production practices, including the use of renewable resources and materials.

## 2 Light-emitting diode (LED) lighting systems

The advent of LEDs has generated a renewed interest in lighting systems in protected cultivation. The primary benefits of LEDs over previous generations of artificial lighting systems are their higher energy-use efficiency and longer life span. The application of LED technologies to plant production has required plant researchers, plant growers, and LED manufacturers to combine efforts to identify the lighting systems that will generate the most benefit to the horticulture industry.

Lighting system trends for plant production can be divided into two primary uses: supplemental and photoperiodic lighting. Supplemental lighting is used to supplement natural sunlight by delivering additional photons to generate photosynthesis. The sun delivers a photosynthetic photon flux density (PPFD) of up to ~2000 μmol $m^{-2}$ $s^{-1}$, while most commercial light system installations deliver only 50–100 μmol $m^{-2}$ $s^{-1}$. Thus, supplemental lighting is primarily used during periods of low ambient sunlight. Photoperiodic lighting is low-intensity lighting, which is used to influence flowering by creating long-day (short-night) conditions that stimulate flowering of long-day plants or inhibit flowering of short-day plants. Typically, a PPFD of 1 to 2 μmol $m^{-2}$ $s^{-1}$ is sufficient to elicit photoperiodic responses in plants.

In addition to the impact that LEDs are having on supplemental and photoperiodic lighting systems, LEDs are creating new opportunities for plant production in vertical farming (warehouse) situations that omit sunlight, that is, sole-source lighting situations. Because LEDs deliver a relatively low level of infrared energy (heat) to the plant canopy, they can be positioned close to the canopy, allowing production benches to be stacked vertically inside warehouses. Finally, LEDs provide the potential to precisely deliver a specific light spectrum to the plant canopy. The resulting possibilities are beyond any manipulation of plant growth and development previously achieved during the long history of agriculture.

### 2.1 Supplemental lighting

The efficiency of a luminaire for converting electrical energy into light for plant growth is described by the photosynthetic photon efficiency (PPE), which is a measure of the photons of visible light (400–700 nm) emitted by a luminaire per unit of energy required to power the luminaire. The unit is μmol/J, and the calculation is:

$$\text{Photon output } (\mu\text{mol/s}) \times \text{Electrical input}(\text{W or J/s}) = \text{Photon efficiency } (\mu\text{mol/J})$$

The current industry standard for greenhouse lighting is the high-pressure sodium (HPS) luminaire. HPS luminaires emit a large amount of near-infrared energy that is directed to the underlying plant canopy. Consequently, the luminaires are typically mounted ≥2 m above the plants to avoid damaging leaves and flowers. Commercially used 400W, single-ended HPS luminaires have a PPE of ~0.9 μmol/J; however, the newer technology of 1000 W, double-ended HPS luminaires (Fig. 1) has increased the efficiency to ~1.7 μmol/J (Runkle and Bugbee, 2017). In a 2014 study, the most efficient double-ended HPS luminaires proved equally efficient to the two most efficient LED luminaires and more efficient than eight of the LED luminaires on the market at that time (Nelson and Bugbee, 2014). Three of the LED luminaires had a PPE <1.0 μmol/J.

LED technology is rapidly developing, as indicated in a 2017 study that reported the PPE range of ten LEDs to be 1.1 to 2.6 μmol/J (Radetsky, 2018). Manufacturers are now marketing LED luminaires reported to have an efficiency of up to 3.2 μmol/J. While the theoretical maximum efficiency is 4.6–5.1 μmol/J, those numbers are not expected to be achieved in the near future.

Maximum efficiencies are achieved by reducing the power delivered to the luminaire. The downside of ramping down the power requirement is that

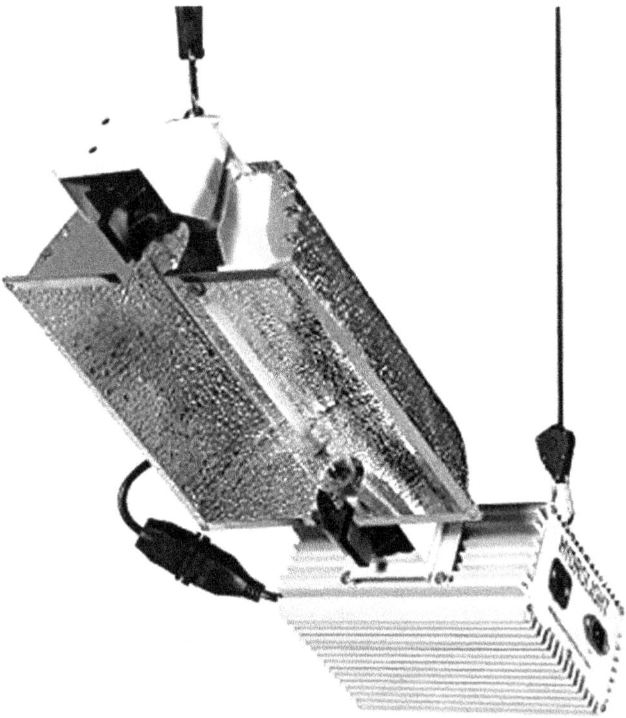

**Figure 1** An example of a double-ended, high-pressure sodium luminaire. Photo: HydroLight.

the photosynthetic photon flux (PPF, μmol/s) emitted by the luminaire also decreases; thus, the number of LEDs in the luminaire must be increased to generate an equivalent PPF. Increased numbers of LEDs in luminaire obviously increases the cost of the LED luminaire and creates the dilemma of how to balance initial capital investment costs with operating costs. For example, growers must decide whether to invest in an expensive, highly efficient luminaire that has lower operating costs or purchase a less expensive, less efficient luminaire that has higher operating costs. This example underscores one of the challenges of the current state of LED technology as manufacturers attempt to balance affordability and energy efficiency.

PPE is not the only measurement to consider when evaluating lighting systems. Photon capture is an additional measure that describes the percentage of photons intercepted by the plant canopy. Photon capture depends on the spatial arrangement of the luminaires as well as the underlying plants, and it can be maximized to greater than 90% in large greenhouses with narrow aisles (Nelson and Bugbee, 2014). Because LED luminaires can be manufactured to emit photons at different angles, the design criterion can be customized to the benching area within a greenhouse. Such an adjustment improves the efficiency of photon capture from LED luminaires in smaller greenhouses or in greenhouses with smaller benches. LED arrays can also be positioned within tall-plant canopies to improve the photosynthetic efficiency of leaves in the lower half of the plant canopy. This technique is termed 'intra-canopy lighting' (Fig. 2) and is being studied primarily in high-wire vegetable production (Kumar et al., 2016). For ornamental crops, intra-canopy lighting may prove useful for cut flower rose production.

Shadows cast by fixtures must also be taken into consideration when evaluating luminaire efficiency. Because LED luminaires provide a lower PPF than HPS luminaires, more LED luminaires are required to provide an equivalent PPF. Typically, three LED luminaires are needed to provide a PPFD to the plant canopy equivalent to a single 1000W HPS luminaire. Consequently, LED luminaire shadows range from 13–55% compared to only 5% for HPS luminaires (Radetsky, 2018). LED luminaires are configured into rectangular panels or linear arrays (Fig. 3) of variable length, and growers can compensate for shadows to some degree by mounting linear arrays directly under greenhouse truss supports.

Unfortunately, the energy-saving potential of LEDs can be overstated when comparisons fail to consider the PPFD provided by each lighting system (Stober et al., 2017). If comparisons in plant growth are made based on providing equal power inputs, then LEDs appear to be superior to HPS luminaires. However, the most important characteristic of lighting systems is the PPFD delivered to the plant canopy. For example, in a comparative analysis, two LED luminaires had a higher PPE efficiency than the HPS luminaires tested; however, when

**Figure 2** An example of intra-canopy lighting with an LED array in a high-wire tomato crop. Photo: Celina Gomez.

equivalent PPFD was provided, both LED luminaires had a higher total power demand than the HPS luminaires (Radetsky, 2018).

The economics of LED use demands consideration of many factors. Double-ended HPS luminaires have a life expectancy of 10 000 h, while LEDs have a predicted life expectancy of 50 000 h; however, the life expectancy of LEDs is reduced when operated at high temperature or high power. The life-cycle cost for ten LED luminaires and two HPS luminaires were compared, and three of the LED luminaires were found to have lower life-cycle costs than

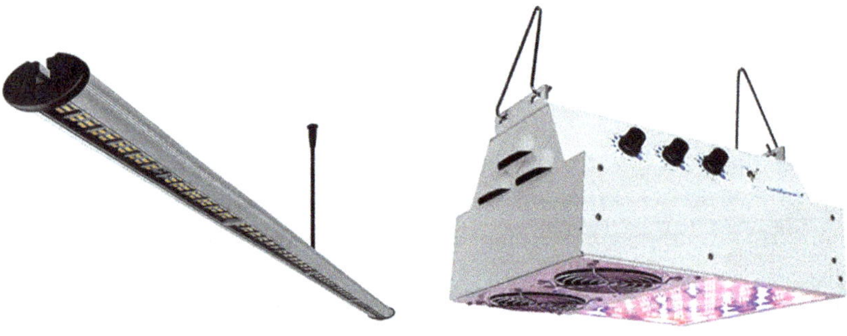

**Figure 3** Examples of LED luminaires. Left: a linear array of white LEDs; right: a rectangular array of blue and red LEDs. Photos: Fluence Bioscience, Lumigrow.

the HPS luminaires tested (Radetsky, 2018). Operating costs for LEDs can be substantially lower than those for HPS, but the initial capital cost per photon can be five to ten times higher than HPS luminaires. The calculated return-on-investment also depends on the price of electricity; LED systems are more attractive when energy prices are higher. Finally, increased legalization of cannabis worldwide is driving investment in indoor LED systems. The unusually high profitability of cannabis has allowed LED prices to remain high, and thus LEDs continue to be cost prohibitive as supplemental lighting for much of the ornamentals industry. Thus far, investment in LEDs by growers of ornamental crops has been limited to situations where the crop is very high value per unit of growing area and where LEDs are expected to significantly reduce crop losses, such as in vertical propagation systems (see Section 5), or where the initial capital investment has been subsidized through government grants or credits issued by the electrical power company.

## 2.2 Sole-source lighting

Sole-source lighting refers to lighting systems used in the absence of sunlight, such as in vertical farming warehouses (Fig. 4). LEDs lend themselves to this use since they can be placed closer to plants without damaging the plant tissues. It is a common misconception, however, that LEDs produce less heat than HPS luminaires. Heat production depends on the efficiency of the luminaires. For example, if we compare HPS and LED luminaires that are equally efficient at generating photons, one 1000W HPS luminaire and three 333W LED luminaires produce similar amounts of heat; however, HPS luminaires direct a higher percentage of that heat directly to the underlying crop, while the heat generated from LEDs is more readily dissipated in an upward or horizontal direction. As a result, the cooling requirements in a warehouse will be similar for equally efficient HPS and LED luminaires delivering similar PPFD. The bottom line is that LEDs lend themselves to stacked systems, that is, vertical farming, yet the cooling requirements for LEDs in these systems remains substantial.

The main focus for LED use as sole-source lighting has been on the production of microgreens, lettuce, and herbs because these crops have a high value per area occupied and are relatively short plants that lend themselves to stacked systems. Ornamentals have also been grown in vertical systems but only in limited ways. Perhaps the greatest success has been in the propagation of high-value species that have a high rate of failure during production. For example, the process of transitioning tissue-cultured plants from the sterile container to the greenhouse propagation bench, which can be a harsh environment, often results in heavy losses of these expensive propagules. In contrast, the environment of an insulated growth room can be maintained at

**Figure 4** Microgreens growing in a vertical farming warehouse with sole-source LED luminaires.

the ideal humidity, temperature, and PPFD. Consistently ideal conditions allow for the successful transition of the tissue-culture and provide a rapid return-on-investment for LEDs (see Section 5).

## 2.3 Spectral responses

Perhaps the most unique feature of LED technology is the precise wavelengths that are emitted, which provides horticulturists with the ability to manipulate the light spectrum in a manner never before possible. LEDs are available in all colors, but the efficiency of different colored LEDs can vary dramatically. Blue and red LEDs are the most efficient at ~1.8 and 1.7 μmol/J, respectively, while green provides just ~0.5 μmol/J and ultraviolet is ~0.2 μmol/J (Nelson and Bugbee, 2014). Thus, blue and red are the primary colors used in LED luminaires. White LEDs are actually blue LEDs covered with a phosphor coating that absorbs blue light and re-emits it at higher wavelengths (Fig. 5). Thus, while white LEDs are less efficient (~1.5 μmol/J) than blue, they are popular in sole-source situations, since humans need white light to properly view the plants.

The light spectrum actually has a relatively small direct effect on photosynthetic rates (Paradiso et al., 2011). Plant morphology, however, can be dramatically affected by the light spectrum, and morphological changes can have a large indirect effect on plant growth as well as the appearance of

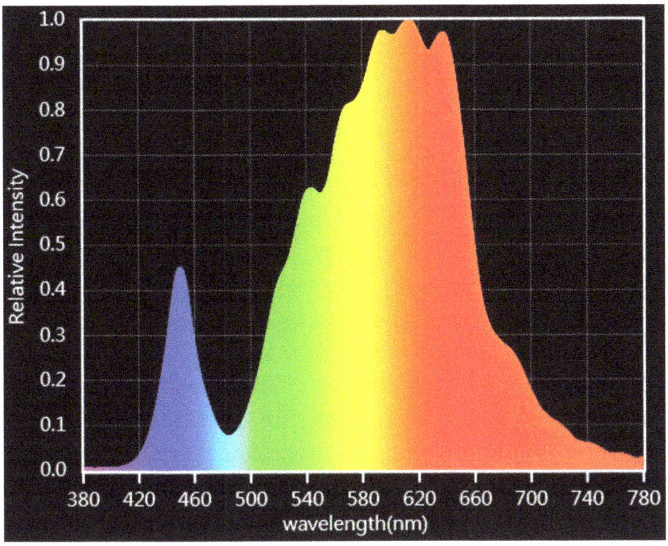


**Figure 5** Spectrum emitted from a white LED luminaire.

ornamentals. Spectral responses become enhanced in a sole-source lighting situation where the photons emitted by the LEDs are not mixed with the full spectrum of photons delivered from sunlight. Following are examples of the effects of light spectrum on plant morphology, growth, and flowering:

- Blue light (400–500 nm) decreases cell expansion and thus reduces leaf area and photon capture (Dougher and Bugbee, 2004). For ornamentals, blue light effectively acts as a growth regulator in sole-source lighting systems by producing more compact plants with smaller leaves, while the effect is less pronounced in greenhouses. Blue light and ultraviolet (300–400 nm) light promote purplish pigmentation development in some species. The amount of blue light emitted from cool-, neutral-, and warm-white LEDs varies, and the subsequent leaf expansion and stem elongation also varies in a similar pattern, that is, smaller leaves with a higher intensity of blue light (Snowden et al., 2016). The percentage of blue light typically found in red and blue LED arrays is 5–20%.
- Green light (500–600 nm) is commonly misperceived as relatively ineffective at generating photosynthesis and at being perceived in photoperiod responses. In reality, nearly 85% of green light is absorbed by leaves while just 5–15% is transmitted or reflected. Green light penetrates deeper into the leaf tissue than red or blue light, which improves its efficiency at driving photosynthesis (Terashima et al., 2009). Studies have shown that green light is as effective at promoting flowering on some long-day

plants as is red plus far-red light (Meng and Runkle, 2019). Green light is also important for human observation of plants in sole-source lighting systems, since the absence of green light makes it difficult to identify plant disorders and creates eyestrain for workers. Most LED luminaires do not contain green LEDs because they are much less efficient at converting energy into photosynthetically active photons compared to red and blue LEDs. White LEDs emit significant levels of green light and are therefore the most economical method of supplying green light to plants and to those who care for them.

- Red light (600–700 nm) is the predominate spectrum provided in most LED arrays, accounting for 75–85% of the PPFD. In sole-source lighting situations, small amounts of blue light must be added to red LEDs to grow plants with a desirable appearance. The amount of red light relative to the amount of far-red light has a strong impact on leaf expansion, stem elongation, and axillary bud development, for example, increasing the red light to far-red light ratio results in shorter plants with smaller leaves and improved lateral branching. Red light is also important for photoperiod perception.
- Far-red light (700–800 nm) enhances cell expansion and thus increases leaf expansion and stem elongation. Far-red light provided at the top of the canopy increases leaf expansion, which can increase overall plant productivity by improving light interception (Zhang et al., 2019). This response is particularly noteworthy for high-wire vegetable production but may also have an application for cutflower rose production. Far-red light has also been shown to be critical for photoperiod perception of long-day plants (Kohyama et al., 2014), especially when daily light integral is low ($\leq$6 mol $m^{-2}$ $d^{-1}$). Consequently, newer versions of LED luminaires often include far-red LEDs among the red and blue LEDs. White LEDs usually have a sufficient far-red spectrum to elicit photoperiod responses (Fig. 5).

Since ornamental production focuses on plant appearance rather than overall growth rates, manipulation of the light spectrum opens up an array of possibilities for growers to enhance plant size, shape, and pigmentation by tweaking wavelengths. Considerable research needs to be done to develop 'light recipes' to allow growers to take full advantage of the possibilities that lie ahead for ornamental production (Fig. 6).

## 2.4 Photoperiodic lighting

Due to low cost and ease of installation, incandescent bulbs (INC) were the standard luminaire for photoperiodic lighting for many decades; however,

**Figure 6** Multi-color LED luminaires at the Controlled-Environment Lighting Laboratory (CELL), Michigan State University.

INC have been removed from most markets due to their low-energy efficiency. Compact fluorescent luminaires (CFL) have replaced INC in recent years, but the lack of far-red light in the CFL spectrum makes them less effective for promoting flowering of some long-day species. White LEDs with a screw-in connector serve as direct replacements for screw-in INC and CFL luminaires. The white LEDs typically have a sufficiently large far-red spectrum to make them useful for growing long-day plants.

Growers use HPS luminaires for photoperiodic lighting in outdoor situations by mounting the luminaires atop tall poles similar to those used for security lighting in parking lots. This configuration has been used to promote early-season flowering of long-day herbaceous perennial species grown outdoors.

Cyclical lighting, a process developed in the 1950s, provides intermittent lighting rather than continuous lighting to save energy during long-night (short-day) treatments. Oscillating reflectors can be attached to HPS luminaires, thus allowing a single luminaire to shine cyclic light on a large area by directing the light sequentially in multiple directions. HPS luminaires have been mounted on irrigation booms to provide cyclical lighting as the boom tracks back and forth above a crop throughout the night. Growers are just beginning to consider the use of LED arrays mounted on irrigation booms (Fig. 7). Limited research on cyclical lighting suggests that light must be provided every 30 min during a 4 h night interruption treatment (Blanchard and Runkle, 2016). Plant response to the light cycle is quantitative, meaning that the number of photons delivered during the light cycle is more important than simply providing a duration of light above a threshold PPF. Blanchard and Runkle (2016) termed this the 'night

**Figure 7** LED luminaires mounted on an irrigation boom to provide cyclical lighting. Photos: John Davies.

interruption lighting integral'. This measurement appears to be more critical for promoting flowering of long-day plants than for inhibiting the flowering of short-day plants.

## 3 Manipulation of plant growth and development

In today's industry, efficient and profitable plant production requires the precise management of the greenhouse light environment, the utilization of automated equipment, and the fine-tuned control of plant growth and flowering through the use of plant growth regulators. This section first explores the recent advancements in the greenhouse industry that allow growers to better understand and control the light environment in greenhouses. Then, the changes in cutting production and propagation systems are described, focusing on how innovations in automation and robotics are redefining where and how young plants are grown. Finally, horticulturists continue to experiment with plant-growth-regulating compounds. Knowledge of how to manipulate the hormonal responses of plants provides a powerful tool in the hands of a skilled and observant grower.

### *3.1 Light management*

Historically, the limitation to accurately managing the amount of light delivered to crops in protected cultivation was the lack of measurement instrumentation. The concept of an integrated light measurement–termed the 'daily light integral' (DLI )–was introduced in the 1980s and has been used since then in plant science research, but the ability to communicate this information in a meaningful way to growers did not exist until affordable and easy-to-use instrumentation was developed. The new instrument is simply a quantum sensor mounted on a data logging device, but this tool allows academics to communicate plant requirements to growers and then growers to monitor and appropriately modify their production environments. Today, growers in the greenhouse industry are comfortable discussing the DLI delivered to their crops, and they understand how to manage DLI with retractable shade curtains, shading compounds applied to the greenhouse glazing, and supplemental lighting from electrical luminaires.

Maps have been developed describing the monthly distribution of DLI across the United States (Faust and Logan, 2018), and a project is currently underway to create DLI maps for Europe and Asia. The maps allow growers to anticipate monthly variations in light delivered to greenhouses. This information is used to estimate the number of days per year that supplemental lighting would be useful and thus provides a means for projecting the return-on-investment for supplemental lighting. Similarly, one can estimate the percentage of shade required for a crop and the number of days per year that shade would be beneficial.

Plants respond to DLI in various ways. The most consistent response across plant species is the promotion of dry matter accumulation as DLI increases up to very high levels, for example, >30 mol $m^{-2}$ $d^{-1}$. While this response is useful for agronomic crops, dry matter is of limited value for most ornamental species. Flower number, stem number (branching), and stem diameter are more vital measurements for ornamental plant productivity and quality, and these measures typically increase at a diminishing rate as DLI increases up to 20–25 mol $m^{-2}$ $d^{-1}$ (Faust et al., 2005) DLIs above 25 mol $m^{-2}$ $d^{-1}$ may demonstrate moderate increases in plant growth; however, the ornamental characteristics are rarely beneficial in protected cultivation because heat stress becomes limiting and may result in reduced flowering and damaged or deformed tissues. Greenhouses frequently lack the cooling capacity to remove the additional energy entering the greenhouse when very high DLI is provided.

Time to flower typically decreases with increasing DLI, although the largest response occurs as DLI changes at the low end of the DLI range. For example, flower initiation can become light-limited as DLI decreases below 5–10 mol $m^{-2}$ $d^{-1}$, while DLI >10 mol $m^{-2}$ $d^{-1}$ usually has little added benefit in terms of time to flower (Faust et al., 2005). Root initiation of cuttings in propagation increases as DLI increases from up to 10 mol $m^{-2}$ $d^{-1}$ (Lopez and Runkle, 2008). DLI >12 mol $m^{-2}$ $d^{-1}$ causes severe wilting of cuttings even under mist, so this limits the DLI delivered to unrooted cuttings in propagation. In general, high-quality ornamental plants can be grown when the DLI is 10–20 mol $m^{-2}$ $d^{-1}$. For crops in which a portion of the crop is harvested or removed from the plant canopy, such as cutflowers, DLI at 15–25 mol $m^{-2}$ $d^{-1}$ should be provided to maximize yield.

Recent engineering advances have created a diffused glass that is essentially glass that has had the inside surface scratched in order to scatter the light. Diffused light supplied to a plant canopy is beneficial because scattered light penetrates the canopy better than direct light, thus allowing the upper leaves to receive a lower PPFD while the lower leaves receive a higher PPFD. Diffused light leads to an overall improvement in whole plant photosynthesis and a reduction in leaf and flower temperature at the top of the canopy (Dueck et al., 2012). Diffused light also creates a more uniform light environment within greenhouses, which allows the crop to grow more uniformly. Higher PPFD can be provided to crops when light is diffused, since the uppermost leaves are less susceptible to leaf burn.

## 3.2 Cutting production and propagation systems

The increasing market for vegetatively propagated species has created a dynamic industry for stock plant and unrooted cutting production. Vegetative propagation demands cuttings derived from disease-free stock plants. The process of growing 'clean' stock begins with tissue-cultured plantlets that have been indexed for viruses and bacteria. A block of 'elite' stock plants is started from the tissue-cultured plantlets, then cuttings are removed to produce large numbers of 'commercial' stock plants from which unrooted cuttings are harvested and shipped to propagators.

The risk associated with introducing pathogens and pests to propagation facilities is so great that extensive disease-prevention protocols must be strictly enforced. Following are examples:

- Full-body protective clothing, including hairnets, latex gloves, and boots are commonplace, while handwashing stations, disinfection solutions, and footbaths are located at the entrance to individual sections of greenhouses.
- Physical partitioning and mandatory worker flow patterns are utilized to minimize the potential for workers to inadvertently vector pathogens or spread pests.
- Water-treatment facilities use redundant technologies to ensure that water supplies are not a source of pathogens.
- Every bag of cuttings, typically containing 105 individual cuttings, is identified with barcodes and/or QR codes to precisely locate the stock plants from which the cuttings were harvested in the rare case that a pathogen is identified later in the production cycle.
- The vents of stock plant greenhouses are covered with insect screening to keep out pests as small as thrips.

Today's propagators demand cuttings that are uniform in size, flower uniformly, do not flower prematurely, do not suffer from postharvest disorders, and have 100% rooting success. Inspectors at ports-of-entry have zero tolerance for pests or diseases observed in the shipments of cuttings. These exacting expectations have driven cutting producers to utilize plant growth regulators, water and nutrient management strategies, the greenhouse environment, stock-plant trimming techniques, and harvesting protocols to produce uniform cuttings with well-toned tissues that will withstand the rigors of the shipping environments (Faust and Grimes, 2005; Samarakoon et al., 2017; Rapaka et al., 2007a).

The time from cutting harvest to delivery to the propagator greenhouse is typically just 48 h, yet postharvest disorders are common when temperature abuse occurs en route. Much organizational effort is exerted to maintain good cold-chain management protocols throughout the postharvest process. While most species are harvested in the morning, when greenhouse temperatures are cool, some ethylene-sensitive species (e.g. lantana and portulaca) are harvested mid-day or in the afternoon, when their carbohydrate levels are high and the cuttings are less sensitive to ethylene in the postharvest environment (Rapaka et al., 2007b).

In contrast to the European Union, the United States does not allow for the importation of roots contained in a propagation medium; therefore, the U.S. market requires unrooted cuttings, while rooted cuttings and seedlings can be imported into the E.U. countries. Sticking cuttings is a laborious task, and the U.S. industry is attempting to address this issue with different approaches.

One method is to use robotic-arm machines that can identify the cutting stem, pick it up, and insert it into the propagation medium (Fig. 8). While the financial investment is substantial, the machines can run 24 h per day. The seasonal nature of some greenhouse production means that the machines are idle in the off-season and unable to meet the high demand in the peak season, so seasonal workers are still necessary.

A second method addresses the labor cost of sticking cuttings by pre-sticking the cuttings into cartridges at the stock plant production location (Fig. 9). Upon arrival at the propagator, the cartridges are loaded into machines that separate the individual cuttings and insert them into the propagation medium.

Once the cuttings are stuck into the propagation medium and placed on the mist bench, an irrigation boom delivers mist intermittently. Most growers still rely on a pre-programmed time-clock controller to trigger mist events; however, mist controllers that link mist frequency to environmental measurements of vapor pressure deficit or solar radiation are becoming more popular.

### 3.3 Novel uses of plant growth regulators

Plant growth regulators (PGR) were first introduced in the 1950s for the purpose of manipulating plant growth and flowering by enhancing or reducing hormone signals. While the number of different active compounds available today is somewhat limited, scientists and professional horticulturists continue to experiment with novel uses of these compounds to improve plant production. These innovations can improve plant appearance, performance,

**Figure 8** Robotic arm used to stick unrooted cuttings into propagation media. Photo: Four Star Greenhouse.

**Figure 9** Top: Unrooted cuttings pre-stuck into cartridges (AutoStix) prior to transport to the propagator. A machine separates the cuttings and inserts them into the propagation media. Bottom: Cuttings inserted into cartridges, rooted without substrate, then shipped to propagators where machines separate and transplant the cuttings (Basewell). Photos: Visser Horti Systems, Dummen-Orange.

and uniformity, thus affording growers more precise control in order to better meet the demands of the current production systems.

Benzyladenine (BA) is a synthetic cytokinin applied to herbaceous perennials, such as *Hosta* and *Echinacea*, to increase bud development from the below-ground crown. This helps the plant fill out the container, which allows it to be marketable at an earlier date. Similarly, improved branching on above-ground stems can occur on crops such as *Osteospermum* and *Sempervivum*. In cutflower rose crops, BA is painted onto dormant buds in the lower canopy in order to stimulate budbreak and shoot formation.

Ethephon can be sprayed onto plants to provide an ethylene signal that elicits a range of responses. The growth of the vegetatively propagated cutting market has been aided by the weekly application of ethephon to stock plants for the purpose of aborting flower buds, promoting branching, and decreasing leaf and stem expansion. All of these responses improve stock plant productivity and cutting quality. The primary negative effect of ethephon is that it can stimulate autocatalytic ethylene production for several weeks after the application. This can result in heightened concentrations of ethylene occurring in the postharvest environment that may result in leaf yellowing, abscission, or senescence during shipping or during the first few days in propagation (Leatherwood et al., 2016). Leaf curling of cuttings recently treated with ethephon causes the base of the stem to be lifted out of the propagation media, which in turn causes plant death unless the cutting is re-inserted. As a result of these potential dangers, stock plant growers typically cease ethephon applications immediately prior to the cutting harvest season or apply relatively low rates throughout the harvest season.

1-methylcyclopropene (1-MCP) is a competitive inhibitor of ethylene action and has proven useful for improving the postharvest performance of many horticultural products. For ornamentals, the primary uses include placing sachets of 1-MCP powder into boxes of unrooted cuttings before closing the box and sending them to the customer (Leatherwood et al., 2016). The hydrophilic powder quickly absorbs moisture from the surrounding air and release 1-MCP gas, which infiltrates the packaging. Another application technique is to place the harvested cuttings into a sealed room, release the 1-MCP overnight, and package and ship the treated cuttings the following day. These techniques are reserved for use on species that are the most ethylene-sensitive, such as *Lantana*, *Portulaca*, *Euphorbia*, and *Pelargonium*. 1-MCP is also used on truck transport of flowering plants such as *Pelargonium* and *Cineraria*. Similarly, 1-MCP is used during the shipping of ethylene-sensitive cutflower species such as carnations, alstroemeria, and roses.

Rooting hormones, such as indole-3-butyric acid (IBA), are typically applied to the base of the stem of cuttings, but this is a time-consuming task that slows down sticking lines. A time-saving alternative is the foliar application. The water-soluble potassium salt form of IBA (K-IBA) is the product of choice for this application, since the alcohol-soluble forms tend to cause more phytotoxicity. K-IBA sprays can be applied quickly to a large area of cuttings on the day following sticking to promote rooting.

Imported cuttings can also be completely submersed in K-IBA solutions prior to sticking. Other products, such as disinfectants, pesticides, and miticides, can be added to the dipping solution to potentially stave off pest and disease problems that might arrive with a shipment of cuttings.

Growers have experimented with an array of tank mixes that combine multiple active ingredients to achieve a unique combination of responses. For example, daminozide and ethephon applied in propagation or shortly after transplant can (1) abort flower buds that may already be initiated, but not visible in cuttings; (2) reduce stem elongation and leaf expansion; and (3) promote branching. Tank mixes of daminozide, ethephon, and BA are also used to promote branching (Fig. 10).

Chemical manufacturers have combined BA and gibberellic acid (GA) into a pre-mixed tank mix that has found many applications. Following are several examples:

- Reduce leaf yellowing and abscission of *Pelargonium* leaves following the shipment of unrooted or rooted cuttings (Currey et al., 2013).
- Reduce leaf yellowing of the lower leaves in potted lilies due to low light and crowded conditions (Funnell and Heins, 1998).
- Promote stem elongation of excessively short plants, especially when the plants have been stunted due to exposure to an excessive amount of the strong gibberellin-synthesis inhibitors, such as paclobutrazol and uniconizole.
- Promote elongation of poinsettia stems immediately prior to providing a mechanical pinch with a shearing device in order to make the shearing more uniformly effective.
- Increase bract expansion during the last week of poinsettia flowering (Blanchard and Runkle, 2008) (Fig. 11) and to create whiter bracts on varieties that naturally have a greenish or yellowish tint.

**Figure 10** Lantana treated with a tank mix of daminozade and ethephon (right) compared to an untreated control (left). Photo: Royal Heins.

**Figure 11** Effect of benzyladenine (BA) and gibberellic acid (GA) tank mix spray on poinsettia bract expansion. Left: untreated, Right: BA+GA spray at 5 ppm.

## 4 Sustainable practices and renewable inputs

The sustainability of an industry is dictated by its ability to adapt. Adaptation can be a change in practices as well as a change in the materials and energy consumed in the production process. The biggest adaptation challenges for the ornamentals industry relate to society's rejection of the usage and reliance upon chemical pesticides and fungicides, non-renewable energy, and plastics. Significant strides have been made to identify sustainable practices and increase use of renewable inputs, but there is much room for further development.

### *4.1 Biological controls*

Biological control of insects has become a standard practice in most greenhouse businesses. This trend is driven by multiple forces. For example, societal pressure to reduce the use of neonicotinoides has been great and has resulted in a *de facto* ban on their use in some markets. While the introduction of new synthetic chemicals into the niche market of ornamentals has been a slow trickle, new biological options continue to be introduced at a growing rate. Pest and disease resistance to chemical pesticides remains a significant issue. Finally, and perhaps most importantly, biocontrol has proven to work well and, in many cases, is more effective than traditional, preventative chemical spray programs.

Biocontrol programs require more skilled management to implement, and product costs are significant, but techniques for using biologicals have proven reliable and effective. The following list shows just a few examples of biological controls that have been integrated into ornamental businesses:

- Entomopathogenic nematodes, *Steinernema feltiae*, applied as a drench in propagation are effective at controlling fungus gnat larvae. Nematode drenches are also useful for controlling thrips by feeding on the pupae stage, which occurs in the growing medium.
- Entomopathogenic fungus, *Beauvaria bassiana*, applied as a foliar spray, controls thrips, whiteflies, and aphids as the spores infect the pest.
- Endophytic fungus, *Trichoderma* sp., applied as a drench to improve root resistance to fungal infection.
- Endophytic bacterium, *Bacillus subtilis*, applied as a foliar spray to control *Botrytis* or as a media drench to control *Rhizoctonia* and *Fusarium*.
- Predatory mites, *Amblyseius swirskii*, reared in sachets are placed within plant canopies to feed on thrips, whitefly, and spider mite larvae. Cattail pollen is blown over crops to provide an additional food source for the predatory mites.
- Predatory mites, *Phytoseleiulus persimilis*, eradicate spider mites in cutflower rose canopies.
- Parasitic wasps, *Aphidius colemani*, attack aphid adults and nymphs. Grasses are grown as 'banker plants' in the greenhouse to provide an additional food source (cereal aphids) for the wasps.
- Pirate bugs, *Orius laevigatus*, feed on thrips. Pepper plants are grown in the greenhouse to provide pollen as an additional source of food for the beneficials.
- Predatory beetles, *Atheta coriaria*, target shore fly larvae in propagation greenhouses.
- Pheromone lures are used to help monitor thrips populations.

While biological products are often purchased from companies that specialize in their production, some growers have invested in on-site laboratories to produce their own beneficial nematodes, fungi, and mites. Often, the rural location of the farms, particularly in developing countries, makes this approach necessary.

## 4.2 Fungicide resistance management

*Botrytis* is a major international pathogen for ornamental crops that has a high risk of fungicide resistance development because it has a short life cycle, high genetic and phenotypic variability, and a high reproduction rate. In situations where fungicide selection pressure is high, isolates have demonstrated simultaneous resistance to six FRAC codes (Muñoz et al., 2019). In one study, *Botrytis* isolates were collected from cutflower roses grown in six different countries in Africa and South, Central and North America, and resistance to several FRAC codes occurred in all of these locations (Muñoz et al., 2018).

Fungicide resistance development in *Botrytis* isolates is particularly of concern in greenhouse cutflower production for several reasons: the same species are produced in the same greenhouses for multiple years, senescing tissues are always present, species such as roses and gerbera daisies are highly susceptible, and condensation occurs nearly every night in the greenhouses in many of the primary geographic locations used for cutflower production. Thus, disease pressure is persistently high, so preventative fungicide applications have been the primary management tool used for many years.

Growers in South America are using several approaches to combat fungicide resistance. First, fungicide resistance screening is performed at in-house labs or at commercial lab testing facilities. The results of these tests are used to identify the best fungicides to use in the upcoming season.

Currently, most cutflower species are transported via airfreight, but the use of sea transport is expected to increase in the coming years. The longer shipping time with sea transport will place additional pressure on growers' ability to prevent *Botrytis* infections in the greenhouse prior to harvest. Much additional research will be needed to address this challenge.

### 4.3 Renewable growing media

Peat and perlite are the primary components of most growing media used in container production. The typical peat to perlite ratio is 80:20. While peat is a natural product, it takes thousands of years to produce, so it can hardly be called a renewable resource. Likewise, perlite is a naturally occurring product, but it is non-renewable, the dust is toxic, and it is a relatively expensive media component. So, there is a continual search for alternative growing media.

Products such as coir (from coconut husks) and rice hulls have been used extensively in hydroponic systems and to a lesser extent in container production. Rockwool is also popular for hydroponic production, but disposal at the end of a crop can be problematic. Recycling programs, such as using rockwool as a component of concrete, address some of the environmental concerns surrounding this product.

Wood fiber is a renewable resource that has been used in Europe for a couple of decades and has recently become popular in the United States. Wood-fiber substrates are derived from whole trees ground into a fibrous material (Fig. 12). Different evergreen species can be used, but pine is most common. Particle size can be controlled during the grinding process to provide a range of physical properties in the media (Altland et al., 2018).

Many growers have replaced the perlite fraction of their growing media with wood fiber to reduce costs, while some growers are also attempting to also replace a percentage of the peat with wood fiber. Replacing peat is a more difficult proposition for a couple of reasons. First, peat is very acidic (pH = ~4.5),

**Figure 12** Examples of wood-fiber substrates derived from pine trees through different manufacturing processes and coloring techniques. The initial feedstock is pictured in the center. Photo: Brian Jackson.

so lime is added to the media to adjust the pH to 6.0. The lime also provides pH buffering so that rapid fluctuations do not occur. Wood fiber, on the other hand, is less acidic, so the lime rates need to be reduced. This results in a lack of pH buffering, so growers can experience rapid changes in the growing media when high percentages of wood fiber are used.

The second issue with wood fiber is that it is a high carbon material, and the microbes feeding on the carbon tie up nitrogen. To address this problem, nitrogen can be impregnated into the fiber during processing or higher nitrogen rates can be provided in the fertilization program.

### 4.4 Renewable energy

The expenditure of fossil fuels to grow herbaceous plants in controlled environments comes at a high carbon cost. The use of renewable energy sources, for example, solar, wind, geothermal, and waste wood, can reduce the annual emission of carbon dioxide and thus reduce the industry's carbon footprint. The availability and economic feasibility of these options varies with geographical location and the availability and cost of competing fuels.

The use of recycled wood products has become popular in areas where wood is readily available. Wood fuel is derived from various waste products and processes, such as pallets, cleared land, wood chips, and sawdust, and the cost can be considerably lower than fuel oil. Adequate storage room can be an issue, and boilers must be purchased or renovated. In the east coast of the United States and in parts of Europe, wood fuel has proven to be an effective, renewable energy resource for greenhouse production and is contributing to future sustainability of the ornamentals industry.

### 4.5 Recycling plastics

The success of plasticulture in the 1960s and 1970s allowed for the rapid expansion of ornamental plant production, and even today plastics are invaluable as a greenhouse covering and for container production. However, the ubiquitous use of plastics has clearly created a sustainability issue.

The low cost, light weight, and versatility of polyethylene cannot be matched by other glazing materials, so the use of polyethylene coverings for greenhouses and as a mulching material has no end in sight. Re-use is not possible due to the degradation of polyethylene under ultraviolet light. Filling landfills with plastic is also not an option. Thus, the focus must be on recycling. Unfortunately, recycling solutions are usually dictated by local programs or businesses that offer recycling solutions. Success stories abound, but the availability of recycling options can be limited depending on one's geographical locale.

Plastic containers are terribly convenient for growing transporting and selling ornamentals, but they continue to be the scourge of growers and consumers. High-quality plastic containers, such as those used in propagation trays, are often disinfected and re-used; however, relatively low-quality plastics are used for producing containerized plants that will eventually be sold to consumers.

Several years ago, a wave of bio-pots was introduced, and the large retailers demanded that wholesale growers use them. The bio-pots consisted of a smorgasbord of biodegradable plastics and other compostable products: rice hulls, polylactic acid, recycled paper and cardboard, soy, corn, bamboo, and poultry feathers. Some containers were 'plantable', meaning the consumer could plant the plant and the pot in their garden, and the roots would penetrate the sides of the container. These were made with coir fiber, wood pulp, paper, cow manure, and rice hulls (Nambuthiri et al., 2015).

Initially, bio-pots appeared to be promising, but after several years of trial and error, the industry gravitated back to plastic containers. Ultimately, bio-pots failed because their cost was higher than and their performance inferior to plastics (Brumfield et al., 2015). Performance was an issue for several reasons: they did not fit well into automated systems, they did not slide on benches, they were too brittle or too pliable, they broke down too quickly or too slowly, and unsightly molds grew on some.

The bottom line is that plastic greenhouse coverings and containers work extremely well for horticulture production, and there are no signs that they will be replaced anytime soon. Therefore, more effort must be exerted to create and implement recycling programs that minimize the industry's environmental impact.

## 5 Case study: Propagation of ornamentals in a vertical propagation area

Stage 3, tissue-cultured plantlets consist of in vitro plantlets with roots that are ready to be transplanted and transitioned (hardened-off) in a greenhouse propagation area. The hardening-off process can be difficult due to the harsh propagation environment, where temperature, light, humidity, and moisture conditions are in continuous flux. A fluctuating propagation environment can cause a high percentage of losses of expensive propagules.

To address this challenge, Battlefield Farms, a 45-acre greenhouse located in Rapidan, Virginia, renovated a cooler with sole-source LED lighting, a fog system, and carbon dioxide injection to provide a stable transition environment. Because the cooler had the capacity to remove the large amount of heat generated by the LEDs, the temperature could be held at 20–22°C year round. The fog system provided a relative humidity of ~95%, and carbon dioxide was provided at 800 ppm.

Blue and red LED linear arrays were mounted horizontally to deliver light to the plants positioned on each shelf of their ten-shelve carts (Fig. 13). The light from the LEDs was also directed horizontally so that each shelf of the cart was illuminated from two directions. Thus, the LEDs were located in fixed positions creating 'stalls' or parking spaces for the carts of young plants, which could be rapidly wheeled in and out of the stalls. Each cart holds 8,100 tissue-cultured plantlets. The cooler has 30 stalls and thus provides the equivalent of 225 $m^2$ of greenhouse propagation space.

The LEDs provided a PPFD of 125–150 μmol $m^{-2}$ $s^{-1}$ throughout a 16 h photoperiod, resulting in a DLI of 7.2–8.6 mol $m^{-2}$ $d^{-1}$. It is easier to provide long-day conditions in the vertical LED system than in the greenhouse, so facultative long-day plants, such as petunias, are started in the vertical system to promote earlier and more uniform flowering.

The vertical propagation system developed for difficult tissue-culture propagated species such as *Heuchera* and *Echinacea* was a tremendous success. The results included dramatic reduction in plant losses as well as improved uniformity of rooting and faster production schedules. Labor needs were also reduced, since the high uniformity and strong root systems reduced the need to cull dead or weak plants. High uniformity made watering easier, and

**Figure 13** Vertical propagation area at Battlefield Farms, Rapidan, Virginia, USA. Photo: Marc Verdel.

the compact growth of plantlets reduced the need for plant growth regulator applications.

After the initial success of the vertical propagation system, its use was expanded to seed propagation. White LEDs were installed to provide a higher PPFD, 200 μmol $m^{-2}$ $s^{-1}$. For example, fall pansy sales require plug propagation during hot summer months when pansies do not germinate and grow well. Since temperatures can be controlled more easily in the warehouse, pansy seeds can be propagated in July and August before being transplanted in the greenhouse.

Battlefield Farms propagate 1.3 million young plants each year in their vertical farming warehouse. The use of the system for various tissue-culture, unrooted cutting and seed-propagated crops allows for 75% occupancy over the course of one year.

## 6 Conclusion and future trends

Over the past two decades, scientific advancement, societal and economic changes, and environmental awareness have propelled the ornamentals industry toward a more sustainable future. The increased use of LEDs, clean stock management practices, biological control methods, automation and robotics, and wood-fiber substrates have improved the grower's ability to maintain a profitable business while meeting the exacting requirements of wholesale customers and retail consumers, and while reducing energy consumption, the use of chemicals for disease and pest control, labor requirements, and carbon dioxide emissions. These are substantial accomplishments, but much work remains.

The industry will benefit greatly from future research focused on furthering our understanding of how to more effectively utilize energy-efficient technologies and renewable materials. Following are possible avenues for future exploration:

- Identify the best applications for LED technology. Today's growers have little idea of how they will be using LEDs in 5, 10, 20 years, and the technology is expensive and changing rapidly. So, there is a hesitancy to invest. Meanwhile, LED manufacturers do not exactly know what configuration of LEDs to create, so they have yet to focus their attention on a specific application. These forces keep the investment cost high and the return-on-investment low for most applications in the ornamentals industry. Once it is clearer where the market will be for LED use, manufacturing will be more efficient and pricing will become more attractive to a broader sector of the industry.
- For example, guidelines need to be developed for using LEDs for cyclical lighting while mounted on irrigation booms. A better understanding of how photoperiodic plants respond to cyclical lighting is needed before growers will invest in LED arrays for this application. Similarly, LED manufacturers need to know the spectrum and PPFD that is most efficient for use in a cyclical lighting application.
- The scope of spectral responses possible through the manipulation of the light spectrum with LEDs provides a second example of a research area with great need. Recipes need to be developed to allow growers to further utilize this powerful tool. Currently, LED luminaires are available that allow the spectrum to be adjusted, but growers do not necessarily know what adjustments to make. Once plant responses to light spectrum are better understood, LEDs may become the PGRs of the future.
- Develop 'smart' systems that aid growers in their decision-making processes or are able to make decisions without human input. For example, weather stations can aid in disease management decisions. Radio frequency identification technology (RFID) is currently used for inventory management in greenhouses where individual chips are attached to individual trays or benches of plants. RFIDs could also be used to communicate to irrigation booms that track overhead, allowing the boom to make species-specific irrigation decisions. Essentially, the boom would act like a drone as is done in outdoor, precision agriculture applications. A second example is current technology being developed

that allows LEDs to adjust the PPF emitted based on real-time ambient PPFD from sunlight. A quantum sensor mounted on top of a luminaire senses the ambient PPFD and then instantaneously adjusts the LED output so that the underlying crops receive a constant PPFD.

- Develop production and post-postharvest practices that allow growers to ship cut flowers, such as roses, via ocean carrier. This will require new techniques that produce more resilient plants and handling procedures that minimize stresses in the postharvest environment. Accomplishment of this goal will have a tremendous impact on the carbon footprint of the international cut flower industry.

## 7 Where to look for further information

*Further reading*

- LED Lighting for Urban Agriculture. 2016. T. Kozai, K. Fujiwara, E. Runkle (Eds). Springer Singapore.
- Light Management in Controlled Environments, 2017. R. Lopez and E. Runkle (Eds), Meister Media Worldwide, Willoughby, OH.

*Conferences and meetings*

- International Society for Horticultural Science Symposia and Workshops on Light, Growing Media, GreenSys, Vertical farming, etc.
- NCERA-101 Committee on Controlled Environment Technology and Use.

*Major research projects*

- *Botrytis* & Thrips Research Initiative, American Floral Endowment.
- Floriculture Crop Resilience, USDA-ARS-Floriculture & Nursery Research Initiative & Floriculture Research Alliance; James Faust, Clemson University; Erik Runkle, Michigan State University; Charlie Hall, Texas A&M University; Paul Fisher, University of Florida; John Erwin, University of Minnesota.
- LED and Vertical Farming Research (focused primarily on vegetable and herb production systems):
  - Controlled Environment Agriculture Center (CEAC)-Murat Kacira, University of Arizona.
  - Controlled-Environment Lighting Laboratory (CELL)-Erik Runkle, Michigan State University.
  - USDA-Specialty Crops Research Initiative. Lighting Approaches to Maximize Profits (LAMP)-Marc van Iersel, University of Georgia.

  - Wageningen University & Research–Leo Marcelis, Horticulture and Product Physiology Group, The Netherlands.
- Wood Fiber Research, Brian Jackson, North Carolina State University Horticultural Substrates Lab, NC, USA; Bruce Bugbee, Utah State University.

## 8 References


Altland, J. E., Owen, J. S., Jackson, B. E. and Fields, J. S. 2018. Physical and hydraulic properties of commercial pink-bark substrate products used. *HortScience* 53(12), 1883–90. doi:10.21273/HORTSCI13497-18.

Blanchard, M. G. and Runkle, E. S. 2008. Increasing stem elongation and bract size of poinsettia 'Freedom Red' with gibberellins and benzyladenine. *Acta Hortic.* 774, 209–16. doi:10.17660/ActaHortic.2008.774.26.

Blanchard, M. G. and Runkle, E. S. 2016. Investigating reciprocity of intensity and duration of photoperiodic lighting to regulate flowering of long-day plants. *Acta Hortic.* 1134, 41–8. doi:10.17660/ActaHortic.2016.1134.6.

Brumfield, R. G., Devincentis, A. J., Wang, X., Fernandez, R. T., Nambuthiri, S., Geneve, R. L., Koeser, A. K., Bi, G., Li, T., Sun, Y., Niu, G., Cochran, D., Fulcher, A. and Stewart, J. R. 2015. Economics of utlizing alternative containers in ornamental crop production systems. *HortTechnology* 25(1), 17–25. doi:10.21273/HORTTECH.25.1.17.

Currey, C. J., Lopez, R. G., Rapaka, V. K., Faust, J. E. and Runkle, E. S. 2013. Exogenous applications of benzyladenine and gibberellic acid inhibit lower-leaf sensescence of geraniums during propagation. *HortScience* 48(11), 1352–7. doi:10.21273/HORTSCI.48.11.1352.

Dougher, T. A. O. and Bugbee, B. 2004. Long-term blue light effects on the histology of lettcue and soybean leaves and stems. *Journal of the American Society for Horticultural Science* 129(4), 467–72. doi:10.21273/JASHS.129.4.0467.

Dueck, T., Janse, J., Li, T., Kempkes, F. and Eveleens, B. 2012. Influence of diffuse glass on the growth and production of tomato. *Acta Hortic.* 956, 75–82. doi:10.17660/ActaHortic.2012.956.6.

Faust, J. E. and Grimes, L. W. 2005. Pinch height of stock plants during scaffold development affects cutting production. *HortScience* 40(3), 650–3. doi:10.21273/HORTSCI.40.3.650.

Faust, J. E., Holcombe, V., Rajapakse, N. C. and Layne, D. R. 2005. The effect of daily light integral on bedding plant growth and flowering. *HortScience* 40(3), 645–9. doi:10.21273/HORTSCI.40.3.645.

Faust, J. E. and Logan, J. 2018. Daily light integral: a research review and high-resolution maps of the United States. *HortScience* 53(9), 1250–7. doi:10.21273/HORTSCI13144-18.

Funnell, K. A. and Heins, R. D. 1998. Plant growth regulators reduce postproduction leaf yellowing of potted asiflorum lilies. *HortScience* 33(6), 1036–7. doi:10.21273/HORTSCI.33.6.1036.

Kohyama, F., Whitman, C. and Runkle, E. S. 2014. Comparing flowering responses of long-day plants under incandescent and two commercial light-emitting diode lamps. *HortTechnology* 24(4), 490–5. doi:10.21273/HORTTECH.24.4.490.

Kumar, K. G. S., Hao, X., Khosla, S., Guo, X. and Bennett, N. 2016. Comparison of HPS lighting and hybrid lighting with top HPS and intra-canopy LED lighting for high-wire mini-cucumber production. *Acta Hortic.* 1134, 111–8. doi:10.17660/ActaHortic.2016.1134.15.

Leatherwood, W. R., Dole, J. M., Bergmann, B. A. and Faust, J. E. 2016. 1-Methylecyclopropene improves ethylene tolerance of unrooted herbaceous cuttings but delays adventitious root development in *Angelonia*, *Calibrachoa*, *Impatiens*, *Portulaca*, *Sutera*, and *Verbena* cultivars. *HortScience* 51(2), 164–70. doi:10.21273/HORTSCI.51.2.164.

Lopez, R. G. and Runkle, E. S. 2008. Photosynthetic daily light integral during propagation influences rooting and growth of cuttings and subsequent development of New Guinea impatiens and petunia. *HortScience* 43(7), 2052–9. doi:10.21273/HORTSCI.43.7.2052.

Meng, Q. and Runkle, E. S. 2019. Regulation of flowering by green light depends on its photon flux density and involves cryptochromes. *Physiologia Plantarum* 166(3), 762–71. doi:10.1111/ppl.12832.

Muñoz, M., Schnabel, G. and Faust, J. 2018. Fungicide resistance in *Botrytis*: an international challenge. *Greenhouse Grower*, May, 62–4.

Muñoz, M., Faust, J. E. and Schnabel, G. 2019. Characterization of *Botrytis cinerea* from commercial cut flower roses. *Plant Disease* 103(7), 1577–83. doi:10.1094/PDIS-09-18-1623-RE.

Nambuthiri, S., Fulcher, A., Koeser, A. K., Geneve, R. and Niu, G. 2015. Moving toward sustainability with alternative containers for greenhouse and nursery crop production: a review and research update. *HortTechnology* 25(1), 8–16. doi:10.21273/HORTTECH.25.1.8.

Nelson, J. A. and Bugbee, B. 2014. Economic analysis of greenhouse lighting: light emitting diodes vs. high intensity discharge fixtures. *PLoS ONE* 9(6), e99010. doi:10.1371/journal.pone.0099010.

Paradiso, R., Meinen, E.,Snel, J. F. H., Marcelis, L. F. M., van Ieperen, W. and Hogenwoning, S. W. 2011. Light use efficiency at different wavelengths in rose plants. *Acta Hortic.* 893, 849–55. doi:10.17660/ActaHortic.2011.893.93.

Radetsky, L. C. 2018. LED and HID horticultural luminaire. Testing report. Rensselaer Polytechnic Institute, Troy, NY.

Rapaka, V. K., Faust, J. E., Dole, J. M. and Runkle, E. S. 2007a. Diurnal carbohydrate dynamics affect postharvest ethylene responsiveness in portulaca (*Portulaca grandiflora* 'Yubi Deep Rose') unrooted cuttings. *Postharvest Biology and Technology* 44(3), 293–9. doi:10.1016/j.postharvbio.2006.12.004.

Rapaka, V. K., Faust, J. E., Dole, J. M. and Runkle, E. S. 2007b. Effect of time of harvest on postharvest leaf abscission in lantana (*Lantana camara* L. 'Dallas Red') unrooted cuttings. *HortScience* 42(2), 304–8. doi:10.21273/HORTSCI.42.2.304.

Runkle, E. and Bugbee, B. 2017. Plant lighting efficiency and efficacy: μmols per joule. *Greenhouse Product News*, July.

Samarakoon, U. C., Faust, J. E. and Dole, J. M. 2017. Quantifying the effects of foliar-applied calcium chloride and its contribution to postharvest durability of unrooted cuttings. *HortScience* 52(12), 1790–5. doi:10.21273/HORTSCI11820-17.

Snowden, M. C., Cope, K. R. and Bugbee, B. 2016. Sensitivity of seven diverse species to blue and green light interactions with photon flux. *PLoS ONE* 11(10), e0163121. doi:10.1371/journal.pone.0163121.

Stober, K., Lee, K., Yamada, M. and Pattison, P. M. 2017. *Energy Savings Potential of SSL in Horticultural Applications*. United States Department of Energy and Office of Energy Efficiency and Renewable Energy, Washington DC.

Terashima, I., Fujita, T., Inoue, T., Chow, W. S. and Oguchi, R. 2009. Green light drives leaf photosynthesis more efficiently than red light in strong white light: revisiting the enigmatic question of why leaves are green. *Plant Cell Physiol.* 50(4), 684-97. doi:10.1093/pcp/pcp034.

Zhang, Y., Zhang, Y., Yang, Q. and Li, T. 2019. Overhead supplemental far-red light stimulates tomato growth under intra-canopy lighting with LEDs. *J. Integr. Agric.* 18(1), 62-9. doi:10.1016/S2095-3119(18)62130-6.

# Chapter 11

# Diseases affecting ornamental geophytes and their control

*Gary A. Chastagner and Andrea R. Garfinkel, Washington State University, USA*



## 1 Introduction

Ornamental geophytes consist of plants that have underground, fleshy storage structures, such as (Cornwell and Giles, 2019):

- bulbs (e.g. tulips, daffodils, hyacinths, grape hyacinths (muscari), and alliums)
- corms (e.g. gladiolus, crocus, and autumn crocus)
- tubers (e.g. caladiums, oxalis, and anemones)
- tuberous roots (e.g. dahlia and peonies)
- rhizomes (e.g. lily-of-the-valley)
- fleshy roots (e.g. daylily)

Geophytes represent a significant segment of the world floriculture industry (De Hertogh et al., 2012; Miller, 2017). According to Miller (2017), the three major uses of these diverse flower crops are (Fig. 1):

http://dx.doi.org/10.19103/AS.2020.0066.15

**Figure 1** Major uses of ornamental geophytes include production of cut flowers, potted plants and their use in landscape plantings (Gary A. Chastagner © 2020. All rights reserved).

- cut flower production (70%)
- production of potted plants (10%)
- use in the landscape (20%)

Although a number of genera are important in the cut flower and potted plant industries, the industry is dominated by seven genera: *Tulipa* (Fig. 2), *Lilium*, *Narcissus*, *Gladiolus*, *Hyacinthus*, *Crocus*, and *Iris* (De Hertogh et al., 2012). Although the Netherlands has historically been and continues to be the worldwide leader in the production of ornamental geophytes, more recent southern hemisphere production regions have made it so that some bulbous species, such as tulips, are now available year-round (De Hertogh et al., 2012; Miller, 2017).

Lakshman et al. (2019) indicate that the diseases on ornamental geophytes can lead to 100% crop losses as a result of cosmetic injury, yield reduction, death, or regulatory action. Even when only a small number of bulbs or flowers are diseased, the cost of sorting out infected plant material is often not practical, and the entire crop has to be destroyed. Overall, it has been estimated that there could be as many as $10 billion USD in annual losses in floriculture crops due to plant diseases alone (McGovern and Elmer, 2018c).

The management of diseases and pests is more complicated in bulb crops than in many other crops due to low consumer tolerance for damage, the

**Figure 2** Field production of tulips – one of the most important ornamental geophytes (Gary A. Chastagner © 2020 All rights reserved).

systemic nature of some infections, and the potential for infestation of planting stock, which is mostly vegetatively propagated (Lakshman et al., 2019). The life cycles of most geophytes also increase the risk and impact that the diseases have on these crops. In landscapes, the common practice of cultivating geophytes, such as tulips, in the same beds year-after-year increases the risk of buildup of certain diseases. While diseases are an issue in the cultivation of ornamental geophytes, diseases can also affect the postharvest storage and quality of cut flowers and propagative structures. Given the globalization of the geophyte industry (De Hertogh et al., 2012; Miller, 2017; McGovern and Elmer, 2018c), the presence of pathogens on flowers and planting stock can limit access to certain markets.

While the focus of this chapter will be on diseases, it is important to note that in addition to being affected by biotic diseases, geophytes are affected by a number of abiotic disorders as well. In addition, insects, such as aphids, are important pests of ornamental geophytes. In some instances, insect pests are important vectors of plant viruses or phytoplasmas. This chapter provides a general overview of the diseases and general approaches that are used to manage the major groups of pathogens on commonly grown ornamental geophytes. Discussed in more detail is a more comprehensive summary of recent changes in our understanding of diseases caused by *Botrytis* species which are considered to be one of the most important groups of fungal pathogens on ornamental geophytes. Additional sources of information relating to diseases of ornamental geophytes and their management are included at the end of the chapter.

## 2 Biotic and abiotic diseases

### *2.1 Fungi*

Fungi cause a number of economically important production and postharvest diseases on geophyte crops (Table 1). Lakshman et al. (2019) point out that soilborne fungi such as the *Fusarium* spp. that cause bulb rot and wilting diseases, and *Rhizoctonia* spp., which causes bulb-, foot-, and root-rots, are among the most difficult to control (Fig. 3). There are also a number of economically important aerially dispersed fungi, such as *Botrytis* spp. that cause leaf spots, foliage blights, bulb rots, and postharvest flower decays (Fig. 4). Blue molds, caused by a number of *Penicillium* spp. (Dugan et al., 2014, 2017), are important storage pathogens of propagative structures (bulbs, rhizomes, corms, etc.) (Fig. 5). If left uncontrolled, fungal diseases can have a devastating impact on the production and marketability of propagative structures and cut flowers.

### *2.2 Oomycetes*

Oomycetes (also referred to as 'water molds') cause a number of economically important diseases of geophytes (Table 1). Oomycetes are fungal-like organisms with cell walls that are composed of cellulose rather than chitin and produce motile zoospores that spread in water. *Pythium* and *Phytophthora* spp. are soilborne oomycetes that frequently cause root rots and/or bulb decays when geophytes are planted in poorly drained soils or where plants are over-irrigated. Downy mildews are foliar diseases that are caused by oomycetes in the Peronosporaceae family. In addition to the direct damage by the members of this group of plant pathogens, *Olpidium brassicae* is a water mold that vectors Tobacco necrosis virus (TNV), which causes 'Augusta' disease in tulips.

### *2.3 Viruses*

Viruses are infectious, submicroscopic, intracellular, obligate parasites that are composed of nucleic acid (mostly RNA) and protein (Loebenstein et al., 1995; Hull, 2002). The vegetatively propagated geophytes are especially prone to the buildup of viruses which cannot reproduce outside of a living host cell (Table 2). Infected plants exhibit a range of symptoms depending on the host, virus, and environmental conditions. Symptoms can include flower break, chlorotic mosaic, streaking, or ringspot patterns on the foliage (Fig. 6). Plant growth is generally reduced, resulting in a significant decline in the productivity of infected stocks. Up to 50% increases in growth and yields have been reported for virus-free planting stock of a number of geophytes (Allen, 1975;

**Table 1** Some common fungal and oomycete diseases of the seven most widely cultivated ornamental geophyte species

| Geophyte species | Diseases | Causal organism | Notes | References |
|---|---|---|---|---|
| **Tulips** | Botrytis Blight/Tulip Fire | *Botrytis tulipae* | *Botrytis tulipae* is one of the most important pathogens of tulip and can be moved with tulip bulbs. See Table 5 for more information. | McGovern and Elmer (2018) |
| | Fusarium Bulb Rot | *Fusarium oxysporum* f. sp. *tulipae* | Although it can cause losses in the field for cut tulip growers, this disease is more important for producers of tulip bulbs as the disease can continue to develop in cold storage of the bulb. | McGovern and Elmer (2018) |
| | Gray Bulb Rot | *Rhizoctonia tuliparum* | *Rhizoctonia tuliparum* has long-lived overwintering structures, called sclerotia, which can be transported in infested soil and bulbs. This pathogen can also infect lily. | McGovern and Elmer (2018) |
| | Penicillium Bulb Rot (aka. Blue Mold) | *Penicillium* spp. | See section on daffodil. | Dugan et al. (2014, 2017) |
| | Pythium Root and Bulb Rot | *Pythium* spp. | Risk of *Pythium* infection is highest in soil with high water content and causes typical root rot symptoms of stunting and chlorosis. | McGovern and Elmer (2018) |
| **Lilies** | Basal Rot | *Fusarium oxysporum* f. sp. *lilii* and *Cylindrocarpon* spp. | The two fungi that cause basal rot are often found co-infecting lily bulbs whose roots and basal plate can be destroyed by the pathogens. | Chastagner et al. (2018) |
| | Botrytis Blight/Fire | *Botrytis elliptica* | *Botrytis elliptica* is one of two major *Botrytis* pathogens of lily, along with *B. cinerea*. See Table 5 for additional *Botrytis* species reported on lily. | Chastagner et al. (2018) |
| | Penicillium Bulb Rot (aka. Blue Mold) | *Penicillium* spp. | See section on daffodil. | Chastagner et al. (2018) |

(*Continued*)

**Table 1** (*Continued*)

| Geophyte species | Diseases | Causal organism | Notes | References |
|---|---|---|---|---|
| | Scale Rot | Fungal species *Fusarium* and *Cylindrocarpon* in association with a bacterial species, *Pseudomonas* | See 'Scale Rot' in Table 3 on bacterial diseases. | Chastagner et al. (2018) |
| | Gray Bulb Rot | *Rhizoctonia tuliparum* | See section on tulip. | Chastagner et al. (2018) |
| | Soft Rot | *Rhizopus* spp. | See section on bulbous iris. | Chastagner et al. (2018) |
| | Southern Blight | *Sclerotium rolfsii* and *S. delphinii* | Although not considered a major pathogen of lilies, *Sclerotium* species attack at the plant crown and can cause complete plant death. | Chastagner et al. (2018) |
| | White Stem Rot | *Sclerotinia sclerotiorum* | *Sclerotinia sclerotiorum* has a wide host range of many ornamental and non-ornamental crops and its presence may be especially important in area where other major host plants are present. | Chastagner et al. (2018) |
| **Gladiolus** | Botrytis Blight | *Botrytis gladiolorum* | *Botrytis gladiolorum* is one of three *Botrytis* species reported on gladiolus. See Table 4 for more information. | Elmer and Kamo (2018) |
| | Curvularia Spot | *Curvularia gladiolus* | *Curvularia gladiolous* can be a highly destructive pathogen of all above-ground parts of gladiolus, however, the economic impact of this pathogen is not well known. | Elmer and Kamo (2018) |
| | Fusarium Yellows and Fusarium Corm Rot | *Fusarium oxysporum* f. sp. *gladioli* | Considered the most destructive diseases of gladiolus, *F. oxysporum* f. sp. *gladioli* causes a corm rot that can cause to up to 80% yield losses. | Elmer and Kamo (2018) |

| | | | | |
|---|---|---|---|---|
| | Penicillium Bulb Rot (aka. Blue Mold) | *Penicillium* spp. | See section on daffodil. | Dugan et al. (2014, 2017), Elmer and Kamo (2018) |
| | Gladiolus Rust | *Uromyces transversalis* | Gladiolus rust is the only rust pathogen considered of major importance to geophyte crops and was once regulated as a quarantine pathogen in the United States. | Elmer and Kamo (2018) |
| | Stromatinia Rot | *Stromatinia gladioli* | *Stromatinia gladioli* is a widely-distributed pathogen of gladiolus corms. | Elmer and Kamo (2018) |
| **Daffodils** | Basal Rot and Neck Rot | *Fusarium oxysporum* f. sp. *narcissi* | Infection of daffodil by *F. oxysporum* f. sp. *narcissi* causes damage to the bulb and base of the leaves and can lead to up to 30% losses in the field. | Hanks and Chastagner (2018) |
| | Fire | *Botrytis polyblastis* | Unlike many *Botrytis* diseases of ornamental geophytes, the sexual stage plays a large role in the development of disease due to *B. polyblastis*. See Table 5 for additional details. | Hanks and Chastagner (2018) |
| | Leaf Scorch | *Didymella curtisii* (formerly *Peyronellaea curtsii*) | *Didymella curtsii* infection can continue to develop on flowers and stems in cold storage, making this pathogen especially damaging in cut flower production. | Hanks and Chastagner (2018) |
| | Penicillium Bulb Rot (aka. Blue Mold) | *Penicillium* spp. | Blue mold is one of the most important storage pathogens of ornamental geophytes. Multiple *Penicillium* species have been found infecting the bulbs/corms/rhizomes of ornamental geophyte species with varying severity. Wounding is considered necessary for infection. | Dugan et al. (2014, 2017), Hanks and Chastagner (2018) |
| | Smolder | *Botrytis narcissicola* | One of two host-specific *Botrytis* species of daffodil, *B. narcissicola* infection often originates from the overwintering structures, called sclerotia, which can survive on bulbs, and infects emerging shoots. See Table 5 for additional details. | Hanks and Chastagner (2018) |

*(Continued)*

**Table 1** (*Continued*)

| Geophyte species | Diseases | Causal organism | Notes | References |
|---|---|---|---|---|
| | Soft Rot | *Rhizopus* spp. | Soft rot due to *Rhizopus* spp. is typically considered to be a problem only under inadequate bulb storage conditions. | Hanks and Chastagner (2018) |
| **Hyacinth** | Botrytis blight | *Botrytis hyacinthi* | *Botrytis hyacinthi* is one of two *Botrytis* species reported on hyacinth. See Table 5 for more information. | Elad et al. (2016) |
| | Penicillium Bulb Rot (aka. Blue Mold) | *Penicillium* spp. | See section on daffodil. | Dugan et al. (2014, 2017) |
| | Root Rot | *Fusarium* spp., *Phytophthora* spp., and *Pythium* spp. | Some reports suggest that *Pythium* species are a greater contribution to root rot than either *Fusarium* or *Phytophthora*. Regardless, reports of all pathogens can be found in the literature. | Byther and Chastagner (1993), Saaltink (1969) |
| **Crocus** | Botrytis Blight | *Botrytis croci* | In addition to being a pathogen of ornamental crocus, *B. croci* has also been reported on the saffron crocus, *Crocus sativus*. See Table 5 for more information. | Elad et al. (2016) |
| | Bulb Rot | *Fusarium* spp. | Two forms of *Fusarium* have been reported on crocus, *F. oxysporum* f. sp. *croci*, a crocus-specific pathogen, and *F. oxysporum* f. sp. *gladioli*. | Baayen et al. (1998) |
| | Gray Bulb Rot | *Rhizoctonia tuliparum* | See section on tulip. | Chastagner et al. (2018) |
| | Penicillium Bulb Rot (aka. Blue Mold) | *Penicillium* spp. | See section on daffodil. | Dugan et al. (2014, 2017) |

| | | | | |
|---|---|---|---|---|
| **Bulbous Iris** | Basal Rot | *Fusarium oxysporum* f. sp. *gladioli* | Although named for its pathogenicity on gladiolus, *F. oxysporum* f. sp. *gladioli* can also infect other members of the Iridaceae family. | Baayen et al. (1998) |
| | Black Slime | *Sclerotinia bulborum* | *S. bulborum* also affectes hyacinth and tulips. | Byther and Chastagner (1993) |
| | Botrytis Gray Mold | *Botrytis convoluta* | *Botrytis convoluta* is one of three species of *Botrytis* reported on iris. See Table 5 for more information. | Elad et al. (2016) |
| | Fire | *Mycosphaerella macrospora* | *Mycosphaerella macrospora* causes severe disease on both bulbous and rhizomatous iris. | Byther and Chastagner (1993) |
| | Gray Bulb Rot | *Rhizoctonia tuliparum* | See section on tulip. | Byther and Chastagner (1993) |
| | Ink Spot | *Bipolaris iridis* | *Bipolaris iridis* affects the foliage of bulbous iris, but can also be found on the bulbs of netted iris (*Iris reticulata*). | Gould (1950) |
| | Neck Rot | *Rhizoctonia solani* | Considered a wide host range pathogen, *Rhizoctonia solani* can also infect gladiolus, lilies, and daffodils. | Byther and Chastagner (1993) |
| | Penicillium Bulb Rot (aka. Blue Mold) | *Penicillium spp.* | See section on daffodil. | Dugan et al. (2014, 2017) |
| | Rust | *Puccinia iridis* | Rust diseases, such as those caused by *Puccinia iridis*, are not typically considered of large economic importance on bulbous iris except for in certain environmental conditions. | Byther and Chastagner (1993) |
| | Southern Blight | *Sclerotium rolfsii* and *S. delphinii* | Southern blight is considered more severe in bulbous iris than in other ornamental geophytes. | Byther and Chastagner (1993) |

**Figure 3** Soilborne diseases such as Fusarium yellows (*Fusarium oxysporum* f. sp. *gladioli*) on gladiolus (left) are among the most difficult diseases to control. Pathogens can be introduced into planting sites on infested planting stock and often produce long-lived survival structures, such as sclerotia (middle-circled) that allow them to survive in soil. Infected plants often fail to emerge or die shortly after emergence (upper right), resulting in patches of diseased plants or bare ground in the field (lower right). Diseases such as gray bulb rot (*Rhizoctonia tuliparum*) on tulips (middle and right images) can have a major impact of the growth of plants in production and landscape sites. (Gary A. Chastagner © 2020 All rights reserved).

**Figure 4** Fungi cause a number of economically important diseases on the foliage and flowers of ornamental geophytes. Examples are Botrytis leaf blight and shoot dieback on peonies (left), Scorch (*Peyronellaea curtisii*) on daffodils (top middle), leaf spot (*Mycosphaerella macrospora*) on bulbous iris (bottom middle), powdery mildew (*Golovinomyces (Erysiphe) cichoracearum*) on dahlia (top right), and Botrytis blight (*Botrytis tulipae*) on tulip flowers (bottom right). (Gary A. Chastagner © 2020 All rights reserved).

**Figure 5** Bulb mold (*Penicillium* spp.) is a common disease of propagative structures, such as tulip bulbs. (Gary A. Chastagner © 2020 All rights reserved).

Asjes, 1990; Sutton et al., 1986). In addition, plant material infected with some viruses cannot be exported, resulting in severe economic damage. Viruses are transmitted from plant to plant by insects, nematodes, mites, sap (i.e. via mechanical transmission), and oomycetes (Loebenstein et al., 1995).

### *2.4 Bacterial and Phytoplasma diseases*

Bacterial and Phytoplasma diseases are not as common as fungal or viral diseases, but a few species have the potential to cause severe damage on some geophytes (Fig. 7 and Table 3). Bacterial pathogens infect through natural openings and wounds. Several *Dickeya* spp. and *Pectobacterium* spp. can cause serious soft rot in some geophytes, such as calla lily. Infection by *Xanthomonas* spp. leads to poor bulb quality, while infection by different *Rhodococcus* spp. leads to soft rot and distortion of the roots, scales, and shoots. Phytoplasmas are organisms that lack cell walls and were formerly called mycoplasma-like organisms (MLO). Phytoplasmas are smaller than bacteria and are sometimes referred to as 'fastidious' bacteria because most of them have not been cultured or can only be grown on complex, specialized media. Aster yellows disease of gladiolus is caused by a phytoplasma that is transmitted by leafhoppers.

### *2.5 Nematodes*

Nematodes are non-segmented roundworms and those which are plant pathogens are obligate parasites and have a sharp, hollow needle-like feeding structure called a stylet used to penetrate plant host tissue. Nematodes have a relatively simple lifecycle in which they develop into a complete worm

**Table 2** Some common virus diseases of the seven most widely cultivated ornamental geophyte species

| Geophyte species | Diseases | Virus genus | Notes | Reference |
|---|---|---|---|---|
| **Tulips** | Arabis mosaic virus | *Nepovirus* | Arabis mosaic virus has been reported in a number of ornamental geophyte hosts worldwide and is transmitted by nematodes in the genus *Xiphinema*. | McGovern and Elmer (2018) |
| | Cucumber mosaic virus | *Cucumovirus* | Cucumber mosaic virus has been reported on the most hosts of any other plant virus including a number of ornamental geophyte hosts worldwide and is transmitted mechanically and by aphids. | Hanks and Chastagner (2018), McGovern and Elmer (2018) |
| | Lily symptomless virus | *Carlavirus* | See section on lily. Causes flower breaking symptoms in tulip. | McGovern and Elmer (2018) |
| | Tobacco necrosis virus | *Necrovirus* | Tobacco necrosis virus has a wide host-range and has been reported to be transmitted by a soilborne fungus. | Hanks and Chastagner (2018), McGovern and Elmer (2018) |
| | Tobacco rattle virus | *Tobravirus* | See section on daffodil. | McGovern and Elmer (2018) |
| | Tulip breaking virus | *Potyvirus* | Tulip breaking virus is the most common virus in tulip and famous for its 'breaking' symptoms on flowers that causes streaks of color that were prized in the Dutch flower trade of tulips piror to the symptoms being identified as being caused by a virus. | Chastagner et al. (2018), McGovern and Elmer (2018) |
| **Lilies** | Arabis mosaic virus | *Nepovirus* | See section on tulips. Lilies infected with Arabis mosaic virus can also be symptomless. | Chastagner et al. (2018) |
| | Cucumber mosaic virus | *Cucumovirus* | See section on tulips. Symptoms on lilies co-infected with Lily symptomless virus are often more severe. | Chastagner et al. (2018) |
| | Lily mottle virus | *Potyvirus* | Lily mosaic virus has a restricted host range as compared with other viruses of ornamental geophytes and is the most common virus in lilies. Lily mosaic virus is aphid-transmitted. | Chastagner et al. (2018) |

| | | | | |
|---|---|---|---|---|
| | Lily symptomless virus | *Carlavirus* | As its name suggests, Lily symptomless virus can remain symptomless, but expresses severe symptoms only under certain environmental conditions. The virus is aphid-transmitted. | Chastagner et al. (2018) |
| | Lily virus X | *Potexvirus* | Lily virus X does not produce symptoms in lily, but can be mechanically transmitted and is suspected to have an insect vector. | Chastagner et al. (2018) |
| | Plantago asiatica mosaic virus | *Potexvirus* | Plantago asiastica mosaic virus is a recently discovered virus on lily with global distribution and wide host range. The virus causes severe symptoms, especially as a co-infection with other lily viruses. | Chastagner et al. (2018) |
| | Strawberry latent ringspot virus | Unassigned (family Secoviridae) | Strawberry latent ringspot virus is symptomless in lily, but transmitted by nematodes in genus *Xiphinema*. | Chastagner et al. (2018) |
| | Tobacco rattle virus | *Tobravirus* | See section on daffodil. | Chastagner et al. (2018) |
| | Tulip breaking virus | *Potyvirus* | While more important on tulips, Tulip breaking virus causes mild to moderate symptoms on the leaves of lily. | Chastagner et al. (2018) |
| **Gladiolus** | Bean yellow mosaic virus | *Potyvirus* | Symptoms of Bean yellow mosaic virus are non-existant to mild and expression is highly-dependent on environmental conditions. Nearly all gladiolus are thought to be infected with this virus. | Elmer and Kamo (2018) |
| | Cucumber mosaic virus | *Cucumovirus* | Cucumber mosaic virus symptoms are usually not evident on gladiolus until flowering. See section on lily for more information. | Elmer and Kamo (2018) |
| **Daffodils** | Arabis mosaic virus | *Nepovirus* | See section on lily. Symptoms not readily evident on daffodil. | Hanks and Chastagner (2018) |

(*Continued*)

**Table 2** (*Continued*)

| Geophyte species | Diseases | Virus genus | Notes | Reference |
|---|---|---|---|---|
| | Narcissus latent virus | *Macluravirus* | Narcissus latent virus has also been found in gladiolus and iris and is aphid-transmitted. | Hanks and Chastagner (2018) |
| | Narcissus mosaic virus | *Potexvirus* | Daffodil is the only known host of Narcissus mosaic virus whose symptoms can be absent to severe. | Hanks and Chastagner (2018) |
| | Narcissus tip necrosis virus | Unassigned (family Tombusviridae) | Daffodil is the only known host of Narcissus tip necrosis virus and symptoms of this virus appear after flowering. | Hanks and Chastagner (2018) |
| | Narcissus white streak virus | Unassigned (family Potyvirdae) | Daffodil is the only known host of Narcissus white streak virus and can cause yield losses of up to 40% over several years. | Hanks and Chastagner (2018) |
| | Narcissus yellow stripe virus | *Potyvirus* | This virus is also referred to as 'Narcissus mosaic virus' in the literature and is aphid-transmitted. | Hanks and Chastagner (2018) |
| | Strawberry latent ringspot virus | Unassigned (family Secoviridae) | See section on tulip. | Chastagner et al. (2018), Hanks and Chastagner (2018) |
| | ***Tobacco rattle virus*** | *Tobravirus* | Tobacco rattle virus has a wide range of hosts, including numerous ornamental geophytes. The virus is vectored by stubby root nematodes. In daffodils, cultivars range from having severe symptoms to symptomless infections to the apparent ability to resist infection. | Hanks and Chastagner (2018) |
| **Hyacinth** | Hyacinth mosaic virus | *Potyvirus* | Hyacinth mosaic virus is aphid transmitted. | Asjes (1985) |

| | | | | |
|---|---|---|---|---|
| | Tobacco necrosis virus | *Necrovirus* | See section on tulip. | Derks et al. (2003) |
| | Tobacco rattle virus | *Tobravirus* | See section on daffodil. | Chastagner et al. (2018) |
| **Crocus** | Arabis mosaic virus | *Nepovirus* | See section on tulip. | Chastagner et al. (2018) |
| | Cucumber mosaic virus | *Cucumovirus* | See section on tulip. | Chastagner et al. (2018) |
| | Tobacco rattle virus | *Tobravirus* | See section on daffodil. | Chastagner et al. (2018) |
| **Bulbous Iris** | Cucumber mosaic virus | *Cucumovirus* | See section on tulip. | Duraisamy and Pokorný (2009) |
| | Iris mild mosaic virus | *Potyvirus* | Iris mild mosaic virus is vectored by aphids and symptom expression is dependent on when in the season the plant becomes infected. | Asjes (1979), van der Vlugt et al. (1988) |
| | Iris severe mosaic virus | *Potyvirus* | Iris severe mosaic virus is vectored by aphids and has also been reported on gladiolus. | Asjes (1979), Elmer and Kamo (2018), van der Vlugt et al. (1988) |
| | Narcissus latent virus | *Macluravirus* | See section on daffodil. | Hanks and Chastagner (2018) |
| | Bean yellow mosaic virus | *Potyvirus* | See section on gladiolus. | Duraisamy and Pokorný (2009), van der Vlugt et al. (1988) |

**Table 3** Some common bacterial and phytoplasma diseases of the seven most widely cultivated ornamental geophyte species

| Geophyte species | Diseases | Causal organism | Notes | Reference |
|---|---|---|---|---|
| **Tulips** | Bacterial Soft Rot | *Pectobacterium carotovorum* | *P. caratovorum* affects the bulb (or corm or rhizome) of a number of widely-cultivated ornamental geophytes and is a problem in both the field and in storage. In tulips, *P. caratovorum* can also infect the foliage and flower tissues. | Boyraz et al. (2006), McGovern and Elmer (2018) |
| | Hell Fire | *Curtobacterium flaccumfaciens pv. oorti* (*Corynebacterium oortii*) | Limited to plants grown outdoors. The torn leaf epidermis that is separated from the underlying tissues is associated with large temperature variation and night frosts. | Conijn (2017) |
| **Lilies** | Scale Rot | *Pseudomonas* spp. (complex with fungal species *Fusarium* and *Cylindrocarpon*) | Scale rot in lilies has mostly been associated with Easter lilies and wounding is typically considered essential for disease to occur. | Chastagner et al. (2018) |
| | Leafy Gall | *Rhodococcus fascians* | This bacterium has a wide host-range and affects a number of herbaceous perennials. This pathogen, carried in the bulb, was previously managed by the addition of formaldahyde to bulb dips, but new strategies are needed as formaldahyde use has been phased out of common practice. Due to the lack of effective treatments, bulbs infested with *R. fascians* should be destroyed. | Putnam and Miller (2007), Chastagner et al. (2018) |
| | Bacterial Soft Rot | *Pectobacterium carotovorum* | Bacterial soft rot of lily bulbs has also been reported to be caused by the bacterium *Pseduomonas marginalis*. | Hahm et al. (2003) |
| **Gladiolus** | Bacterial Scab, Stem and Neck Rot | *Burkholderia gladioli* pv. *gladioli* | While this bacterium does not cause mortality, corms infected with *Burkholderia gladioli* pv. *gladioli* should not be planted or used in propagation. | Elmer and Kamo (2018) |
| | Aster Yellows | '*Candidatus* Phytoplasma asteria' | Aster yellows are common in a number of ornamental and non-ornamental plants. '*Candidatus* Phytoplasma asteria' is not a true bacterium, rather a non-cultureable, bacteria-like organism. | Elmer and Kamo (2018) |

| | | | | |
|---|---|---|---|---|
| | Bacterial Soft Rot | *Pectobacterium carotovorum* | Although *P. caratovorum* has been isolated from gladiolus bulbs, this pathogen does not seem to be as important as it is in other bulb crops, such as calla lily (*Zantedeschia*). | Taghinasb et al. (2008) |
| **Daffodils** | Bacterial Soft Rot | *Pectobacterium carotovorum* subsp. *carotovorum* | *Pectobacterium carotovorum* subsp. *carotovorum* is not as commonly associated with daffodils as for other bulb crops, however, has been seen on the bulbs of tezetta cultivars. | Hanks and Chastagner (2018) |
| | Bacterial Streak | *Pseudomonas* sp. | Reports of a *Pseudomonas* sp. on daffodil originate only from the Pacific Northwest of the United States and appear to affect daffodils both in the field and in cold storage. | Gould (1957), Hanks and Chastagner (2018) |
| **Hyacinth** | Bacterial Soft Rot | *Pectobacterium carotovorum* and subspecies | Disease caused by *P. caratavorum* on hyacinth is not typically a large production problem, but widely reported. | Gardan et al. (2003) |
| | Bacterial Wilt | *Dickeya* spp. | Soft rot of hyacinth by *Dickeya* species causes significant losses in bulb production and causes flower abortion and pre-flowering flower drop. | Jafra et al. (2009), Slootweg et al. (2015), van Doorn et al. (2008) |
| | Hyacinth Yellows | *Xanthomonas hyacinthi (X. c. pv. hyacinthi)* | Hyacinth yellows is considered the most important disease of cultivated hyacinth, and a number of resistant cultivars have been identified to manage this disease (see reference). | van Tuyl and Toxopeus (1980) |
| **Crocus** | Bacterial Soft Rot | *Pectobacterium corotovorum* | Although P. caratovorum has been isolated from crocus bulbs, it appears to be minor pathogen compared with other bulb crops. | van Doorn et al. (2008) |
| **Bulbous Iris** | Bacterial Blight | *Xanthomonas campestris* pv. *tardicrescens* | *Xanthomonas campestris pv. tardicrescens* affects the foliage of bulbous iris, but not the rhizome. | Okubo and Sochacki (2013) |
| | Bacterial Soft Rot | *Pectobacterium corotovorum* | Bacteria in the genus *Dickeya can also infect bearded iris.* | Okubo and Sochacki (2013), van Doorn et al. (2008) |

**Figure 6** Common symptoms associated with virus diseases include flower breaks and distortion, chlorotic mosaic, streaking, and ringspot patterns on the foliage of many hosts. *Tulip breaking virus* on tulips (left), *Narcissus late season yellows virus* on daffodil (middle), *Dahlia mosaic virus* on dahlia (top right), and *Tobacco rattle virus* on peony (bottom right). (Gary A. Chastagner © 2020 All rights reserved).

**Figure 7** Hell fire on tulips, caused by *Curtobacterium flaccumfaciens* pv. *oortii* (*Corynebacterium oortii*) (Gary A. Chastagner © 2020 All rights reserved).

within eggs prior to hatching and then move through four larval stages before they become reproductive adults. There are two groups of plant-parasitic nematodes: ectoparasites, which can move from plant to plant, and endoparasites, which remain on or in the same plant. Within these groups are both migratory and sedentary species, the former of which moves through host tissue and the latter of which does not move once the initial infection has been initiated. Nematode infections can lead to serious reductions in plant yield and quality and are most often associated with root problems (Table 4). Soilborne nematodes include *Pratylenchus* spp., which cause root lesions on several bulb crops. The stem and bulb nematode (called the stem nematode in the UK), *Ditylenchus dipsaci,* can be a serious pest on geophytes. There are different biological races, such as the tulip race, of this pathogen. *D. dipsaci* and specific races are quarantine pests in a number of countries. There is one genus of nematodes, *Aphelenchoides* spp., that do not infect roots. Instead, the so-called 'foliar nematodes' cause leaf spots and growth distortion (Fig. 8). In addition to being pathogens, a few migratory ectoparasitic nematodes are also vectors of viruses.

### *2.6 Abiotic disorders*

Non-infectious or abiotic disorders affecting geophytes are sometimes also referred to as physiological disorders (De Hertogh and Le Nard, 1993a). Abiotic disorders are generally grouped into disorders that are caused by exposure to environmental conditions such as extreme temperature, moisture, light and nutrient availability, or exposure to harmful chemicals such as ethylene, air pollutants, herbicides, or salt. Some disorders, such as upper leaf necrosis of lilies (Fig. 9) and topple of tulips and bulbous iris (both caused by a calcium deficiency), and gummosis of tulips (caused by ethylene exposure) are well understood, but the causes of other disorders, such as a chocolate spot on daffodils, are poorly understood.

## 3 General disease management principles

Chastagner et al. (2012) and Lakshman et al. (2019) provide comprehensive overviews of recent advances relating to the integrated management of diseases on ornamental geophytes. A number of additional references that address specific management practices for diseases on geophytes are mentioned at the end of this chapter.

General disease management principles are centered around the concept of integrated pest management, often referred to simply as 'IPM'. Concerns resulting from increased instances of resistance to pesticides, the effect of pesticides on the environment, and the sustainability of agricultural production systems led to the concept of IPM in 1967 (Smith and van den Bosch, 1967). Initially the focus was on the implementation of IPM programs to reduce the impact of pesticides on the environment and was later followed by efforts to reduce energy consumption and costs associated with crop production (Rudd-Jones, 1975; van Julsingha, 1980). Cook (2000) indicates that the goal of IPM is to implement an economically, environmentally, and socially acceptable plant health management system.

**Table 4** Some common nematodes of the seven most widely cultivated ornamental geophyte species

| Geophyte species | Nematodes (common name) | Latin name | Notes | References |
|---|---|---|---|---|
| **Tulips** | Stem and Bulb Nematode | *Ditylenchus dipsaci* | *Ditylenchus dipsaci* is an important pathogen on nearly all ornamental geophyte crops. Symptoms include swelling and distortion of leaves and stems, and necrosis or rotting of the base of the stems or bulb/tuber/rhizome. Infected bulbs are more susceptible to postharvest infection by fungi and bacteria. In tulip, the foliage and flowers may also have spots or localized swellings. | Chastagner et al. (2018), McGovern and Elmer (2018) |
| | Stubby Root Nematode | *Trichodorus* spp. and *Paratrichodorus* spp. | Subby root nematodes feed on plant roots and cause stunting and swelling. Due to root damage and subsequent decrease in uptake of water and nutrients, above-ground symptoms include chlorosis and wilting. Some stubby root nematodes are also vectors of important viruses of ornamental geophytes. | Chastagner et al. (2018), McGovern and Elmer (2018) |
| | Foliar Nematode | *Aphelenchoides* spp. | See lily section on *Aphelenchoides*. Two *Aphelenchoides* species have been found associated with tulip. | McCuiston et al. (2007) |
| | No common name reported | *Aphelenchoides subtenuis* | See description under daffodil section. | Hanks and Chastagner (2018) |
| **Lilies** | Root Lesion Nematode | *Pratylenchus penetrans* | See daffodil section for description. Tulips infected by *P. penetrans* may fail to emerge. | Chastagner et al. (2018) |
| | Foliar Nematode | *Aphelenchoides* spp. | Foliar nematodes are common in a number of ornamental geophyte crops. Foliage nematodes enter and migrate through leaf tissue, causing discoloration and distortion. Foliar nematodes have been found on shipments of geophytes, therefore, can represent a route of spread. | Chastagner et al. (2018) |
| | Stem and Bulb Nematode | *Ditylenchus dipsaci* | See symptoms above in tulip section. Like in other ornamental geophytes, *Ditylenchus dipsaci* is an important pathogen of lily. | Chastagner et al. (2018) |
| | Root-Knot Nematode | *Meloidogyne* spp. | Several species of root knot nematodes infect a wide range of ornamental and non-ornamental hosts. Swellings or 'galls' can be seen on fine feeder roots. Above-ground symptoms can include wilting or chlorosis. | Chastagner et al. (2018) |
| | Needle Nematode | *Longidorus* spp. | Needle nematode is especially damaging to seedlings, causing stunting and galling at root tips. | Chastagner et al. (2018) |
| **Gladiolus** | Root-Knot Nematode | *Meloidogyne* spp. | See symptoms above in tulip section. Research done in the 1970s suggests an association between the root knot nematode and bacterial scab of gladiolus. | El-Goorani et al. (1974). |

| | | | | |
|---|---|---|---|---|
| **Daffodils** | Stem and Bulb Nematode | *Ditylenchus dipsaci* | See symptoms above in tulip section. *Ditylenchus dipsaci* is an important quarantine organism for the daffodil bulb trade and is managed by hot water dips. See reference for full description. | Hanks and Chastagner (2018) |
| | No common name reported | *Aphelenchoides subtenuis* | Unlike most *Aphelenchoides* species which attack foliage, *A. subtenuis* feeding occurs on the roots. See reference for full description. | Hanks and Chastagner (2018) |
| | Root Lesion Nematode | *Pratylenchus penetrans* | The root lesion nematode is limited to the roots of daffodil (it does not attack the bulb) and causes small lesions on root tissue and above-ground symptoms of stunting and chlorosis. | Hanks and Chastagner (2018) |
| | Foliar Nematode | *Aphelenchoides* spp. | See description in lily section. | Kohl (2011) |
| **Hyacinth** | Stem and Bulb Nematode | *Ditylenchus dipsaci* | See description in tulip section. | McGovern and Elmer (2018) |
| **Crocus** | Foliar Nematode | *Aphelenchoides* spp. | See description in lily section. | McCuiston et al. (2007) |
| | Iris Nematode (aka. Potato Tuber Nematode) | *Ditylenchus destructor* | See description in bulbous iris section. | Okubo and Sochacki (2013) |
| | No common name reported | *Aphelenchoides subtenuis* | See description in daffodil section. | Hanks and Chastagner (2018) |
| **Bulbous Iris** | Stem and Bulb Nematode | *Ditylenchus dipsaci* | See description in tulip section. | McGovern and Elmer (2018) |
| | Iris Nematode (aka. Potato Tuber Nematode) | *Ditylenchus destructor* | The iris nematode, *Ditylenchus destructor*, is also an important pathogen of potatoes tubers. Although typically associated with disease in iris, *D. destructor* also has the ability to infect gladiolus, hyacinth, and tulip. | Goodey (1951), Jeger et al. (2016) |
| | Root-Knot Nematode | *Meloidogyne* spp. | See description in lily section. | Okubo and Sochacki (2013) |
| | Foliar Nematode | *Aphelenchoides* spp. | See description in lily section. | McCuiston et al. (2007) |
| | No common name reported | *Aphelenchoides subtenuis* | See description in daffodil section. | Hanks and Chastagner (2018) |

**Figure 8** Foliar nematodes (*Aphelenchoides* spp.) can cause leaf spots, such as this spot on peony which is delimited by the leaf veins. (Gary A. Chastagner © 2020 All rights reserved).

**Figure 9** Upper leaf necrosis, which is caused by calcium deficiency can be a serious abiotic problem on greenhouse grown lilies. (Gary A. Chastagner © 2020 All rights reserved).

Although the reduction in pesticide usage is a potential benefit of IPM programs, Kogan (1998) indicates that in order to maintain viable agricultural production, it may be necessary, in special instances, to use more, not less, pesticides. The goal is that pesticide use is guided by the principles of IPM, that is, selective use after maximizing the effectiveness of cultural and natural controls. Kogan (1998) provides an excellent historical perspective on the development of IPM which also includes the concept of a 'key pest' (Jacobsen, 1997; Smith and van den Bosch, 1967). A key pest is 'a serious, perennially occurring species that dominates control practices because in the absence of deliberate control by man, the pest population usually remains above the economic injury levels'. Chastagner et al. (2012) suggest that aphid-transmitted viruses; soilborne fungi, such as *Fusarium oxysporum* and *Rhizoctonia* spp.; bulb and stem nematodes; bacterial soft rot pathogens; and pathogens that cause foliage diseases, such as those caused by *Botrytis* species, are all examples of 'key pests' on geophytes.

Additionally, understanding the 'disease triangle' concept helps inform the methods of infectious disease control. The disease triangle concept involves an understanding that disease development requires the co-occurrence and interaction of three factors: a susceptible host, a pathogen, and a favorable environment (Agrios, 2005). The application of IPM principles requires an additional understanding of factors within the disease cycle that affects the epidemiology of the disease, such as inoculum production, spread, and survival; host susceptibility; infection; symptom development; and disease progression.

Accurate diagnosis is the first step in any successful disease management program. Implicit is scouting to monitor the health of the crop being grown and the effectiveness of any disease management treatment. Although in-field diagnostic skills increase with experience, accurate identification often requires that samples can be sent to a laboratory. In addition to the traditional culturing and microscopy laboratory approaches, significant advances have been made in the use of serological tests, such as the enzyme-linked immunosorbent assay (ELISA), and nucleic acid-based tests, such as polymerase chain reaction (PCR), to identify causal agents. The development of PCR-detection techniques has also provided a means to identify specific taxa of pathogens and assess infection and colonization of plant tissue and quantify inoculum survival in plants and soil (de Boer, 2011; van Doorn et al., 2011; Chiocchetti et al., 1999; Lees et al., 2002).

Following an accurate diagnosis, successful disease management requires an integrated approach that often includes combining diverse tactics such as:

- Exclusion to prevent the introduction of invasive pathogens and vectors and the use of certified pathogen-free planting stock. Exclusion measures through the Plant Quarantine Act from 1919 to 1926 were essential in the attempt to prevent the introduction of the stem and bulb nematode on daffodil in the United States.
- Avoidance of planting sites and times when environmental conditions favor the pathogen or coincide with periods of increased susceptibility of the host, avoiding wounding or injury that provides entry sites for pathogens such as bacteria, and manipulation of the environment to reduce conditions that are favorable to pathogens. One of the most important methods of preventing foliar diseases of geophytes involves avoiding overhead irrigation, which produces free moisture on the leaves in which fungal spores can germinate (Fig. 10).

**Figure 10** Extended periods of free moisture favor the sporulation and infection of pathogens like *Botrytis*. (Gary A. Chastagner © 2020 All rights reserved).

- Eradication to reduce the carryover of inoculum from one year to another via rotation to nonhost crop, removal of weed hosts, destruction of infected host residues, application of some types of pesticides and disinfectants that target pathogens, heat treatments, solarization, steam, soil fumigation, hot water treatments, rogueing out infected plants, destruction of vectors, and thermotherapy and/or meristem culture to produce pathogen-free propagation material. Hot water dips of many bulbous ornamentals, including lilies, daffodils, and tulips, is a critical management strategy for eradicating nematodes from propagative material and is a standard practice, especially in bulbs destined for international export.
- Sanitation is an important approach that is used to minimize the risk of certain diseases. Sanitation involved: hand or chemical rogueing of symptomatic plants in the field, such as tulips exhibiting symptoms of tulip break virus; removal of diseased plant material during the growing season and during the processing of bulbs; and planting stock and cut flowers on packing lines (Fig. 11). Sanitizers are often used as spray or dip applications to sanitize bulbs, trays, pots, work surfaces, and tools (Moore et al., 1979; Gould, 1957). Growers now have a much broader choice of oxidizers to sanitize objects and tools (Copes, 2018). For example, chlorine dioxide ($ClO_2$) is an effective replacement for the use of formaldehyde to reduce the spread of *Fusarium* during the hot water dip treatment of daffodil bulbs, and controlled-release systems to produce $ClO_2$ gas for a number of uses in the horticultural industry are being developed by a number of companies (Chastagner and Riley, 2002; Copes et al., 2014; Copes, 2018). This technology and the direct application of other oxidizers to plants has the potential to reduce postharvest diseases on bulbs and cut flowers

**Figure 11** Scouting and removal of diseased plants from a field of hyacinths. (Gary A. Chastagner © 2020 All rights reserved).

**Figure 12** Fungicides applications (right) are an effective way of protecting lilies against 'fire', which is caused by *Botrytis elliptica* (left). (Gary A. Chastagner © 2020 All rights reserved).

(Chastagner and Copes, 2002; Chastagner and Riley, 2005; Macnish et al., 2008; Macnish et al., 2010).

- Protection of the crop involves the use of biocontrol agents, application of some types of pesticides, such as fungicides that provide protective barriers to infection (Fig. 12); control of plant disease vectors; the use of antibiotics or antagonists; mandated crop-free periods to limit the spread of viruses; barrier crops; reflective mulches; cross protection; and induced host resistance plants. The reduction of chemical pesticide use in ornamental geophytes remains as one of the major goals for the industry and is discussed at length in the following sections.

Another tactic in disease management is disease-resistant varieties. Although most geophytes that are used in the production of bulbs, cut flowers and potted plants are selected for various horticultural characteristics and customer appeal, the selection and utilization of plants having disease and pest resistance is an increasingly important objective of many ornamental plant-breeding programs. Breeding for resistance is particularly important for key pests and pathogens such as *Botrytis* that currently require high-fungicide inputs and for bacteria, phytoplasmas, viruses, and a few soilborne fungi for which there are no or few effective control measures. Conventional and molecular approaches are being used to genetically manipulate host resistance, and natural and chemical defense elicitors are being used to promote the induction of natural plant defense mechanisms. Induced resistance against several bacterial and fungal pathogens has been reported in a number of ornamental geophytes (Chastagner et al., 2012). While promising, additional work is needed to improve the performance of elicitors, particularly under field conditions (Klessig et al., 2018).

## 4 Approaches to reduce chemical fungicide usage

Even though pesticides have continued to play a predominant role in the management of some ornamental geophyte diseases, a number of approaches are being used to reduce the amount of pesticides used and reduce the development of fungicide resistance.

One of the ways that chemical pesticide use has reduced is the development of products that are effective at lower concentrations. Prior to the introduction of the dithiocarbamate fungicides in the 1940s, fungicides were often applied at very high rates to control foliar diseases. Bordeaux mixture was one of the earliest fungicides (Millardet, 1933). Depending on the concentration of fungicide needed, growers would mix 4–10 pounds of copper sulfate with an equal amount of hydrated lime (calcium hydroxide) in 100 gallons of water and spray this mixture onto their crop to control diseases caused by pathogens such as *Botrytis*. The introduction of dithiocarbamates eliminated the need for growers to mix ingredients, reduced the risk of phytotoxicity to their crop, and reduced the amount of fungicide needed to obtain disease control (Brandes, 1953). Newer, reduced-risk fungicides, which according to the US Environmental Protection Agency have a low impact on human health, low toxicity to non-target organisms, low potential for groundwater contamination, and are effective at low use rates, have led to significant additional reductions in the amount of fungicides that are needed to control plant diseases (Morton and Staub, 2008). Morton and Staub indicate that between 1944 and 2002, the total amount of fungicides applied to crops in the United States was reduced by about 60% as more effective and selective fungicides were introduced into the market. On geophytes, mancozeb, a fungicide introduced in 1961, is still recommended to control diseases such as *Botrytis* on tulips at a rate of 1.2 lbs active ingredient (ai)/100 gallons. On the other hand, the newer fungicide fludioxonil is applied at a rate of only 1–2 oz ai/100 gallons. Similarly, the in-furrow application rates

**Figure 13** Disease forecasting models have been used to reduce the number of applications of fungicides that are applied to control white mold (*Ramularia vallisumbrosae*) on daffodils in the UK. (Gary A. Chastagner © 2020 All rights reserved).

of newer products like flutolanil, fludioxonil, and trifloxystrobin used to control gray bulb rot (*Rhizoctonia tuliparum*) and crown rot (*Sclerotium delphinii*) on tulips, iris, and lilies are about 80% lower than the amount of the older fungicide pentachloronitrobenzene (PCNB) required for control (Chastagner et al., 2012).

Reducing application frequency has also been shown as a possible method of reducing pesticide use. Growers have historically used a calendar-based spray schedule, particularly in the management of diseases such as *Botrytis* on field-grown crops like tulips and lilies. Ongoing research indicates that switching to a crop phenology-based program where the majority of sprays are applied early in the growing season to keep inoculum from building up and protecting the plant during periods of rapid growth and the development of flower buds have the potential to reduce the number of applications of fungicide needed to control *Botrytis* on tulips and lilies by 40–60% (Chastagner, unpublished).

A number of weather-based models have been developed that provide an additional approach to reduce the number of pesticide applications during the production of ornamental geophytes. The BoWaS: Botrytis Warning System (Bastiaansen et al., 1997) resulted in a potential 30–80% reduction in fungicide use to control *Botrytis* leaf blight in lily and has been commercially available in the Netherlands to lily, tulip, and gladiolus growers since 1999 (van den Ende et al., 2000). Similar forecasting models, when combined with newer fungicides, have reduced the number of applications needed to control smolder (*Botrytis narcissicola*) and white mold (*Ramularia vallisumbrosae*) on daffodil in the United Kingdom by 50% (Fig. 13) (Hanks et al., 2003; O'Neill et al., 2004).

While many challenges still exist (West et al., 2003), the use of precision application technologies to ensure targeted applications of pesticides is another approach that is being used to reduce fungicide usage. The amount of fungicides needed to control *Botrytis* leaf blight on lilies was reduced by at least 21% using a canopy density sprayer that employed a sensor to quantify crop canopy combined with variable rate-spray technologies (De Boer, 2009; Van der Zande et al., 2010).

There is an increasing interest in the use of biopesticides to control diseases and arthropod pests on many ornamentals, including geophytes. Although there is some variation in their classification, biopesticides are generally classified as natural products (microorganism-, higher plant-, and animal-derived products) and microorganism-based products (viruses, bacteria, fungi, nematodes, and protozoa) (Copping and Menn, 2000; Yoon et al., 2013). The number of registered products is increasing (Daughtrey and Buitenhuis, 2019), but challenges still exist to effective implementation of these products due to variable control with some products, particularly on field-grown geophytes (Chastagner and DeBauw, 2014, 2017). Borges et al. (2018) have suggested that the research relating to the development of controlled-release formulations of botanicals (essential oils and plant extracts) that contain a multitude of bioactive substances that are active against fungi, bacteria, and nematodes may have the potential to improve the efficacy of botanicals against plant pathogens.

## 5 Managing fungicide resistance

While the introduction of fungicides with single-site modes of actions (MOA) in the 1960s has contributed to the overall reduction in the rates of fungicides used to control diseases, they have increased the risk of fungicide resistance development within populations of pathogens (Morton and Staub, 2008). Benzimidazole fungicides were introduced in the 1960s and early 1970s. Benzimidazole fungicides proved effective at low-use rates against a broad spectrum of fungi and were systemic with post-infection activity that allowed for extended spray intervals. In addition to being widely used during the production of crops, they were also highly effective as postharvest treatments. In 1969, one year after their introduction, the first case of resistance to benzimidazoles occurred in powdery mildew in greenhouses. By 1984, resistance had been reported on many of the pathogens against which benzimidazoles are active (Smith, 1988). A broad range of plant pathogens have developed resistance to fungicides since the 1960s, including pathogens that cause Botrytis diseases, Penicillium blue mold, powdery and downy mildews, and Pythium root rots and decays. Given the global movement of planting stock, there is a risk that resistant individuals or populations of pathogens can be easily introduced into other production regions, thus complicating growers' disease and resistance management programs.

As a result of the increasing problems with fungicide resistance and questions relating how best to manage it, the Fungicide Resistance Action

Committee (FRAC) was developed in 1980–1981 to 'provide fungicide resistance management guidelines to prolong the effectiveness of "at-risk" fungicides and to limit crop losses should resistance occur' (FRAC, 2019). Each year, the FRAC publishes a FRAC code list that summarizes the modes of action, cross resistance patterns, and resistance risk for commercial fungicides and the most important bactericides. In 2019, the FRAC code list contained products listed under 78 code numbers, 35 of which were identified as requiring resistance management.

Reducing the risk of resistance development should be an integral component of every grower's disease management program. Approaches to manage fungicide resistance development include limiting the number of applications of at-risk fungicides and rotating or combining at-risk fungicides with chemicals from different FRAC group designations. In some cases, registrants are formulating products that contain a combination of chemistries with different modes of action in an effort to maintain the usefulness of highly effective at-risk active ingredients. In addition to indicating a fungicide's FRAC code, most fungicide labels now provide specific resistance management recommendations to help reduce the development of fungicide resistance. If fungicide resistance does develop, the use of the fungicide in question and other fungicides within the same FRAC code should be stopped until it is confirmed that the population of resistant strains of the pathogen is no longer present. If the fungicide still needs to be used to control other pathogens, it should only be used in combination with one or more different FRAC codes that are known to control the resistant pathogen.

## 6 *Botrytis* infection of ornamental geophytes

Diseases caused by species of *Botrytis* are considered some of the most important fungal pathogens of ornamental geophytes. Genus *Botrytis* is currently composed of approximately 35 officially described species, although this number constantly increases with the advancement of molecular detection technologies, which allow for easy discrimination between species (discussed more below). Many of the early-discovered species of *Botrytis* were originally found associated with ornamental geophytes and were named accordingly, such as *B. tulipae*, named for its first-discovered host, the tulip (*Tulipa*), or *B. narcissicola*, named after its host, the daffodil (*Narcissus*). There are 17 different species of *Botrytis* that have been described on ornamental geophytes (Table 5), some of which display a wide host range, such as the generalist pathogen, *B. cinerea*, and others which have been reported to have a narrow host range, such as *B. polyblastis* whose only known host is daffodil. Although some species are named after their original host, the pathogen's host range is not always limited to its namesake. Such is the case with *B. tulipae* which has

**Table 5** *Botrytis* species and their ornamental geophyte host plants

| | Geophyte Host | | |
|---|---|---|---|
| *Botrytis* species** | Common name | Species | Reference |
| *B. aclada* | Ornamental allium | *Allium caeruleum* (syn. *A. azureum)* | Elad et al. (2016) |
| *B. cinerea** | Amaryllis | *Amaryllis belladona* | Elad et al. (2016) |
| | Autumn crocus | *Colchicum* spp. | Elad et al. (2016) |
| | Calla lily | *Zantedeschia* spp. | Elad et al. (2016) |
| | Canna lily | *Canna* spp. | Elad et al. (2016) |
| | Daffodil | *Narcissus* spp. | Elad et al. (2016) |
| | Dahlia | *Dahlia* spp. | Elad et al. (2016) |
| | Daylily | *Hemerocallis spp. hybrids* | Elad et al. (2016) |
| | Gladiolus | *Gladiolus* spp. and hybrids | Elad et al. (2016) |
| | Hyacinth | *Hyacinthus orientalis* | Elad et al. (2016) |
| | Hymenocallis spider lily | *Hymenocallis* spp. | Elad et al. (2016) |
| | Iris | *Iris* spp. | Farr and Rossman (2019) |
| | Lily | *Lilium* spp. and hybrids | Elad et al. (2016) |
| | Lily of the valley | *Convallaria majalis* | Elad et al. (2016) |
| | Peace lily | *Spathiphyllum* spp. | Elad et al. (2016) |
| | Peony | *Paeonia* spp. | Elad et al. (2016) |
| | Persian cyclamen | *Cyclamen persicum* | Elad et al. (2016) |
| | Peruvian lily | *Alstroemeria* spp. | Elad et al. (2016) |
| | Tubcrose (aka. polyanthus lily) | *Polianthes tuberosa* | Elad et al. (2016) |
| | Tuberous begonia | *Begonia* x *tuberhybrida* | Elad et al. (2016) |
| | Tulip | *Tulipa* spp. | Elad et al. (2016) |
| *B. convoluta* | Bearded iris | *I. flavescens* | Elad et al. (2016) |
| | Dutch iris | *Iris × hollandica* | Elad et al. (2016) |
| | Reticulated iris | *I. reticulata* | Elad et al. (2016) |
| *B. croci* | Spring crocus | *Crocus versicolor* | Farr and Rossman (2019) |
| *B. deweyae* | Daylily | *Hemerocallis spp. hybrids* | Grant-Downton et al. (2014) |
| *B. elliptica** | Autumn crocus | *Colchicum autumnale* | Elad et al. (2016) |
| | Daylily | *Hemerocallis* spp. hybrids | Elad et al. (2016) |
| | Easter lily | *Lilium longiflorium* | Elad et al. (2016) |
| | Gladiolus | *Gladiolus* spp. and hybrids | Elad et al. (2016) |
| | Lily | *Lilium* spp. and hybrids | Elad et al. (2016) |
| | Toad lily | *Tricyrtis hirta* | Elad et al. (2016) |

| *Botrytis* species** | Geophyte Host<br>Common name | <br>Species | Reference |
|---|---|---|---|
| | Tuberose (aka. polyanthus lily) | *Polianthes tuberosa* | Elad et al. (2016) |
| | Yellow avalanche lily | *Erythronium grandiflorum* | Elad et al. (2016) |
| *B. euroamericana* | Herbaceous peony | *Paeonia lactiflora* and hybrids | Garfinkel et al. (2017) |
| *B. galanthina* | Opal flower | *Lachenalia aloides* | Elad et al. (2016) |
| | Snowdrop | *Galanthus nivalis* | Elad et al. (2016) |
| *B. gladiolorum** | Freesia | *Freesia* spp. hybrids | Farr and Rossman (2019) |
| | Gladiolus | *Gladiolus* spp. and hybrids | Elad et al. (2016) |
| *B. hyacinthi* | Grape hyacinth | *Muscari armeniacum* | Elad et al. (2016) |
| | Hyacinth | *Hyacinthus orientalis* | Elad et al. (2016) |
| | Lily | *Lilium* spp. and hybrids | Elad et al. (2016) |
| | Pineapple lily | *Eucomis vandermerwei* | Elad et al. (2016) |
| *B. narcissicola** | Autumn crocus | *Colchicum autumnale* | Elad et al. (2016) |
| | Daffodil | *Narcissus* spp. | Elad et al. (2016) |
| | Iris | *Iris* spp. | Elad et al. (2016) |
| | Polyanthus | *Polianthes* spp. | Farr and Rossman (2019) |
| | Rhododendron | *Rhododendron* spp. | Farr and Rossman (2019) |
| *B. paeoniae* | Lily of the valley | *Convallaria majalis* | Elad et al. (2016) |
| | Opal flower | *Lachenalia aloides* | Elad et al. (2016) |
| | Peony | *Paeonia* spp. | Elad et al. (2016) |
| | Solomon's seal | *Polygonatum* spp. | Farr and Rossman (2019) |
| *B. polyblastis** | Daffodil | *Narcissus pseudonarcissus* and *N. tazetta* | Elad et al. (2016) |
| *B. ranunculi** | Giant spider lily | *Crinum augustum* | Farr and Rossman (2019) |
| *B. sphaerosperma* | Lily | *Lilium spp. and hybrids* | Chastagner et al. (2018) |
| *B. squamosa** | Lily | *Lilium* spp. | Farr and Rossman (2019) |
| *B. tulipae* | Autumn crocus | *Colchicum autumnale* | Elad et al. (2016) |
| | Lily | *Lilium* spp. | Elad et al. (2016) |
| | Tulip | *Tulipa* spp. | Elad et al. (2016) |

*Species with a known teleomorph (sexual) (*Botryotinia*) stage (Terhem, 2015).
**Additional *Botrytis* spp. not included in the table include: *B. byssoidea**, *B. calthae**, *B. californica*, *B. caroliniana*, *B. eucalypti*, *B. fabae**, *B. fabiopsis*, *B. ficariarum**, *B. fragariae*, *B. globosa**, *B. medusae*, *B. pelargonii**, *B. polyphyllae*, *B. porri**, *B. prunorum*, *B. pseudocinerea*, *B. sinoallii*, *B. sinoviticola*, *B. sphaeroperma**. *B. aclada*, *B. byssoidea*, *B. globosa*, B. *porri*, *B. sinoallii*, and *B. squamosa* all have non-ornamental geophyte hosts in the genus *Allium*.

also been reported on geophyte species *Colchicum* and *Lilium*. See Table 5 for a list of *Botrytis* species that have been reported to have ornamental geophyte hosts.

In general, *Botrytis* species are considered as aggressive pathogens which can quickly and completely kill their host under ideal conditions for infection. Several diseases caused by *Botrytis* in ornamental geophyte cut flower crops are referred to as 'fire' (or in the case of *B. narcissicola* on daffodil, 'smoulder') due to their ability to cause rapid, widespread damage up to complete losses in the field or greenhouse, not unlike the effects of a fire. *Botrytis* species of ornamental geophytes can cause disease on both above- and below-ground tissues of their host and attack both in the field and during cold storage, making the management of *Botrytis* in ornamental geophytes essential for producers of these crops (Figs. 14 and 15).

There are numerous publications that describe *Botrytis* management on ornamental geophyte crops, many of which are discussed in the 'Additional Resources' section below. Therefore, this section is limited to two specific topics as they relate to *Botrytis*: the diversity of *Botrytis* species and the ability of *Botrytis* to colonize plant tissue in a non-pathogenic, endophytic relationship. The impact of these two recent developments on disease management is yet to be fully determined, but initial investigations show that studying *Botrytis* on ornamental geophytes can provide some valuable insight into this pathogen and its biology.

**Figure 14** *Botrytis cinerea* and a number of host-specific *Botrytis* spp. attack foliage, flowers, and propagative structures, causing a number of pre- and postharvest diseases on ornamental geophyte such as peonies (top left and top middle), tulips (top right and bottom left), cyclamen (bottom middle), and lilies (bottom right). (top left: Andrea R. Garfinkel and others: Gary A. Chastagner © 2020 All rights reserved).

**Figure 15** Smoulder, caused by *Botrytis narcissicola*, can cause a decay and death of emerging daffodil foliar and flowers (left) and diagnostic one-sided lesions that result in shepherd crooked leaves (right) (Gary A. Chastagner © 2020 All rights reserved).

## 7 Diversity of *Botrytis* species

Molecular advancements in the biological sciences, especially involving PCR and DNA sequencing, have allowed for important insight into issues of diversity in genus *Botrytis*. Two main areas of study could be especially important concerning *Botrytis* management in ornamental geophytes, namely, the recognition of new species and the intra-specific diversity (or population genetics) within species.

Although the genus *Botrytis* was first described in 1729, there has been an explosion in only the last decade in the number of species that have been described. Gene sequencing and phylogenetic analysis have allowed for the discovery of the so-called 'cryptic' *Botrytis* species that previously went undetected due to the difficulty in traditional, morphological delineation among species. At least 12 species have been officially described in the last 8 years (Ferrada et al., 2016; Garfinkel et al., 2017; Grant-Downton et al., 2014; Li et al., 2012; Liu et al., 2016; Rupp et al., 2017; Saito et al., 2016; Walker et al., 2011; Zhang et al., 2010a,b, 2016; Zhong et al., 2019), and at least a dozen additional species have been provisionally identified using molecular techniques alone (Garfinkel et al., 2019; Shaw et al., 2016; Shipunov et al., 2008). Some of these new species have been identified on ornamental geophytes, including *B. deweyae* on daylily (*Hemerocallis*) (Grant-Downton et al., 2014) and *B. euroamericana* (Garfinkel et al., 2017), as well as more than 10 yet undescribed, but likely new species of *Botrytis* on peony (*Paeonia*) (Garfinkel et al., 2019).

The question of new species is more than purely scientific; while many *Botrytis* species on ornamental geophytes have similar epidemiology, it has been documented that some species of *Botrytis* have unique lifecycles (Hanks and Chastagner, 2018), susceptibilities to fungicides (Walker et al., 2011), propensities to develop fungicide resistance in the field (Plesken et al., 2015), and host ranges which ultimately affect the efficacy of disease management strategies. For example, the primary inoculum of fire on daffodils are the sexual spores, ascospores, which are produced by apothecia that arise from overwintered sclerotia of *B. polyblastis*. Ascospore infection is limited to flowers. The asexual stage of this pathogen then sporulates on infected flowers and the spores, conidia, are spread via wind to the foliage. Additional conidia are produced on the infected foliage, resulting in rapid disease buildup and death of the foliage (Hanks and Chastagner, 2018). Unlike most *Botrytis* diseases, foliage can be protected from infection with a single application of fungicide applied right after flowering (Chastagner, 1983). Other species of *Botrytis*, such as *B. paeoniae*, are not known to have a sexual cycle and therefore all of the infection is considered to initiate from conidia, and multiple sprays are needed for effective control. A greater understanding of the diversity of *Botrytis* species in ornamental geophytes and their life cycle may lead to additional insight on how to manage the diseases caused by this group of pathogens in the future as more effective and ecologically sustainable methods are sought.

Molecular advancements have also given insight into the intra-species diversity in *Botrytis*, which has allowed for making inferences into the biology of *Botrytis* species that affect ornamentals. Population genetics studies on *B. elliptica*, a pathogen that is primarily known for causing disease on lilies (*Lilium* spp.), in the United States and Taiwan, highlighted the importance of asexual reproduction and spread in the Pacific Northwest of the United States (PNW), likely facilitated through the movement of bulbs. This study helped demonstrate the potential importance of the bulb trade in the movement of *Botrytis*. The same study revealed that both asexual and sexual recombination is likely occurring within local populations in Taiwan, despite the rare sighting of sexual reproductive structures in the field (Huang et al., 2001). As described above, sexual recombination can have important impacts on disease management and genetic diversity of the pathogen. Lastly, this study showed some slight host-specialization of different genotypes of *B. elliptica* on different cultivars of Oriental lily (Huang et al., 2001), potentially suggesting that host resistance can be employed as a disease management strategy.

Although there are few examples of population genetics studies in ornamental geophytes, population genetics data to evaluate sexual recombination and discriminate among populations have been critical in other ornamental industries in identifying the routes of spread and limiting the exchange of plant pathogens (Goss, 2015). Additional population genetics studies in *Botrytis* have allowed researchers to identify temporal differences within *Botrytis* populations (Johnston et al., 2014), differences in populations exposed to different fungicide pressures (Ajouz et al., 2010), and even differences among populations exposed to different storage conditions (Baraldi et al., 2002). Together, these studies have elucidated unique patterns of distribution, fungicide resistance, host specificity, and mating behavior in *Botrytis* that can help inform disease management decisions. Although yet not widely utilized in *Botrytis* of ornamental geophytes, population genetics analysis of *B. paeoniae* in peony suggests that sexual reproduction does not likely occur in peony populations in the PNW (unpublished data). Additional population genetics studies may be warranted as the global trade of ornamental geophyte propagative materials increases, disease management strategies evolve, and breeding of new cultivars continues.

## 8 *Botrytis* as an endophyte

While *Botrytis* species have historically been characterized as aggressive and destructive necrotrophs (organisms that kill their host), research over the past decade has shown that a number of species within this genus also have the potential to live complex and cryptic lifestyles as endophytes while maintaining their virulence as pathogens. This surprising behavior has been recorded in various plants, such as lettuce (*Lactuca sativa*) (Sowley et al., 2010), primrose (*Primula × polyanta*) (Barnes and Shaw, 2002, 2003), common dandelion (*Taraxacum officinale*) (Shaw et al., 2016), and spotted knapweed (*Centaurea stoebe*) (Shipunov et al., 2008), and the endophytic growth of *Botrytis* has been induced and studied in the lab in *Arabadopsis thaliana* (Shaw et al., 2016; Emmanuel et al., 2018).

Although no studies have confirmed this behavior in ornamental geophytes, this cryptic lifecycle is speculated for a newly discovered *Botrytis* species in *Hemerocallis*, *B. deweyae*. Given its pattern of limited sporulation in culture and on host tissue, variations in virulence among isolates, and the fact that *B. deweyae* has been found associated only with cultivated *Hemerocallis* and not its native progenitors, the authors of this new species speculate that *B. deweyae* may be a native endophyte of *Hemerocallis* that is undergoing a transition towards increasing pathogenicity (Grant-Downton et al., 2014). The authors suggest that this transition could be facilitated by the breeding (and subsequent loss of genetic diversity and potential defense genes) and movement of *Hemerocallis* cultivars out of their native range (Grant-Downton et al., 2014). Subsequent clonal propagation of *Hemerocallis*, as is common in most ornamental geophyte species, may be helping to rapidly increase the prevalence of susceptible hosts and virulent strains of the pathogen.

Although this hypothesis is purely speculative on the part of the authors, the ability of other fungal plant pathogens to behave as an endophyte or pathogen depending on host genotype and the apparent prevalence of endophytic *Botrytis* species in nature suggests that researchers studying *Botrytis* on geophytes and other clonally propagated plants should be carefully identifying the *Botrytis* species causing the disease. This may be especially true for plants grown outside of their native or historical ranges. A recent study on *Botrytis* in new plantings of ornamental peonies (*Paeonia*) in Alaska revealed upwards of 10 putatively new *Botrytis* species on this one crop alone, some of which appear to have also been reported as endophytes in *C. stoebe* (Garfinkel et al., 2019), further exemplifying the importance of tracking *Botrytis* species on geophytes.

## 9 Conclusions

It is unclear what impact the diversity of *Botrytis* species on ornamental geophytes will have on grower management practices. Although a number of practices, such as crop rotation, irrigation, and manipulation of the environment are used to help manage *Botrytis* diseases, the management of *Botrytis* disease in ornamental geophytes is currently reliant on, and dominated by, the (often-intensive) use of chemical fungicides. The chemical management of *Botrytis* pathogens is complicated by the high propensity for species in this genus to develop resistance to fungicides and the lack of effective bio-products against this pathogen. The 'Where to look for further information' section in this chapter outlines a number of publications that discuss the management of *Botrytis* diseases on specific ornamental geophyte hosts.

Tremendous progress has been made in the integrated management of diseases of ornamental geophytes, particularly in greenhouse production systems. While advances in breeding for host resistance and the management of some soilborne, virus, and bacterial pathogens has lagged, significant advances have been made and are being made in areas such as molecular-based diagnostic and detection systems, metagenomics, biotechnology, disease and pest forecasting, application technology, induced resistance, and biocontrol (Chastagner et al., 2012; Lakshman et al., 2019). In addition, new highly efficacious pesticides with reduced human and environmental risks have replaced many historical pesticides (Chastagner et al., 2012; Lakshman et al., 2019; Morton and Staub, 2008; Palmer and Vea, 2018).

As indicated by Daughtrey and Buitenhuis (2019), approaches used to disseminate information to growers are changing at a rapid pace. For growers to improve their disease management programs, the key to success is the transfer of locally relevant, research-based information that provides growers with practical, economical, commercially available solutions relating to the management of diseases on their crops in a timely manner. In instances where new growers have limited commercial horticultural experience, information relating to basic crop production and disease management principles is needed. This includes a basic understanding of the concepts relating to the disease triangle so that growers have the ability to identify factors that affect disease development and understand the different approaches that are used to manage diseases. Growers should know the diseases that commonly occur on their crops, be able to identify the symptoms and signs associated with them, understand the host range of the pathogens, understand how the pathogens survive from one season to the next, know what conditions favor inoculum survival and production of primary and secondary inoculum, and understand how these pathogens are commonly spread. In short, they need to understand the disease and life cycles of the pathogens to be able to optimize their disease management program.

## 10 Where to look for further information

As indicated by Lakshman et al. (2019), a number of other comprehensive books and book chapters relating to the management of diseases of ornamental geophytes and ornamentals have also been recently published. Chastagner et al. (2012) reviewed advances in the integrated management of diseases, arthropods, and weeds on ornamental geophytes. Daughtrey and Buitenhuis (2019) provide a summary of key pests and diseases of flowering potted plants and cut flower crops that occur on greenhouse-grown crops. Daughtrey and Buitenhuis (2019) also include lists of biologicals, fungicides, and insecticides/miticides that are available for use against common greenhouse ornamental diseases and pests. Conijn (2017) provides a comprehensive overview of diseases on tulips. McGovern and Elmer's (2018d) Handbook of Florists' Crops Diseases includes chapters with specific information about the management of diseases on caladium (Bowman et al., 2018), cyclamen (Elmer and Daughtrey, 2018), gladiolus (Elmer and Kamo, 2018), lily (Chastagner et al., 2018), daffodil (Hanks and Chastagner, 2018), peonies (Garfinkel and Chastagner, 2018), and tulips (McGovern and Elmer, 2018a). Other chapters address postharvest diseases (Reid and Jiang, 2012; Darras, 2018), mineral nutrition and nutritional disorders (Datnoff and Elmer, 2018; Raudales, 2018), physiological disorders (De Hertogh and Le Nard, 1993b), and the management of insect vectors that spread plant pathogens (Cloyd, 2018). In addition to providing information on diseases and pests, Okubo and Sochacki (2012) provide an overview of the propagation and production of crocus, Dutch iris, gladiolus, hyacinth, lily, daffodil, and tulip.

There are a number of additional publications which summarize advances in diagnostics (Schubert et al., 2018), breeding for disease resistance (van Tuyl et al., 2012; Deng, 2018), environmental modification (Gullino and Garibaldi,

2018), fungicides and biocontrols (Palmer and Vea, 2018), sanitation and soil/media disinfestation (Copes, 2018; Rosskopf et al., 2018), modification of traits via biotechnology (Kamo et al., 2012; Lütken et al., 2012), and the induction of systemic acquired resistance (Klessig et al., 2018). Older, but still very relevant publications include ADAS (1986), Asjes (1974), Baker (1993), Chastagner and Byther (1985), de Best and Zwart (2000), Lane (1984), Loebenstein et al. (1995), Lorbeer et al. (2007), Moore et al. (1979), Rees (1992), Strider (1985), Van Aartrijk et al. (2000), Van Rijn and Pfaff (1995), and De Hertogh and Le Nard's (1993a,b) book that includes chapters that provide a comprehensive overview of diseases on flower bulb crops (Byther and Chastagner, 1993) as well as detailed information on the production of common diseases of 20 different genera of ornamental geophytes. Because of regional differences in the availability and registration of disease control products, specific recommendations relating to products are not generally included in these publications. However, there are a number of regional disease management guides, such as the PNW Disease Management Handbook (https://pnwhandbooks.org/plantdisease), that are updated on a regular basis and do provide specific recommendations of products that are registered for use on ornamental geophytes in specific areas.

## 11 References


ADAS. 1986. *Control of Diseases of Bulbs*. Booklet 2524. MAFF Publications, Alnwick, UK.

Agrios, G. N. 2005. *Plant Pathology* (5th edn.). Elsevier-Academic Press, San Diego, CA.

Ajouz, S., Decognet, V., Nicot, P. C. and Bardin, M. 2010. Microsatellite stability in the plant pathogen *Botrytis cinerea* after exposure to different selective pressures. *Fungal Biology* 114(11–12), 949–54. doi:10.1016/j.funbio.2010.09.004.

Allen, T. C. 1975. Viruses of lilies and their control. *Acta Horticulturae* 47, 69–76. doi:10.17660/ActaHortic.1975.47.11.

Asjes, C. J. 1974. Soil-borne virus diseases in ornamental bulbous crops and their control in the Netherlands. *Agriculture and Environment* 1(3), 303–15. doi:10.1016/0304-1131(74)90029-0.

Asjes, C. J. 1979. Viruses and virus diseases in Dutch bulbous irises (*Iris hollandica*) in the Netherlands. *Netherlands Journal of Plant Pathology* 85(6), 269–79. doi:10.1007/BF01977599.

Asjes, C. J. 1985. Control of field spread of non-persistent viruses in flower-bulb crops by synthetic pyrethroid and pirimicarb insecticides, and mineral oils. *Crop Protection* 4(4), 485–93. doi:10.1016/0261-2194(85)90054-7.

Asjes, C. J. 1990. Production for virus freedom of some principal bulbous crops in the Netherlands. *Acta Horticulturae* 266, 517–30. doi:10.17660/ActaHortic.1990.266.69.

Baayen, R. P., Förch, M. G., Waalwijk, C., Bonants, P. J. M., Löffler, H. J. M. and Roebroeck, E. J. A. 1998. Pathogenic, genetic and molecular characterisation of *Fusarium oxysporum* f. sp. *lilii*. *European Journal of Plant Pathology* 104(9), 887-94. doi:10.1023/A:1008676729143.

Baker, J. R. 1993. Insects. In: De Hertogh, A. A. and Le Nard, M. (Eds), *The Physiology of Flower Bulbs*. Elsevier, pp. 101-53.

Baraldi, E., Bertolini, P., Chierici, E., Trufelli, B. and Luiselli, D. 2002. Genetic diversity between *Botrytis cinerea* isolates from unstored and cold stored kiwi fruit. *Journal of Phytopathology* 150(11-12), 629-35. doi:10.1046/j.1439-0434.2002.00809.x.

Barnes, S. E. and Shaw, M. W. 2002. Factors affecting symptom production by latent *Botrytis cinerea* in *Primula* × *polyantha*. *Plant Pathology* 51(6), 746-54. doi:10.1046/j.1365-3059.2002.00761.x.

Barnes, S. E. and Shaw, M. W. 2003. Infection of commercial hybrid *primula* seed by *Botrytis cinerea* and latent disease spread through the plants. *Phytopathology* 93(5), 573-8. doi:10.1094/PHYTO.2003.93.5.573.

Bastiaansen, C., Koster, A. J., Van der Meer, L. J., Van den Ende, J. E., Pennock, M. E. and Buurman, F. M. P. 1997. A disease-forecasting system of Botrytis blight (fire) in lily. *Acta Horticulturae* 430, 657-60 doi:10.17660/ActaHortic.1997.430.104.

Borges, D. F., Lopes, E. A., Fialho Moraes, A. R., Soares, M. S., Visôtto, L. E., Oliveira, C. R. and Moreira Valente, V. M. 2018. Formulation of botanicals for the control of plant-pathogens: a review. *Crop Protection* 110, 135-40 doi:10.1016/j.cropro.2018.04.003.

Bowman, H. W., Polston, J. E. and McGovern, R. L. 2018. Diseases of caladium. In: McGovern, R. J. and Elmer, W. H. (Eds), *Handbook of Florists' Crops Diseases*. Springer International Publishing, pp. 1073-98.

Boyraz, N., Bastas, K. K., Maden, S. and Yasar, A. 2006. Bacterial leaf and peduncle soft rot caused by *Pectobacterium carotovorum* on tulips in Konya, Turkey. *Phytoparasitica* 34(3), 272-80. doi:10.1007/BF02980956.

Brandes, G. A. 1953. The history and development of the ethylene bisdithiocarbamate fungicides. *American Potato Journal* 30(6), 137-40. doi:10.1007/BF02861938.

Byther, R. S. and Chastagner, G. A. 1993. Diseases. In: De Hertogh, A. A. and Le Nard, M. (Eds), *The Physiology of Flower Bulbs*. Elsevier, pp.71-99.

Chastagner, G. A. 1983. Narcissus fire: prevalence, epidemiology and control in western Washington. *Plant Disease* 67(12), 1384-6. doi:10.1094/PD-67-1384.

Chastagner,G. S. and Byther, R. S. 1985. Bulbs - narcissus, tulips, and iris. In: Strider, D. L. (Ed.), *Diseases of Floral Crops*. Praeger Scientific, New York, NY, pp. 447-506.

Chastagner, G. A. and Copes, W. E. 2002. Potential use of chlorine dioxide to control diseases in ornamental plant production systems. *Combined Proceedings - International Plant Propagators' Society* 51, 275-9.

Chastagner, G. A. and DeBauw, A. 2014. The effectiveness of reduced-risk and new biocontrol products in controlling fire on lilies. *Acta Horticulturae* 1027, 231-9. doi:10.17660/ActaHortic.2014.1027.24.

Chastagner, G. A. and DeBauw, A. 2017. Integration of bio-fungicides in *Botrytis* disease management programs on field-grown ornamental geophytes. *Acta Horticulturae* 1171, 331-40. doi:10.17660/ActaHortic.2017.1171.43.

Chastagner, G. A. and Riley, K. L. 2002. Potential use of chlorine dioxide to prevent the spread of *Fusarium* basal rot during the hot water treatment of daffodil bulbs. *Acta Horticulturae* 570, 267-73. doi:10.17660/ActaHortic.2002.570.34.

Chastagner, G. A. and Riley, K. L. 2005. Sensitivity of pathogen inocula to chlorine dioxide gas. *Acta Horticulturae* 673, 355-9. doi:10.17660/ActaHortic.2005.673.45.

Chastagner, G. A., Hanks, G. R., Daughtrey, M. L., Yedidia, I., Miller, R. W. and Pappu, H. R. 2012. Sustainable production and integrated management: environmental issues. In: Kamenetsky, R. and Okubo, H. (Eds), *Ornamental Geophytes: From Basic Science to Sustainable Production*. CRC Press, Boco Raton, FL, pp. 363–419.

Chastagner, G. A., van Tuyl, J. M., Verbeek, M., Miller, W. B. and Westerdahl, B. B. 2018. Diseases of lily. In: McGovern, R. J. and Elmer, W. H. (Eds), *Handbook of Florists' Crops Diseases*. Springer International Publishing, pp. 1229–88.

Chiocchetti, A., Bernardo, I., Daboussi, M. J., Garbaldi, A., Gullion, M. L., Langin, T. and Migheli, Q. 1999. Detection of *Fusarium oxysporum* f. sp. *dianthi* in carnation tissue by PCR amplification of transposon insertions. *Phytopathology* 89(12), 1169–75. doi:10.1094/PHYTO.1999.89.12.1169.

Cloyd, R. A. 2018. Insect management for disease control in florists' crops. In: McGovern, R. J. and Elmer, W. H. (Eds), *Handbook of Florists' Crops Diseases*. Springer International Publishing, pp. 69–85.

Conijn, C. 2017. *Tulip Disease Shown for Each Growth Stage*. RoodBont Publishers B.V., the Netherlands, p. 150.

Cook, R. J. 2000. Advances in plant health management in the twentieth century. *Annual Review of Phytopathology* 38, 95–116. doi:10.1146/annurev.phyto.38.1.95.

Copes, W. E. 2018. Sanitation for management of florists' crops diseases. In: McGovern, R. J. and Elmer, W. H. (Eds), *Handbook of Florists' Crops Diseases*. Springer International Publishing, pp. 201–36.

Copes, W. E., Barbeau, B. and Chastagner, G. A. 2014. Chlorine dioxide for irrigation water treatment. In: Hong, C., Moorman, G. W., Wohanka, W. and Buttner, C. (Eds), *Biology, Detection, and Management of Plant Pathogens in Irrigation Water*. APS Press, St. Paul, MN, pp. 251–66.

Copping, L. G. and Menn, J. J. 2000. Biopesticides: a review of their action, applications and efficacy. *Pest Management Science* 56(8), 651–76. doi:10.1002/1526-4998(200008)56:8<651::AID-PS201>3.0.CO;2-U.

Cornwell, R. and Giles, F. 2019. Bulbs and other rootling structures. University of Illinois Extension. Available at: https://extension.illinois.edu/bulbs/bulbbasics.cfm (accessed on 02 March 2019).

Darras, A. I. 2018. Postharvest disease management. In: McGovern, R. J. and Elmer, W. H. (Eds), *Handbook of Florists' Crops Diseases*. Springer International Publishing, pp. 253–79.

Datnoff, L. E. and Elmer, W. H. 2018. Mineral nutrition and florists' crops diseases. In: McGovern, R. J. and Elmer, W. H. (Eds), *Handbook of Florists' Crops Diseases*. Springer International Publishing, pp. 237–51.

Daughtrey, M. and Buitenhuis, R. 2019. Ornamentals. In: Gulllino, M. L., Albajes, R. and Nicot, P. (Eds), *Integrated Pest and Disease Management in Greenhouse Crops* (2nd edn.). Plant Pathology in the 21st Century Series. Springer (in press). Chapter 23.

De Best, A. L. I. C. and Zwart, M. J. (Eds). 2000. *Ziekten en afwijkingen bij bolgewassen. Deel 1: Liliaceae* (3rd edn.). Laboratorium and Voor Bloembollenonderzoek, Lisse, the Netherlands, 194p.

De Boer, M. 2009. Effectieve vuurbestrijding in lelie: minder middel - minder kosten. *Bloembollenvisie* (24 September), 24–5.

De Boer, M. 2011. Producing bulbs and perennials; sustainable control of diseases, pests and weeds. *Acta Horticulturae* 886, 59–67. doi:10.17660/ActaHortic.2011.886.6.

De Hertogh, A. A. and Le Nard, M. 1993a. Physiological disorders. In: De Hertogh, A. A. and Le Nard, M. (Eds), *The Physiology of Flower Bulbs*. Elsevier, pp. 155–60.

De Hertogh, A. A. and Le Nard, M. (Eds). 1993b. *The Physiology of Flower Bulbs*. Elsevier, 811p.

De Hertogh, A. A., van Scheepen, J., LeNard, M., Okubo, H. and Kamenetsky, R. 2012. Globalization of the flower bulb industry. In: Kamenetsky, R. and Okubo, H. (Eds), *Ornamental Geophytes: From Basic Science to Sustainable Production*. CRC Press, Boco Raton, FL, pp. 1–16.

Deng, Z. 2018. Breeding for disease resistance in florists' crops. In: McGovern, R. J. and Elmer, W. H. (Eds), *Handbook of Florists' Crops Diseases*. Springer International Publishing, pp. 87–117.

Derks, T., Pham, K., Lemmers, M. and Blom-Barnhorn, G. 2003. Research virus. TNV in hyacinth and lily (in Dutch). *BloembollenVisie* 6 maart, nummer 5, 18–9.

Dugan, F. M., Lupien, S. L., Vahling-Armstrong, C. M., Chastagner, G. A. and Schroeder, B. K. 2014. Host ranges of North American isolates of *Penicillium* causing blue mold of bulb crops. *Crop Protection* 64, 129–36. doi:10.1016/j.cropro.2014.06.013.

Dugan, F. M., Lupien, S. L., Vahling-Armstrong, C. M., Chastagner, G. A. and Schroeder, B. K. 2017. Host ranges of *Penicillium* species causing blue mold of bulb crops in Washington State and Idaho. *Crop Protection* 96, 265–72. doi:10.1016/j.cropro.2017.03.002.

Duraisamy, G. S. and Pokorný, R. 2009. Survey of virus pathogens in gladiolus, iris and tulips in the Czech Republic. *Acta Universitatis Agriculturae et Silviculturae Mendelianae Brunensis* 57(5), 79–86. doi:10.11118/actaun200957050079.

Elad, Y., Pertot, H., Cotes Prado, A. M. and Stewart, A. 2016. Plant hosts of *Botrytis* spp. In: Fillinger, S. and Elad, Y. (Eds), *Botrytis-the Fungus, the Pathogen and Its Management in Agricultural Systems*. Springer International Publishing, Switzerland, pp. 413–86.

El-Goorani, M. A., Abo-El-Dahab, M. K. and Mehiar, F. F. 1974. Interaction between root knot nematode and *Pseudomonas marginata* on gladiolus corms. *Phytopathology* 64(2), 271–2. doi:10.1094/Phyto-64-271.

Elmer, W. H. and Daughtrey, M. L. 2018. Diseases of cyclamen. In: McGovern, R. J. and Elmer, W. H. (Eds), *Handbook of Florists' Crops Diseases*. Springer International Publishing, pp. 1099–128.

Elmer, W. H. and Kamo, K. K. 2018. Diseases of gladiolus. In: McGovern, R. J. and Elmer, W. H. (Eds), *Handbook of Florists' Crops Diseases*. Springer International Publishing, pp. 1289–312.

Emmanuel, C. J., van Kan, J. A. L. and Shaw, M. W. 2018. Differences in the gene transcription state of *Botrytis cinerea* between necrotic and symptomless infections of lettuce and *Arabidopsis thaliana*. *Plant Pathology* 67(9), 1865–73. doi:10.1111/ppa.12907.

Farr, D. F. and Rossman, A. Y. 2019. Fungal databases. U.S. National Fungus Collections, ARS, USDA. Retrieved June 2019, from https://nt.ars-grin.gov/fungaldatabases/.

Ferrada, E. E., Latorre, B. A., Zoffoli, J. P. and Castillo, A. 2016. Identification and characterization of *Botrytis* blossom blight of Japanese plums caused by *Botrytis cinerea* and *B. prunorum* sp. nov. in Chile. *Phytopathology* 106(2), 155–65. doi:10.1094/PHYTO-06-15-0143-R.

FRAC. 2019. Fungicide resistance action committee. Available at: https://www.frac.info/home (assessed on 24 July 2019).

Gardan, L., Gouy, C., Christen, R. and Samson, R. 2003. Elevation of three subspecies of *Pectobacterium carotovorum* to species level: *Pectobacterium atrosepticum* sp. nov., *Pectobacterium betavasculorum* sp. nov. and *Pectobacterium wasabiae* sp. nov. *International Journal of Systematic and Evolutionary Microbiology* 53(2), 381–91. doi:10.1099/ijs.0.02423-0.

Garfinkel, A. R. and Chastagner, G. A. 2018. Diseases of peonies. In: McGovern, R. J. and Elmer, W. H. (Eds), *Handbook of Florists' Crops Diseases*. Springer International Publishing, pp. 663–92.

Garfinkel, A. R., Lorenzini, M., Zapparoli, G. and Chastanger, G. A. 2017. *Botrytis euroamericana*, a new species found on peony and grape in North America and Europe. *Mycologia* 109(3), 495–507. doi:10.1080/00275514.2017.1354169.

Garfinkel, A. R., Coats, K. P., Sherry, D. L. and Chastagner, G. A. 2019. Genetic analysis reveals unprecedented diversity of a globally-important plant pathogenic genus. *Scientific Reports* 9(1), Article Number 6671. doi:10.1038/s41598-019-43165-y.

Goodey, J. B. 1951. The potato tuber nematode, *Ditylenchus destructor* Thorne, 1945; the cause of eelworm disease in bulbous iris. *Annals of Applied Biology* 38(1), 79–90. doi:10.1111/j.1744-7348.1951.tb07790.x.

Goss, E. M. 2015. Genome-enabled analysis of plant-pathogen migration. *Annual Review of Phytopathology* 53, 121–35. doi:10.1146/annurev-phyto-080614-115936.

Gould, C. J. 1950. Diseases of bulbous iris. Extension Bulletin No. 424. State College of Washington, Pullman, WA.

Gould, C. J. 1957. Narcissus. Fungal diseases. In: *Handbook on Bulb Growing and Forcing for Bulbous Iris, Easter Lilies, Hyacinths, Narcissus, Tulips*. Northwest Bulbgrowers' Association, Mt. Vernon, WA, pp. 117–26.

Grant-Downton, R. T., Terhem, R. B., Kapralov, M. V., Mehdi, S., Rodriguez-Enriquez, M. J., Gurr, S. J., van Kan, J. A. L. and Dewey, F. M. 2014. A novel *Botrytis* species is associated with a newly emergent foliar disease in cultivated Hemerocallis. *PLoS ONE* 9(6), e89272. doi:10.1371/journal.pone.0089272.

Gullino, M. L. and Garibaldi, A. 2018. Environmental modification for disease management. In: McGovern, R. J. and Elmer, W. H. (Eds), *Handbook of Florists' Crops Diseases*. Springer International Publishing, pp. 119–36.

Hahm, S. S., Han, K. S., Shim, M. Y., Park, J. J., Kwon, K. H. and Park, J. E. 2003. Occurrence of bacterial soft rot of lily bulb caused by *Pectobacterium carotovorum* subsp. *carotovorum* and *Pseudomonas marginalis* in Korea. *The Plant Pathology Journal* 19(1), 43–5. doi:10.5423/PPJ.2003.19.1.043.

Hanks, G. R. and Chastagner, G. A. 2018. Diseases of daffodils (*Narcissus*). In: McGovern, R. J. and Elmer, W. H. (Eds), *Handbook of Florists' Crops Diseases*. Springer International Publishing, pp. 1129–228.

Hanks, G. R., Kennedy, R. and O'Neill, T. M. 2003. Narcissus leaf diseases: forecasting and control of white mould and smoulder. Report on Project BOF 41. HDC, East Malling, UK.

Huang, J., Hsieh, T., Chastagner, G. A. and Hsiang, T. 2001. Clonal and sexual propagation in *Botrytis elliptica*. *Mycological Research* 105(7), 833–42. doi:10.1017/S0953756201004051.

Hull, R. 2002. *Matthews' Plant Virology* (4th edn.). Academic Press, New York, NY, 1001p.

Jacobsen, B. J. 1997. Role of plant pathology in integrated pest management. *Annual Review of Phytopathology* 35, 373–91. doi:10.1146/annurev.phyto.35.1.373.

Jafra, S., Przysowa, J., Gwizdek-Wiśniewska, A. and Van Der Wolf, J. M. 2009. Potential of bulb-associated bacteria for biocontrol of hyacinth soft rot caused by *Dickeya zeae*. *Journal of Applied Microbiology* 106(1), 268–77. doi:10.1111/j.1365-2672.2008.04000.x.

Jeger, M., Bragard, C., Caffier, D., Candresse, T., Chatzivassiliou, E., Dehnen-Schmutz, K., Gilioli, G., Grégoire, J., Jaques Miret, J. A., MacLeod, A., Navajas Navarro, M., Niere, B., Parnell, S., Potting, R., Rafoss, T., Rossi, V., Van Bruggen, A., Van Der Werf, W., West, J., Winter, S., Mosbach-Schulz, O. and Urek, G. 2016. Risk to plant health of *Ditylenchus destructor* for the EU territory. *EFSA Journal* 14(12) doi:10.2903/j.efsa.2016.4602.

Johnston, P. R., Hoksbergen, K., Park, D. and Beever, R. E. 2014. Genetic diversity of *Botrytis* in New Zealand vineyards and the significance of its seasonal and regional variation. *Plant Pathology* 63(4), 888–98 doi:10.1111/ppa.12143.

Kamo, K., Krens, F. A. and Ziv, M. 2012. Biotechnology for the modification of horticultural traits in geophytes. In: Kamenetsky, R. and Okubo, H. (Eds), *Ornamental Geophytes: From Basic Science to Sustainable Production*. CRC Press, Boco Raton, FL, pp. 159–95.

Klessig, D. F., Choi, H. W. and Dempsey, D. A. 2018. Systemic acquired resistance and salicylic acid: past, present, and future. *Molecular Plant-Microbe Interactions: MPMI* 31(9), 871–88. doi:10.1094/MPMI-03-18-0067-CR. PubMed.

Kogan, M. 1998. Integrated pest management: historical perspectives and contemporary developments. *Annual Review of Entomology* 43, 243–70. doi:10.1146/annurev.ento.43.1.243.

Kohl, L. M. 2011. Foliar nematodes: a summary of biology and control with a compilation of host range. *Plant Health Progress* 12(1), 23. doi:10.1094/PHP-2011-1129-01-RV.

Lakshman, D. K., Cloyd, R. A. and Chastagner, G. A. 2019. Integrated management of diseases and pests on ornamental geophytes: challenges and progress. *Acta Horticulturae* 1237(1237), 13–32. doi:10.17660/ActaHortic.2019.1237.3.

Lane, A. 1984. *Bulb Pests* (7th edn.). Reference Book 51. HMSO, London, UK.

Lees, A. K., Cullen, D. W., Sullivan, L. and Nicolson, M. J. 2002. Development of conventional and quantitative real-time PCR assays for the detection and identification of *Rhizoctonia solani* AG-3 in potato and soil. *Plant Pathology* 51(3), 293–302. doi:10.1046/j.1365-3059.2002.00712.x.

Li, X., Kerrigan, J., Chai, W. and Schnabel, G. 2012. *Botrytis caroliniana*, a new species isolated from blackberry in South Carolina. *Mycologia* 104(3), 650–8. doi:10.3852/11-218.

Liu, Q. L., Li, G. Q., Li, J. Q. and Chen, S. F. 2016. *Botrytis eucalypti*, a novel species isolated from diseased *Eucalyptus* seedlings in South China. *Mycological Progress* 15(10–11), 1057–79. doi:10.1007/s11557-016-1229-1.

Loebenstein, G., Lawson, R. H. and Brunt, A. A. 1995. *Virus and Virus-Like Diseases of Bulb and Flower Crops*. John Wiley & Sons, 543p.

Lorbeer, J. W., Seub, A. M., deBoer, M. and van den Ende, J. E. 2007. *Botrytis* species on bulb crops. In: Elad, Y., Williamson, B., Tudzynski, P. and Delen, N. (Eds), *Botrytis: Biology, Pathology and Control*. Springer, pp. 273–94.

Lütken, H., Clarke, J. L. and Müller, R. 2012. Genetic engineering and sustainable production of ornamentals: current status and future directions. *Plant Cell Reports* 31(7), 1141–57 doi:10.1007/s00299-012-1265-5. PubMed.

Macnish, A. J., Leonard, R. T. and Nell, T. A. 2008. Treatment with chlorine dioxide extends the vase life of selected cut flowers. *Postharvest Biology and Technology* 50(2-3), 197-207. doi:10.1016/j.postharvbio.2008.04.008.

Macnish, A. J., Morris, K. L., de Theije, A., Mensink, M. G. J., Boerrigter, H. A. M., Reid, M. S., Jiang, C. and Woltering, E. J. 2010. Sodium hypochlorite: a promising agent for reducing *Botrytis cinerea* infection on rose flowers. *Postharvest Biology and Technology* 58(3), 262-7. doi:10.1016/j.postharvbio.2010.07.014.

McCuiston, J. L., Hudson, L. C., Subbotin, S. A., Davis, E. L. and Warfield, C. Y. 2007. Conventional and PCR detection of *Aphelenchoides fragariae* in diverse ornamental host plant species. *Journal of Nematology* 39(4), 343-55.

McGovern, R. J. and Elmer, W. H. 2018a. Diseases of tulip. In: McGovern, R. J. and Elmer, W. H. (Eds), *Handbook of Florists' Crops Diseases*. Springer International, pp. 1313-37. doi:10.1007/978-3-319-39670-5_49.

McGovern, R. J. and Elmer, W. H. 2018b. Florists' crops: global trends and disease impact. In: McGovern, R. J. and Elmer, W. H. (Eds). *Handbook of Florists' Crops Diseases, Handbook of Plant Disease Management*, pp. 1-9, doi:10.1007/978-3-319-39670-5_8.

McGovern, R. J. and Elmer, W. H. 2018c. Global trends and disease impact. In: McGovern, R. J. and Elmer, W. H. (Eds), *Handbook of Florists' Crops Diseases*. Springer International Publishing, pp. 1-9.

McGovern, R. J. and Elmer, W. H. 2018d. *Handbook of Florists' Crops Diseases*. Springer International Publishing, p. 1365.

Millardet, P. M. A. 1933. *Phytopathological Classics - the Discovery of Bordeaux Mixture*. American Phytopathological Society, St. Paul, MN, 25p.

Miller, W. B. 2017. Flower bulbs worldwide: perspectives on the production chain and research. *Acta Horticulturae* 1171(1171), 1-8 doi:10.17660/ActaHortic.2017.1171.1.

Moore, W. C. and Dickens, J. S. W. (Eds), Brunt, A. A., Price, D. and Rees, A. R. (revisers) 1979. *Diseases of Bulbs* (2nd edn.). Reference Book HPD 1. HMSO, London, UK.

Morton, V. and Staub, T. 2008. A short history of fungicides. *APSnet Features*. Online. doi:10.1094/APSnetFeature-2008-0308.

Okubo, H. and Sochacki, D. 2012. Botanical and horticultural aspects of major ornamental geophytes. In: Kamenetsky, R. and Okubo, H. (Eds), *Ornamental Geophytes: From Basic Science to Sustainable Production*. CRC Press, Boco Raton, FL, pp.77-121.

Okubo, H. and Sochacki, D. 2013. Botanical and horticultural aspects of major ornamental geophytes. In: Kamenetsky, R. and Okubo, H. (Eds), *Ornamental Geophytes: From Basic Science to Sustainable Production*. CRC Press Taylor and Francis Group LLC, Boca Raton, FL, pp. 77-122.

O'Neill, T. M., Hanks, G. R. and Wilson, D. W. 2004. Control of smoulder (*Botrytis narcissicola*) in narcissus with fungicides. *Annals of Applied Biology* 145(1), 129-37. doi:10.1111/j.1744-7348.2004.tb00368.x.

Palmer, C. L. and Vea, E. 2018. Fungicides and biocontrols for management of florists' crops diseases. In: McGovern, R. J. and Elmer, W. H. (Eds) *Handbook of Florists' Crops Diseases*. Springer International Publishing, pp.136-66.

Plesken, C., Weber, R. W. S., Rupp, S., Leroch, M. and Hahn, M. 2015. *Botrytis pseudocinerea* is a significant pathogen of several crop plants but susceptible to displacement by fungicide-resistant *B. cinerea* strains. *Applied and Environmental Microbiology* 81(20), 7048-56. doi:10.1128/AEM.01719-15.

Putnam, M. L. and Miller, M. L. 2007. *Rhodococcus fascians* in Herbaceous Perennials. *Plant Disease* 91(9), 1064–76. doi:10.1094/PDIS-91-9-1064.

Raudales, R. E. 2018. Nutritional disorders of florists' crops. In: McGovern, R. J. and Elmer, W. H. (Eds), *Handbook of Florists' Crops Diseases*. Springer International Publishing, pp. 41–67.

Rees, A. R. 1992. *Ornamental Bulbs, Corms and Tubers*. CAB International, 220p.

Reid, M. S. and Jiang, C.-Z. 2012. Postharvest: cut flowers and potted plants. In: Kamenetsky, R. and Okubo, H. (Eds), *Ornamental Geophytes: From Basic Science to Sustainable Production*. CRC Press, Boco Raton, FL, pp.333–61.

Rosskopf, E. N., Kokalis-Burelle, K., Fennimore, S. A. and Wiler, C. A. 2018. Soil/media disinfestation for management of florists' crops diseases. In: McGovern, R. J. and Elmer, W. H. (Eds), *Handbook of Florists' Crops Diseases*. Springer International Publishing, pp.167–99.

Rudd-Jones, D. 1975. Introduction. *Acta Horticulturae* 47, 13–5.

Rupp, S., Plesken, C., Rumsey, S., Dowling, M., Schnabel, G., Weber, R. W. S. and Hahn, M. 2017. *Botrytis fragariae*, a new species causing gray mold on strawberries, shows high frequencies of specific and efflux-based fungicide resistance. *Applied and Environmental Microbiology* 83(9), e00269-17. doi:10.1128/AEM.00269-17.

Saaltink, G. J. 1969. Root rot of hyacinths caused by species of *Pythium*. *Netherlands Journal of Plant Pathology* 75(6), 343–54. doi:10.1007/BF01975505.

Saito, S., Margosan, D., Michailides, T. J. and Xaio, C. L. 2016. *Botrytis californica*, a new cryptic species in the *B. cinerea* species complex causing gray mold in blueberries and table grapes. *Mycologia* 108(2), 330–43. doi:10.3852/15-165.

Schubert, T., Jeyaptakash, A. and Harmon, C. 2018. Fundamentals and advances in plant problem diagnostics. In: McGovern, R. J. and Elmer, W. H. (Eds), *Handbook of Florists' Crops Diseases*. Springer International Publishing, pp.13–39.

Shaw, M. W., Emmanuel, C. J., Emilda, D., Terhem, R. B., Shafia, A., Tsamaidi, D., Emblow, M. and van Kan, J. A. L. 2016. Analysis of cryptic, systemic *Botrytis* infections in symptomless hosts. *Frontiers in Plant Science* 7, 625 doi:10.3389/fpls.2016.00625.

Shipunov, A., Newcombe, G., Raghavendra, A. K. H. and Anderson, C. L. 2008. Hidden diversity of endophytic fungi in an invasive plant. *American Journal of Botany* 95(9), 1096–108. doi:10.3732/ajb.0800024.

Slootweg, C., Dijkema, M., van der Wolf, J., Meekes, E. and Westerhof, J. 2015. *Bacterieziekten in de bloemisterij* (in Dutch). Wageningen UR, Wageningen, the Netherlands.

Smith, C. M. 1988. History of benzimidazole use and resistance. In: Delp, C. J. (Ed.), *Fungicide Resistance in North America*. American Phytopathological Society, St. Paul, MN, pp. 23–24.

Smith, R. F. and van den Bosch, R. 1967. Integrated control. In: Kilgore, W. and Doutt, R. (Eds), *Pest Control: Biological, Physical, and Selected Chemical Methods*. Academic Press, New York, NY, pp. 295–340.

Sowley, E. N. K., Dewey, F. M. and Shaw, M. W. 2010. Persistent, symptomless, systemic, and seed-borne infection of lettuce by *Botrytis cinerea*. *European Journal of Plant Pathology* 126(1), 61–71. doi:10.1007/s10658-009-9524-1.

Strider, D. L. 1985. *Diseases of Floral Crops*. Praeger Scientific, New York, NY.

Sutton, M. W., Dixon, G. R. and Willock, M. 1986. Virus-tested *Narcissus*: progress with field evaluation in Scotland. *Acta Horticulturae* 177(177), 221–6. doi:10.17660/ActaHortic.1986.177.29.

Taghinasb, M., Khademlou, E. and Rahimian, H. 2008. Identification of *Pectobacterium carotovorum* casual agent of gladiolus bulb and stalk rot in Mahallat greenhouses. *Proceedings of the 18th Iranian Plant Protection Congress*, Hamadan, Iran, 465 (Abstract).

Terhem, R. B. 2015. Sexual development of *Botrytis* species. Ph.D. Thesis. Wageningen University, Wageningen, the Netherlands, p. 188.

van Aartrijk, J., van den Ende, J. and Peeters, I. J. 2000. *Ziekten en afwijkingen bij bolgewassen. Deel 1: Liliaceae*. Laboratorium and voor Bloembollenonderzoek, Lisse, the Netherlands, 193p.

van den Ende, J. E., Pennock-Vos, M. G., Bastiaansen, C., Koster, A. T. H. J. and van der Meer, L. J. 2000. BoWaS: a weather-based warning system for the control of *Botrytis* blight in lily. *Acta Horticulturae* 519(519), 215–20 doi:10.17660/ActaHortic.2000.519.22.

van der Vlugt, C. I., Derks, A. F. L. M., Dijkstra, J. and Goldbach, R. 1988. Towards a rapid and reliable detection method for iris severe mosaic virus in iris bulbs. *VII International Symposium on Virus Diseases of Ornamental Plants*. ISHS, Sanremo, Italy, pp. 191–8.

van der Zande, J. C., Achten, V. T. J. M., Schepers, H. T. A. M., van der Lans, A., Kempenaar, C., Michielsen, J. M., Stallinga, H. and van Velde, P. 2010. Precision disease control in bed-grown crops. In: Oerke, E. C., et al. (Eds), *Precision Crop Protection - the Challenge and Use of Heterogeneity*. Springer Science+Business Media BV, pp. 403–15.

van Doorn, J., Vreeburg, P. J. M., van Leeuwen, P. J. and Dees, R. H. L. 2008. The presence and survival of soft rot (*Erwinia*) in flower bulb production systems. *X International Symposium on Flower Bulbs and Herbaceous Perennials*, Lisse, the Netherlands. ISHS, pp. 365–79.

van Doorn, J., Vreeburg, P. J. M., van Leeuwen, P. J. and Dees, R. H. L. 2011. The presence and survival of soft rot (*Erwinia*) in flower bulb production systems. *Acta Horticulturae* 886, 365–79. doi:10.17660/ActaHortic.2011.886.51.

van Julsingha, E. B. 1980. Developments in bulb research. *Acta Horticulturae* 109(109), 19–24. doi:10.17660/ActaHortic.1980.109.2.

van Rijn, J. F. A. T. and Pfaff, H. G. M. (Eds). 1995. *Ziekten en afwijkingen bij bolgewassen. 2. Amaryllidaceae, Araceae, Begoniaceae, Cannaceae, Compositae, Iridaceae, Oxalidaceae, Ranunculaceae* (2nd edn.). Informatie en Kennis Centrum Landbouw, Laboratorium voor Bloembollenonderzoek, Lisse, the Netherlands.

van Tuyl, J. M. and Toxopeus, S. J. 1980. Breeding for resistance to yellow disease of hyacinths. I. Investigations on F1's from diallel crosses. *Euphytica* 29(3), 555–60. doi:10.1007/BF00023202.

Van Tuyl, J. M., Arens, P. and Marasek-Ciolakowska, A. 2012. Breeding and genetics of ornamental geophytes. In: Kamenetsky, R. and Okubo, H. (Eds), *Ornamental Geophytes: From Basic Science to Sustainable Production*. CRC Press, Boco Raton, FL, pp. 131–58.

Vol. I and Vol. II, 579p.

Walker, A. S., Gatier, A. L., Confais, J., Martinho, D., Viaud, M., Le Pêcheur, P., Dupont, J. and Fournier, E. 2011. *Botrytis pseudocinerea*, a new cryptic species causing gray mold in French vineyards in sympatry with *Botrytis cinerea*. *Phytopathology* 101(12), 1433–45. doi:10.1094/PHYTO-04-11-0104.

West, J. S., Bravo, C., Oberti, R., Lemaire, D., Moshou, D. and McCartney, H. A. 2003. The potential of optical canopy measurement for targeted control of field crop diseases. *Annual Review of Phytopathology* 41(1), 593–614 doi:10.1146/annurev.phyto.41.121702.103726. PubMed.

Yoon, M. Y., Cha, B. and Kim, J. C. 2013. Recent trends in studies on botanical fungicides in agriculture. *The Plant Pathology Journal* 29(1), 1–9 doi:10.5423/PPJ.RW.05.2012.0072. PubMed.

Zhang, J., Zhang, L., Li, G.-Q., Yang, L., Jiang, D.-H., Zhuang, W.-Y. and Huang, H.-C. 2010a. *Botrytis sinoallii*: a new species of the grey mould pathogen on *Allium* crops in China. *Mycoscience* 51(6), 421–31. doi:10.1007/S10267-010-0057-4.

Zhang, J., Wu, M. D., Li, G. Q., Yang, L., Yu, L., Jiang, D. H., Huang, H. C. and Zhuang, W. Y. 2010b. *Botrytis fabiopsis*, a new species causing chocolate spot of broad bean in central China. *Mycologia* 102(5), 1114–26. doi:10.3852/09-217.

Zhang, J., Yang, H., Yu, Q. Y., Wu, M. D. and Yang, L. 2016. *Botrytis pyriformis* sp. nov., a novel and likely saprophobic species of *Botrytis*. *Mycologia* 108, 982–696.

Zhong, S., Zhang, J. and Zhang, G. Z. 2019. *Botrytis polyphyllae*: a new *Botrytis* species causing gray mold on *Paris polyphylla*. *Plant Disease* 103(7), 1721–7. doi:10.1094/PDIS-07-18-1284-RE.

# Index

www.ingramcontent.com/pod-product-compliance
Ingram Content Group UK Ltd.
Pitfield, Milton Keynes, MK11 3LW, UK
UKHW021825100926
8556IPUK00003B/158

9 781786 763280